The Sticky Synapse

Michael Hortsch · Hisashi Umemori
Editors

The Sticky Synapse

Cell Adhesion Molecules and Their Role in Synapse Formation and Maintenance

Springer

Editors
Michael Hortsch
Department of Cell & Developmental Biology
University of Michigan
Medical School
109 Zina Pitcher Place
Ann Arbor MI 48109
Biomedical Sciences
Research Bldg.
USA
hortsch@umich.edu

Hisashi Umemori
Molecular and Behavioral Neuroscience Institute and Department of Biological Chemistry
109 Zina Pitcher Place
Ann Arbor, MI 48109
USA
umemoh@umich.edu

Cover illustrations: Developing Synapses - Synapses are formed at points of contact between axons and their targets. From left, Drosophila neuromuscular junctions (motor axons, red; muscles, green), mouse neuromuscular junctions (motor axons, green; neuromuscular junctions, pink), and mouse cerebellar synapses in culture (pontine axons, blue; cerebellar granule cell dendrites, pink; synapses, green).
Courtesy of Carrero-Martinez and Chiba (Drosophila) and Harris and Umemori (mouse).

ISBN 978-0-387-92707-7 e-ISBN 978-0-387-92708-4
DOI 10.1007/978-0-387-92708-4
Springer Dordrecht Heidelberg London New York

Library of Congress Control Number: 2009929373

Printed on acid-free paper

Springer is part of Springer Science+Business Media (www.springer.com)

Preface

The molecular mechanisms, which are responsible for the functional differences between the various types of neuronal synapses, have become one of the central themes of modern neurobiology. It is becoming increasingly clear that a misregulation of synaptogenesis and synaptic remodeling and dysfunctional neuronal synapses are at the heart of several human diseases, both neurological disorders and psychiatric conditions. As synapses present specialized cellular junctions between neurons and their target cells, it may not come as a surprise that neural cell adhesion molecules (CAMs) are of special importance for the genesis and the maintenance of synaptic connections. Genes encoding adhesive molecules make up a significant portion of the human genome, and neural CAMs even have been postulated to be a major factor in the evolution of the human brain. These are just some of the many reasons why we thought a book on neural CAMs and their role in establishing and maintaining neuronal synapses would be highly appropriate for summarizing our current state of knowledge. Without question, over the near future, additional adhesive proteins will join the ranks of synaptic CAMs and our knowledge, and how these molecules enable neurons and their targets to communicate effectively will grow. We hope that this book will provide a comprehensive and timely synopsis of the role of CAMs at synaptic connections and will encourage other researchers to join this exciting field of neuroscience, which has the promise not only to yield new insights into the functioning of our brain but also to shed light on some devastating human diseases.

Ann Arbor, MI

Michael Hortsch
Hisashi Umemori

Contents

Contributors

Michael V.L. Bennett Dominick P Purpura Department of Neuroscience, Albert Einstein College of Medicine, 1410 Pelham Parkway South, Room 704, Bronx, NY 10804, USA, mbennett@aecom.yu.edu

Laurent Bogdanik The Jackson Laboratory, 600 Main St., Bar Harbor, ME 04609, USA, laurent.bogdanik@jax.org

Thomas Bourgeron Human Genetics and Cognitive Functions Unit, Department of Neuroscience, Institut Pasteur, 25 rue du Docteur Roux, 75724 Paris Cedex 15, France, thomasb@pasteur.fr

Mario Rosario Buffelli Dipartimento di Scienze Neurologiche e della Visione, Sezione di Fisiologia, Universita' di Verona, Strada Le Grazie 8, 37134 Verona, Italy, mario.buffelli@univr.it

Robert W. Burgess The Jackson Laboratory, 600 Main St., Bar Harbor, ME 04609, USA, rburgess@jax.org

Franklin A. Carrero-Martínez Department of Biology, University of Puerto Rico, Mayagüez, Mayagüez, Puerto Rico 00681-9012, franklin.carrero@upr.edu

Akira Chiba Department of Biology, University of Miami, 234 Cox Science Center, 1301 Memorial Drive, Coral Gables, FL 33124, USA, akira.chiba@miami.edu

Alexander A. Chubykin The Picower Institute for Learning and Memory, Massachusetts Institute of Technology, 77 Massachusetts Avenue, 46-3301, Cambridge, MA 02139, USA, chubykin@mit.edu

James C. Clemens Department of Biochemistry, Purdue University, 175 S. University St., West Lafayette, IN 47907, USA, jclemens@purdue.edu

Staffan Cullheim Department of Neuroscience, Karolinska Institutet, SE-171 77 Stockholm, Sweden, staffan.cullheim@ki.se

Michael A. Fox Department of Anatomy and Neurobiology, Virginia Commonwealth University Medical Campus, Box 980709, Richmond, VA 23298-0709, USA, mafox@vcu.edu

Juan Mauricio Garré Dominick P Purpura Department of Neuroscience, Albert Einstein College of Medicine, 1410 Pelham Parkway South, Room 704, Bronx, NY 10804, USA

Andrew M. Garrett Department of Biology, Graduate Program in Neuroscience, The University of Iowa, Iowa City, IA 52242, USA

Tanja Angela Godenschwege Department of Biological Sciences, Florida Atlantic University, Sanson Science Building 1/209, 777 Glades Road, Boca Raton, FL 33431, USA, tanjag@biology.fau.edu

Y. Kate Hong Program in Neuroscience, Harvard Medical School, Boston MA, 02115, USA; Department of Molecular and Cellular Biology, Harvard University, Fairchild Bldg, 7 Divinity Ave., Cambridge, MA 02138, USA, yhong@fas.harvard.edu

Michael Hortsch Department of Cell and Developmental Biology, University of Michigan, BSRB, 109 Zina Pitcher Place, Ann Arbor, MI 48109, USA, hortsch@umich.edu

Hitesh Kathuria Department of Biochemistry, Purdue University, 175 S. University St., West Lafayette, IN 47907, USA

Lars V. Kristiansen Research Laboratory for Stereology and Neuroscience, H. S. Bispebjerg University Hospital, Copenhagen, Denmark, lkri0062@bbh.regionh.dk

Catherine E. Krull Biologic and Materials Sciences, University of Michigan, 5211 Dental, 1011 N. University Ave., Ann Arbor, MI 48109-1078, USA, krullc@umich.edu

Daniel J. Liebl Miller School of Medicine, University of Miami, Miami Project to Cure Paralysis, P.O. Box 016960 R-48, Miami, FL 33101, USA, dliebl@miami.edu

Devi Majumdar Department of Biological Sciences, Vanderbilt Kennedy Center for Research on Human Development, Vanderbilt University, 465 21st Avenue South, Nashville, TN 37232, USA

Hisakazu Ogita Division of Molecular and Cellular Biology, Department of Biochemistry and Molecular Biology, Kobe University Graduate School of Medicine, 7-5-1 Kusunoki-cho, Chuo-ku, Kobe 650-0017 Japan

Sylwia Owczarek Protein Laboratory, Department of Neuroscience and Pharmacology, University of Copenhagen, Faculty Of Health Sciences, Blegdamsvej 3B, DK-2200 Copenhagen, Denmark

Suzanne Paradis Department of Biology, Brandeis University, P.O. Box 549110, Waltham, MA 02454-9110, USA, paradis@brandeis.edu

Bibiana Scelfo Dipartimento di Neuroscienze – Sezione di Fisiologia, Istituto Nazionale di Neuroscienze, Universita' di Torino, Corso Raffaello 30, 10125 Torino, Italy

Dietmar Schreiner Department of Biology, The University of Iowa, Iowa City, IA 52242, USA

Kang Shen Department of Biological Sciences, Stanford University, 371 Serra Mall, Gilbert 109, Stanford, CA 94305-5020, USA, kangshen@stanford.edu

Yoshimi Takai Division of Molecular and Cellular Biology, Department of Biochemistry and Molecular Biology, Kobe University Graduate School of Medicine, 7-5-1 Kusunoki-cho, Chuo-ku, Kobe 650-0017 Japan, ytakai@med.kobe-u.ac.jp

Sebastian Thams Department of Neuroscience, Karolinska Institutet, SE-171 77 Stockholm, Sweden, sebastian.thams@ki.se

Hideru Togashi Division of Molecular and Cellular Biology, Department of Biochemistry and Molecular Biology, Kobe University Graduate School of Medicine, 7-5-1 Kusunoki-cho, Chuo-ku, Kobe 650-0017 Japan

Hisashi Umemori Molecular and Behavioral Neuroscience Institute, University of Michigan, BSRB, 109 Zina Pitcher Place, Ann Arbor, MI 48109, USA, umemoh@umich.edu

Smitha Babu Uthaman Department of Biological Sciences, Florida Atlantic University, Sanson Science Building 1/209, 777 Glades Road, Boca Raton, FL 33431, USA

Philip Y. Wang Laboratory of Neurochemistry, National Institute on Deafness and Other Communication Disorders, National Institutes of Health, Bethesda, MD 20892, USA; Department of Biology, College of Chemical and Life Sciences and Neuroscience and Cognitive Science Program, University of Maryland, College Park, MD 20742, USA

Donna J. Webb Department of Biological Sciences and Vanderbilt Kennedy Center for Research on Human Development, Vanderbilt University, VU station B, Box 35-1634, Nashville, TN 37235, USA, donna.webb@vanderbilt.edu

Joshua A. Weiner Department of Biology, Graduate Program in Neuroscience, The University of Iowa, Iowa City, IA 52242, USA, joshua-weiner@uiowa.edu

Robert J. Wenthold Laboratory of Neurochemistry, National Institute on Deafness and Other Communication Disorders, National Institutes of Health, Bethesda, MD 20892, USA, wenthold@nidcd.nih.gov

Peter S. Walmod Protein Laboratory, Department of Neuroscience and Pharmacology, Faculty of Health Sciences, University of Copenhagen, Blegdamsvej 3B, DK-2200 Copenhagen, Denmark, psw@sund.ku.dk

Masahito Yamagata Department of Molecular and Cellular Biology, Harvard University, Fairchild Bldg, 7 Divinity Ave., Cambridge, MA 02138, USA, yamagatm@mcb.harvard.edu

Masahiro Yasuda Molecular and Behavioral Neuroscience Institute, University of Michigan, BSRB, 109 Zina Pitcher Place, Ann Arbor, MI 48109, USA

Chapter 1
A Short History of the Synapse – Golgi Versus Ramón y Cajal

Michael Hortsch

The history of the synapse started not only as a struggle between two ideas but also as a feud between the two founding fathers of modern neuroscience, the Italian Camillo Golgi (1843–1926) and the Spaniard Santiago Ramón y Cajal (1851–1934). Preceding their groundbreaking portrayals of the nervous system structure, Robert Remak (1815–1865), Theodore Schwann (1810–1882), Otto Friedrich Karl Deiters (1834–1863), and others had published only rudimentary histological descriptions of nerves and of some other parts of the nervous system. However, the limited resolution of the microscopic analysis at that time did not allow them to elucidate the cellular details and the functional relationships between individual nervous system components. In 1872, Joseph von Gerlach (1820–1896) formulated the first theory to explain the cellular organization of the nervous system (Gerlach 1872). His model, the Reticular Theory, postulated that the nervous system consists of a continuous syncytial network or reticulum. Nerve fibers, dendrites, and neuronal cells would be directly connected to each other by cytoplasmic bridges with the neuronal cell bodies providing only nourishment support.[1] Over the following years, Joseph von Gerlach together with Camillo Golgi became the major proponents of the initially widely accepted Reticular Theory. Ironically, it was a fortuitous discovery by Camillo Golgi that ultimately led to its demise.

[1] J. Gerlach J (1872) Von dem Rückenmark. In: Stricker S (eds) Handbuch der Lehre von den Geweben des Menschen und der Thiere. Verlag von Wilhelm Engelmann, Leipzig on page 684: "...the finest divisions of the protoplasmic processes take part in the formation of the fine nerve fiber network, which I consider to be an essential constituent of the gray matter of the spinal cord. ...(T)he neuronal and cytoplasmic extensions of the cells in the gray matter are therefore connected in two ways with the nerve fibers of the spinal cord. First, by means of the nerve process...and secondly through the finest branches of the protoplasmic processes, which become a part of the fine nerve fiber net of the gray matter."

M. Hortsch (✉)
Department of Cell and Development Biology, University of Michigan,
Ann Arbor, MI 48109-2200, USA
e-mail: hortsch@umich.edu

M. Hortsch, H. Umemori (eds.), *The Sticky Synapse*,
DOI 10.1007/978-0-387-92708-4_1,

In 1873, Camillo Golgi reported a novel histological staining procedure, which selectively highlights a small number of neuronal cells at random while leaving most other neurons unstained (Golgi 1873). This effect is achieved by impregnating fixed neuronal tissues with potassium dichromate and silver nitrate. All stained cells are entirely filled with a brown or black precipitate of silver chromate, revealing even slender dendritic and axonal processes. In 1887, Santiago Ramón y Cajal learned about this novel histological method and developed it further to reveal even minute details of neuronal structures (Fig. 1.1). Over the following years, both Ramón y Cajal and Golgi used this staining technique for a detailed survey of many neuronal tissues. From his results, Santiago Ramón y Cajal concluded that the nervous system is not a continuous network, but rather consists of separate, discontinuous units or cells.

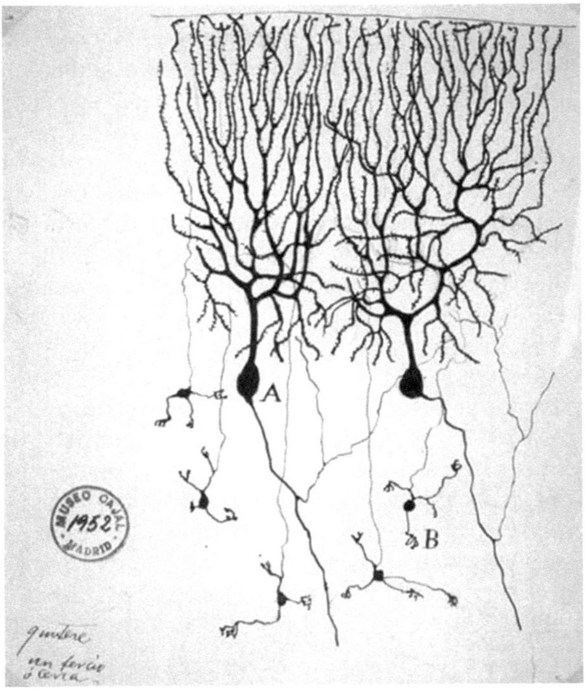

Fig. 1.1 Drawing of Purkinje (**A**) and granule cells (**B**) from an adult pigeon cerebellum by Santiago Ramón y Cajal (Golgi method), 1899. Instituto Santiago Ramón y Cajal, Madrid, Spain

Feeling scientifically isolated at his position as professor of histology and pathological anatomy in Barcelona, Ramón y Cajal traveled to the October 1889 meeting of the German Anatomical Society, which was held at the University of Berlin (Ramón y Cajal 1937). There he made the acquaintance of Rudolph Albert von Kölliker (1817–1905), Wilhelm His (1831–1904), Heinrich Wilhelm Gottfried von Waldeyer-Hartz (1836–1921), Arthur van Gehuchten (1861–1914), and other eminent histologists. After viewing Ramón y Cajal's preparations, Albert von Kölliker in particular encouraged him to publish his findings more widely and later even confirmed and extended them with his own work.

Based on Santiago Ramón y Cajal's conclusions and the results of other researchers, Wilhelm von Waldeyer-Hartz in 1891 published a paper, in which he outlined an alternative theory, the Neuron Doctrine of the nervous system (Waldeyer-Hartz 1891), which subsequently received overwhelming support throughout the scientific community. In his publication, von Waldeyer-Hartz used for the first time the term "neuron" (Greek "νευρων" for sinew or tendon) to describe the separate cellular subunit that is common to all neuronal tissues. At that time, it had become clear that most neuronal cells consist of three different subcellular domains: the neuronal cell body or soma, fine tree-like cytoplasmic processes, and a single long fiber-like extension. Inspired by their branch-like structure and after the Greek word "δεντρο" for tree, Wilhelm His in 1889 had suggested the use of the phrase "dendrites" for the finer cytoplasmic neuronal processes (His 1889). Later in 1896, Albert von Kölliker added the term "axon" (Greek "αξον" for axle or axis) for the long, fiber-like extension (von Koelliker 1896). Over the following years, Santiago Ramón y Cajal in Spain and Arthur van Gehuchten in Belgium independently modified and extended the Neuron Doctrine by adding the Law of Dynamic Polarization, which states that neuronal signals only travel in one direction in a neuron, from dendrites and cell bodies to axons (Berlucchi 1999).

However, as the acceptance of the Neuron Doctrine grew, it raised a new problem. Neither von Waldeyer-Hartz's hypothesis nor Ramón y Cajal's morphological analysis offered an explanation of how a neuronal signal would be transferred from one neuronal cell to the next. Although specialized contact regions between neurons were soon suspected to be responsible for this process, no mechanistic explanation would be forthcoming for a considerable time. When preparing the 6th edition of his *Handbook of Human Physiology*, Sir Michael Foster (1836–1907) secured the assistance of his student Sir Charles Scott Sherrington (1857–1952) for writing the chapter on the Central Nervous System (Foster and Sherrington 1897). They both felt that a proper term for describing these special contact points between neurons was lacking and requested the help of Arthur Woolgar Verrall (1851–1912), a classical Greek scholar at the Trinity College in Cambridge (Tansey 1997). Verrall suggested the term "synapse" from the Greek "συν" (syn meaning together) and "απτειν" (haptein meaning to clasp), which was adapted by Foster and Sherrington and thereby introduced as the scientific term for describing neuronal contacts.

In 1906, the accomplishments of Camillo Golgi and Santiago Ramón y Cajal were jointly recognized with the Nobel Prize for Physiology or Medicine, the first of many to honor discoveries in the field of neuroscience (Table 1.1). The committee awarded the prize to both scientists "in recognition of their work on the structure of the nervous system" (Grant 2007). In his acceptance speech, given December 12, 1906, in Stockholm, Santiago Ramón y Cajal summarized his extensive histological work and that of other scientists, which argued against

Table 1.1 Nobel Prizes for Physiology or Medicine, which have been awarded for basic neuroscience discoveries

1906	Camillo Golgi and Santiago Ramón y Cajal "in recognition of their work on the structure of the nervous system"
1932	Sir Charles Sherrington and Lord Edgar Douglas Adrian "for their discoveries regarding the functions of neurons"
1936	Sir Henry Halett Dale and Otto Loewi "for their discoveries relating to chemical transmission of nerve impulses"
1944	Joseph Erlanger and Herbert Spencer Gasser "for their discoveries relating to the highly differentiated functions of single nerve fibers"
1957	Daniel Bovet "for his discoveries relating to synthetic compounds that inhibit the action of certain body substances, and especially their action on the vascular system and the skeletal muscles"
1961	Georg von Békésy "for his discoveries of the physical mechanism of stimulation within the cochlea"
1963	Sir John Eccles, Alan Lloyd Hodgkin, and Andrew Fielding Huxley "for their discoveries concerning the ionic mechanisms involved in excitation and inhibition in the peripheral and central portions of the nerve cell membrane"
1967	Ragnit Granit, Haldan Keffer Hartline, and George Wald "for their discoveries concerning the primary physiological and chemical visual processes in the eye"
1970	Sir Bernard Katz, Ulf von Euler, and Julius Axelrod "for their discoveries concerning the humoral transmittors in the nerve terminals and the mechanism for their storage, release, and inactivation"
1977	Roger Guillemin and Andrew Viktor Schally "for their discoveries concerning the peptide hormone production of the brain" and Rosalyn Yalow for "for the development of radioimmunoassays of peptide hormones"
1981	Roger W. Sperry "for his discoveries concerning the functional specialization of the cerebral hemispheres" and David H. Hubel and Torsten N. Wiesel "for their discoveries concerning information processing in the visual system"
1986	Stanley Cohen and Rita Levi-Montalcini "for their discoveries of growth factors"
1991	Erwin Neher and Bert Sakmann "for their discoveries concerning the function of single ion channels in cells"
1997	Stanley B. Prusiner "for his discovery of Prions – a new biological principle of infection"
2000	Arvid Carlsson, Paul Greengard, and Eric R. Kandel "for their discoveries concerning signal transduction in the nervous system"
2004	Richard Axel and Linda B. Buck "for their discoveries of odorant receptors and the organization of the olfactory system"

the Reticular Theory and in support of the Neuron Doctrine[2] (Ramón y Cajal 1967). He acknowledged that in the future, novel techniques might reveal new structures and mechanisms and how neuronal cells are connected. However,

[2] S. Ramón y Cajal, Nobel Prize Lecture (1967): "From the whole of these facts, the neuronal doctrine of His and of Forel, accepted by many neurologists and physiologists, is derived as an inevitable postulate... The irresistible suggestion of the reticular conception, of which I have spoken to you has led several physiologists and zoologists to object to the doctrine of the propagation of nerve currents by contact or at a distance. All their allegations are based on the findings by incomplete methods showing far less than those which have served to build the imposing edifice of the neuronal conception."

from the data, which were available to him, he rejected a continuous neuronal network and therefore the Reticular Theory. Much to his chagrin, Camillo Golgi in his Nobel lecture, which he had delivered the previous day, presented a diametric opposite view and a scathing rejection of the Neuron Doctrine[3] (Golgi 1967). In his autobiography, Santiago Ramón y Cajal describes Camillo Golgi's Nobel lecture as self-serving and his attitude as arrogant (Ramón y Cajal 1937). He accuses him of ignoring the experimental results of other researchers and of "worship of his own ego."[4] Certainly no love was lost between these two pioneers of neuroscience. Until his death in 1926, Camillo Golgi remained an ardent supporter of the Reticular Theory.

First insights into the mechanism and the chemical nature of synaptic signals came at the beginning of the 20th century, mainly from the laboratory of John Newport Langley (1852–1925) at Cambridge University in England. In 1904, his student Thomas Renton Elliott (1877–1961) discovered that adrenaline from the adrenal gland mimics the effect of sympathetic nerve innervation on various muscles and glands (Elliott 1905). Adrenaline had previously been recognized as a small bioactive molecule derived from the adrenal medulla; its structure had been determined and it had just been chemically synthesized. Although he mistakenly assumed that adrenaline, rather than noradrenaline, might be released from the peripheral sympathetic nerve endings, Thomas Elliott laid the conceptual foundation for the activity of neurotransmitters as small chemical molecules that bridge the synaptic gap between nerve endings and their targets (Elliott 1904). The identification of the first genuine neurotransmitter can be credited to another former student of Langley, Sir Henry Halett Dale (1865–1968) (Tansey 2006). Together with his colleague Arthur James Ewins (1882–1957) at the Wellcome Physiological Research Laboratories he identified and isolated acetylcholine from a bacterial contamination of the cereal fungus ergot and characterized its physiological activity (Dale 1914, Ewins 1914). However, the final proof of its physiological significance fell to his friend and 1936 fellow Nobel laureate (Table 1.1), the physiologist Otto Loewi (1873–1961). Otto Loewi's experiments on explanted frog hearts established that signaling across most synapses is mediated by small chemical compounds, now referred to as neurotransmitters (Loewi 1921). Nevertheless, it took a considerable time until it was generally accepted that synaptic signal transduction usually is based on a chemical and not on a bioelectrical mechanism. Even in 1937, Sir John Eccles (1903–1997), one of the 1963 Nobel laureates

[3] C. Golgi, Nobel Prize Lecture (1967): "I shall ... confine myself to saying that, while I admire the brilliancy of the (neuron) doctrine, which is a worthy product of the high intellect of my illustrious Spanish colleague, I cannot agree with him on some points of an anatomical nature."

[4] S. Ramón y Cajal, Recollections of My Life (1937): "Contrary to what we all expected, instead of pointing out the valuable facts, which he (Golgi) had discovered, he attempted in it to refloat his almost forgotten theory of interstitial nerve nets. Likewise he considered it unnecessary to correct any of his old theoretical errors, or of his lapses of observation."

for his work on the ionic mechanisms of nerve cell excitation and inhibition (Table 1.1), still favored an electrical transmission model (Eccles 1937). Only later he converted to Henry Dale's view of a chemical-centered signal transmission at synapses. Over the next decades, a number of additional neurotransmitters were identified. For example, a student of Henry Dale, Ulf Svante von Euler (1905–1983), demonstrated in 1946 that noradrenalin is the major neurotransmitter of the sympathetic nervous system (von Euler 1946). Also the first mechanistic details about the process of synaptic transmission began to emerge. At the beginning of the 1950s, Sir Bernard Katz (1911–2003) and his coworkers showed that neurotransmitter molecules were released from the presynaptic termini in discrete quantal amounts (Fatt and Katz 1952, Del Castillo and Katz 1954), and Julius Axelrod (1912–2004) and his research group demonstrated that secreted neurotransmitters were not just rapidly degraded by enzymes, but also taken up and recycled by the surrounding cells (Whitby et al. 1961). In 1961, their contributions to the understanding of synaptic processes were also recognized by the Nobel Prize committee (Table 1.1).

Although physiological and biochemical experiments settled the chemical nature of synaptic signal transmission, a new microscopic technique was needed to elucidate the fine structure of synaptic organization and to demonstrate how transmitters are released into the synaptic cleft. In 1933, Ernst August Friedrich Ruska (1906–1988) had developed the first electron microscope, and at the beginning of the 1950s, this technology was used to investigate the subcellular organization of many biological tissues including neuronal cells. These initial studies by Eduardo de Robertis (1913–1988), J. David Robertson (1922–1995), Fritiof S. Sjöstrand (born 1912), and others provided the final morphological proof for the central hypothesis of the Neuron Doctrine, the existence of a discontinuity or gap between the pre- and the postsynaptic cell (Robertson 1953, Estable, Reissig and De Robertis 1954, Sjöstrand 1958). The superior magnification and resolution of the electron microscope also revealed additional structural details, which had not been seen using other techniques. One such revelation was the presence of small secretory vesicles in the presynaptic terminus (De Robertis and Bennett 1955, Palay 1956). These membrane vesicles were soon postulated to contain neurotransmitters and thus provided an explanation for the quantal release of neurotransmitters, which had been observed by Sir Bernard Katz and his group. Early electron microscopic analyses also reported an electron-dense region at the membrane of the postsynaptic neuron, now referred to as the postsynaptic density (Akert et al. 1969). Despite this wealth of new structural information about the general subcellular organization of synaptic connections, electron microscopic studies alone were unable to identify the molecular components and proteins that form them.

Over the last 40 years, genetic, biochemical, molecular biological and genomic approaches have finally revealed a plethora of protein components, which constitute the synaptic apparatus. Among these synaptic proteins are components of the secretory pathway, which are responsible for vesicle transport, polypeptides involved in membrane vesicle docking and fusion, neurotransmitter receptors and ion channels, enzymes responsible for neurotransmitter processing, inactivation and uptake, cytoskeletal elements and scaffolding proteins, extracellular matrix components, cellular signaling proteins, and also cell adhesion molecules (CAMs). As synapses are special contact points between neurons and their targets it may not be surprising that CAMs are important components of synaptic connections. However, it was somewhat unexpected that many CAMs, which have been found at synapses, also have important non-synaptic functions in neuronal cell and in tissues outside the nervous system, such as during neuronal differentiation, axonal pathfinding, cell migration, or epithelial stability. Only relatively few adhesive molecules appear to have an exclusive synaptic function. Several general characteristics of CAMs appear to be of special relevance for their functional role at synapses. Synaptic contacts contain not only homophilic CAMs but also heterophilic CAMs, which interact with a heterologous binding partner on the pre- or postsynaptic cell surface. Such heterophilic pairs of adhesive molecules or pre- versus postsynaptic differences in the expression of CAM-interacting proteins might play a role in the differential organization of pre- and postsynaptic membranes. Besides their extracellular adhesive specificities, many CAMs also exhibit evolutionarily well-conserved, cytoplasmic binding activities to different cytoskeletal elements. These interactions appear to be of special importance in integrating different structural and functional aspects of the synaptic apparatus. More recently, it has become increasingly clear that many adhesive proteins directly or indirectly influence various cellular signaling processes. This is relevant not only during synapse formation but also during synaptic functioning and remodeling. In turn, cellular signaling processes, especially those involving protein phosphorylation and proteolysis as well as interactions with the cytoskeleton are known to regulate the adhesive ability of many CAMs. For synaptic CAMs, this may be important for facilitating synaptic plasticity, when existing synaptic connections are weakened or severed. Therefore, synaptic CAMs may be directly involved in processes like long-term potentiation and depression and synaptic remodeling. Almost all of the major CAM families have one or more representatives that are expressed at synaptic contacts, and different classes of synapses appear to have specific subsets of adhesive proteins. Although all chemical synapses share some general characteristics, this variety of CAMs is certainly part of the structural and functional diversity between different types of synaptic contacts. While our knowledge of how different CAM families contribute to synapse formation and functioning is still incomplete, the available data support some general themes, which are summarized above and in the following chapters. In the coming years, our understanding of the crucial role of CAMs at synapses will certainly deepen and possibly new adhesive molecules will join the list of known synaptic CAMs that are discussed in this book.

Today the term synapse is used in connection with three different types of cellular junctions (Yamada and Nelson 2007). It describes contact points not only between neuronal cells but also between immune cells and epithelial cells. An immunological synapse is the interface between antigen-presenting cells (e.g., macrophages, dendritic or activated B cells) and lymphocytes (Grakoui et al. 1999). Adhesion complexes, such as tight and adherent junctions, between epithelial cells are sometimes referred to as epithelial synapses (Yamada and Nelson 2007). However, usually the term synapse alludes to neuronal synapses. The majority of neuronal synapses are chemical based, as presumed in the preceding part of this chapter. More recently, evidence for an alternative type of neuronal synapse has emerged, which uses an electrical mode of signal transduction. These electrical synapses are formed by connexin/pannexin-containing gap junctions, which allow the direct propagation of the action potential from one neuronal cell to the next without the need for a chemical transmitter intermediate (Connors and Long 2004). As gap junctions form small cytoplasmic connections between neighboring cells, the existence of electrical synapses might be viewed as a partial exoneration of Camillo Golgi's old idea that neuronal cells are directly linked to each other. The relative importance of electrical versus chemical synapses currently remains unclear. Obviously, the structural and functional interactions between neuronal cells and their targets have grown increasingly intricate and multifaceted. As Santiago Ramón y Cajal pointed out in 1906 "Unfortunately, nature seems unaware of our intellectual need for convenience and unity, and very often takes delight in complication and diversity. Besides, we believe that we have no reason for scepticism. While awaiting the work of the future, let us be calm and confident in the future of our work" (Ramón y Cajal 1967).

References

Akert K, Moor H, Pfenninger K et al. (1969) Contributions of new impregnation methods and freeze etching to the problems of synaptic fine structure. In: Akert K and Waser PG (eds) Prog Brain Res. Elsevier Publishing Company, Amsterdam/London/New York

Berlucchi G (1999) Some aspects of the history of the law of dynamic polarization of the neuron. From William James to Sherrington, from Cajal and van Gehuchten to Golgi. J Hist Neurosci 8:191–201

Connors BW and Long MA (2004) Electrical synapses in the mammalian brain. Annu Rev Neurosci 27:393–418

Dale HH (1914) The action of certain esters and ethers of choline. J Pharmacol 6:147–190

De Robertis ED and Bennett HS (1955) Some features of the submicroscopic morphology of synapses in frog and earthworm. J Biophys Biochem Cytol 1:47–58

Del Castillo J and Katz B (1954) Quantal components of the end-plate potential. J Physiol 124:560–573

Eccles JC (1937) Synaptic and neuro-muscular transmission. Physiol Rev 17:538–555

Elliott TR (1904) On the action of adrenalin. J Physiol 31(suppl.):XX–XXI

Elliott TR (1905) The action of adrenalin. J Physiol 32:401–467

Estable C, Reissig M and De Robertis E (1954) Microscopic and submicroscopic structure of the synapsis in the ventral ganglion of the acoustic nerve. Exp Cell Res 6:255–262

Ewins AJ (1914) Acetylcholine, a new active principle of ergot. Biochem J 8:44–49

Fatt P and Katz B (1952) Spontaneous subthreshold activity at motor nerve endings. J Physiol 117:109–128

Foster M and Sherrington CS (1897) A Text Book of Physiology. Macmillan and Co., New York
Gerlach J (1872) Von dem Rückenmark. In: Stricker S (eds) Handbuch der Lehre von den Geweben des Menschen und der Thiere. Verlag von Wilhelm Engelmann, Leipzig
Golgi C (1873) Suella struttura della sostanza grigia del cervello [On the structure of the brain grey matter]. Gazzetta Medica Italiana Lombarda 33:244–246
Golgi C (1967) The Neuron Doctrine – Theory and Facts (Lecture delivered Decmber 11, 1906). In: (eds) Nobel Lectures, Physiology or Medicine 1901–1921. Elsevier Publishing Company, Amsterdam-London-New York
Grakoui A, Bromley SK, Sumen C et al. (1999) The immunological synapse: a molecular machine controlling T cell activation. Science 285:221–227
Grant G (2007) How the 1906 Nobel Prize in Physiology or Medicine was shared between Golgi and Cajal. Brain Res Rev 55:490–498
His W (1889) Die Neuroblasten und deren Entstehung im Embryonalen Mark. Abhandlungen der mathematisch-physischen Classe der Königl. Sächsischen Gesellschaft der Wissenschaften 15:311–372
Loewi O (1921) Über humorale Übertragbarkeit der Herznervenwirkung. I. Mitteilung. Pflüger's Arch ges Physiol 189:239–242
Palay SL (1956) Synapses in the central nervous system. J Biophys Biochem Cytol 2:193–202
Ramón y Cajal S (1937) Recollections of my Life. M.I.T. Press, Cambridge, MA
Ramón y Cajal S (1967) The structure and connexions of neurons (Lecture delivered December 12, 1906). In: (eds) Nobel Lectures, Physiology or Medicine 1901–1921. Elsevier Publishing Company, Amsterdam-London-New York
Robertson JD (1953) Ultrastructure of two invertebrate synapses. Proc Soc Exp Biol Med 82:219–223
Sjöstrand FS (1958) Ultrastructure of retinal rod synapses of the guinea pig eye as revealed by three-dimensional reconstructions from serial sections. J Ultrastruct Res 2:122–170
Tansey EM (1997) Not committing barbarisms: Sherrington and the synapse, 1897. Brain Res Bull 44:211–212
Tansey EM (2006) Henry Dale and the discovery of acetylcholine. C R Biol 329:419–425
von Euler US (1946) A specific sympathomimetric ergone in adrenergic nerve fibres (Sympathin) and its relations to adrenaline and nor-adrenaline. Acta Physiol Scand 12:73–97
von Koelliker A (1896) Handbuch der Gewebelehre des Menschen. Engelmann, Leipzig
Waldeyer-Hartz W (1891) Ueber einige neuere Forschungen im Gebiete der Anatomie des Centralnervensystems. Deutsche medicinische Wochenschrift 17:1213–1218, 1244–1216, 1287–1219, 1331–1212 and 1350–1216
Whitby LG, Axelrod J and Weil-Malherbe H (1961) The fate of H^3-norepinephrine in animals. J Pharmacol Exp Ther 132:193–201
Yamada S and Nelson WJ (2007) Synapses: sites of cell recognition, adhesion, and functional specification. Annu Rev Biochem 76:267–294

Chapter 2
Cell Adhesion Molecules at the *Drosophila* Neuromuscular Junction

Franklin A. Carrero-Martínez and Akira Chiba

Abstract A major goal in neuroscience is the understanding of organizational principles underlying cellular communication and the ensuing molecular integrations that lead to a functional nervous system. The establishment of neuromuscular connections (junctions) is a complex process that requires enumerable cellular and molecular interactions. There are many known and well-characterized molecular events involved in every aspect of neuromuscular junction (NMJ) formation. For instance, at the presynaptic side the motoneuron must differentiate, polarize, undergo dendrogenesis and axogenesis, and extend its processes out to the muscle field. This requires axon guidance, pathfinding, and finally synaptogenesis. At the postsynaptic side, the muscle cell must differentiate and find its correct place in the embryonic body plan to receive motor axons. There are many molecules known to play essential roles during each step in these self-organizational processes. Genetic and biochemical studies have identified molecules that facilitate accurate synaptic target recognitions, as well as those responsible for pre- and postsynaptic specializations. Cell adhesion molecules (CAMs) are known to play an essential role in establishing the NMJ. In this chapter, we begin by exploring *Drosophila* and its NMJ as a model system for glutamatergic synapses in the mammalian central nervous system. We continue by discussing selected CAMs, with known roles in *Drosophila* NMJ formation. We also explore the role these CAMs play in establishing the basic cytoarchitecture that ultimately results in functional neuromuscular synapses. We then examine the role CAMs play in synapse formation and plasticity. We conclude by providing an integrative model for CAMs function during synapse formation.

Keywords *Drosophila* · Filopodia · Myopodia · Cell adhesion molecule (CAM) · Capricious (Caps) · Connectin (Con) · Down syndrome cell

F.A. Carrero-Martínez (✉)
Department of Biology, University of Puerto Rico, Mayagüez, Puerto Rico 00681-9012
e-mail: franklin.carrero@upr.edu

M. Hortsch, H. Umemori (eds.), *The Sticky Synapse*,
DOI 10.1007/978-0-387-92708-4_2,

adhesion molecule (Dscam) · Fasciclin II (FasII) · Fasciclin III (FasIII) · Integrin · N-Cadherin · Neuroglian (Nrg) · Toll

2.1 Introduction

Considering the number of neurons (billions in the human brain) and the connections among them (trillions), the study of how neuronal networks emerge is a daunting task. Even with available modern tools, addressing this fundamental question is difficult and appears virtually impossible. While animals display seemingly endless variations of different developmental strategies, the underlying molecular mechanisms of assembling a functional neuromuscular network are surprisingly well conserved between chordate and arthropod species.

For this reason, the use of simpler model organisms such as the fruit fly *Drosophila melanogaster* has allowed the identification, cloning, and functional assessment of genes at the molecular, cellular, and organism levels. This model organism offers a well-characterized repertoire of genetic tools, a relatively short life span, a rapid reproduction rate, a panel of efficient molecular techniques, and a completely sequenced and mapped genome (Adams et al. 2000). In addition, due to a high degree of evolutionary conservation, the analysis of gene functions in *Drosophila* yields information that is usually relevant for and applicable to more complex organisms, such as mice and humans.

The vertebrate nervous system is divided into two main systems: central nervous system (CNS) and peripheral nervous system (PNS). The CNS is composed of the spinal cord and the brain, while the PNS is composed of sensory neurons and the neurons that connect them to the brain. In *Drosophila*, the nervous system is divided into two systems as well: CNS and PNS. The fly CNS is composed of a series of neuronal cell bodies grouped into clusters, called ganglia. These ganglia are connected to each other by parallel connectives along the ventral midline axis of the organism as well as perpendicular commissures, giving rise to the characteristic ladder-like organization of the ventral nerve cord (VNC). Motor neurons send their axons away from the VNC forming an anterior and posterior fascicle, also known as intersegmental nerve and segmental nerve, respectively. The PNS is formed by sensory input neurons (multiple dendritic neurons, external sensory organs, and chordotonal organs), which carry information to the CNS using the anterior and posterior fascicles (Hartenstein 1993).

The *Drosophila* neuromuscular network has been established as a standard genetic and cell biological model by several pioneers such as Corey Goodman, Michael Bate, Haig Keshishian, and many others. Developmental processes can be analyzed in *Drosophila* at the level of a single gene or a single cell, an ability that is essential for studying the underlying fundamental organizational principles of complex self-organizing cellular networks (Hoang and Chiba 2001).

Motor neurons in the developing CNS and their muscle cell targets are experimentally accessible during embryonic development and follow a stereotypic pattern in each segment (Landgraf et al. 1997, Schmid et al. 1999), which persists through larval development. Individual neuron lineages, axon pathways, synaptic target muscles, and the types of synaptic boutons axons develop have all been documented (Chiba 1999, Schmid et al. 1999, Landgraf et al. 2003). In each half-segment, a total of 34 neurons, including 2 which are bilaterally innervating ventral unpaired median (VUM) motoneurons, make up the motor neuron pool that innervates 30 embryonic muscle cells by the end of embryogenesis. This means that muscle and neuronal cells are each uniquely identifiable with numbers considerably smaller than in vertebrate nervous systems. This innervation ratio, together with a stereotypical spatial arrangement, means that a given neuron/muscle synaptic pair can be reproducibly accessed for analysis during well-defined embryonic developmental stages (Fig. 2.1). A diagram of the stereotypical neuronal and muscle cells localization is provided in Fig. 2.2. Table 2.1 provides a convenient conversion for the two existing muscle nomenclature systems.

The *Drosophila* NMJ is glutamatergic and thus often considered as a convenient model for studying regulatory mechanisms for mammalian central glutamatergic synapses (Johansen et al. 1989, Budnik 1996, Keshishian et al. 1996, Davis and Goodman 1998, Chiba 1999). Thus, the underlying general

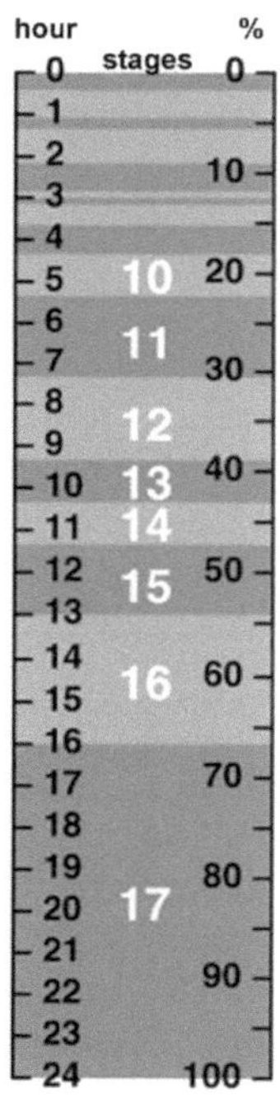

Fig. 2.1 ***Drosophila*** **embryonic development**. Wild-type embryonic development at 25°C has been characterized in different scales such as (*left*) hours after egg laying (AEL), (*center*) morphological and developmental events defining stages (Campos-Ortega and Hartenstein 1985), and (*right*) completed development as a percentage function

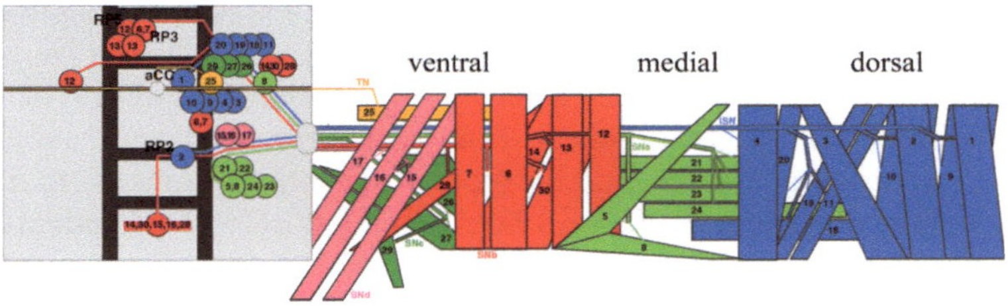

Fig. 2.2 Schematic representation of *Drosophila* neuromuscular network. Synaptic matchmaking between motoneurons (*left*) and embryonic muscles (*right*) is color coded according to the innervating nerve branch. Neuronal cell body localization is presented with the muscle number they innervate. Neurons commonly referenced throughout (RP5, RP3, aCC, RP2) are specifically named. Axons of the intersegmental nerve (ISN) and their partner muscles are shown in *blue*, while the transverse nerve (TN) is shown in *orange*. The segmental nerve (SN) branches are shown as follows: SNa (*green*), SNb (*red*), SNc (*green*), SNd (*pink*). There are two different naming conventions for *Drosophila* embryonic muscles. In this diagram we used the muscle numbering convention. Please refer to Table 2.1 for the corresponding name in the muscle location convention. For reference, the anteroposterior axis of the *Drosophila* embryo is always presented top to bottom, while the dorsolateral (ventral) axis is from right to left. That is, CNS is located to the left of the muscle field

Table 2.1 Muscle nomenclature conversion table

Number	Name	Number	Name
1	Dorsal acute 1 (DA1)	16	Ventral oblique 5 (VO5)
2	Dorsal acute 1 (DA2)	17	Ventral oblique 6 (VO6)
3	Dorsal acute 3 (DA3)	18	Dorsal transverse 1 (DT1)
4	Lateral longitudinal 1 (LA1)	19	Dorsal oblique 4 (DO4)
5	Lateral oblique 1 (LO1)	20	Dorsal oblique 5 (DO5)
6	Ventral longitudinal 3 (VL3)	21	Lateral transverse 1 (LT1)
7	Ventral longitudinal 4 (VL4)	22	Lateral transverse 2 (LT2)
8	Segmental border muscle (SMB)	23	Lateral transverse 3 (LT3)
9	Dorsal oblique 1 (DO1)	24	Lateral transverse 4 (LT4)
10	Dorsal oblique 2 (DO2)	25	Ventral transverse 1 (VT1)
11	Dorsal oblique 3 (DO3)	26	Ventral acute 1 (VA1)
12	Ventral longitudinal 1 (VL1)	27	Ventral acute 2 (VA2)
13	Ventral longitudinal 2 (VL3)	28	Ventral oblique 3 (VO3)
14	Ventral oblique 1 (VO1)	29	Ventral acute 3 (VA3)
15	Ventral oblique 4 (VO4)	30	Ventral oblique 2 (VO2)

There are two existing naming conventions for the embryonic and larval musculature. Throughout this chapter we use the muscle numbering nomenclature (Bate and Rushton 1993); however, since some references use the muscle location naming convention (Crossley 1978), we provide this table to ease cross-referencing.

principles presented here may apply to other systems. Ultimately (ignoring the specific identities of the cells discussed in this chapter), the fundamental question is (reduced to) why and how two cells decide to connect (synapse), remodel that connection (synaptic plasticity), or abnormally end their interaction (neurodegeneration).

2.2 CAMs at the NMJ

CAMs play critical roles in every single developmental stage leading up to the formation of functional NMJs. The study of CAMs has provided us with a functional explanation for the observed explicit cell motilities and required molecular integration within the emerging NMJ network. Here we provide a short list of cell-specific membrane-spanning CAMs that have been identified as target recognition molecules in the *Drosophila* neuromuscular system. Figure 2.4 provides a visual representation of the expression pattern of the molecules discussed below.

2.2.1 Capricious

Capricious (Caps) is a single-span transmembrane protein with 14 leucine-rich repeats (LRRs) in its extracellular portion. Caps is regulated by the transcription factor Krüppel and necessary for proper defasciculation of SNb axons (Abrell and Jackle 2001). Presynaptically, Caps is expressed in the anterior corner cell (aCC), RP2, U, and RP5 motoneurons. These cells innervate their Caps-positive muscle partners, muscles 1, 2, 9, 10, and 12. Caps-positive muscles innervated by Caps-negative neurons are muscles 14, 28, 15, 16, and 17 (Shishido, Takeichi and Nose 1998). Caps loss-of-function (LOF) results in muscle 12's motor axons miswiring to muscle 13. In muscles, Caps intracellular domain mediates target recognition, but not in neurons (Taniguchi et al. 2000). However, when Caps is overexpressed in all muscles, RP5 initially contacts muscle 12, before innervating muscle 13 (Shishido et al. 1998, Taniguchi et al. 2000). Taken together, these results suggest a mechanism by which upstream molecular events mediate Caps expression.

2.2.2 Connectin

Drosophila connectin (Con) is a cell surface protein with ten LRRs thought to mediate homophilic attractive adhesion (Nose et al. 1997). Starting at embryonic stage 12, Con is expressed in ventral and lateral muscles and on the intersegmental nerve (ISN) and segmental nerve (SN) axonal tracts that innervate them (Nose et al. 1992). This protein is proposed to play a dual role at NMJs, where it specifies (a) muscle pattern formation and (b) synapse formation. For instance, in muscles 18 and 21–24 an accumulation of Con protein to high levels is observed at muscle–muscle boundaries. In Con null mutants, gaps between these muscles are visible, while other Con-negative muscles develop normally (Raghavan and White 1997). Con gain-of-function (GOF) conditions, which are induced with pan-muscular promoters, do not result in major CNS, PNS, or muscle defects (Nose et al. 1992, 1997). At the presynaptic side, the protein is negatively regulated by the engrailed gene product (Siegler and Jia 1999). Con is also expressed on the surface of glial cells PG1, PG3, and glial-like cell PG4 (Nose, Mahajan and Goodman 1992), which are thought to provide guidance cues for motoneuron axons. Given the dual roles in muscle pattern formation and synaptogenesis, we propose Con to play a general role in target selection at the NMJ.

2.2.3 Down Syndrome Cell Adhesion Molecule

Down syndrome cell adhesion molecule (Dscam, see also Chapter 9) is the *Drosophila* homologue of human Down syndrome cell adhesion molecule (DSCAM) and participates in NMJ presynaptic cell (motor neuron) pattern formation. It has been proposed that it may modulate the actin cytoskeleton through activation of the adaptor proteins Pak and Dock. The fly gene encodes a transcript that can be alternatively spliced to generate more than 38,000 predicted protein isoforms. These protein isoforms usually contain ten immunoglobulin (Ig) and six fibronectin (FN) type III extracellular domains. Dscam mutants are lethal during early larval development and exhibit a mild disorganization of the connective and commissural tracts within the ventral nerve cord (Schmucker et al. 2000). This protein is expressed in all muscles and all motor neurons; however, expression patterns of alternatively spliced isoforms are not known. This information may lead to a better understanding of adhesive regulation and activation of intracellular events.

2.2.4 Fasciclin II

Fasciclin II (FasII) is a homophilic CAM known to be important for the development, maintenance, and plasticity of the NMJ pattern. This protein contains five Ig and two FN type III domains (Grenningloh et al. 1991) and is considered as the fly ortholog of the mammalian neuronal cell adhesion molecule (NCAM). All motoneuron axon pathways and their growth cones express this protein from axonal outgrowth to synapse formation (van Vactor et al. 1993). The protein is also expressed at low levels in all muscle cells (Davis et al. 1997). Genetic increase in presynaptic FasII results in fusion of motoneuron axons, while genetic decrease leads to a complete or partial defasciculation of motor axon pathways (Lin et al. 1994, Lin and Goodman 1994). In pioneer axons such as aCC/RP2, FasII has been demonstrated to be required and sufficient to facilitate guidance of follower axons (Sanchez-Soriano and Prokop 2005). This suggests that FasII plays an essential role in the establishment of the presynaptic cell pattern. Postsynaptically, FasII is necessary for the postsynaptic accumulation of various proteins, including the scaffolding protein Discs large (Dlg), glutamate receptor subunits GluRIIA and GluRIIB, and FasII itself (Kohsaka et al. 2007). Hypomorphic mutant alleles (in which FasII levels were reduced by 50%) show a significant increase in presynaptic bouton numbers, but not in synaptic transmission (Schuster et al. 1996b, a). Furthermore, a transient increase in FasII levels in specific muscle cells results in the formation of new ectopic functional synapses in those muscle cells (Davis et al. 1997). There is considerable evidence that FasII is able to activate intracellular signaling events through its interactions with PDZ (*P*ostsynaptic density protein (PSD95), *Drosophila* disc large tumor suppressor (DlgA), and *Z*onula occludens-1 protein (zo-1)) domain-containing proteins such as Dlg (Kohsaka et al. 2007) and dX11/Mint/Lin-10 (Ashley et al. 2005). These experiments focused on embryonic and larval developmental stages and raise the possibility of a developmentally regulated choice between various PDZ scaffolding proteins as their interacting molecules, which either initiate synapse formation or modify NMJs during later developmental stages. For additional regulatory mechanisms involving FasII, refer to Packard et al. (2003). Taken together, all this evidence, together with its expression pattern (Fig. 2.4d), suggests that FasII plays essential roles in pattern formation and synapse initiation and maintenance, but not in target selection.

2.2.5 Fasciclin III

Fasciclin III (FasIII) is a single-span transmembrane immunoglobulin superfamily (IgCAM) member with three extracellular Ig domains that mediate homophilic adhesion and a PDZ-binding cytoplasmic domain that mediates interaction with postsynaptic Dlg (Woods et al. 1996). FasIII is normally expressed in muscle 6 and muscle 7 and the RP motoneuron axons, including RP3 which is part of the SNb nerve branch. LOF results in a failure of RP3 axons to innervate their normal target, while GOF experiments show that RP3 mistargets neighboring muscles misexpressing FasIII. In wild-type embryos, both aCC motoneurons, which are part of the ISN, and muscle 2 are FasIII negative. However, when FasIII is misexpressed in both aCC motoneurons and muscle 2, aCC axons misinnervate muscle 2 (Chiba et al. 1995). Furthermore, when a cytoplasmically truncated form of FasIII, which maintains its homophilic interacting domain, is misexpressed in all neurons, axon–muscle adhesion is observed. However, whether or not this leads to successful synaptogenesis is still unknown (Rose et al. unpublished). FasIII's cell-specific expression pattern may dictate its function as a positive target recognition molecule.

2.2.6 Integrins

Integrins are part of a large family of heterodimeric transmembrane proteins with five α subunits and a single β subunit in *Drosophila*. These six subunits generate five different integrin heterodimers. In addition to many other roles during embryonic development, it has been suggested that integrins play a role in linking the presynaptic partner axon with the postsynaptic muscle cell. *Drosophila* embryonic muscles express α_{PS1}/β_{PS} (PS1) and α_{PS2}/β_{PS} (PS2) heterodimeric integrins. PS1, encoded by the gene *myospheroid*, is reported to bind to the ECM component laminin, while PS2 (encoded by the gene *inflated*) has RGD-binding activity (Gotwals et al. 1994). α_{PS1} and α_{PS2} integrin knockout mutations lead to widespread miswiring and reduced synaptogenesis (Hoang and Chiba 1998). That is, axonal fasciculation appears normal, but embryonic motoneuron axons overshoot their target muscles. Neuronal expression of an integrin transgene into the knockout greatly reduces axonal misguidance, but still fails to rescue synaptogenesis (Hoang and Chiba 1998). In embryos lacking postsynaptic α_{PS} integrins, NMJ adhesion is affected, but presynaptic synaptotagmin accumulation occurs at wild-type levels (Prokop et al. 1998). β_{PS} null mutant animals exhibit a muscle fiber twitch, even after the characteristic detached phenotype (refer to Section 2.3.2), suggesting that synaptic transmission still occurs in this altered NMJ (Prokop et al. 1998). However, at the larval NMJ synaptic arborization is greatly reduced (Beumer et al. 1999). This observation may be explained by integrins' ability to recruit essential postsynaptic components such as Dlg and FasII to the postsynaptic membrane (Beumer et al. 2002). These observations suggest that *Drosophila* integrins play multiple roles during NMJ formation and their postembryonic development.

2.2.7 N-Cadherin

N-cadherin (N-Cad) is an evolutionarily conserved classical cadherin with a large, complex extracellular domain that is composed of 15 cadherin repeats, a Fcc box (fly classic cadherin box), 2 cysteine-rich domains, and a laminin A globular segment. In addition it contains a catenin-binding cytoplasmic domain (Salinas and Price 2005, Suzuki and Takeichi 2008). Identification of 12 developmentally regulated alternative splice variants highlights a role of classical cadherins in synaptogenesis (reviewed in Halbleib and Nelson 2006). A common molecular architecture among splice variants, with differences in their extracellular and membrane-spanning domains, has been described (Yonekura et al. 2006). N-Cad regulates axonal pattern formation, presumably by regulating axonal fasciculation in the developing embryo (Iwai et al. 1997). However, a new study highlights the importance of these splice variants at the onset of synaptogenesis as they are differentially expressed in either presynaptic neuronal cells or the postsynaptic muscle cells (Hsu et al. unpublished). Identification of splice variants expression pattern is an essential step toward the understanding of how an organism fine-tunes its cellular connectivity.

2.2.8 Neuroglian

Neuroglian (Nrg) contains six Ig-like domains and five FN type III domains and participates in homophilic interactions (Hortsch 2000). Alternative splicing generates 2 isoforms with an identical extracellular domain and 53 additional amino acid residues in the cytoplasmic domain of the neuronally expressed protein isoform (Hortsch et al. 1990). Protein expression pattern is negatively regulated by engrailed (Siegler and Jia 1999), with a shorter protein form expressed in body wall muscles, trachea, and gut and the longer form expressed in CNS and peripheral nervous system (PNS) neurons and their processes (Hall and Bieber 1997). The neuron-specific isoform is expressed in RP1, RP2, RP3, aCC, and pCC (posterior corner cell) motoneuron axonal projections and glial cells associated with them as they exit the CNS (Bieber et al. 1989). The muscle-specific isoform is expressed at high levels in muscles 7, 6, 13, 12, and 4 and at lower levels in other muscle cells and accumulates at the future site of synaptogenesis. Nrg null mutant analysis revealed motoneuron axon misprojections and stalling close to the target postsynaptic muscle cell. These mutants show complete embryonic development, but fail to hatch (Hall and Bieber 1997). The fact that Nrg accumulates at the future site of synaptogenesis raises the possibility that Nrg plays an essential at the NMJ. As proposed below, it will be interesting to investigate Nrg distribution within the myopodia and myopodial cluster.

2.2.9 Toll

Toll is a member of the LRR family of transmembrane proteins. This protein contains 15 extracellular LRR domains and is expressed in the embryonic muscles but preferentially accumulates at muscle–muscle contact. Toll displays a dynamic spatiotemporal expression pattern during axon targeting and exerts an inhibitory influence on motoneuron axons (Rose et al. 1997, Rose and Chiba 1999, Suzuki et al. 2000). Genetic misexpression of Toll in muscle 12 beyond hour 15 AEL results in RP5 motoneuron stalling just before its partner muscle. It has been proposed that Toll spatiotemporal regulation is crucial for its role in development, specifically the local inhibition of synaptogenesis of specific motoneurons (Rose et al. 1997).

In general, the CAMs reviewed here have specific expression patterns. In some cases there is a general expression pattern in both neurons and muscles, while the expression of other CAMs is restricted to a subset of neurons and/or muscle cells. Further studies addressing splice variants and their developmental regulation will lead to a better understanding of the affinity-based selection process in support of NMJ pattern formation and connectivity. The expression patterns of individual CAMs are presented in Fig. 2.4.

2.3 CAMs and Neuromuscular Network Formation

In this section we look at the essential roles that CAMs play in the establishment of the neuromuscular circuits at the stereotypical locations characteristic of the *Drosophila* NMJ (Fig. 2.2). Starting at around embryonic stage 12, myoblasts fuse and motor axons start to navigate out of the ventral nerve chord. Muscle development occurs independently of motoneuron innervation and innervation occurs at the correct muscle partner cell even if position and/or morphology of its partner muscle are altered (Cash et al. 1992, Broadie and Bate 1993). Although both of these CAM-mediated events occur almost simultaneously we look at them separately to facilitate discussion.

2.3.1 Presynaptic Cell Pattern Formation

Drosophila neuronal network pattern formation is a critical, developmentally regulated process, in which CAMs and guidance cues help the axon to navigate the muscle field in search of its synaptic partner. CAMs play a critical role in establishing the neuronal network pattern by regulating three distinct types of adhesion: axon–extracellular matrix (ECM), axon–axon (i.e., fasciculation/defasciculation), and axon–muscle adhesion. In this section we cover axon–ECM and axon–axon adhesion events.

2.3.1.1 CAMs and Axon–ECM Adhesion

During embryonic development, motor axons navigate out to the periphery in search of their postsynaptic partners in a process known as axon pathfinding. All of these CNS axons must navigate and sort through many non-partner cells before contacting their respective synaptic targets. During this process, interactions with the ECM play a critical role for NMJ development and pattern formation (Ackley et al. 2003). At embryonic *Drosophila* stage 12 before muscle formation, mesoblasts intermingle with somatic mesoderm and start the deposition of collagen type IV (Mirre et al. 1988). Immunostaining confirmed the presence of this ECM component at this early developmental stage and showed collagen sheaths enveloping muscles, CNS, and other structures (Lunstrum et al. 1988). In general, integrin-mediated cell adhesion to the ECM provides a way for cells to adhere to a substrate in support of axon navigation toward its postsynaptic partner without engaging in a direct cell–cell interaction. This may account for the observation that integrin LOF mutants show severe patterning defects (Brown 2000). In this context, attachment of motoneuron axons to the ECM is a crucial and essential step to provide the mechanical stability that is essential for continued navigation. This principle has been demonstrated through surgical axotomy in a live, undissected embryo. When the developing aCC axon is cut, the resulting ends slowly recoil away from each other. This slow recoil suggests adhesion to the surrounding ECM (Siechen et al. unpublished). Dynamic regulation of these ECM interactions may be provided through matrix metalloproteinases (MMPs). MMPs are a large family of conserved proteases with two representatives in the *Drosophila* genome (Page-McCaw 2008). They are strongly expressed starting at embryonic stage 14 (Miller, Page-McCaw and Broihier

2008) and are able to degrade the basement membrane proteins fibronectin and type IV collagen and the ECM (Llano et al. 2000), which has led to the hypothesis that they clear ECM materials for supporting axonal growth cone pathfinding (McFarlane 2003). This type of cell adhesion may provide physical stability as the axon further explores the peripheral muscle field in search of its synaptic partner, even in the presence of a moving target (see below).

2.3.1.2 CAMs and Axon–Axon Adhesion

Axons that exit the CNS at the anterior fascicle eventually form the ISN, while those exiting the CNS at the posterior fascicle form the segmental nerves a–d (Hartenstein 1993). CAMs play an essential role in axon pattern formation (please refer to Fig. 2.2). For example, FasII is expressed on all motoneurons during embryogenesis and is necessary to maintain adhesion between axons in a process called fasciculation (van Vactor et al. 1993, Lin and Goodman 1994). When this FasII is removed, axonal growth cones do not extend properly and fail to fasciculate (Grenningloh et al. 1991). Fasciculation and defasciculation must be spatiotemporally regulated in order to allow for the formation of the highly stereotypical pattern of motor axons at the embryonic NMJ. It has recently been shown that MMPs may not be required for motoneuron axon extension, but instead promote FasII-mediated motor axon fasciculation and antagonize the semaphorin signaling pathway (Miller et al. 2008). The semaphorin pathway is essential for motor axon defasciculation. FasII or Con LOF mutations suppress Semaphorin LOF phenotypes, indicating that defasciculation of motoneuron axons occur through interference with axon–axon adhesion (Winberg et al. 1998, Yu et al. 2000). Other studies show that FasII is required to facilitate guidance of follower axons (Sanchez-Soriano and Prokop 2005), therefore playing an essential role in the establishment of the neuronal pathway. Taken together, these observations suggest that the right amount of cellular adhesion must be present or at least dynamically regulated in order for motoneuron axons to fasciculate/defasciculate at choice points en route to their synaptic targets.

2.3.1.3 CAMs and Axon–Muscle Adhesion

ECM deposition begins during early embryonic stages (Lunstrum et al. 1988, Mirre et al. 1988), even before the muscle cell pattern is established (see below). It is therefore likely that the interaction of a growth cone with the muscle surface is mediated by the ECM. Therefore, axon–muscle adhesion may not directly contribute to the establishment of the presynaptic cell pattern formation, but instead ECM interaction plays a larger role than previously considered. However, recent observations provide a novel role for axon–muscle adhesion in support of synaptogenesis. Please refer to Section 2.7.1 below for a discussion of these findings.

2.3.2 Postsynaptic Cell Pattern Formation

During embryonic development, myoblasts fuse to give rise to a multinucleated single syncytial muscle cell. In *Drosophila* this process involves neuronal CAM (NCAM/FasII), sticks and stones, and dumbfounded (also called Kirre) (Dworak and Sink 2002), all of which are members of the Ig superfamily. By the end of embryonic stage 13, the characteristic muscle cell pattern (Fig. 2.3) starts to emerge. The establishment of the embryonic muscle

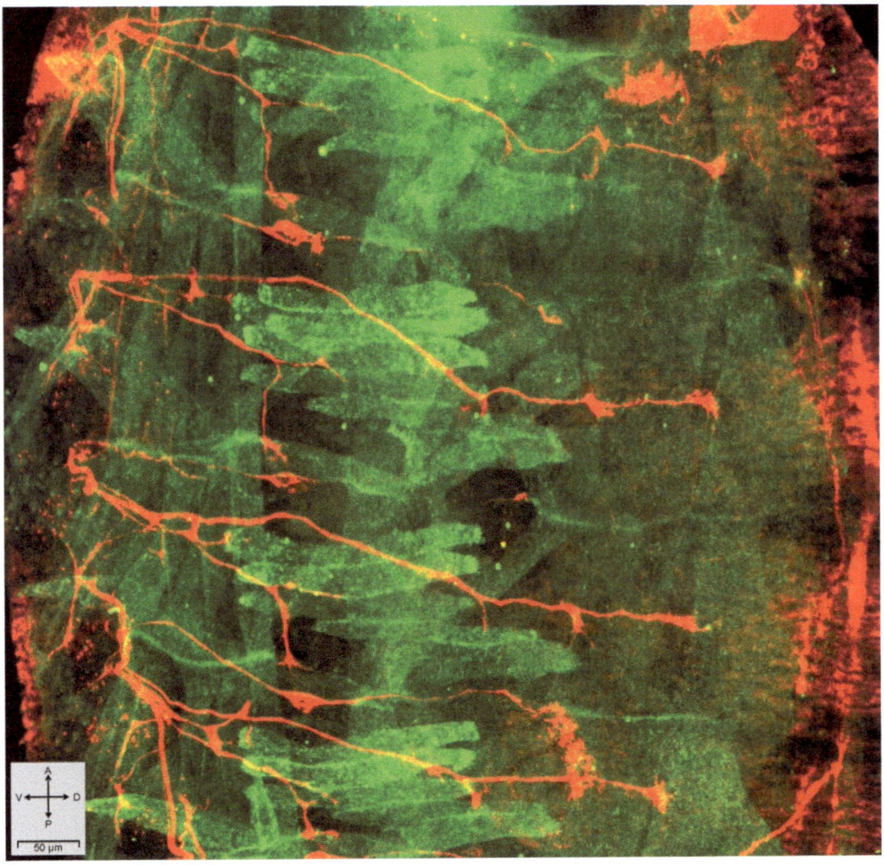

Fig. 2.3 Stereotypic embryonic neuromuscular cellular pattern. Laser scanning confocal micrograph projection reveals the highly stereotypic musculature (*green*; membrane-targeted GFP expression driven by the pan-muscle driver line Gal4-24B) and axonal (*red*; HRP antibody staining) pattern. The distinct stereotypical cellular pattern formed by each uniquely identifiable muscle and neuron, together with powerful genetic tools, makes the *Drosophila* NMJ an ideal system to dissect both cellular and molecular interactions under in vivo conditions. In this image, the anteroposterior axis runs from top to bottom and the dorsoventral axis is from right to left. This notation will be used throughout unless otherwise indicated

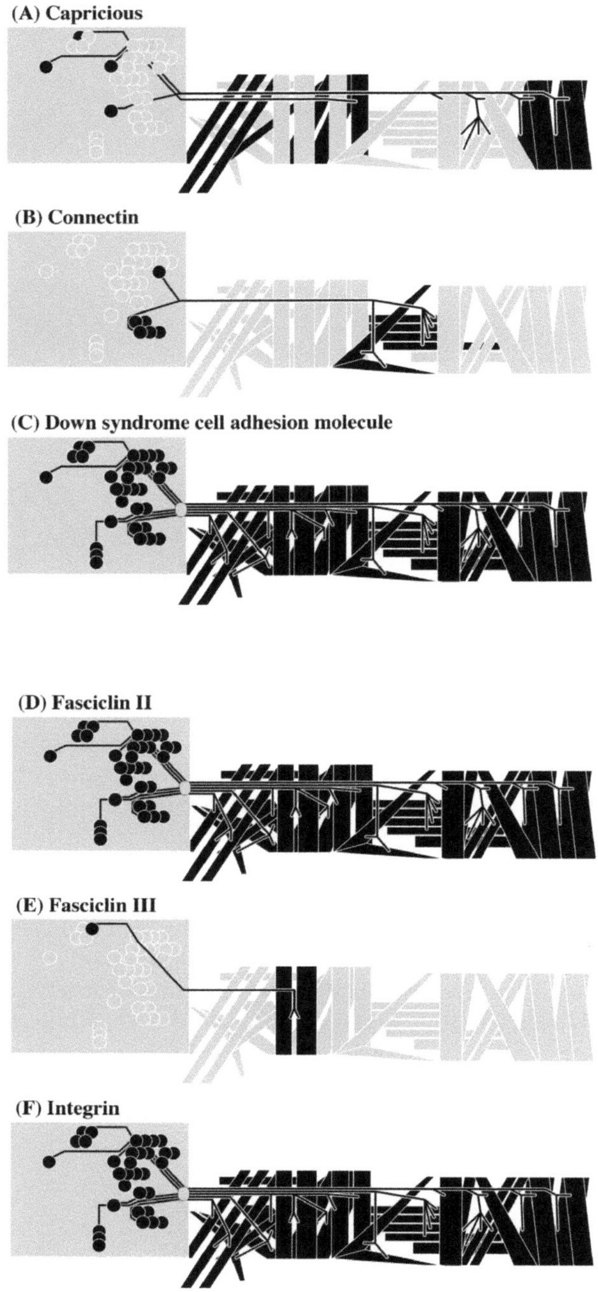

Fig. 2.4 CAM expression pattern at the embryonic NMJ. Diagrammatic representation of membrane-spanning adhesion molecules at the *Drosophila* neuromuscular junction. (**A**) capricious, (**B**) connectin, (**C**) Down syndrome cell adhesion molecule, (**D**) fasciclin II, (**E**) fasciclin III, (**F**) integrin, (**G**) N-cadherin, (**H**) neuroglian, and (**I**) Toll

Fig. 2.4 (continued)

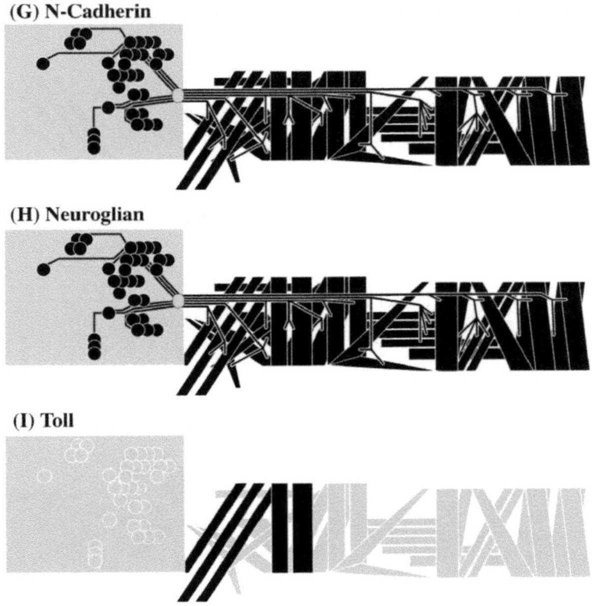

pattern is governed in part by the direct apposition of integrins. α_{PS1}/β_{PS} (PS1) and α_{PS2}/β_{PS} (PS2) integrin complexes in both the epidermis and the muscle surface facilitate muscle insertion at apodemes of the lateral epidermis (Fessler and Fessler 1989). Removal of these molecules results in the embryonic lethal *myospheroid* phenotype. Myospheroid is characterized by muscle detachment from their respective insertion sites as soon as contractions begin, resulting in rounded muscle fibers (Wright 1960, Bokel and Brown 2002). Another essential PS2 function during myogenesis is the recruitment of non-muscle myosin II, which is also required for proper embryonic muscle pattern development (Bloor and Kiehart 2001). Integrins are also involved in the organization of the actin–myosin contractile structure into sarcomeres (Brown 2000, Bloor and Kiehart 2001). This is an important postsynaptic structure that will eventually support muscle function after NMJ formation. Taken together, these observations highlight the roles that CAMs play in establishing the muscle cellular pattern and in the maturation of the postsynaptic cell in preparation for its contractile function.

2.4 CAM-Mediated Intracellular Signaling Activation at the NMJ

The list of required proteins to successfully assemble a synapse is fairly extensive and continues to grow as research efforts are directed toward the functional characterization of CAMs at the *Drosophila* NMJ. These research efforts have identified roles for a number of *Drosophila* CAMs in the activation of intracellular pathways on either side of the neuromuscular synapse. This activation may be mediated either directly by the CAM or indirectly through intermediary/linker proteins. Here we examine several examples in which CAMs facilitate the activation of intracellular events at the NMJ.

It has recently been shown that postsynaptic Dlg recruitment is mediated by FasII (Kohsaka et al. 2007), hinting at a CAM-mediated sequence of events at the onset of neuromuscular synapse formation. Dlg, the *Drosophila* representative of the mammalian PSD-95 family of scaffolding proteins, is essential for synaptogenesis of the embryonic NMJ (Carrero-Martínez et al. unpublished). It has been postulated that the multiple PDZ protein–protein interaction domains facilitate the molecular organization of the postsynaptic density (Thomas et al. 2000). Among the known postsynaptic molecular proteins interacting with Dlg are glutamate receptors, Shaker potassium channels, and many other essential postsynaptic proteins, including the CAMs FasIII, Nrg (Woods et al. 1996), and FasII (Thomas et al. 1997). Through these adhesive interactions, Dlg directly apposes its molecular counterparts on the presynaptic side (Thomas et al. 1997, Thomas et al. 2000), facilitating its role as a postsynaptic membrane coordinator.

Another example of CAM-mediated signaling at the NMJ involves *Drosophila* Ankyrin2 (Ank2). Ankyrins are adaptor proteins that mediate attachment of integral membrane proteins to the underlying actin cytoskeleton. *Drosophila* Ank2 is a neuron-specific ankyrin (Bouley et al. 2000), which is required for the proper organization of Nrg and FasII at the NMJ and for synaptic maintenance and functionality (Koch et al. 2008, Pielage et al. 2008). Currently, it is unknown whether Nrg is the Ank2-binding partner at the NMJ, but as Nrg is the major membrane-associated ankyrin2 ligand in *Drosophila* neurons (Bouley et al. 2000) it is an excellent candidate for interacting with Ank2 at the NMJ. Ank2 removal from the presynaptic cell results in disassembly of the larval NMJ due to the retraction of the microtubule-based cytoskeleton (Koch et al. 2008, Pielage et al. 2008). Ank2 is also associated with β-spectrin and when β-spectrin is removed from the presynaptic cell, FasII and Nrg become disorganized and eventually lose their NMJ localization. However, when spectrin is removed postsynaptically, similar defects were not observed (Pielage et al. 2005). These observations may explain why spectrin knockout flies have morphologically normal neurons at embryonic stages with reduced neurotransmission (Featherstone et al. 2001). At the larval NMJ, α- and β-spectrin RNAi knockout leads to synaptic disassembly and elimination (Pielage et al. 2005). Interestingly, the formation of the embryonic NMJ is not affected in Ank2 mutants, but as development progresses, motoneuron axons innervating posterior segments show a decreased bouton number, which manifests as posterior paralysis in the fly. Further dissection of these Ank2-mediated synaptic disintegration may prove relevant in the context of human motor neuron conditions (Koch et al. 2008).

CAMs provide the critical point-of-contact interactions that facilitate transduction of adhesive events intracellularly at both ends of the developing synapse. These CAM-mediated events are either directly or indirectly coupled to cytoplasmic proteins, which coordinate additional intracellular events, including essential interactions to the underlying cytoskeleton. The question of which protein initiates NMJ formation remains to be answered. Are CAMs the *first* molecules to set a signaling cascade in motion that leads to the activation of intracellular events? Or do intracellular events result in the clustering of CAMs at the site of synapse formation? According to the model we present below, a combination of both events may contribute to the successful genesis of the *Drosophila* neuromuscular synapse.

2.5 CAMs Mediate FORCES

By mutual and exclusive adhesion, CAMs are the ideal candidates to initiate intracellular signaling after the correct synaptic partner has been found. Recently, we have observed that muscles start to contract even before they become innervated (Siechen et al. unpublished). This is consistent with prior observations which established that muscle contractions precede the formation of the sarcomere in embryos (Volk et al. 1990, Bloor and Brown 1998). This type of movement is thought to originate either myogenically and/or due to the contraction of neighboring muscles sharing a common insertion point (Carrero-Martínez and Chiba unpublished). The timing of synaptogenesis initiation for each muscle, as assessed by axon–muscle contacts, varies slightly from muscle to muscle and is highly reproducible. Therefore, at the onset of neuromuscular synaptogenesis in vivo, the motoneuron is presented with a mechanically rich environment from which to select its correct synaptic target muscle.

While the significance of muscle contractions in the context of the developing NMJ remains to be addressed, we propose that CAMs mediate FORCES. *C*ell *A*dhesion *M*olecules mediate *F*orce-*O*rchestrated *R*etrograde *C*ommunication to *E*nhance the *S*ynapse and support its continued development. Preliminary observations support this model in which sufficient cellular adhesion is provided to withstand the muscle contractions while providing a way to transduce the axon–muscle adhesion event to both synaptic partners. While addressing this question we found that when a motoneuron axon is surgically cut at different developmental time points, different retraction/recoiling rates are observed. Prior to synaptogenesis, the resulting ends slowly recoil, but after the synapse is established, exogenously applied forces, which exceed physiological conditions, are insufficient to separate both synaptic partners (Siechen et al. unpublished). We concluded that adhesive molecules progressively accumulate as the NMJ develops into a functional structure. Taken together, these observations suggest that during axon navigation the source of adhesive stability may be provided by interactions with the surrounding ECM. As development proceeds, adhesive forces provided by precisely matched synaptic partners increase the mechanical stability of the developing synapse. Stay tuned as we and other research groups continue to dissect this novel role for CAMs at the NMJ.

2.6 CAMs in NMJ Plasticity

During the life cycle of *Drosophila*, there are two distinct classes of NMJs: the embryonic/larval NMJs (reviewed by Keshishian et al. 1996) and the adult NMJs (reviewed by Patricia K. Rivlin 2004). A popular and versatile model for studying NMJ development and plasticity has been established in a set of motoneurons found in abdominal segments. In the context of metamorphosis, adult muscle cells develop from twist-expressing myoblasts, which are present in the larva (Currie and Bate 1991). These cells are closely associated with neural cells and continue this association throughout metamorphosis, to at least until 51 hours after puparium formation (APF) when the adult muscle pattern has been completed (Currie and Bate 1991). In specific cases, surgical denervation at the onset of metamorphosis impairs muscle development, but not its characteristic distribution pattern (Currie and Bate 1995). A genetically induced reduction of FasII affects the morphology of adult NMJs by reducing the area, but not the size of individual synaptic boutons. Conversely, a genetically encoded increase of FasII results in increased numbers of boutons and presynaptic area (Hebbar et al. 2006).

Taken together, these results suggested a role of adhesion molecules in maintaining the muscle–neuron contact even during the dramatic reorganization that takes place during metamorphosis and that this interaction is essential for the development of adult muscle morphology and thus for adult fly function and behavior. The astounding NMJ cellular reorganization that takes place during the process of metamorphosis makes this system an excellent choice to study the process of synaptic plasticity.

2.7 A Two-Step Model for CAM-Mediated NMJ Formation

The body of knowledge, which has been generated during the last 20 years in the fields of adhesion molecules and synapse formation, has provided us with a staggering amount of molecular players, cellular events, and their respective interactions, which are essential for the formation of a functional NMJ synapse. Here we propose an integrative model, which takes into consideration earlier observations and recent work describing the appearance of postsynaptic filopodia and their role in NMJ formation. This two-step model puts the role that CAMs play during the process of synaptogenesis into a new perspective. The first step involves the presentation of CAMs at the tips of myopodia to facilitate attractive or repulsive growth cone recognition. The second step occurs locally at the postsynaptic cell surface and involves the CAM-mediated creation of a subcellular space that enhances the chance of molecular complexing, which is required for the establishment of a functional neuromuscular synapse.

2.7.1 *Myopodia Brings CAMs Closer to Navigating Motor Axons*

Using three- and four-dimensional analyses of undissected, live *Drosophila* embryos expressing GFP-based bioprobes, we have observed highly motile actin-based microprocesses called *myopodia* (Fig. 2.5) that extend from muscle cells before neuromuscular synaptogenesis is initiated in *Drosophila* (Ritzenthaler et al. 2000). Appearance of myopodia coincides with motoneuron outgrowths in both *Drosophila* (Ritzenthaler et al. 2000) and mouse embryos (Uhm et al. 2001, Misgeld et al. 2002). In vivo, these myopodia reach up to 30 μm in length translating to nearly the full length of a segment in a *Drosophila* embryo (Ritzenthaler et al. 2000). When considered in context with previous reports showing that neurofilopodia reach up to 15 μm in length (Johansen et al. 1989), it is conceivable that prior to synaptogenesis all synaptic partners in a given segment fall within direct reach of each other. Furthermore, the rapid protrusion/retraction rates of each individual myopodia may serve as a mechanism to break through ECM (please refer to Section 2.3.1.1 for

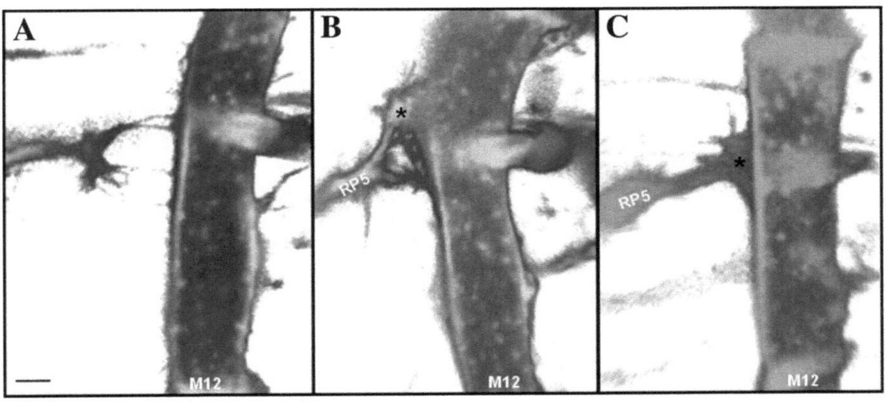

Fig. 2.5 Myopodial behavior at the onset of synaptogenesis. Myopodia are dynamic filopodia-like structures emerging from the postsynaptic muscle cell surface prior to synaptogenesis. (**A**) As development proceeds (hour 14:00 AEL), a precisely matched synaptic partner (RP5 motor neuron) starts its interaction with muscle 12's myopodia. (**B**) Within 30 min of initial contact, presynaptic and postsynaptic filopodia form a myopodial cluster (* in **B** and **C**). (**C**) This new subcellular space serves as a signaling hot spot where the postsynaptic density is organized at around hour 15:00. Live confocal imaging of membrane-targeted GFP expressed in muscle 12 and in neurons (b–d). Scale bar = 10 μm

additional details), making this an ideal structure for the presentation of 'recognition molecules', such as CAMs.

If we consider that every muscle presents its myopodia and every growth cone displays a number of filopodia, then the potential combinatorial interactions among pre- and postsynaptic filopodia are staggering. However, two developmentally regulated events greatly restrict the number of possible interactions. First, initial outgrowth of pioneer ISN axons takes place before muscle fibers have formed (Campos-Ortega and Hartenstein 1985). Second, with few exceptions myopodia have a strong tendency to aggregate toward the ventral nerve cord (Ritzenthaler and Chiba 2003). Thus, lateral and ventral muscles are not able to interact with pioneer axons which have navigated dorsally by the time these muscles are formed and start extending their myopodia. This spatiotemporal restriction is important and may explain why certain CAMs, which are simultaneously expressed by different muscles (i.e., Caps), support selective NMJ formation.

Since myopodia formation is a transient and unique embryonic event, which has only recently been described (Ritzenthaler et al. 2000, Ritzenthaler and Chiba 2003), many studies have not taken into consideration its role in CAM presentation and CAM-mediated synaptic enhancement (see FORCES above). However, in recent studies, FasII (Kohsaka et al. 2007) and Caps (Nose 2008) have been shown to be present in myopodia. Caps is expressed by both dorsal and ventral muscles (Shishido et al. 1998) and is enriched at the tip of myopodia. When mutated, the initial neurofilopodia–myopodia contact fails to stabilize (Nose 2008). These observations suggest a mechanism in which localized regions of the postsynaptic membrane guide and facilitate the formation of the NMJ by bringing appropriate CAMs and other signaling components in direct contact with their presynaptic partners, well before synaptogenesis is initiated. Thereby, precisely matched synaptic partners present CAMs through filopodial (i.e., neurofilopodia–myopodia) interactions to support axonal growth cone guidance, pathfinding, and synaptogenesis (Fig. 2.6). In addition, these CAMs may provide the necessary adhesive force to keep synaptic partners connected even in the presence of a moving target.

2.7.2 CAM-Mediated Postsynaptic Signaling Hub

We have shown that every neuromuscular synapse in live embryos is preceded by a transient stabilization/clustering of presynaptic and postsynaptic filopodia from matched partners upon contact. These interactions result in the creation of a new postsynaptic subcellular space, called *myopodial cluster* (Fig. 2.5d). In a prospero mutant background, which has severe motor neuron axon outgrowth delays from the CNS, myopodial cluster fails to form. Furthermore, filopodial interactions between non-partner cells, though extensive and essential for proper axon guidance, do not result in myopodial clustering. Myopodial cluster

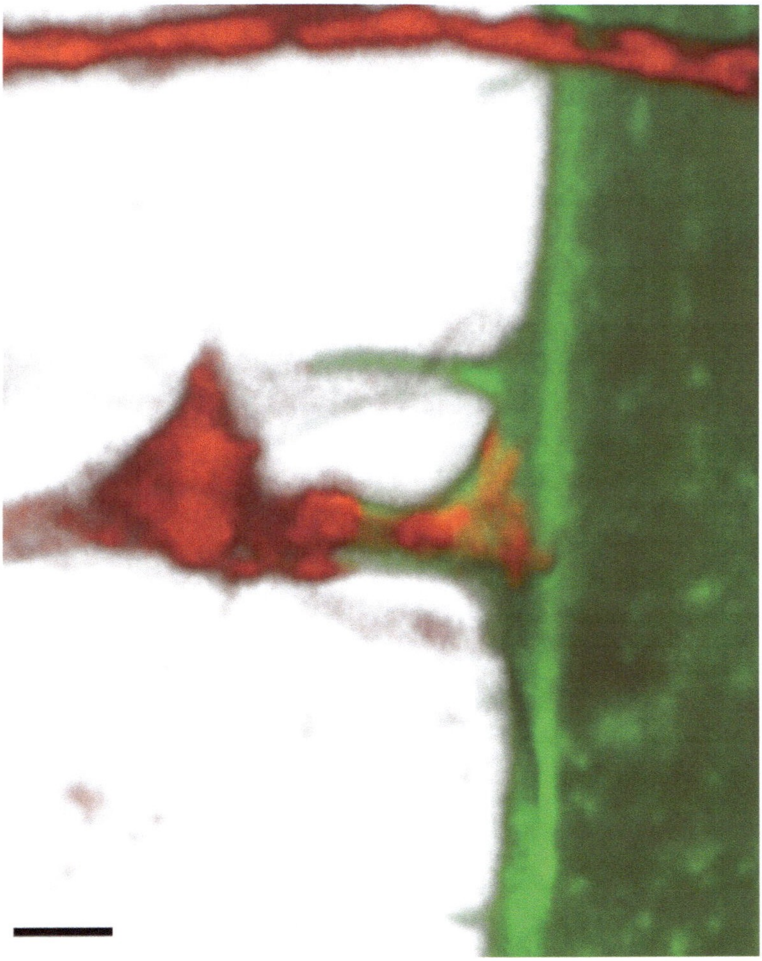

Fig. 2.6 Synaptogenesis. Muscle 12 expressing membrane-targeted GFP (*green*) and RP5 motor axon, as revealed by HRP antibody staining (*red*), in an intact (i.e., undissected) embryo fixed at hour 14:30 AEL. Scale bar = 10 μm

formation is also impaired in α_{PS2} integrin or Toll misexpression mutants, which exhibit abnormal axon targeting (Ritzenthaler et al. 2000, Ritzenthaler and Chiba 2003). These observations suggest that myopodial clustering only occurs in response to correctly matched synaptic binding partner.

Myopodial structures have not been observed at later larval developmental stages when new synaptic boutons are continuously added to the synaptic terminal. This may provide an explanation for earlier observations that inappropriate synaptic connections could be stabilized when FasII is increased in specific muscles during embryogenesis, but not when a similar increase is induced during larval stages (Davis et al. 1997). A recent research report may provide us with a potential cellular and molecular mechanism that may underlie this effect. FasII aggregates at the postsynaptic myopodial cluster and is necessary for postsynaptic Dlg accumulation (Kohsaka et al. 2007). Functional genetic dissection of Dlg reveals that it is essential for NMJ formation, but is not required for myopodial cluster formation. Furthermore, when myopodial clustering is genetically inhibited by a dominant-negative form of ezrin, synaptogenesis still proceeds, but the resulting NMJs are morphologically altered (Carrero-Martínez et al. unpublished). These observations suggest a model in which presynaptic FasII facilitates the recruitment and clustering of postsynaptic FasII, which in turn recruits Dlg molecules and initiates the formation of a functional synapse (Fig. 2.7). Taken together, these results highlight the role for CAMs in the development of the embryonic postsynaptic subcellular space, which facilitates and perhaps even initiates intracellular signaling events that are essential for the development of the embryonic NMJ (Fig. 2.8).

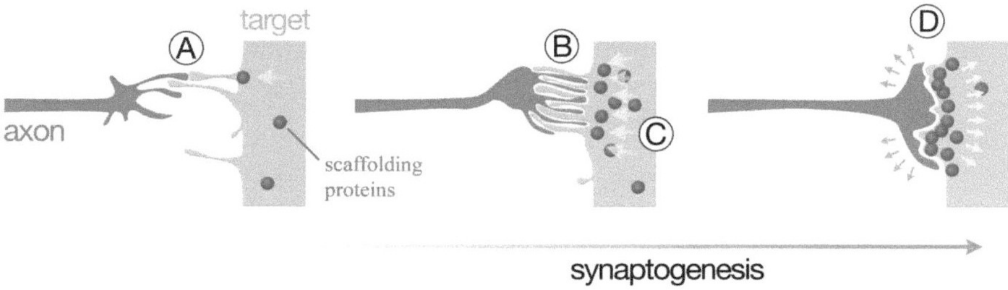

Fig. 2.7 Sequence of events leading to the successful formation of the NMJ. CAMs are presented at a distant site from the muscle surface (**A**). This interaction allows the muscle cell to start the assembly process of the postsynaptic specialization as the synaptic partner approaches. Increased and continued interaction between precisely matched synaptic partners results in the formation of a new subcellular space, called myopodial cluster (**B**). The creation of this space serves as s signaling hot spot to which several proteins such as the scaffolding protein Dlg are recruited (**C**). This space facilitates molecular interactions and the formation of signaling complexes, which are required for the formation of the neuromuscular synapse (**D**)

2.8 CAMs: The Cellular Glue that Holds Our Thoughts Together

Over the last decade, progress has been moderate in terms of identifying new CAMs that are involved in the formation of the *Drosophila* NMJ. Instead, attention in the field has shifted to more contextual issues such as filopodia. However, pioneer work in CAM and NMJ research was performed before postsynaptic filopodia had been described. Therefore, the early analyses did not take into consideration myopodia as active players for CAMs presentation

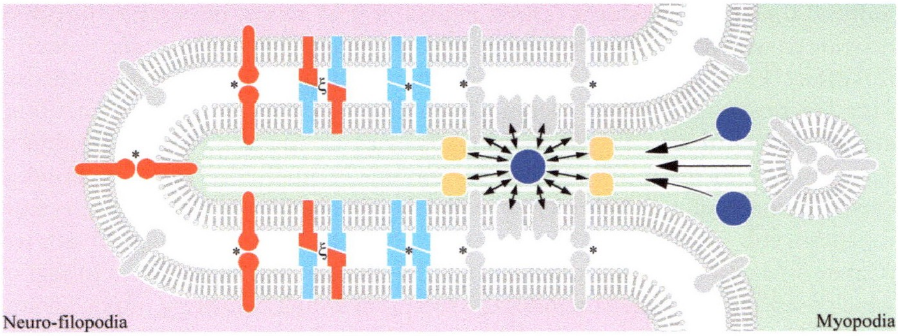

Fig. 2.8 CAMs facilitate and sustain NMJ formation. We propose that membrane-spanning proteins (such as CAMs) presented by developmentally regulated filopodia in the presynaptic (i.e., neurofilopodia; *left*) and postsynaptic (i.e., myopodia; *right*) cells may provide a mechanism that facilitates axonal pathfinding and target selection (Fig. 2.7). This is the first step toward the successful formation of the neuromuscular junction (see Section 2.7.1). As development of the NMJ progresses, myopodia aggregates to form a myopodial cluster (see Section 2.7.2). The increased surface area presented at the myopodial cluster/neurofilopodial interface locally increases the chance for trans-synaptic interaction among homophilic (*) and heterophilic (ξ) CAMs and initiate trans-synaptic signaling (*double-headed arrows*) between both synaptic partners. The narrow cytoplasmic space within each individual myopodia facilitates interaction between membrane-spanning CAMs and cytoplasmic scaffolding proteins such as Dlg, Ankyrin2, Pak, Dock, and other proteins (*circles* and *squares*). Polarization of the cytoskeleton may facilitate the recruitment of vesicles packed with other components essential for the development of the NMJ

during the process of embryonic synaptogenesis. This has led us to propose a two-step model in which CAMs facilitate the formation of the NMJ. The first step occurs during the process of growth cone extension and pathfinding. At this developmental stage CAMs are presented by both neurofilopodia and postsynaptic filopodia. This means that synaptic partner recognition may take place further away from the site of synaptogenesis than previously considered. According to our model, this ensures that both synaptic partners could form a CAM-mediated stable interaction and are able to withstand the forces generated by non-myogenic muscle contraction. At the same time, axons that are not appropriately matched with their corresponding synaptic partners will not be able to withstand the intercellular mechanical tension produced by these early muscle contractions and thus fail to activate appropriate postsynaptic signal transduction events. This activation of postsynaptic signaling events is the second step in our model. The early interactions are eventually transformed into the myopodial cluster, which serves as a signaling hot spot for the transformation of the presynaptic filopodia into synaptic boutons, a process that is concluded by the end of embryogenesis. Our model only accounts for the generation of the embryonic neuromuscular network pattern, which remains largely intact through larval stages. This is because myopodia are only transient structures, which uniquely respond to the mutual recognition by synaptic partners at the onset of embryonic synaptogenesis.

Identification of new synaptic CAMs may provide additional insights into how the *Drosophila* NMJ is established and maintained. Furthermore, identification of CAM splice variants, their developmental regulation, and localization may provide us with additional insights into how the *Drosophila* NMJ is fine-tuned. For now, the existence and importance of these isoforms in NMJ development remain largely unknown.

Myopodia, myopodial clustering, and their interaction with presynaptic filopodia may offer an opportunity to further dissect the molecular integrations of this glutamatergic synapse in an in vivo model. If we consider that flies which are heterozygous for a specific integrin mutation have short-term memory defects (Grotewiel et al. 1998) and that FasII-mediated adhesion may be involved in long-term memory processes (Cheng et al. 2001), we might hypothesize that CAMs are the cellular glue that holds our thoughts together.

Acknowledgments We thank Dr. Julie Dutil and Grissell Carrero-Martínez for editorial comments and suggestions on the manuscript. We also thank Dr. O'Neil Guthrie for helpful scientific discussions and comments and Dr. Daniel P. Kiehart for access to valuable resources and critical comments. F.A.C.-M. is an ASCB MAC visiting scholar with D.P.K. This award is supported by a MARC grant from the NIH NIGMS to the American Society for Cell Biology Minorities Affairs Committee.

References

Abrell S and Jackle H (2001) Axon guidance of Drosophila SNb motoneurons depends on the cooperative action of muscular Kruppel and neuronal capricious activities. Mech Dev 109:3–12

Ackley BD, Kang SH, Crew JR et al. (2003) The basement membrane components nidogen and Type XVIII collagen regulate organization of neuromuscular junctions in Caenorhabditis elegans. J Neurosci 23:3577–3587

Adams MD, Celniker SE, Holt RA et al. (2000) The genome sequence of Drosophila melanogaster. Science 287:2185–2195

Ashley J, Packard M, Ataman B et al. (2005) Fasciclin II signals new synapse formation through amyloid precursor protein and the scaffolding protein dX11/Mint. J Neurosci 25:5943–5955

Bate M and Rushton E (1993) Myogenesis and muscle patterning in Drosophila. C R Acad Sci III 316:1047–1061

Beumer K, Matthies HJ, Bradshaw A et al. (2002) Integrins regulate DLG/FAS2 via a CaM kinase II-dependent pathway to mediate synapse elaboration and stabilization during postembryonic development. Development 129:3381–3391

Beumer KJ, Rohrbough J, Prokop A et al. (1999) A role for PS integrins in morphological growth and synaptic function at the postembryonic neuromuscular junction of Drosophila. Development 126:5833–5846

Bieber AJ, Snow PM, Hortsch M et al. (1989) Drosophila neuroglian: a member of the immunoglobulin superfamily with extensive homology to the vertebrate neural adhesion molecule L1. Cell 59:447–460

Bloor JW and Brown NH (1998) Genetic analysis of the Drosophila alphaPS2 integrin subunit reveals discrete adhesive, morphogenetic and sarcomeric functions. Genetics 148:1127–1142

Bloor JW and Kiehart DP (2001) zipper Nonmuscle Myosin-II functions downstream of PS2 integrin in Drosophila myogenesis and is necessary for myofibril formation. Dev Biol 239:215–228

Bokel C and Brown NH (2002) Integrins in development: moving on, responding to, and sticking to the extracellular matrix. Dev Cell 3:311–321

Bouley M, Tian MZ, Paisley K et al. (2000) The L1-type cell adhesion molecule neuroglian influences the stability of neural ankyrin in the Drosophila embryo but not its axonal localization. J Neurosci 20:4515–4523

Broadie K and Bate M (1993) Muscle development is independent of innervation during Drosophila embryogenesis. Development 119:533–543

Brown NH (2000) Cell-cell adhesion via the ECM: integrin genetics in fly and worm. Matrix Biol 19:191–201

Budnik V (1996) Synapse maturation and structural plasticity at Drosophila neuromuscular junctions. Curr Opin Neurobiol 6:858–867

Campos-Ortega JA and Hartenstein V (1985) The Embryonic Development of Drosophila melanogaster. Springer-Verlag, Berlin, New York

Cash S, Chiba A and Keshishian H (1992) Alternate neuromuscular target selection following the loss of single muscle fibers in Drosophila. J Neurosci 12:2051–2064

Cheng Y, Endo K, Wu K et al. (2001) Drosophila fasciclinII is required for the formation of odor memories and for normal sensitivity to alcohol. Cell 105:757–768

Chiba A (1999) Early development of the Drosophila neuromuscular junction: a model for studying neuronal networks in development. Int Rev Neurobiol 43:1–24

Chiba A, Snow P, Keshishian H et al. (1995) Fasciclin III as a synaptic target recognition molecule in Drosophila. Nature 374:166–168

Crossley CA (1978) The morphology and development of the Drosophila muscular system. In: Ashburner M and Wright TRF (eds) The Genetics and Biology of Drosophila. Academic Press, New York

Currie DA and Bate M (1991) The development of adult abdominal muscles in Drosophila: myoblasts express twist and are associated with nerves. Development 113:91–102

Currie DA and Bate M (1995) Innervation is essential for the development and differentiation of a sex-specific adult muscle in Drosophila melanogaster. Development 121: 2549–2557

Davis GW and Goodman CS (1998) Genetic analysis of synaptic development and plasticity: homeostatic regulation of synaptic efficacy. Curr Opin Neurobiol 8:149–156

Davis GW, Schuster CM and Goodman CS (1997) Genetic analysis of the mechanisms controlling target selection: target-derived Fasciclin II regulates the pattern of synapse formation. Neuron 19:561–573

Dworak HA and Sink H (2002) Myoblast fusion in Drosophila. Bioessays 24:591–601

Featherstone DE, Davis WS, Dubreuil RR et al. (2001) Drosophila alpha- and beta-spectrin mutations disrupt presynaptic neurotransmitter release. J Neurosci 21:4215–4224

Fessler JH and Fessler LI (1989) Drosophila extracellular matrix. Annu Rev Cell Biol 5:309–339

Gotwals PJ, Fessler LI, Wehrli M et al. (1994) Drosophila PS1 integrin is a laminin receptor and differs in ligand specificity from PS2. Proc Natl Acad Sci USA 91:11447–11451

Grenningloh G, Rehm EJ and Goodman CS (1991) Genetic analysis of growth cone guidance in Drosophila: fasciclin II functions as a neuronal recognition molecule. Cell 67:45–57

Grotewiel MS, Beck CD, Wu KH et al. (1998) Integrin-mediated short-term memory in Drosophila. Nature 391:455–460

Halbleib JM and Nelson WJ (2006) Cadherins in development: cell adhesion, sorting, and tissue morphogenesis. Genes Dev 20:3199–3214

Hall SG and Bieber AJ (1997) Mutations in the Drosophila neuroglian cell adhesion molecule affect motor neuron pathfinding and peripheral nervous system patterning. J Neurobiol 32:325–340
Hartenstein V (1993) Atlas of Drosophila development. Cold Spring Harbor Laboratory Press, Plainview, NY
Hebbar S, Hall RE, Demski SA et al. (2006) The adult abdominal neuromuscular junction of Drosophila: a model for synaptic plasticity. J Neurobiol 66:1140–1155
Hoang B and Chiba A (1998) Genetic analysis on the role of integrin during axon guidance in Drosophila. J Neurosci 18:7847–7855
Hoang B and Chiba A (2001) Single-cell analysis of Drosophila larval neuromuscular synapses. Dev Biol 229:55–70
Hortsch M (2000) Structural and functional evolution of the L1 family: are four adhesion molecules better than one? Mol Cell Neurosci 15:1–10
Hortsch M, Bieber AJ, Patel NH et al. (1990) Differential splicing generates a nervous system-specific form of Drosophila neuroglian. Neuron 4:697–709
Iwai Y, Usui T, Hirano S et al. (1997) Axon patterning requires DN-cadherin, a novel neuronal adhesion receptor, in the Drosophila embryonic CNS. Neuron 19:77–89
Johansen J, Halpern ME and Keshishian H (1989) Axonal guidance and the development of muscle fiber-specific innervation in Drosophila embryos. J Neurosci 9:4318–4332
Keshishian H, Broadie K, Chiba A et al. (1996) The drosophila neuromuscular junction: a model system for studying synaptic development and function. Annu Rev Neurosci 19:545–575
Koch I, Schwarz H, Beuchle D et al. (2008) Drosophila Ankyrin 2 is required for synaptic stability. Neuron 58:210–222
Kohsaka H, Takasu E and Nose A (2007) In vivo induction of postsynaptic molecular assembly by the cell adhesion molecule Fasciclin2. J Cell Biol 179:1289–1300
Landgraf M, Bossing T, Technau GM et al. (1997) The origin, location, and projections of the embryonic abdominal motorneurons of Drosophila. J Neurosci 17:9642–9655
Landgraf M, Jeffrey V, Fujioka M et al. (2003) Embryonic origins of a motor system: motor dendrites form a myotopic map in Drosophila. PLoS Biol 1:E41
Lin DM, Fetter RD, Kopczynski C et al. (1994) Genetic analysis of Fasciclin II in Drosophila: defasciculation, refasciculation, and altered fasciculation. Neuron 13:1055–1069
Lin DM and Goodman CS (1994) Ectopic and increased expression of Fasciclin II alters motoneuron growth cone guidance. Neuron 13:507–523
Llano E, Pendas AM, Aza-Blanc P et al. (2000) Dm1-MMP, a matrix metalloproteinase from Drosophila with a potential role in extracellular matrix remodeling during neural development. J. Biol. Chem. 275:35978–35985
Lunstrum GP, Bachinger HP, Fessler LI et al. (1988) Drosophila basement membrane procollagen IV. I. Protein characterization and distribution. J Biol Chem 263:18318–18327
McFarlane S (2003) Metalloproteases: carving out a role in axon guidance. Neuron 37:559–562
Miller CM, Page-McCaw A and Broihier HT (2008) Matrix metalloproteinases promote motor axon fasciculation in the Drosophila embryo. Development 135:95–109
Mirre C, Cecchini JP, Le Parco Y et al. (1988) De novo expression of a type IV collagen gene in Drosophila embryos is restricted to mesodermal derivatives and occurs at germ band shortening. Development 102:369–376
Misgeld T, Burgess RW, Lewis RM et al. (2002) Roles of neurotransmitter in synapse formation: development of neuromuscular junctions lacking choline acetyltransferase. Neuron 36:635–648
Nose A (2008) Personal communication and laboratory website (bio.phys.s.u-tokyo.ac.jp/lab_page/english/englishresearch.html). 7/20/2008 3:54 PM
Nose A, Mahajan VB and Goodman CS (1992) Connectin: a homophilic cell adhesion molecule expressed on a subset of muscles and the motoneurons that innervate them in Drosophila. Cell 70:553–567

Nose A, Umeda T and Takeichi M (1997) Neuromuscular target recognition by a homophilic interaction of connectin cell adhesion molecules in Drosophila. Development 124:1433–1441
Packard M, Mathew D and Budnik V (2003) FASt remodeling of synapses in Drosophila. Curr Opin Neurobiol 13:527–534
Page-McCaw A (2008) Remodeling the model organism: matrix metalloproteinase functions in invertebrates. Semin Cell Dev Biol 19:14–23
Rivlin PK, Ryan, St Clair RM, Vilinsky I, Deitcher DL (2004) Morphology and molecular organization of the adult neuromuscular junction of *Drosophila*. J Comp Neurol 468:596–613
Pielage J, Cheng L, Fetter RD et al. (2008) A presynaptic giant ankyrin stabilizes the NMJ through regulation of presynaptic microtubules and transsynaptic cell adhesion. Neuron 58:195–209
Pielage J, Fetter RD and Davis GW (2005) Presynaptic spectrin is essential for synapse stabilization. Curr Biol 15:918–928
Prokop A, Martin-Bermudo MD, Bate M et al. (1998) Absence of PS integrins or laminin A affects extracellular adhesion, but not intracellular assembly, of hemiadherens and neuromuscular junctions in Drosophila embryos. Dev Biol 196:58–76
Raghavan S and White RA (1997) Connectin mediates adhesion in Drosophila. Neuron 18:873–880
Ritzenthaler S and Chiba A (2003) Myopodia (postsynaptic filopodia) participate in synaptic target recognition. J Neurobiol 55:31–40
Ritzenthaler S, Suzuki E and Chiba A (2000) Postsynaptic filopodia in muscle cells interact with innervating motoneuron axons. Nat Neurosci 3:1012–1017
Rose D and Chiba A (1999) A single growth cone is capable of integrating simultaneously presented and functionally distinct molecular cues during target recognition. J Neurosci 19:4899–4906
Rose D, Zhu X, Kose H et al. (1997) Toll, a muscle cell surface molecule, locally inhibits synaptic initiation of the RP3 motoneuron growth cone in Drosophila. Development 124:1561–1571
Salinas PC and Price SR (2005) Cadherins and catenins in synapse development. Curr Opin Neurobiol 15:73–80
Sanchez-Soriano N and Prokop A (2005) The influence of pioneer neurons on a growing motor nerve in Drosophila requires the neural cell adhesion molecule homolog FasciclinII. J Neurosci 25:78–87
Schmid A, Chiba A and Doe CQ (1999) Clonal analysis of Drosophila embryonic neuroblasts: neural cell types, axon projections and muscle targets. Development 126:4653–4689
Schmucker D, Clemens JC, Shu H et al. (2000) Drosophila Dscam is an axon guidance receptor exhibiting extraordinary molecular diversity. Cell 101:671–684
Schuster CM, Davis GW, Fetter RD et al. (1996a) Genetic dissection of structural and functional components of synaptic plasticity. I. Fasciclin II controls synaptic stabilization and growth. Neuron 17:641–654
Schuster CM, Davis GW, Fetter RD et al. (1996b) Genetic dissection of structural and functional components of synaptic plasticity. II. Fasciclin II controls presynaptic structural plasticity. Neuron 17:655–667
Shishido E, Takeichi M and Nose A (1998) Drosophila synapse formation: regulation by transmembrane protein with Leu-rich repeats, CAPRICIOUS. Science 280: 2118–2121
Siegler MVS and Jia XX (1999) Engrailed negatively regulates the expression of cell adhesion molecules connectin and neuroglian in embryonic Drosophila nervous system. Neuron 22:265–276
Suzuki E, Rose D and Chiba A (2000) The ultrastructural interactions of identified pre- and postsynaptic cells during synaptic target recognition in Drosophila embryos. J Neurobiol 42:448–459

Suzuki SC and Takeichi M (2008) Cadherins in neuronal morphogenesis and function. Dev Growth Differ 50 Suppl 1:S119–130

Taniguchi H, Shishido E, Takeichi M et al. (2000) Functional dissection of drosophila capricious: its novel roles in neuronal pathfinding and selective synapse formation. J Neurobiol 42:104–116

Thomas U, Ebitsch S, Gorczyca M et al. (2000) Synaptic targeting and localization of discs-large is a stepwise process controlled by different domains of the protein. Curr Biol 10:1108–1117

Thomas U, Kim E, Kuhlendahl S et al. (1997) Synaptic clustering of the cell adhesion molecule fasciclin II by discs-large and its role in the regulation of presynaptic structure. Neuron 19:787–799

Uhm CS, Neuhuber B, Lowe B et al. (2001) Synapse-forming axons and recombinant agrin induce microprocess formation on myotubes. J Neurosci 21:9678–9689

van Vactor DV, Sink H, Fambrough D et al. (1993) Genes that control neuromuscular specificity in Drosophila. Cell 73:1137–1153

Volk T, Fessler LI and Fessler JH (1990) A role for integrin in the formation of sarcomeric cytoarchitecture. Cell 63:525–536

Winberg ML, Noordermeer JN, Tamagnone L et al. (1998) Plexin A is a neuronal semaphorin receptor that controls axon guidance. Cell 95:903–916

Woods DF, Hough C, Peel D et al. (1996) Dlg protein is required for junction structure, cell polarity, and proliferation control in Drosophila epithelia. J Cell Biol 134:1469–1482

Wright TR (1960) The phenogenetics of the embryonic mutant, lethal myospheroid, in Drosophila melanogaster. J Exp Zool 143:77–99

Yonekura S, Ting C-Y, Neves G et al. (2006) The variable transmembrane domain of Drosophila N-cadherin regulates adhesive activity. Mol Cell Biol 26:6598–6608

Yu HH, Huang AS and Kolodkin AL (2000) Semaphorin-1a acts in concert with the cell adhesion molecules fasciclin II and connectin to regulate axon fasciculation in Drosophila. Genetics 156:723–731

Chapter 3
Development of the Vertebrate Neuromuscular Junction

Michael A. Fox

Abstract The precise alignment of nerve terminals to postsynaptic specializations suggests that trans-synaptic cues direct synapse formation. As with much of our understanding of synaptic function, initial insight into both the presence and the identity of these synaptogenic cues was derived from studies at the vertebrate neuromuscular junction (NMJ), a synapse formed between motoneurons and skeletal muscle fibers. Unlike central synapses, the wide synaptic cleft of the NMJ contains a network of cell-associated extracellular glycoproteins in the form of a specialized basal lamina (BL). The discovery that components of this synaptic BL direct pre- and postsynaptic differentiation has fueled three decades of intense research on the molecular signals regulating NMJ formation. Here, in addition to describing the organization and morphological development of the vertebrate NMJ, the roles of these extracellular adhesion molecules in the formation, maturation, and maintenance of this synapse are discussed.

Keywords Synapse formation · Basal lamina · Laminin · Agrin · Collagen IV · Nidogen

3.1 Vertebrate Neuromuscular Junction: A Model Synapse

The vertebrate neuromuscular junction (NMJ) has been extensively studied for over 160 years. Much of our initial understanding of synaptic organization, function, and formation was derived from this synapse. To demonstrate its value as a model synapse, a few fundamental synaptic properties discovered from studies at the NMJ warrant mention:

M.A. Fox (✉)
Department of Anatomy and Neurobiology, Virginia Commonwealth University, Richmond, VA 23298-0709, USA
e-mail: mafox@vcu.edu

M. Hortsch, H. Umemori (eds.), *The Sticky Synapse*,
DOI 10.1007/978-0-387-92708-4_3,

(i) **Synaptic partners are individual entities and are not continuous:** In the late 1800s, many believed in Gerlach's reticular theory, which described the nervous system as a continuous anastomosis of cells. At the NMJ, two variations of this reticular theory prevailed. The first claimed motor nerve terminals fused with muscle fibers allowing the direct flow of cytoplasm from nerve to muscle. In the mid-1800s, Claude Bernard (1813–1878) performed a series of studies on the effects of the paralytic neurotoxin curare on muscle contraction that challenged this theory. Curare-bathed muscle could not be stimulated to contract by untreated motor nerves, whereas curare-bathed nerves retained the ability to induce contractions in untreated muscle (Bernard 1856). The action of curare on muscle but not nerve indicated that nerve and muscle were not continuous. The second reticular theory posited that nerves did not terminate on muscle but rather anastomosed with sensory afferent fibers that returned to the spinal cord. Wilhelm Kühne (1837–1900) discounted this theory with detailed microscopic observations demonstrating nerves terminating at NMJs in numerous vertebrate species (Kühne 1862, 1887). Kühne's conclusion that NMJs were the sites of neurotransmission between separate cells predated similar theories concerning synaptic connections between neurons in the central nervous system (Kühne 1888).

(ii) **Chemical neurotransmission:** At the turn of the twentieth century, the notion emerged that synaptic transmission may result from the secretion of neurochemicals rather than electric coupling. While it has been debated as to who first speculated on the existence of chemical neurotransmitters, Otto Loewi (1873–1961) first demonstrated that stimulated nerve terminals release neurochemicals and Henry Dale (1875–1968) first identified acetylcholine (ACh) as the neurotransmitter at vertebrate NMJs (Loewi 1921, Dale et al. 1936, Brown et al. 1936, reviewed in Valenstein 2002).

(iii) **Molecular components of synapses:** Early studies at the NMJ identified the first molecular components of synapses. Three examples warrant attention. As described above, ACh was first described and rigorously tested at the vertebrate NMJ. Second, Marnay and Nachmansohn identified acetylcholinesterase (AChE), a protein concentrated and retained at synaptic sites (Marnay and Nachmansohn 1937). Those who argued against the chemical nature of synapses claimed chemical transmission could not account for the short duration of a single synaptic event, i.e., the time required for a chemical to diffuse out of the cleft and no longer stimulate postsynaptic receptors exceeded the duration of measured synaptic responses. The discovery of AChE, an enzyme that rapidly hydrolyzes and inactivates ACh, provided an explanation as to how the duration of a chemical signal was tightly regulated. Finally, the first neurotransmitter receptor to be identified, electrophysiologically studied, biochemically characterized, and molecularly cloned was the nicotinic acetylcholine receptor (AChR) (reviewed in Duclert and Changeux 1995).

(iv) **Quantal and vesicular theories of neurotransmission:** After the discovery that synaptic transmission was chemical in nature, it remained unclear how a neurotransmitter was secreted from nerve terminals. Bernard Katz and colleagues discovered small, spontaneous changes in the postsynaptic membrane potential (termed miniature endplate potentials [mEPPs]) while recording from NMJs. Noting little change in mEPP amplitude, they proposed the 'quantal release hypothesis,' which posited that ACh is packaged and released in specific quantities or quanta. Thus, mEPPs result from the release of a single quantum of ACh molecules (Fatt and Katz 1950, 1952). This led to the question of how ACh was packaged in the nerve terminal. After viewing electron micrographs demonstrating vesicles accumulated in motor nerve terminals (Robertson 1956a, b), Katz and Jose del Castillo proposed the 'vesicular hypothesis,' which stated that ACh is packaged into synaptic vesicles and released quantally from nerve terminals (del Castillo and Katz 1956).

(v) **Active zones are sites of neurotransmitter release:** After electron microscopy was applied to synaptic structures, and particularly motor nerve terminals, it was apparent that vesicles closely surround specialized, thickened portions of the presynaptic membrane. It was posited that these specializations were sites of neurotransmitter release, however, definitive proof that exocytosis occurred at these sites remained elusive for more than a decade after the proposal of the 'vesicular hypothesis' (Birks et al. 1960). Finally in 1970, Couteaux and Pecot-Dechavassine captured images of synaptic vesicle exocytosis at active zones in motor nerve terminals (Couteaux and Pecot-Dechavassine 1970).

(vi) **Molecular signals drive synapse formation:** Of particular importance to the subject matter of this book, studies at the NMJ provided the first clues that molecular signals were passed between synaptic partners to orchestrate synaptogenesis. This will be discussed in detail later in this chapter.

Having highlighted some of the most significant synaptic discoveries originating from studies at the NMJ, it is important to discuss features of this synapse that have made it amenable to diverse experimental paradigms. Most notably, neuromuscular synapses offer anatomical advantages over other peripherally and centrally located synapses. The vertebrate NMJ is large (up to ~50 μm in diameter in mammals), relatively isolated from other synapses, and peripherally located (i.e., accessible in comparison with synapses within the cranial vault) (Fig. 3.1). While other synapses do share some of these advantageous characteristics (e.g., calyx of Held synapses are large, autonomic smooth muscle synapses are isolated, and peripheral ganglionic synapses are accessible), it is the combination of features that has made the vertebrate NMJ such a useful experimental model. In addition to size and location, its simple postsynaptic geometry (i.e., lack of dendrites) and physiological robustness further contributed to making the vertebrate NMJ accessible to early electrophysiological approaches.

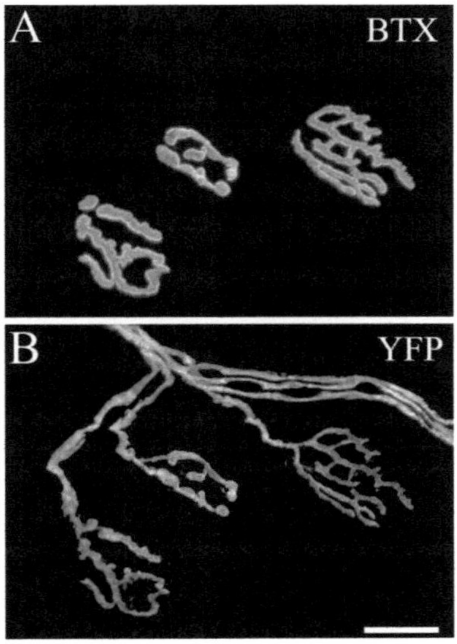

Fig. 3.1 The vertebrate NMJ. (**A**) Postsynaptic specializations of the muscle fiber are labeled with fluorescently conjugated bungarotoxin (BTX). (**B**) Motor axons and nerve terminals are labeled by their expression of a yellow derivative of green fluorescent protein (YFP). Note the large size of each NMJ and their relative isolation from each other. Scale bar is 25 μm

In addition to these exceptional 'intrinsic' properties, at least two 'extrinsic' tools have greatly contributed to the study of vertebrate NMJs. First, several toxins that inhibit neuromuscular transmission have been experimentally applied to studies at the NMJ. Venoms from many elapid snakes (i.e., cobra, mamba, krait) and hydrophid snakes (sea snakes) contain neurotoxins that block neurotransmission; however, it is the active component of Taiwanese banded krait (*Bungarus multicinctus*) venom, α-bungarotoxin, that has been most widely applied to the vertebrate NMJ. First identified by Chun-Chang Chang and Chen-Yuan Lee in the 1960s, α-bungarotoxin binds selectively and quasi-irreversibly to nicotinic AChRs to inhibit neuromuscular transmission (Chang and Lee 1963, Lee et al. 1967). Various derivatives of this small neurotoxin have been used to identify, purify, quantify, and characterize nicotinic AChRs (Miledi et al. 1971, Changeux et al. 1971, Berg et al. 1972, Fertuck and Salpeter 1976, Anderson and Cohen 1977). Even today, those who study the vertebrate NMJ universally apply α-bungarotoxin to label AChRs within the postsynaptic membrane.

While the relative isolation of individual vertebrate NMJs is advantageous for microscopists and electrophysiologists, it is a major limitation to biochemists and molecular biologists. The sparseness and small size of NMJs in comparison with the large muscle fiber (<0.1% of a muscle fiber's surface area is synaptic) makes isolation of synaptic components from muscle difficult. Therefore, a second extrinsic factor that has facilitated biochemical and molecular studies at the NMJ is the electric organ, or electroplax, of the marine ray *Torpedo*. The molecular composition of *Torpedo* electric organ resembles a hypertrophied NMJ (Cartaud et al. 2000). Structural and molecular similarities exist between neuromuscular and electroplax synapses because electric organs are derived from and innervated in a manner similar to embryonic branchial muscle. An important distinction, however, is that myotubes in the embryonic electric organ fail to elongate, lose their contractile apparatus, and differentiate into flattened electrocytes. Whereas a nerve terminal apposes only a minor portion of each muscle fiber, cholinergic nerve terminals contact one entire surface of each electrocyte (Sheridan 1965, Israel et al. 1976). Thus, the ratio of synaptic to non-synaptic components in electric organ is orders of magnitude higher than that of vertebrate skeletal muscle. Because of these features, many components of the NMJ were initially isolated from *Torpedo* electric organ (e.g., AChR [Miledi et al. 1971, Changeux et al. 1971], AChE [Lwebuga-Mukasa et al. 1976, Schumacher et al. 1986], agrin [Nitkin et al. 1987], muscle-specific kinase [MuSK, Jennings et al. 1993], rapsyn [Porter and Froehner 1983], dystrobrevin [Carr et al. 1989], and vesicle-associated membrane protein 1 [VAMP-1, Trimble et al. 1988]).

3.2 Vertebrate Neuromuscular Junction: The Basics

While morphology varies greatly between vertebrate species (Fig. 3.2A), all NMJs are composed of four basic elements (Fig. 3.3): (1) a presynaptic motor nerve terminal (Fig. 3.3B); (2) a postsynaptic specialization of the muscle fiber plasma membrane (Fig. 3.3C); (3) a cohort of perisynaptic Schwann cells capping the motor nerve terminal (Fig. 3.3D); and (4) a basal lamina (BL) separating nerve from muscle (Fig. 3.3E). All three cells and the BL are each highly specialized for synaptic function. Here, a brief description of these elements and how they contribute to synaptic function is provided.

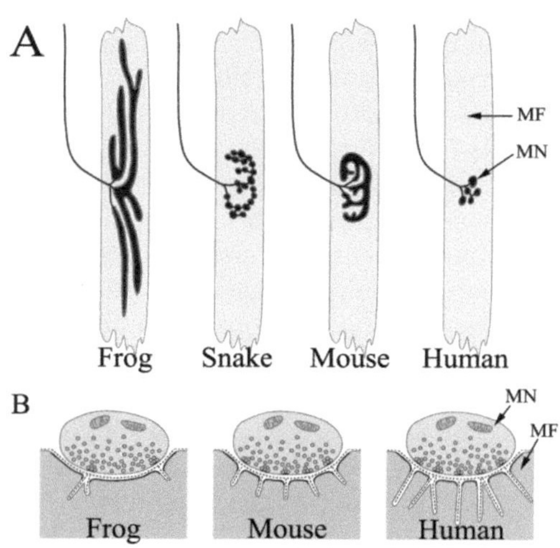

Fig. 3.2 NMJ structure differs between vertebrate species. (**A**) Differences in the size and morphology of frog, snake, mouse, and human NMJs. (**B**) Ultrastructural differences exist between vertebrate NMJs. The most striking difference is the depth and number of postsynaptic folds. MF, muscle fiber; MN, motor nerve terminal

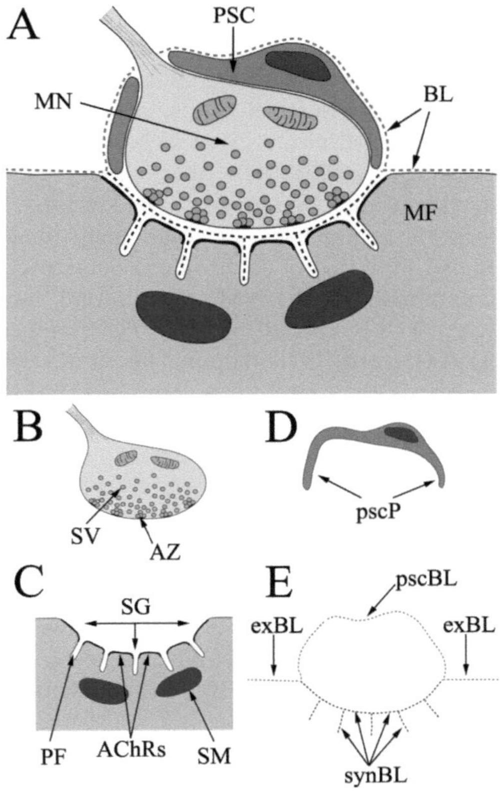

Fig. 3.3 Structure of the vertebrate NMJ (A). Neuromuscular synapses are composed of four main components: a motor nerve terminal (**B**), a postsynaptic specialization of the muscle fiber (**C**), a cohort of non-myelinating, perisynaptic Schwann cells (**D**), and a set of specialized basal laminas (**E**). AChRs, acetylcholine receptors; AZ, active zone; BL, basal lamina; exBL, extrasynaptic basal lamina; MF, muscle fiber; MN, motor nerve terminal; PF, postsynaptic fold; PSC, perisynaptic Schwann cell; pscBL, perisynaptic Schwann cell basal lamina; pscP, perisynaptic Schwann cell process; SG, synaptic gutter; SM, subsynaptic myonuclei; SV, synaptic vesicle; synBL, synaptic basal lamina

3.2.1 Motor Neurons and Their Presynaptic Terminals

In mammals, motor neurons residing in the ventral spinal cord or the brainstem motor nuclei extend myelinated axons into the periphery to innervate skeletal muscles. As motor axons approach the central, synaptogenic region of skeletal muscle fibers (termed the endplate band; see Fig. 3.4) they lose their myelin sheath and send an array of terminal branches that contact the surface of a target muscle fiber. With few exceptions, adult muscle fibers are innervated by a single motor neuron. A single motor axon, however, branches intramuscularly to innervate many muscle fibers (Fig. 3.5A, B). All muscle fibers contacted by branches of the same motor axon function in concert and constitute the fundamental unit of muscle contraction. Therefore, a motor neuron and all of the muscle fibers it innervates are termed a 'motor unit.'

At the distal end of each motor axon branch lies a nerve terminal, which is anatomically specialized to ensure reliable and reproducible neurotransmitter release. The most prominent features of nerve terminals are hundreds of thousands of small (~50 nm), spherical vesicles packed with the neurotransmitter

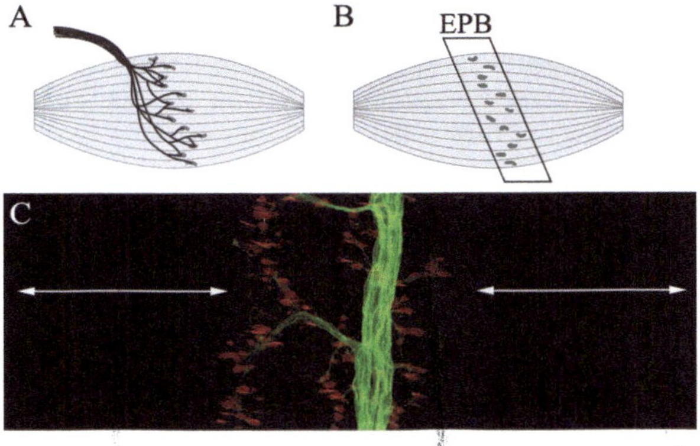

Fig. 3.4 Innervation of muscle by motor axons. Despite the large size of myofibers, all neuromuscular synapses are confined to a central region of muscle – termed the endplate band (EPB). (**A**) Schematic representation of motor axons (*black*) contacting postsynaptic specializations (*grey*) in the endplate region of muscle. (**B**) Schematic representation showing postsynaptic specializations tightly clustered in the EPB. Box highlights region of EPB. (**C**) Confocal image of mouse diaphragm muscle stained for AChRs (*red*) and neurofilament within motor axons (*green*). Both axons and AChR clusters are confined to the central region of muscle. Arrows demonstrate regions of muscle lacking synaptic specializations

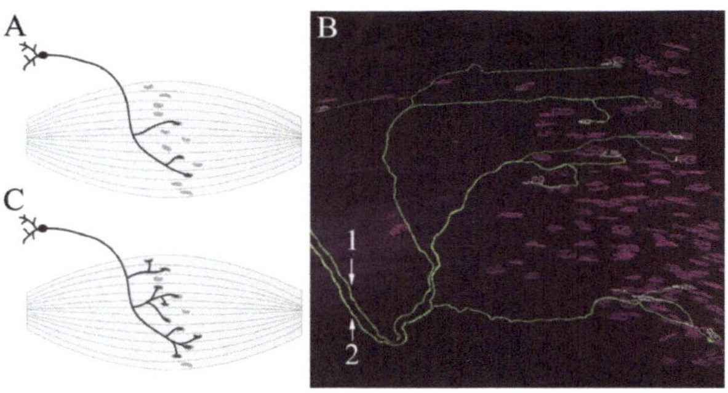

Fig. 3.5 Motor axons branch to innervate multiple muscle fibers within the same muscle. (**A**) Schematic representation of an adult motor unit. (**B**) Confocal reconstruction of two motor units (labeled 1 and 2) in the levator palpebrae superioris muscle from a mouse expressing green fluorescent protein (GFP) in a small subset of motor neurons (Feng et al. 2000). *Green* – motor axons are labeled with GFP; *Magenta* – AChRs labeled with bungarotoxin. (**C**) Schematic representation of an immature motor unit. See text for details

ACh. Since a single synaptic vesicle may contain up to 10,000 ACh molecules, the amount of ACh within a single nerve terminal is staggering (Kuffler and Yoshikami 1975). In addition to ACh, synaptic vesicle-associated proteins essential for vesicle transport and neurotransmitter release are abundant in nerve terminals. Based on their relative absence from motor axons and dense accumulation at synaptic sites, families of synaptic vesicle-associated proteins (such as VAMP-1, synaptophysins, synaptic vesicle protein 2 (SV2), synapsins, synaptobrevins, and synaptotagmins; for review see Südhof 2004) are often used as reliable markers of motor nerve terminals (Fig. 3.6).

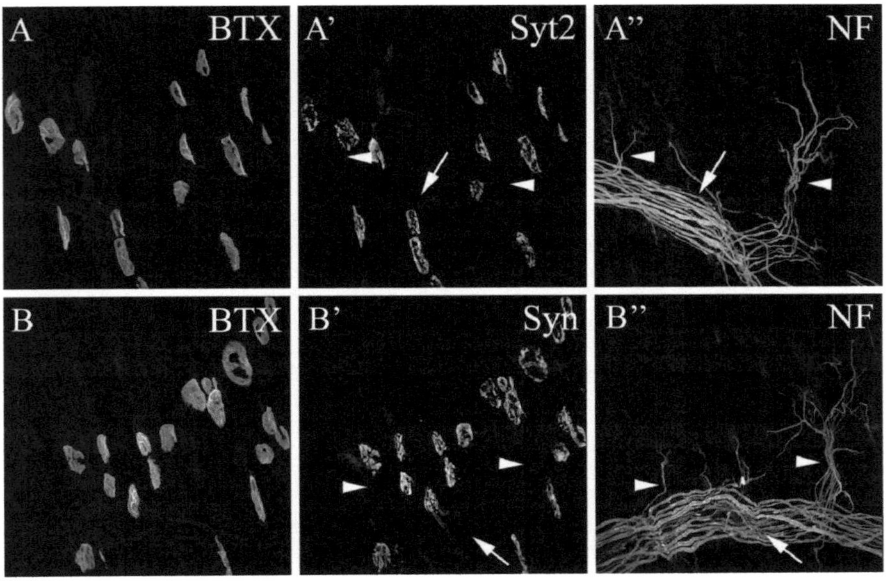

Fig. 3.6 Synaptic vesicle-associated proteins are reliable markers of the motor nerve terminal. Three-week-old mouse muscle triple labeled with a postsynaptic marker (BTX), an axonal marker (neurofilament [NF]), and one of two synaptic vesicle-associated protein markers – synaptotagmin 2 (Syt2; **A**) or synapsin (Syn; **B**). Syt2 and Syn are enriched in nerve terminals apposed to AChR clusters, but are excluded from intramuscular motor axons (*arrows*) and pre-terminal portions of axons (*arrowheads*)

Synaptic vesicles are not evenly distributed within a motor nerve terminal. Vesicles are scarce in regions of the nerve terminal abutting perisynaptic Schwann cells but densely populate regions closest to the synaptic cleft. Clusters of vesicles near the synaptic cleft and apposed to folds in the postsynaptic membrane (see below) are 'docked' at active zones, specialized sites for neurotransmitter exocytosis (Fig. 3.3B). A human motor nerve terminal may have over 200 such sites, while other mammalian nerve terminals may have closer to 1000 (Slater et al. 1992). In the absence of a nerve impulse, single quanta of ACh are spontaneously exocytosed, giving rise to small postsynaptic depolarizations or mEPPs. mEPPs do not sufficiently depolarize the muscle membrane to generate a postsynaptic action potential. Rather, the generation of an action potential requires an orchestrated release of ACh by numerous vesicles at multiple active zones. This requires specialized molecular machinery that resides both in the cytomatrix of the active zone and the presynaptic membrane. Detailed reviews of the molecular components and function of active zones have recently been published (Schoch and Gundelfinger 2006, Jin and Garner 2008, see also Juranek et al. 2006). At the vertebrate NMJ, presynaptic membranes associated with active zones are rich in voltage-dependent calcium channels (P/Q type) and potassium channels (Robitaille et al. 1990, Robitaille et al. 1993a, b, Urbano et al. 2002). As action potentials reach the motor nerve terminals, P/Q type calcium channels are activated allowing the influx of calcium. Increased cytosolic calcium stimulates synaptic vesicle-associated proteins and components of the active zone cytomatrix to fuse vesicles with the presynaptic membrane, thereby releasing ACh into the synaptic cleft.

A single nerve impulse triggers the release of between 20 and 200 quanta of ACh at the vertebrate NMJ. The number of quanta released is correlated with both the surface area occupied by the nerve terminal (i.e., the larger the contact area between muscle and nerve, the more synaptic vesicles released by a single action potential) and the number of active zones per terminal (Wood and Slater 2001, Slater 2003). The amount of ACh released by an adult nerve terminal in response to a single nerve impulse far exceeds the amount required for the generation of a postsynaptic action potential and subsequent muscle fiber contraction. This excessive release of transmitter ensures that neuromuscular transmission does not fail during repetitive firing (i.e., prolonged activity). Thus, it has been proposed that a 'safety factor' of 3–5 exists at the NMJ. This means that three to five times more quanta are released from motor nerve terminals than the minimal number required to generate a postsynaptic action potential (for review see Wood and Slater 2001).

3.2.2 The Postsynaptic Apparatus

As stated above, only a small region of each adult myofiber (or muscle fiber) is in direct contact with and innervated by a single nerve terminal. Although this specialized region, termed the postsynaptic apparatus, is small in comparison with extrasynaptic regions of the myofiber, it is morphologically, topographically, and molecularly distinct (reviewed in Sanes and Lichtman 2001). The postsynaptic membrane is recessed into the myofiber, creating a synaptic gutter into which the nerve terminal sinks. Strikingly, the muscle membrane within the synaptic gutter contains numerous invaginations, each precisely aligned with active zones in the corresponding nerve terminals (Fig. 3.3A,C). These invaginations, termed postsynaptic folds (or secondary synaptic folds), are unique to the NMJ. The depths of these folds differ greatly between vertebrate species (Fig. 3.2B, Slater 2003). In general, postsynaptic fold depth is inversely proportional to the surface area of the postsynaptic membrane.

In order to respond reliably to the neurotransmitter released by nerve terminals, the most critical component of the postsynaptic apparatus is the presence of neurotransmitter receptors. Nicotinic AChRs are highly concentrated in the postsynaptic membrane and sparsely distributed in extrasynaptic muscle membrane. In addition to differences in their synaptic and extrasynaptic distribution, AChRs are differentially distributed within the postsynaptic membrane. The crests and upper portions of postsynaptic folds are rich in AChRs, whereas few AChRs are present in the depths of postsynaptic folds (Fertuck and Salpeter 1974). Other ion channels, such as voltage-gated sodium channels (VGSC), are present in the depths of postsynaptic folds, an arrangement that has been suggested to enhance the efficacy of neuromuscular transmission (Flucher and Daniels 1989, Slater 2003).

The generation and maintenance of the postsynaptic apparatus relies on specialized cytoskeletal elements underlying the postsynaptic membrane. Several components of postsynaptic cytoskeletal matrix critical for synaptic organization and function have been identified (including rapsyn, utrophin, dystrobrevin, syntrophin, dystrobrevin, ankyrin, PSD-95, MAGI, for reviews see Sanes and Lichtman 1999, 2001, Banks et al. 2003, Lai and Ip 2003). Just as ion channels are differentially distributed within the postsynaptic membrane, components of the cytoskeletal matrix underlying the crests and the depths of postsynaptic folds are distinct (Sanes and Lichtman 1999). For example, rapsyn is enriched in the regions underlying the crests of postsynaptic folds where it binds the cytoplasmic tails of AChRs to coordinate their clustering (Burden et al. 1983, Sealock et al. 1984, Flucher and Daniels 1989). Rapsyn is largely absent from regions underlying the depths of postsynaptic folds where few AChRs reside.

A final specialization of the postsynaptic apparatus is the presence of transcriptionally specialized myonuclei directly under the postsynaptic membrane (Fig. 3.3C). These subsynaptic myonuclei allow synaptic proteins to be synthesized locally rather than being generated at a distance and trafficked to synaptic sites. Not only do subsynaptic myonuclei express synaptic genes, but synaptic gene expression is silenced at non-synaptic myonuclei (Duclert and Changeux 1995, Sunesen and Changeux 2003). Enrichment and silencing of gene expression by synaptic and non-synaptic myonuclei, respectively, is at least partly responsible for the enrichment of AChRs at synaptic sites during NMJ formation.

3.2.3 Non-myelinating Perisynaptic Schwann Cells

Typically, three to five process-bearing glial cells, termed perisynaptic Schwann cells (or terminal Schwann cells), closely overlie the non-synaptic portions of motor nerve terminals (Fig. 3.3D). Unlike Schwann cells associated with motor axons, perisynaptic Schwann cells do not generate or ensheath axons with lipid-rich myelin. Instead, they are process-bearing cells more intimately associated with motor nerve terminals than nerve terminals are with the postsynaptic membrane (Fig. 3.3D). Despite this intimate arrangement, considerably less is known about the role these cells play at the NMJ compared with the synaptic roles of muscle and nerve. That being said, perisynaptic Schwann cells are known to synthesize and secrete trophic factors that are required for motor neuron survival, guiding regenerating motor axons, phagocytosing and removing debris following injury, and modulating synaptic function (for review see Kang et al. 2003, Koirala et al. 2003, Auld et al. 2003, Werle and VanSaun 2003, Colomar and Robitaille 2004, Feng et al. 2005). Although perisynaptic Schwann cells do not appear essential for the initial formation of neuromuscular synapses, they are necessary for its maturation and maintenance (Reddy et al. 2003).

3.2.4 The Synaptic Cleft and Basal Lamina

The synaptic cleft at the vertebrate NMJ is a 50–100 nm gap separating the motor nerve terminal from the postsynaptic membrane. Synaptic clefts at vertebrate NMJs not only are substantially wider than clefts associated with central synapses but also contain a dense network of cell-associated extracellular glycoproteins and proteoglycans in the form of a basal lamina (Fig. 3.3E). In many tissues, cell-associated BLs associate with a collagen fiber-rich reticular lamina and together form a basement membrane (BM). While a BM surrounds extrasynaptic portions of muscle fibers, synaptic sites lack reticular laminas. Therefore, a BL but not a BM occupies the synaptic cleft of the NMJ (Sanes 2003).

Molecularly, BLs are composed of four families of core proteins (collagen IV, laminins, nidogens [also called entactins], and heparan sulfate proteoglycans [HSPGs]) and lack fibrillar collagens (e.g., collagens I, II, III, and V). Collagen IV and laminins each form independent networks within BLs, which are molecularly bridged by their ability to bind nidogens. Although BL networks are known to add structural support to tissue, it has become apparent that these matrices are bioactive and capable of binding cell-surface receptors that are critical for developmental and regenerative processes (Adams 2002, Kalluri 2003, Yurchenco et al. 2004, Berrier and Yamada 2007). Moreover, BLs are rich in other bioactive ECM molecules, enzymes, and growth factors important for cell–cell and cell–ECM signaling.

The concept that a network of matrix proteins occupies the synaptic cleft is somewhat surprising given that pre- and postsynaptic elements are precisely aligned for rapid neurotransmission. One might presume that the presence of a dense BL may slow the diffusion of ACh across the synaptic cleft. Moreover, embedded at high concentrations within the synaptic BL is AChE – an enzyme that rapidly and efficiently catalyzes the hydrolysis of ACh and terminates neurotransmission. So, not only is there a physical barrier potentially hindering the passage of ACh, but this barrier is exceptionally efficient at inactivating

ACh. Remarkably, however, neurotransmission is not inhibited by synaptic BL. In fact, synaptic transmission is as fast, if not faster, at NMJs than at central synapses where neurotransmitters have less distance to diffuse before activating postsynaptic receptors (Edmonds et al. 1995a, b). Moreover, comparison of kinematic models of neuromuscular transmission with empirical recordings suggests that ACh diffuses through the neuromuscular synaptic cleft relatively unimpeded by either the BL or the AChE (Land et al. 1984, Tai et al. 2003).

What then is the role of a BL at the NMJ? One role for synaptic BL is to retain AChE. Several forms of AChE exist, but at the NMJ an asymmetric form, not a globular form, of AChE predominates. Two genes code for the 15 subunits required to synthesize, secrete, and restrict asymmetric AChE to the synaptic cleft of the NMJ. One gene, *ache*, generates the 12 catalytic subunits. The other, *ColQ*, generates a collagenous Q domain, three of which trimerize to form a triple helical collagenous tail. Once in the synaptic cleft, the collagenous ColQ tail tethers asymmetric AChE to the BL through its interactions with perlecan and MuSK (Rotundo 2003, Rotundo et al. 2008, Cartaud et al. 2004). As expected, the genetic removal of ColQ removes asymmetric AChE from the synaptic BL (Feng et al. 1999).

A second role for synaptic BL is to act as a means of communication for pre- and postsynaptic elements. The distance between pre- and postsynaptic partners at the NMJ is too great for conventional transmembrane molecules to act as trans-synaptic signals. Thus, molecular cues necessary for synapse formation, maintenance, or function are deposited in the synaptic BL by muscle, nerve, and perisynaptic Schwann cells. Since BLs surround the entire muscle fiber how specific might such synaptic organizing cues be? To understand this it is essential to highlight three important features of synaptic BL:

(i) **Synaptic BL differs in molecular composition from extrasynaptic BL:** This is not to say that the families of core components differ between synaptic and extrasynaptic BLs, but rather specific isoforms differ (Fig. 3.7, for reviews see Sanes 2003, Patton 2003). Synaptic BL bears distinct isoforms of each core BL component: collagens α3–α6(IV); laminins α4, α5, and β2; nidogen-2; and the proteoglycan agrin. Conversely, some molecules present extrasynaptically are excluded from synaptic BL (e.g., laminin β1) and still others are common to all skeletal muscle BL (nidogen-1, perlecan, fibronectin, collagens α1[IV] and α2[IV], and laminins α2 and γ1). In addition to core BL components, other ECM molecules, such as AChE, specifically associate with synaptic BL.

(ii) **Synaptic BL is heterogeneous in its molecular composition:** Within the synaptic cleft some BL components are evenly distributed, while others are confined to subdomains of the synaptic BL. Synaptic collagen IV, nidogen-2, laminin β2 and α5, and AChE are present in the BL and are associated with both the primary synaptic cleft and postsynaptic folds. Agrin, neuregulin, and laminin α4, however, are confined to the primary

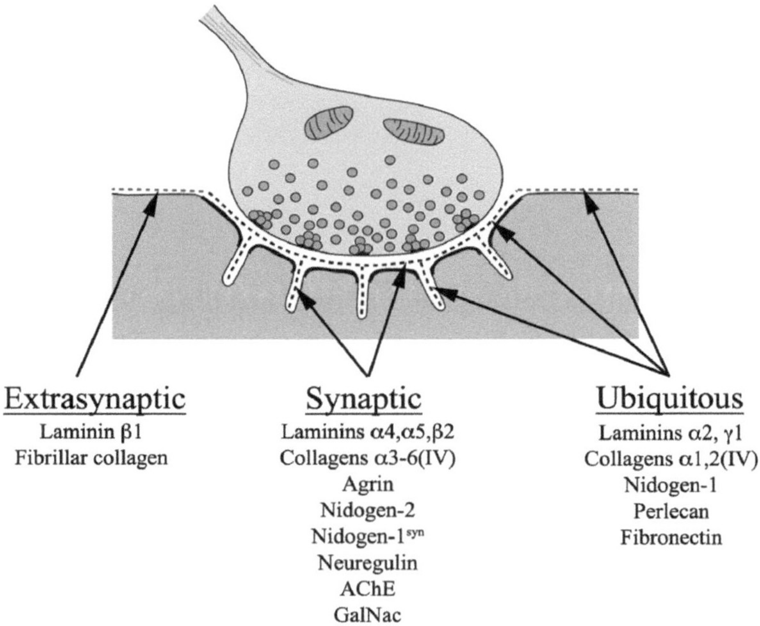

Fig. 3.7 Molecular composition of synaptic and extrasynaptic basal laminas. Immunohistochemical studies have identified numerous components of muscle BLs (see text for details and references). Some extracellular factors are unique to either synaptic or extrasynaptic muscle BL; others are present in both although their expression in one BL may be higher (i.e., expression of perlecan is higher in synaptic BL). In at least one case, synapse-specific modifications have been identified on ubiquitously distributed BL molecules (nidogen-1^{syn}). Collagenous fibers are absent from synaptic sites, but associate with extrasynaptic BL to form a basement membrane

synaptic cleft, postsynaptic folds, or regions flanking presynaptic active zones (i.e., periactive zones), respectively (Patton 2003).

(iii) **Specific glyco-epitopes are present in the synaptic BL:** Components of all BLs and ECMs are highly glycosylated. Evidence that glyco-epitopes differ in synaptic and extrasynaptic BL is derived from studies demonstrating that some plant lectins, proteins that bind carbohydrate chains, specifically label synaptic sites in skeletal muscle (Sanes and Cheney 1982, Scott et al. 1988). Most prominently, carbohydrate chains terminating in *N*-acetylgalactosamine (GalNAc), which are detected by the lectins *Vicia villosa* agglutinin B4 (VVA-B4) and *Dolichos biflorus* agglutinin (DBA), are present in the synaptic BL and absent extrasynaptically (Sanes and Cheney 1982). Moreover, the enzyme responsible for these modifications, GalNAc transferase, is enriched in the synaptic BL (Scott et al. 1990). Not only have modifications been identified on synapse-specific components of muscle BL, but synapse-specific glycosylations of ubiquitously expressed BL components have also been identified (Chui and Ho 1994, reviewed in Martin 2003). While carbohydrate modification of synaptic BL components appears necessary for signaling at the NMJ (Martin and Sanes 1995, Martin et al. 1999, Martin 2003), it remains unclear how synapse-specific modifications of ubiquitous BL molecules alter their functions within the synaptic cleft.

3.3 Morphological Development of the Vertebrate NMJ

3.3.1 Synaptic Differentiation

Much attention has recently focused on determining whether the muscle or the nerve initiates the signaling cascades that lead to NMJ formation. Until a decade ago, the majority of studies supported neuronal control of this process (for review see Sanes and Lichtman 1999), i.e., factors secreted from motor axons first induce AChR clustering, then subsequent muscle-derived signals induce the transformation of a growth cone into a nerve terminal. Early evidence supporting this theory came from co-cultures of muscle and motor neurons. Although cultured myotubes spontaneously cluster AChRs in the absence of innervation, growing neurites from motor neurons do not preferentially contact these pre-existing clusters. Instead, neurites induce the formation of new AChR clusters at sites of nerve–muscle contact (Anderson and Cohen 1977, Frank and Fischbach 1979). These studies led to the hypothesis that nerve-derived signals orchestrate the initial aspects of postsynaptic differentiation. The discovery of several motoneuron-derived factors that induced the synthesis and clustering of AChRs supported this theory (see below, Jessell et al. 1979, Fontaine et al. 1986, Usdin and Fischbach 1986, Godfrey et al. 1984, Nitkin et al. 1987).

Initial challenges to this neurocentric view of postsynaptic differentiation came from observations that AChRs clustered in mutant muscles lacking motor axons (Fig. 3.8A) (Yang et al. 2000, 2001, Lin et al. 2001, reviewed in Ferns and Carbonetto 2001, Kummer et al. 2006). Moreover, similar clusters formed in the absence of agrin, a nerve-derived signal thought to induce the initial clustering of AChRs, agrin (Fig. 3.8B) (Nitkin et al. 1987, Gautam et al. 1996, Lin et al. 2001). Remarkably, even in the absence of innervation or agrin, aneural AChR clusters appear in the central region of muscle, the presumptive site of the endplate band. The presence of these aneural, pre-existing AChR clusters evoked an interesting new question – Do pre-patterned AChR clusters guide motor axons to appropriate synaptic targets within the endplate band in vivo? Although cultured motor axons are not preferentially guided to aneural AChR clusters (see above), pioneering motor axons in zebrafish embryos do appear to be guided by pre-existing AChR (Fig. 3.9, Flanagan-Steet et al. 2005, Panzer

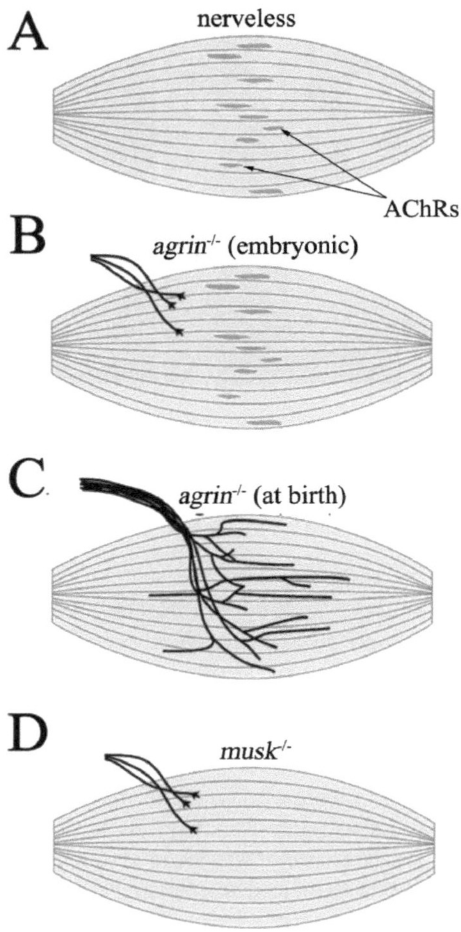

Fig. 3.8 Clustering of AChRs in the absence of nerves and nerve-derived cues. (**A**, **B**) Aneural AChR clusters (*grey ellipses*) form in embryonic mutant muscle lacking motor nerves (**A**) or agrin (**B**). (**C**) Although agrin is dispensable for the initial formation and patterning of aneural AChR clusters, it is required for their reshaping and maintenance. AChR clusters disperse by birth in agrin-deficient mutants and motor axons grow promiscuously outside of the endplate regions of muscle. (**D**) While aneural AChR clusters form in the absence of agrin, they fail to form in embryonic MuSK-deficient muscle

et al. 2005, 2006). Moreover, at least some aneural clusters contacted by these pioneering motor axons are incorporated into functional NMJs.

Specific cues associated with pre-existing, aneural AChR clusters that guide motor axons remain unclear. One possible guidance cue is muscle-specific kinase (MuSK), known best as a putative postsynaptic co-receptor for agrin-induced AChR clustering. MuSK is pre-patterned with AChRs in the endplate region of muscle prior to the entry of motor axons and is necessary and sufficient for the formation of these aneural clusters (Fig. 3.8D) (Lin et al. 2001, Yang et al. 2001, Kim and Burden 2008). MuSK may also play a direct role in guiding motor axons. Expression of MuSK in non-neural cells is sufficient to halt motor axon outgrowth upon contact in vitro (Dimitropoulou and Bixby 2005). Supporting such a direct role for MuSK on motor axons, motor axons branch excessively in muscle and are not confined to the endplate region

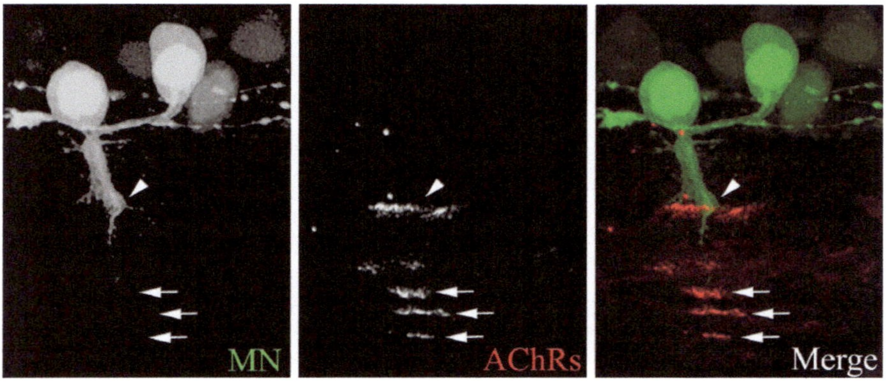

Fig. 3.9 Aneural AChRs precede and guide motor axons. Confocal image of a primary motor neuron (MN, *green*) in *hb9*:GFP transgenic zebrafish extending its axon toward pre-patterned AChRs (*red*). Arrowhead indicates an AChR cluster contacted by a growing motor axon; *arrows* indicate aneural clusters not yet contacted by a motor axon (for details see Flanagan-Steet et al. 2005)

of muscles lacking MuSK (DeChiara et al. 1996). Alternatively, MuSK may also bind other extracellular components, which guide motor axons. A zebrafish homolog of MuSK (*unplugged*) has been proposed to induce changes in muscle ECM that are critical for motor axon guidance (Zhang et al. 2004). The possibilities of MuSK playing direct or indirect roles on motor axons are not mutually exclusive. Changes in the extracellular matrix (ECM) associated with aneural clusters may guide motor axons and direct interactions with MuSK may halt their outgrowth. In support of this multifunctional view of MuSK, overexpression of MuSK in vivo induces ectopic AChR clustering outside of the endplate band and, surprisingly, causes motor axons to find and innervate these ectopic clusters (Kim and Burden 2008).

Are sites of neuromuscular synapses solely determined by the muscle? Despite evidence that some aneural AChR clusters are incorporated into synapses by motor axons, not all evidence supports the notion that these pre-existing clusters dictate where NMJs form. In both mouse and zebrafish muscle, AChR clusters have been identified that are not formed aneurally but instead arise only after contact with motor axons (Flanagan-Steet et al. 2005, Lin et al. 2008). Thus, although muscle pre-patterning might guide motor axons toward the endplate band and some pre-existing AChR clusters are incorporated into synapses, it remains possible that nerve-derived cues, such as agrin, are necessary to induce the formation of some synaptic sites.

Regardless of whether the muscle or the nerve starts the process of synaptic differentiation, major transformations occur in both to ultimately form a functioning neuromuscular synapse. As eluded to above, one such major transformation of the muscle is the clustering of AChRs. Shortly after myofiber formation and before the arrival of motor axons, genes coding for embryonic isoforms of AChRs are activated. Although initially inserted uniformly into muscle membrane (Bevan and Steinbach 1977), AChRs on at least some myotubes aggregate into aneural clusters prior to the arrival of motor axons. After nerve–muscle contact, AChR receptors become highly concentrated synaptically ($>10,000/\mu m^2$) and largely excluded from extrasynaptic sites ($<10/\mu m^2$) (Salpeter and Loring 1985). At these initial stages of postsynaptic development, AChR clusters are small and elliptically shaped, referred to as 'plaques' (Fig. 3.10B,C). Intracellular specializations of the postsynaptic apparatus begin to occur at these early stages of postsynaptic differentiation, including both the assembly of specialized cytoskeletal scaffolds and the enrichment of synaptic gene transcription by subsynaptic myonuclei (Sanes and Lichtman 1999, Sunesen and Changeux 2003). Other elements common to the adult postsynaptic membrane, such as postsynaptic folds, are absent from immature synapses.

The transformation of a growth cone into a nerve terminal is equally as striking as AChR clustering (Fig. 3.10A). Although growth cones contain synaptic vesicles and are capable of releasing small quantities of neurotransmitter, they are mainly mechanosensory organelles constructed to propel and guide axons (Young and Poo 1983). Their motile ability requires a dynamic cytoskeleton, rich in microtubules, and actin. This is clearly distinct from the organization of an adult motor nerve terminal which is highly stable, bears an abundance of synaptic vesicles and neurotransmitter release sites, and is relatively devoid of microtubules or intermediate filaments (i.e., neurofilament, see Fig. 3.10C). Unlike the rapid clustering of AChRs, nerve terminal formation is a protracted process (Fig. 3.10A). Immature nerve terminals are initially bulbous, have a sparse collection of synaptic vesicles, and lack recognizable active zones. As one might assume, these immature terminals are less capable of generating postsynaptic action potentials than adult terminals. While embryonic nerve terminals enlarge, accumulate more synaptic vesicles, and develop more active zones, their quantal content increases and synaptic transmission becomes more reliable (reviewed in Sanes and Lichtman 1999). In mammals, immature nerve terminals begin to more closely resemble their adult counterparts shortly after birth, roughly 1 week after initial contact with their target muscle (Fig. 3.10C). At this time, the immature nerve terminal is polarized with an abundance of synaptic vesicles clustered in the synaptic side of the terminal, recognizable active zones, and docked vesicles clustered around active zones (Fig. 3.10A, Takahashi et al. 1987, Lupa and Hall 1989, Buchanan et al. 1989).

The above description of the developing motor nerve terminal may paint a picture that nerve terminals are individually matched with postsynaptic partners at their conception. This is not the case. Nerve terminals from multiple motoneurons initially converge and interdigitate on single postsynaptic sites of the muscle fiber (Fig. 3.10A,C). The convergence of multiple inputs onto a single muscle fiber is not due to an excess of motor neurons or fewer muscle fibers at this stage in development. Rather, individual motor axons have more intramuscular branches and contact more muscle fibers at young ages than in

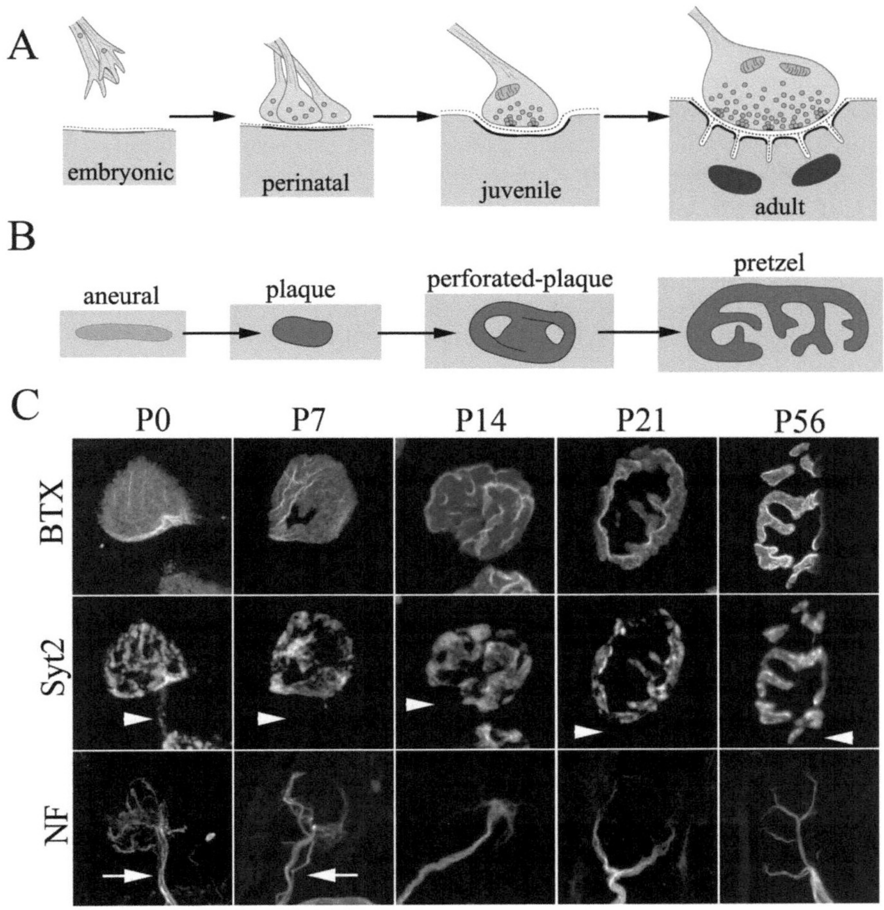

Fig. 3.10 Development of the vertebrate NMJ. (**A**) Ultrastructural changes associated with nerve terminals and the postsynaptic apparatus during the initial formation and postnatal maturation of the NMJ. Approximate ages of mouse development are inset in muscle fibers. Structures are as they were defined in Fig. 3.3. (**B**) Morphological changes in the postsynaptic membrane corresponding to the stages of NMJ development depicted in (**A**). Aneural AChR clusters have a lower density of AChRs per unit area and are therefore depicted by a lighter shade of grey. (**C**) Confocal images of mouse NMJs at various stages of development (postnatal day 0 [P0], P7, P14, P21, and P56). The postsynaptic membrane is labeled with bungarotoxin (BTX); synaptic vesicles and nerve terminals are labeled with anti-synaptotagmin 2 (Syt2); and motor axons (and portions of the nerve terminal cytoskeleton) are labeled with anti-neurofilament (NF). *Arrowheads* demonstrate that the clustering and restriction of synaptic vesicles occurs early in NMJ development. *Arrows* demonstrate multiple motor axons innervating immature postsynaptic sites

adult muscle. Thus, motor units are much larger in immature muscle than in adult muscle (Fig. 3.5C).

While nerves and muscles undergo the initial steps of synaptic differentiation they begin to contribute synapse-specific ECM components into the basal lamina that separates them. Three notable BL components are present and restricted to synaptic clefts of immature NMJs: z-agrin, laminin β2, and AChE (Fallon et al. 1985, Hoch et al. 1993, Patton et al. 1997). Other components present in adult synaptic BL are present at immature synapses, but are not restricted to synaptic sites (Patton et al. 1997, Fox et al. 2007b). Enrichment of these factors in synaptic BL and their concomitant downregulation in extra-synaptic BL occur postnatally as synapses mature. How and why the distribution of synaptic BL components are differentially regulated during NMJ is still under investigation.

3.3.2 Synaptic Maturation and Maintenance

The maturation process that transforms an immature NMJ into an adult NMJ involves three significant morphological changes. First, supernumerary nerve terminals innervating a single muscle fiber are eliminated leaving a single, branched nerve terminal at each synaptic site. Like the initial formation of nerve terminals, the withdrawal of extra nerve terminals, termed synaptic elimination, is a protracted process taking more than 2 weeks in rodents (Redfern 1970, Rosenthal and Taraskevich 1977, Keller-Peck et al. 2001). Initially, some of the intermingled nerve terminals detach and withdraw leading to the segregation of nerve terminals into non-overlap regions (Fig. 3.11).

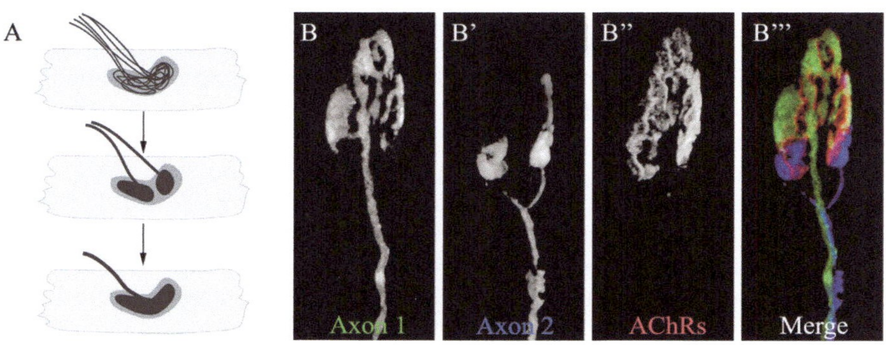

Fig. 3.11 Postnatal pruning of supernumerary nerve terminals at a single postsynaptic site. (**A**) Schematic depicting the pruning of multiple, interdigitated terminals to two segregated terminals and finally a single terminal. (**B**) Confocal image from a transgenic mouse expressing different variants of fluorescent proteins in subsets of neurons (see Feng et al. 2000, Lichtman and Sanes 2003). Notice two segregated nerve terminals (*green* and *blue*) apposed to the same postsynaptic apparatus (*red*)

Ultimately all motor nerves, save one, retract from the synapse. After the withdrawal of excess nerve terminals, the remaining nerve terminal expands and branches to occupy the denervated sites left behind by retracting nerves. Many mechanisms have been proposed to drive synapse elimination, but few involve cell adhesion molecules (CAMs). Thus these studies will not be discussed in this chapter (but see Sanes and Lichtman 1999, Wyatt and Balice-Gordon 2003).

Second, as supernumerary inputs to a single muscle fiber retract, perforations appear within the immature, elliptical AChR-rich postsynaptic membrane. Many, but not all, of these perforations correspond to AChR-rich regions vacated by withdrawing axons. Perforations within AChR clusters not associated with vacating motor nerves are likely the result of a mismatch between muscle and nerve terminal expansion. In either case, without the necessary signals from the apposing nerve terminal AChRs are dispersed and/or endocytosed. As myofibers continue to grow, small perforations expand, transforming plaque-like AChR clusters into pretzel-like, mature postsynaptic membranes (Fig. 3.10B,C) (Steinbach 1981, Slater 1982, Balice-Gordon and Lichtman 1993). It is at this point that the nerve terminal and postsynaptic membrane become precisely apposed.

Third, the topography of the postsynaptic membrane changes drastically in the maturing NMJ. While the postsynaptic membrane is flush with extrasynaptic muscle membrane at immature synapses, the postsynaptic membrane and overlying nerve terminal sink into the muscle fiber during maturation (Fig. 3.10A). In addition, postsynaptic folds are established during NMJ maturation (Fig. 3.10A). One possible mechanism for the formation of postsynaptic folds is a differential expansion of pre- and postsynaptic membranes (Marques et al. 2000). Since core components of the synaptic BL bind pre- and postsynaptic receptors, thereby linking nerve terminal to endplate, the uniform expansion of postsynaptic membrane with little change in presynaptic membrane likely causes the postsynaptic membrane to buckle and fold. This may partly explain differences in postsynaptic fold depth among mammalian species. Mammals with smaller nerve terminals (i.e., humans) have much deeper postsynaptic folds than those with larger terminals on similarly sized myofibers (i.e., mice and rats; Fig. 3.2B) (Slater et al. 1992, Wood and Slater 2001).

NMJs grow throughout life in parallel to muscle fiber growth. However, their unique shape and topography generated during this postnatal maturation period are maintained. Synaptic morphology is retained since growth occurs by either intercalary addition (i.e., new synaptic components are inserted between pre-existing components) or by circumferential, asymmetric addition (Akaaboune et al. 2002, Kummer et al. 2004). Eventually, however, signals that maintain the organization of the NMJ are lost with old age, leading to a deterioration in synaptic architecture. Age-associated NMJ deterioration includes the loss of synaptic adhesion, partial or complete denervation, fragmentation of AChR clusters, diminished concentration of AChRs, nerve terminal sprouting, and/or invasion of the synaptic cleft by perisynaptic Schwann cell processes (Rosenheimer and Smith 1985, Andonian and Fahim 1989, Prakash and Sieck 1998, Hodges et al. 2004, Wokke et al. 1990). Recently, age-associated changes in synaptic BL composition have also been identified and correlated with these morphological defects (J. Tapia, J.R. Sanes and J.W. Lichtman submitted), suggesting that specific extracellular adhesion molecules within the synaptic BL are necessary for not only synapse formation and maturation but also long-term maintenance.

3.4 Trans-synaptic Cues Direct NMJ Formation and Maintenance

3.4.1 Historical Perspective

Pioneering studies aimed at identifying trans-synaptic cues responsible for the formation of synapses were performed at the vertebrate NMJ. A student of Ramon y Cajal, Fernando Tello, first suggested that muscle-derived factors direct the transformation of a growth cone into a nerve terminal at specific sites on a muscle fiber (Cajal 1928). Two observations led Tello to this hypothesis: (1) Motor axons formed synapses only on muscle fibers, despite contacting numerous other cell types on their path from the ventral spinal cord to target muscle. (2) Severed motor axons reformed synaptic contacts with muscle near previously innervated sites (Tello 1907). Thus, not only must muscles produce a chemotactic signal to induce the transformation of a motor axon into a nerve terminal, but only certain regions of muscle fibers, i.e., the postsynaptic apparatus, secrete or present synaptogenic cues.

Noting that muscle and nerve were separated by a basal lamina (Couteaux 1944, 1946) U.J. McMahan and colleagues surmised that these synaptogenic cues might reside within the synaptic BL. To test this hypothesis, they not only denervated skeletal muscle, as Tello had done, but also induced muscle fiber degeneration. The withdrawing axon and degenerating muscle fiber left behind a lone BL. The previous synaptic portions of this BL 'ghost' remained identifiable either by finger-like projections that had previously occupied postsynaptic folds or by the presence of synaptic antigens such as AChE. Remarkably, when the muscle was permitted to regenerate in the absence of nerves, AChRs reclustered only at regions of BL previously associated with synaptic sites (Burden et al. 1979). In the opposite set of experiments, in which nerves regenerated but not muscles, presynaptic terminals precisely formed on previously synaptic portions of BL 'ghosts' (Sanes et al. 1978). In fact, active zones even reformed precisely aligned with the finger-like BL protrusions that had once occupied the cleft within postsynaptic folds (Glicksman and Sanes 1983). Together these studies were the first to demonstrate that molecules embedded within the synaptic BL, i.e., extracellular adhesion molecules, are sufficient to induce pre- and postsynaptic differentiation.

3.4.2 Synaptogenic Molecules Within Synaptic BL

Identifying and characterizing components of the synaptic BL necessary for NMJ formation and maintenance has been an active and fruitful avenue of research for the past three decades. Surprisingly, many of the most well-studied synaptogenic factors have been found to be synapse-specific isoforms of structural BL components: agrin, laminins, and collagen IV. Thus, at synaptic sites these macromolecules play dual roles in structural integrity of the BL and in directing synaptogenesis. Moreover, in several cases a single BL component binds and signals to both pre- and postsynaptic sides of the NMJ. Brief descriptions of some of the roles that these multifaceted extracellular adhesion molecules play in regulating synapse formation are provided below. Additional details as to roles that agrin and laminins play in synapse formation are provided in Chapter 20 of this volume. By focusing more on the role of trans-synaptic signals in forming and maintaining the vertebrate NMJ, details concerning the intracellular mechanisms responsible for synaptogenesis have been omitted. Readers interested in such pathways are referred to several recent, comprehensive reviews (Witzemann 2006, Lai and Ip 2003, Luo et al. 2003, Nelson et al. 2003).

3.4.3 Agrin

After identifying that components of the synaptic BL were sufficient to induce postsynaptic differentiation of regenerating myofibers at previous synaptic sites, McMahan and colleagues pursued the biochemical isolation and identification of BL components capable of clustering AChRs. BL can be biochemically purified based upon its insolubility compared with cellular tissue. However, as described above, synaptic BL represents only a minor fraction of muscle BL, therefore, McMahan's group purified BL fractions from the synapse-rich *Torpedo* electric organ. One fraction capable of inducing AChR clustering in cultured myotubes contained a heparan sulfate proteoglycan, agrin (from the Greek 'ageirein' meaning to 'assemble'), present in the synaptic BL of vertebrate NMJs (Godfrey et al. 1984, Wallace et al. 1985, Fallon et al. 1985, Nitkin et al. 1987). This agrin is produced by motor neurons, transported down motor axons, deposited into the synaptic cleft, and stably associated with synaptic BL. This led to the 'agrin hypothesis,' which postulates that nerve-derived agrin organized the differentiation of the postsynaptic membrane (McMahan 1990). Supporting this hypothesis, genetic deletion of agrin results in severe defects in postsynaptic differentiation (Gautam et al. 1996).

Despite mounting evidence in support of the agrin hypothesis, challenges to the agrin hypothesis eventually arose. First, myofibers also synthesize and secrete agrin. Second, and perhaps more detrimental to the 'agrin hypothesis,' muscle-derived agrin is not selectively associated with synaptic BL, but is also present extrasynaptically as synapses form (Fallon and Gelfman 1989, Bowe and Fallon 1995). How can nerve-derived agrin 'control' where AChRs cluster if muscle releases agrin indiscriminately? The answer to this question came with the discoveries that motor neurons synthesize an alternatively spliced isoform of agrin (termed z-agrin) that is far superior in its ability to stimulate the clustering of AChRs in vitro (Cohen and Godfrey 1992, Ruegg et al. 1992, Ferns et al. 1993, Hoch et al. 1993). Neuromuscular synapses in mice lacking only nerve-derived isoforms of agrin resembled those in mice lacking all forms of agrin, confirming that nerve-derived forms of agrin are both necessary and sufficient for the postsynaptic assembly of AChRs at the NMJ (Burgess et al. 1999).

A more recent challenge to the agrin hypothesis came from the observation that some AChR clusters formed in the absence of nerves and/or agrin, as described earlier (Fig. 3.8, Yang et al. 2000, 2001, Lin et al. 2001). In mutant mice lacking agrin, aneural AChR clusters form but are not properly maintained and disperse by birth (Lin et al. 2001, 2005, Misgeld et al. 2005). Thus, agrin is necessary for maintaining pre-existing AChR clusters, but not for forming them. Although agrin appears dispensable for the generation of prepatterned AChRs, its roles in the initial clustering of receptors in vivo is still under debate. Not all NMJs appear to form pre-existing AChR clusters (Flanagan-Steet et al. 2005, Lin et al. 2008) leaving open the possibility that nerve-derived agrin stimulates the initial formation of AChR clusters on some muscle fibers.

3.4.4 Laminins

While McMahan pursued the biochemical purification of postsynaptic organizers, Sanes and Hall took a wholly different approach to identify putative components of the synaptic BL. They screened antisera for antibodies that selectively stained synaptic sites (Sanes and Hall 1979). The antigen of one antibody that selectively labeled synapses was identified as the β2 subunit of laminin (Hunter et al. 1989). Laminins are large, extracellular heterotrimeric glycoprotein composed of α, β, and γ chains and are present in all BLs. However, a differential distribution of laminin isoforms exists in synaptic and extrasynaptic muscle BL: synaptic laminins contain the β2 subunit, whereas extrasynaptic laminins contain the β1 subunit (Patton 2003, Sanes 2003). Remarkably, despite differential localization, both laminin β1 and β2 subunits are synthesized and secreted into muscle BL by myofibers (Green et al. 1992). At least one mechanism for their differential distribution rests in their primary sequences. The amino-terminal sequence of laminin β1 promotes its binding to extrasynaptic collagen fibers. A 16 amino acid sequence within the carboxy-terminal region of laminin β2 subunits restricts its localization to synapses, perhaps by binding voltage-gated calcium channels (VGCCs) on the motor nerve terminal (Martin et al. 1995, Hunter et al. 1989, 1991, Nishimune et al. 2004).

Expression of laminin β2 and its ability to bind motor neurons suggests it is a prime candidate to drive nerve terminal formation and maintenance. In vitro, laminin β2 halts the growth of motor axons and induces the clustering of VGCCs and synaptic vesicles, two hallmark features of a presynaptic organizing cue (Porter et al. 1995, Nishimune et al. 2004). Unlike agrin mutants, which die at birth due to neuromuscular failure, mice lacking laminin β2 are viable at birth. Synapses in these mutants appear normal at birth, but by the first week of postnatal life mutant NMJs have few active zones, improperly clustered synaptic vesicles, few postsynaptic folds, Schwann cell processes within the synaptic cleft and a reduced capability of releasing neurotransmitters (Noakes et al. 1995, Knight et al. 2003, Nishimune et al. 2004, Fox et al. 2007). Defects in at least active zone formation and/or maintenance have recently been attributed to disrupting direct interactions between laminin β2 and the VGCCs flanking active zones (Nishimune et al. 2004). Thus, laminin β2 is a muscle-derived cue that is necessary for the maturation of nerve terminals (Fig. 3.12) and specifically the assembly and maintenance of active zones.

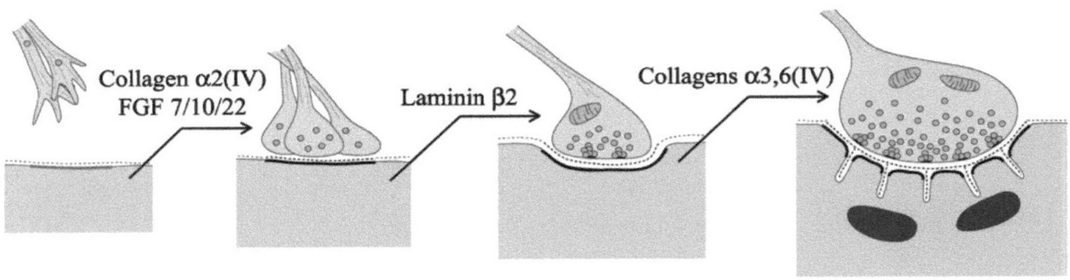

Fig. 3.12 Nerve terminal formation, maturation, and maintenance require the sequential expression of several muscle-derived cues. The FGF 7 subfamily (7/10/22) and collagen α2(IV) pattern the initial formation of motor nerve terminals. Laminin β2 directs the maturation of the nerve terminal by, at least in part, binding presynaptic calcium channels and enhancing active zone formation and/or maintenance. Collagens α3(IV) and α6(IV) are required for the continued maintenance of the nerve terminal. See text for explanations and references

Several lines of evidence suggest that laminins play other roles in forming and maintaining neuromuscular synapses in addition to laminin β2 binding presynaptic VGCCs. First, in vitro assays have demonstrated that laminin, like agrin, is sufficient to induce AChR clustering (Colognato and Yurchenco 2000, Vogel et al. 1983, Kummer et al. 2004). Second, as noted above, laminin β2 associates with three different α subunits (α2 in laminin 221, α4 in laminin 421, and α5 in laminin 521), each of which is differentially distributed in muscle BL (Fig. 3.7), suggesting that each may exhibit its own unique activities at synaptic sites (Patton et al. 1997, Miner et al. 1997, Patton 2003). The most restricted α chain in the synaptic cleft is laminin α4, which resides in periactive zone regions of the BL (Patton et al. 1997). The α2 and α5 laminin chains are both distributed throughout the entire synaptic cleft. However, while the distribution of α5 chains is restricted to synaptic sites, the α2 chain associates with β1 laminin subunits extrasynaptically. To identify specific roles of these α chains in the formation, maturation, and maintenance of the NMJ, synaptic morphology has been examined in mutants lacking each α chain. Genetic removal of the α4 chain results in synaptic fragmentation and misalignment of active zones to postsynaptic folds (Patton et al. 2001). Synapses deficient for α2 chains lack postsynaptic folds (Gilbert et al. 1973, Banker et al. 1979, Patton et al. 1997). In contrast to laminin α4 mutants, nerve terminals and active zones appear normal in laminin α2-deficient mutants. Mice lacking the α5 chains die at birth due to unrelated defects (Miner et al. 1998). Conditional removal of α5 chains from skeletal muscle results in the incomplete apposition of nerve terminals to postsynaptic membranes (Nishimune et al. 2005). Thus, these three sets of mutants demonstrate that synaptic laminins – laminin 221, 421, and 521 – each have unique functions in NMJ development. It remains unclear which of these functions are exerted directly on pre- or postsynaptic receptors or which are due to laminin networks indirectly clustering other synaptogenic factors.

3.4.5 Collagen IV

In addition to laminins and heparan sulfate proteoglycans, collagen IV is a major core component of basal laminas and is present within the synaptic cleft on the NMJ. Collagen IV is a triple helical ECM molecule assembled from three separate polypeptide (α) chains, each containing long collagenous Gly-X-Y repeating peptides – the defining feature of the collagen family of proteins (Myllyharju and Kivirikko 2004, Kadler et al. 2007). Unlike fibril-forming collagens (such as collagens I, II, III, and V) that assemble into collagen fibers, collagen IV is a non-fibril-forming collagen and is assembled into mesh-like networks essential for the formation and maintenance of basement membranes and basal laminas.

Like the other core BL components, several isoforms of collagen IV exist. In mammals, six genes code for α chains of collagen IV (Hudson et al. 2003). Although in principle the chains generated by these genes could assemble into >50 different homo- and heterotrimers (called protomers), only 3 combinations of α(IV) chains are known to exist: α1α1α2, α3α4α5, and α5α5α6 (Fig. 3.13). In muscle, α1α1α2 protomers are present in both synaptic and extrasynaptic BL throughout development of the vertebrate NMJ (Sanes et al. 1990, Miner and Sanes 1994, Fox et al. 2007). The other protomers are absent from muscle BL early in development. However, α3α4α5 and α5α5α6 protomers appear at synaptic sites postnatally, as NMJs mature (Miner and Sanes 1994, Fox et al. 2007). Thus, mature synaptic sites contain all six collagen α(IV) chains.

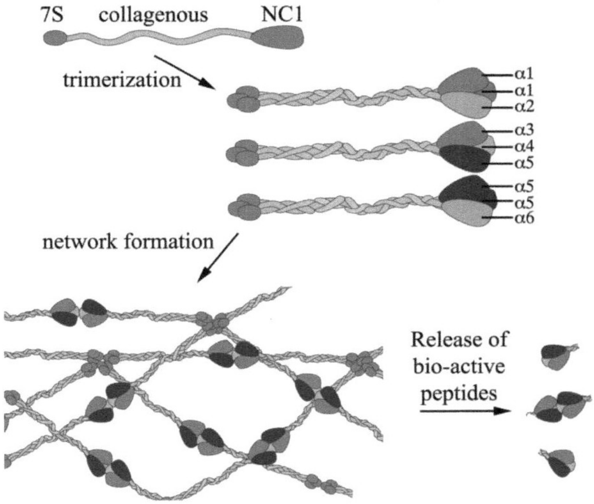

Fig. 3.13 Structure, assembly, and extracellular cleavage of collagen IV. Collagen IV molecules are composed of long collagenous domains flanked by two non-collagenous domains – the 7S and the NC1 domains. Prior to secretion, collagen α(IV) chains trimerize to form a protomer. Despite the presence of six α(IV) chains in mammalian genomes, only three protomers are known to exist (α1α1α2, α3α4α5, α5α5α6). Once secreted protomers are assembled into a collagen IV network. Degradation by extracellular proteases (such as MMPs) releases collagen IV NC1 domains from this network. In addition to the structural roles of collagen IV networks numerous studies have demonstrated novel bioactive roles for cleaved NC1 domains

Progress in understanding potential roles of collagen IV at the NMJ has proceeded slowly. Early studies initially suggested a role for collagen IV in the clustering of AChRs. Both, inhibiting collagen secretion or applying function-blocking collagen IV antibodies, decreased AChR clustering in cultured myotubes (Kalcheim et al. 1982a, b, Bixby 1995). More recently, it has been suggested that these affects are likely due to collagen IV binding and retaining other synaptogenic factors and not a direct activity of collagen IV (Smirnov et al. 2005).

Studies in non-neural cells, however, have demonstrated that collagen IV is not merely a structural molecule but also binds cell-surface receptors and acts as a CAM (Heino 2007). Whether collagen IV plays a direct role in NMJ formation or maintenance remained unclear until recently. In a screen seeking to identify muscle-derived organizers of the NMJ, synaptic BL was purified from *Torpedo* electric organs, separated chromatographically, and fractions were sought capable of inducing presynaptic differentiation in vitro. The active component of one such fraction was a non-collagenous domain of *Torpedo* collagen IV (Fox et al. 2007).

Vertebrate collagen IV contains collagenous domains flanked by two non-collagenous domains (Fig. 3.13). The amino-terminus of collagen IV contains a short 7S domain that is important for network formation, and the carboxy-terminus contains a longer non-collagenous (NC1) domain necessary for both chain trimerization and network formation (Hudson et al. 2003). Outside of the nervous system, NC1 domains have been shown to be proteolytically shed from collagen IV after which they acquire unique bio-activities that are unrelated to the structural role of full-length collagen IV (Ortega and Werb 2002, Kalluri 2003). Likewise, the presynaptic organizing activity of *Torpedo* collagen IV resides in the NC1 domain. Similarly, mammalian α2, α3, and α6(IV) NC1 domains all harbor presynaptic organizing activity in vitro (Fox et al. 2007). Thus, in vitro, both the structural and cell adhesive activities of collagen IV contribute to synaptic differentiation: collagen IV scaffolds stabilize postsynaptic organizers and collagen IV NC1 domains directly regulate presynaptic development.

Testing the role of specific collagen IV and their NC1 domains is difficult in vivo. Genetic removal of a single α chain leads to the inability of a cell to assemble the corresponding protomer. Thus, the absence of a single α chain results in the absence of others from the basal lamina (Hudson et al. 2003). Moreover, genetic truncation of collagen IV lacking carboxy-terminal NC1 domains similarly results in an inability to assemble protomers and thus the loss of several α chains from the basal lamina (Miner and Sanes 1996). Therefore, at present, in vivo analysis has been limited to analyzing collagen IV protomer function in NMJ formation and maintenance.

Of the collagen α(IV) chains capable of inducing presynaptic differentiation in motor axons, only collagen α2(IV) is present as synapse form. Targeted mutants lacking α2(IV) die before synapses form (Poschl et al. 2004), but mice with reduced expression levels survive (Gould et al. 2005). At birth, motor nerve terminals in these mice have improperly clustered synaptic vesicles and extend sprouts beyond their postsynaptic partners (Fox et al. 2007). Thus, collagen α2(IV) appears necessary for proper motor nerve terminal formation. The other collagen IV chains, which are capable of inducing presynaptic differentiation in vitro (α3[IV] and α6[IV]), appear at synaptic sites postnatally. NMJ morphology in mice lacking either α3(IV) or α6(IV) chains, which appear at synaptic sites postnatally, appears normal (Miner and Sanes 1994, Fox et al. 2007). However, mice lacking the α5(IV) chain, which also lack both α3(IV) and α6(IV) chains since neither α3α4α5 nor α5α5α6 protomers can assemble, have improperly maintained NMJs (Fox et al. 2007). Thus, although all three mammalian NC1 domains exhibit similar functions in vitro their roles in vivo differ: protomers containing α2(IV) chains are necessary for nerve terminal formation and protomers containing α3(IV) and α6(IV) chains are necessary for actively maintaining nerve terminals. In combination with studies on synaptic laminins and growth factors (see below), these studies demonstrate that multiple developmentally relevant muscle-derived organizers regulate sequential steps in nerve terminal formation and maintenance (Fig. 3.12).

3.4.6 Nidogens

Nidogens (also previously called entactins) are sulfated monomeric glycoproteins and constitute the fourth core component of BLs. Like laminins, collagen IV, and HSPGs, nidogens play both structural and bioactive roles in BL assembly and function. Structurally, nidogens bridge the laminin and collagen IV networks within the BL and bind HSPGs and chondroitin sulfated proteoglycans to stabilize them within the BL (Ho et al. 2008). Nidogens also bind an array of transmembrane adhesion receptors, such as integrins, and therefore are capable of acting as extracellular adhesion molecules mediating cell–ECM adhesion and signaling (Ho et al. 2008).

Two nidogens are present in vertebrates: nidogen-1 and nidogen-2. While both nidogens are ubiquitously present in most BLs, nidogen isoforms exhibit different distributions in muscle BL. Specifically, nidogen-1 is ubiquitously present in both synaptic and extrasynaptic muscle BL, whereas nidogen-2 is restricted to synaptic BL (Fig. 7, Fox et al. 2007b). In addition to the restricted distribution of nidogen-2 at synapses, a unique glycosylated form of nidogen-1 is restricted to synaptic BL (Chiu and Ko 1994). What roles these synapse-specific nidogens play at the vertebrate NMJ remains to be determined. However, it is noteworthy that a direct role for nidogen in synapse formation has been identified in invertebrates. In *Caenorhabditis elegans*, which expresses a single nidogen gene, nidogen organizes the formation of the NMJ. In the absence of invertebrate nidogen function, neuromuscular synapses are elongated, inappropriately distributed, and have reduced cholinergic transmission (Ackley et al. 2003). Nidogen is not restricted to synaptic sites in *C. elegans*; however, putative nidogen receptors, leukocyte common antigen-related (LAR)-like receptor protein tyrosine phosphatases (RPTPs), are enriched on motor nerve terminals (Ackley et al. 2005). Defects in LAR-like RPTPs mimic those observed in the absence of nidogen, suggesting that nidogen directly regulates presynaptic differentiation by interacting with nerve terminal receptors (Ackley et al. 2005). Although it is unclear whether LAR-like RPTPs are present on vertebrate motor nerve terminals, other putative nidogen receptors are, such as integrins. For example, nidogen-2 has a high affinity for $\alpha3\beta1$ integrins, which flank active zones at the vertebrate NMJ (Cohen et al. 2000). It remains to be seen whether such direct nidogen-2 interactions with pre- or postsynaptic membranes are required for proper vertebrate NMJ organization or function.

3.4.7 Other BL Components Contributing to NMJ Formation and Maintenance

Besides laminins, collagen IV, heparan sulfate proteoglycans, and nidogens, BLs bear a wide array of other extracellular proteins that contribute to its specific function. Although they have received less attention than synapse-specific core

BL components, three other components of synaptic BL have been implicated in the formation and maintenance of the vertebrate NMJ: other matrix molecules, matrix-degrading enzymes, and growth factors.

(i) **Matrix molecules**: In addition to collagen IV, at least two other non-fibril-forming collagenous molecules reside in the synaptic cleft: ColQ and collagen XIII. ColQ, the collagenous subunit of asymmetric AChE, has been well characterized for its role in anchoring AChE to synaptic sites by binding BL components such as the heparan sulfate proteoglycan perlecan (Peng et al. 1999, Rotundo 2003, Feng et al. 1999). Although AChE is absent from NMJs in targeted mouse mutants lacking ColQ, neurotransmission initially appears robust in these mutants. Thus, structural defects that appear during later stages of NMJ development in ColQ mutants suggest that ColQ may play a role in the maturation of NMJs separate from its enzymatic activity. In addition to interacting with perlecan in the synaptic BL, ColQ binds cell-surface receptors on the postsynaptic membrane, including MuSK (Cartaud et al. 2004). Preliminary studies suggest that ColQ–MuSK interactions modulate the expression of AChRs and other components of the postsynaptic apparatus (Sigoillot et al. 2007), thereby providing a potential mechanism for its roles in modulating the organization of neuromuscular synapses.

The third synaptogenic collagen within the neuromuscular synaptic cleft is muscle-derived collagen XIII. This unconventional collagen is a type-II transmembrane molecule whose extracellular collagenous domain can be cleaved in a furin-dependent mechanism to be incorporated into matrix (Vaisanen et al. 2004). Two findings support a role for collagen XIII in NMJ formation: (1) The application of the extracellular domain of collagen XIII onto cultured myotubes enhances the maturation of AChR clusters and (2) genetic removal of collagen XIII impairs the normal progression from plaque- to pretzel-like postsynaptic morphology (Fox et al. 2006). Remarkably, collagen XIII is synthesized by muscles, suggesting that the postsynaptic maturation of the NMJ may be coordinated or modulated in an autocrine fashion.

(ii) **Matrix-degrading enzymes:** A key component of BLs is a cohort of resident matrix-degrading enzymes. Although initially characterized for their roles in degrading ECM macromolecules and thus ECM turnover, it is now well established that extracellular proteases sever cell–ECM and cell–cell interactions, release bioactive cryptic domains from structural ECM molecules, and release and activate growth factors embedded within the matrix (Ortega and Werb 2002, Kalluri 2003, Mott and Werb 2004). To talk solely about the role of extracellular adhesion molecules at the synapse would be incomplete without mentioning these matrix-degrading enzymes.

The presence of proteases within the synaptic cleft has long been appreciated. Early studies characterized increases in synaptic protease activity following muscle stimulation or the direct application of ACh onto isolated muscles (Poberai et al. 1972, Poberai and Savay 1976, O'Brien et al. 1978). Concurrent with these observations, it was noted that the developmental refinement of polyneuronal innervation to a single muscle fiber coincided with increased muscle activity (Redfern 1970). Together these findings suggested that local activity-dependent increases in proteolytic activity might sever some cell–ECM adhesions leading to the detachment of select motor nerve terminals during synapse elimination. Indeed, the application of several protease inhibitors delays the normal developmental withdrawal of supernumerary inputs at the NMJ (O'Brien et al. 1984, Connold et al. 1986, Liu et al. 1994, Zoubine et al. 1996). At present, little is known about which cell–matrix interactions are cleaved by proteases to cause nerve terminal withdrawal. Moreover, why the interactions of one remaining nerve terminal with the synaptic BL is resistant to protease activity remains a mystery.

The findings that both collagen IV NC1 domains and growth factors (see below) are capable of inducing synaptic differentiation suggest that matrix-degrading enzymes may play a role in the release and activation of these factors from synaptic BL. One large family of proteases capable of liberating both collagen IV NC1 domains and growth factors from BL scaffolds are matrix metalloproteinases (MMPs) (Mott and Werb 2004, Sternlicht and Werb 2001). Several MMPs and their regulators have been identified at NMJs, including MMP2, MMP3, MMP7, MMP9, tissue inhibitors of metalloproteinases 2 (TIMP-2), and reversion-inducing cysteine-rich protein with kazal motifs (RECK) (Demestre et al. 2005, Kherif et al. 1998, Kawashima et al. 2008, Lluri et al. 2006, Werle and Vansaun 2003). The loss of some of these enzymes leads to structural changes in the organization of the vertebrate NMJ (for example, see Lluri et al. 2006). However, in all but one case the substrates that these enzymes act upon at synapses remain unclear. The one exception is MMP3, a perisynaptic Schwann cell-derived enzyme that degrades and removes agrin from the synaptic cleft (Vansaun and Werle 2000). In the absence of MMP3, excess agrin leads to an increase in both AChR density and postsynaptic fold depth (Vansaun et al. 2003, Werle and Vansaun 2003).

(iii) **Growth factors:** The final residents of the synaptic BL which influence synaptogenesis are growth factors. One growth factor that has received much attention at the vertebrate NMJ is neuregulin (for reviews see Falls 2003, Rimer 2003, 2007, Loeb 2003). The search for nerve-derived factors that are capable of inducing postsynaptic differentiation at the endplate identified one factor that enhanced AChR expression in cultured myotubes – 'AChR-inducing activity' factor (ARIA). Molecular cloning identified ARIA as a growth factor-like peptide encoded by the neuregulin-1 gene (Falls et al. 1993). Simultaneously, several other growth factor-like peptides (glial growth factor [GGF], heregulin [HRG], and Neu differentiation factor [NDF]) were also identified as alternatively spliced isoforms of neuregulin-1. Numerous studies identified two major roles for neuregulin-1 at the

NMJ: (1) Motoneuron-derived neuregulin-1 is secreted into the synaptic cleft where it can activate postsynaptic EGF receptors (ErbB2,3,4) and induce synapse-specific AChR gene transcription; (2) neuregulin-1 is required for the proliferation and survival of terminal Schwann cells (reviewed in Rimer 2003). Like the agrin hypothesis, roles for neuregulin-1 at the NMJ have been challenged by several unexpected findings in mutant mice. For example, mice lacking postsynaptic EGF receptors have only a modest postsynaptic phenotype (Escher et al. 2005). Furthermore, conditional inactivation of neuregulin from neurons, although lethal, does not alter the distribution or concentration of AChRs at synaptic sites (Jaworski and Burden 2006). Thus, at present it appears more likely that neuregulins are essential for terminal Schwann cell development, but are dispensable for NMJ formation (Rimer 2007, Kummer et al. 2006).

Many of the other growth factor-like molecules present at the neuromuscular synapses are thought to play largely 'trophic' roles for the nerve terminal. Such factors, which include neurotrophins, fibroblast growth factors (FGFs), ciliary neurotrophic factors (CNTFs), insulin-like growth factors (IGFs), and glial-derived growth factors (GDNFs), have been implicated as target-derived cues supporting neuronal survival, regeneration, nerve terminal sprouting, neuron–glial interactions, and even modulating synaptic function (English 2003, Todd et al. 2007). One family of growth factors, however, has been directly implicated in the formation of neuromuscular synapses: FGFs.

FGFs are a well-characterized, large family (22 members in humans and mice) of secreted signaling molecules implicated in almost all aspects of development (for review see Ornitz and Itoh 2001). FGFs bind and activate alternatively spliced forms of four tyrosine kinase receptors, called FGF receptors (FGFRs). Muscle and nerve both express a wide array of FGFs and FGFRs. The first evidence supporting synaptic organizing roles for FGFs came from a set of studies demonstrating their ability to promote presynaptic differentiation in cultured motor neurons (Dai and Peng 1995, Umemori et al. 2004). During NMJ formation, muscle expresses a developmentally regulated subfamily of FGFs with known presynaptic organizing activity (FGF7, 10, and 22; Umemori et al. 2004; Fox and Umemori 2006) and nerve terminals express their specific receptor – FGFR2b (Fox et al. 2007). These FGFs are therefore poised to coordinate the initial formation of motor nerve terminals, but are likely dispensable for synaptic maturation and maintenance since they are downregulated prior to these steps in NMJ development. Genetic disruption of FGF signaling in motor neurons confirmed this hypothesis since synaptic vesicles failed to properly aggregate in nerve terminals in the absence of these signals (Fox et al. 2007). These studies, together with studies on the presynaptic roles of laminin β2 and collagen IV, suggest that multiple muscle-derived cues are necessary for distinct developmental steps in presynaptic differentiation at the NMJ (Fig. 3.12).

3.4.8 Transmembrane Adhesion Molecules Contributing to NMJ Formation and Maintenance

Although it is unlikely that conventional transmembrane molecules embedded within the pre- and postsynaptic membranes directly interact to promote nerve–muscle adhesion, they do act as receptors for synaptogenic cues embedded in the synaptic BL. Many adhesion receptors have been identified in the pre- and postsynaptic membranes of the NMJ; however, only a select few have been determined to be necessary for NMJ formation in vivo (Sanes et al. 1998). Thus, at present little is known about the role that transmembrane cell adhesion receptors play in mediating signals from the synaptic BL.

Perhaps the most likely candidates to mediate signals from BL components are integrins, a large family of cell adhesion receptors capable of both structurally linking ECM molecules with intracellular cytoskeletal scaffolds and of initiating signaling cascades that alter cellular function or gene expression. Integrins are heterodimers composed of α and β subunits; so far over 24 distinct integrin heterodimers have been identified in vivo (Hynes 2002). The combination of α and β subunits dictates the binding specificity of integrins. Many heterodimers exist capable of binding a majority of BL components, including synaptic laminins, collagen IV, HSPGs, and nidogens. Although integrins capable of binding these synaptic BL components are present in pre- and postsynaptic membranes at the vertebrate NMJ, only a role for integrins in postsynaptic organization has been identified in vivo. Specifically, muscle-derived β1-containing integrin heterodimers are essential for NMJ formation (Schwander et al. 2004). Synaptic defects in the absence of muscle-derived β1-containing integrins resemble defects in agrin-null mutants, suggesting that integrins may mediate agrin-induced postsynaptic differentiation. Indeed, inhibition of β1 integrin subunits blocks the ability of agrin to stimulate AChR clustering on cultured myotubes (Martin and Sanes 1997).

Despite this role for integrins in mediating agrin signaling, the preponderance of data suggests that agrin-induced signals are mediated by MuSK, a receptor tyrosine kinase expressed by muscle during synaptogenesis whose distribution is restricted to the postsynaptic membrane (Valenzuela et al. 1995, Glass et al. 1996). MuSK is a single pass transmembrane molecule with an intracellular kinase domain, a PSD-95/Dlg/ZO-1 (PDZ)-binding domain, and four extracellular immunoglobulin-like (Ig) domains. Three initial findings supported an interaction between agrin and MuSK in directing AChR clustering: (1) agrin fails to induce AChR clustering on MuSK-deficient myotubes; (2) soluble MuSK–Ig domains neutralize the ability of agrin to stimulate AChR clustering on wild-type myotubes (Glass et al. 1996, 1997); and (3) synaptic defects in mice lacking MuSK resemble agrin-deficient synapses since AChR clusters are absent from muscle at birth (DeChiara et al. 1996). Despite extensive efforts, however, MuSK has never been shown to directly bind agrin. Thus, it has been proposed that MuSK associates with an agrin receptor (called

MASC) to transduce agrin-induced signaling. At present, the bulk of the studies, which investigate intracellular signaling pathways and scaffolds responsible for AChR clustering at the NMJ, focuses on this model of agrin binding to a MuSK/MASC receptor complex (see Strochlic et al. 2005).

Another postsynaptic protein capable of binding and mediating synaptogenic cues from the BL is dystroglycan. Two subunits of dystroglycan are transcribed from a single gene and a proteolytic cleavage and glycosylation generate α- and β-dystroglycan. Extracellular α-dystroglycan and transmembrane β-dystroglycan complex with utrophin, dystrophin, dystrobrevin, sarcoglycan, sarcospan, and syntrophins to form the dystrophin–glycoprotein complex (DGC) (Ervasti and Campbell 1991). Both agrin and laminins bind α-dystroglycan so the DGC was an initial candidate to mediate agrin- and laminin-induced AChR clustering (Ferns et al. 1993, Campanelli et al. 1994, Gee et al. 1994, Douville et al. 1988, Ibraghimov-Beskrovnaya et al. 1992, Fallon and Hall 1994). In the absence of α-dystroglycan, agrin retains its ability to induce AChR clustering suggesting that the DGC is dispensable for agrin-induced postsynaptic differentiation (Jacobson et al. 2001). Laminin, however, is incapable of inducing AChR clustering in the absence of α-dystroglycan, suggesting that the DGC is necessary for laminin-induced AChR clustering (Jacobson et al. 2001). Neuromuscular synapses in chimeric mice lacking both forms of dystroglycan are severely disrupted (Cote et al. 1999). Remarkably, neuromuscular synapses do initially form in the absence of dystroglycan, thereby hinting that laminin–dystroglycan interactions are not necessary for the initial clustering or reorganization of AChRs (which is presumably mediated through agrin–MuSK signaling) but are necessary for their maturation and maintenance.

The final transmembrane adhesion molecule at the NMJ that warrants mentioning is neural cell adhesion molecule (NCAM). NCAM is a member of the Ig superfamily of adhesion molecules known best for mediating homophilic interaction between neurons in the CNS. In addition to its roles in CNS development, NCAM is expressed by motoneurons, muscle fibers, and perisynaptic Schwann cells during NMJ development (Goridis and Brunet 1992, Sanes et al. 1986, Covault and Sanes 1985, 1986). While homophilic NCAM interactions likely mediate adhesion between Schwann cells and motor nerve terminals, based on the width of neuromuscular clefts it is unlikely to promote synaptic adhesion. Furthermore, the highest concentrations of NCAM are found at a distance from motor nerve terminals and are associated with the membranes of postsynaptic folds (Covault and Sanes 1986). Thus, pre- and postsynaptic NCAM likely interact with components of the synaptic BL components rather than with each other. One potential ECM-binding partner of NCAM at synaptic sites is agrin, although additional ligands are presumed to exist (Storms et al. 1996). The role of NCAM–ECM interactions within the synaptic cleft of the NMJ has yet to be elucidated; however, genetic removal of NCAM leads to a delay in both pre- and postsynaptic maturation (Moscoso et al. 1998, Rafuse et al. 2000).

3.5 Vertebrate Neuromuscular Junction: Concluding Remarks

The aim of this chapter was to provide an overview of the development of the vertebrate NMJ with special emphasis on the role of extracellular adhesion molecules. While our current knowledge regarding synapse formation remains greatest at the vertebrate NMJ, a comprehensive understanding of NMJ development is far from complete. Many of the molecular studies aimed at characterizing receptors for extracellular cues and the subsequent signaling cascades responsible for synapse formation have focused largely on postsynaptic development. In contrast, relatively little is known about the receptors, intracellular scaffolds, and signals responsible for presynaptic differentiation at the NMJ. Furthermore, investigations into NMJ development, maintenance, and function have largely been based on the assumption that synapses on typical skeletal muscles are all created equally. However, this notion has been challenged by the discovery of at least two subgroups of muscles with distinct temporal and spatial patterns of synaptogenesis (Pun et al. 2002, Santos and Caroni 2003). Confirming that not all NMJs may be created equally, genetic deletion or manipulation of several synaptic organizers affects synaptic structure in some muscles more than others (e.g., Pun et al. 2002, Fox et al. 2007, Ksiazek et al. 2007). These studies therefore raise the interesting possibility that distinct sets of trans-synaptic signals are required for NMJ organization and function on different muscles.

Finally, much of the knowledge gained at the large and accessible vertebrate NMJ over the past 200 years has been applicable to central synapses. However, striking differences between the NMJ and its central counterparts lie in the cleft separating nerve from muscle. The neuromuscular cleft is at least twice as wide as central synapses and contains a basal lamina. This has led some to suggest that while extracellular adhesion molecules direct synaptogenesis at the NMJ, they may not be relevant to central synapses. Indeed, many of the first synaptic organizers in the brain have been transmembrane CAMs or diffusible growth factors. However, an increasing number of studies have identified extracellular matrix molecules and matrix-degrading enzymes that are both present at central synapses and contribute to synaptic organization in the CNS (e.g., Libby et al. 1999, Christopherson et al. 2005, Egles et al. 2007, Stevens et al. 2007, Gawlak et al. 2009, Wilczynski et al. 2008). Thus, it appears that another discovery made at the NMJ – i.e., that extracellular cell adhesion molecules deposited into the synaptic cleft by synaptic partners direct synapse formation and maintenance – is applicable to central synapses.

References

Ackley, B. D., Kang, S. H., Crew, J. R., Suh, C., Jin, Y., and Kramer, J. M. (2003). The basement membrane components Nidogen and type XVIII collagen regulate organization of neuromuscular junctions in Caenorhabditis elegans. J Neurosci 23, 3577–3587

Ackley, B. D., Harrington, R. J., Hudson, M. L., Williams, L., Kenyon, C. J., Chisholm, A. D., and Jin, Y. (2005). The two isoforms of the Caenorhabditis elegans leukocyte-common antigen related receptor tyrosine phosphatase PTP-3 function independently in axon guidance and synapse formation. J Neurosci 25, 7517–7528

Adams, J. C. (2002). Molecular organisation of cell-matrix contacts: essential multiprotein assemblies in cell and tissue function. Expert Rev Mol Med 4, 1–24

Akaaboune, M., Grady, R. M., Turney, S., Sanes, J. R., and Lichtman, J. W. (2002). Neurotransmitter receptor dynamics studied in vivo by reversible photo-unbinding of fluorescent ligands. Neuron 34, 865–876

Anderson, M. J., and Cohen, M. W. (1977). Nerve-induced and spontaneous redistribution of acetylcholine receptors on cultured muscle cells. J Physiol 268, 757–773

Andonian, M. H., and Fahim, M. A. (1989). Nerve terminal morphology in C57BL/6NNia mice at different ages. J Gerontol 44, B43–51

Auld, D. S., Colomar, A., Belair, E. L., Castonguay, A., Pinard, A., Rousse, I., Thomas, S., and Robitaille, R. (2003). Modulation of neurotransmission by reciprocal synapse-glial interactions at the neuromuscular junction. J Neurocytol 32, 1003–1015

Balice-Gordon, R. J., and Lichtman, J. W. (1993). In vivo observations of pre- and postsynaptic changes during the transition from multiple to single innervation at developing neuromuscular junctions. J Neurosci 13, 834–855

Banker, B. Q., Hirst, N. S., Chester, C. S., and Fok, R. Y. (1979). Histometric and electron cytochemical study of muscle in the dystrophic mouse. Ann N Y Acad Sci 317, 115–131

Banks, G. B., Fuhrer, C., Adams, M. E., and Froehner, S. C. (2003) The postsynaptic submembrane machinery at the neuromuscular junction: requirement for rapsyn and the utrophin/dystrophin-associated complex. J Neurocytol 32, 709–726

Berg, D. K., Kelly, R. B., Sargent, P. B., Williamson, P., and Hall, Z. W. (1972). Binding of bungarotoxin to acetylcholine receptors in mammalian muscle (snake venom-denervated muscle-neonatal muscle-rat diaphragm-SDS-polyacrylamide gel electrophoresis). Proc Natl Acad Sci U S A 69, 147–151

Bernard, C. (1856). Analyse physiologique des propriétés des systèmes musculaires et nerveux au moyen du curare. CR Acad Sci, 825–829

Berrier, A. L., and Yamada, K. M. (2007). Cell-matrix adhesion. J Cell Physiol 213, 565–573

Bevan, S., and Steinbach, J. H. (1977). The distribution of alpha-bungarotoxin binding sites of mammalian skeletal muscle developing in vivo. J Physiol 267, 195–213

Birks, R., Huxley, H. E., and Katz, B. (1960). The fine structure of the neuromuscular junction of the frog. J Physiol 150, 134–144

Bixby, J. L. (1995). Collagen synthesis inhibition reduces clustering of heparan sulfate proteoglycan and acetylcholine receptors but not agrin or p65, at neuromuscular contacts in vitro. J Neurobiol 26, 262–272

Bowe, M. A., and Fallon, J. R. (1995). The role of agrin in synapse formation. Annu Rev Neurosci 18, 443–462

Brown, G. L., Dale, H. H., and Feldberg, W. (1936). Reactions of the normal mammalian muscle to acetylcholine and to eserine. J Physiol 87, 394–424

Buchanan, J., Sun, Y. A., and Poo, M. M. (1989). Studies of nerve-muscle interactions in Xenopus cell culture: fine structure of early functional contacts. J Neurosci 9, 1540–1554

Burden, S. J., Sargent, P. B., and McMahan, U. J. (1979). Acetylcholine receptors in regenerating muscle accumulate at original synaptic sites in the absence of the nerve. J Cell Biol 82, 412–425

Burden, S. J., DePalma, R. L., and Gottesman, G. S. (1983). Crosslinking of proteins in acetylcholine receptor-rich membranes: association between the beta-subunit and the 43 kd subsynaptic protein. Cell 35, 687–692

Burgess, R. W., Nguyen, Q. T., Son, Y. J., Lichtman, J. W., and Sanes, J. R. (1999). Alternatively spliced isoforms of nerve- and muscle-derived agrin: their roles at the neuromuscular junction. Neuron 23, 33–44

Cajal, S.R.Y. (1928). Degeneration and Regeneration of the Nervous System. (London, Oxford University Press)

Campanelli, J. T., Roberds, S. L., Campbell, K. P., and Scheller, R. H. (1994). A role for dystrophin-associated glycoproteins and utrophin in agrin-induced AChR clustering. Cell 77, 663–674

Carr, C., Fischbach, G. D., and Cohen, J. B. (1989). A novel 87,000-Mr protein associated with acetylcholine receptors in Torpedo electric organ and vertebrate skeletal muscle. J Cell Biol 109, 1753–1764

Cartaud, J., Cartaud, A., Kordeli, E., Ludosky, M. A., Marchand, S., and Stetzkowski-Marden, F. (2000). The torpedo electrocyte: a model system to study membrane-cytoskeleton interactions at the postsynaptic membrane. Microsc Res Tech 49, 73–83

Cartaud, A., Strochlic, L., Guerra, M., Blanchard, B., Lambergeon, M., Krejci, E., Cartaud, J., and Legay, C. (2004). MuSK is required for anchoring acetylcholinesterase at the neuromuscular junction. J Cell Biol 165, 505–515

Chang, C. C., and Lee, C. Y. (1963). Isolation of neurotoxins from the venom of bungarus multicinctus and their modes of neuromuscular blocking action. Arch Int Pharmacodyn Ther 144, 241–257

Changeux, J. P., Meunier, J. C., and Huchet, M. (1971). Studies on the cholinergic receptor protein of Electrophorus electricus. I. An assay in vitro for the cholinergic receptor site and solubilization of the receptor protein from electric tissue. Mol Pharmacol 7, 538–553

Chiu, A. Y., and Ko, J. (1994). A novel epitope of entactin is present at the mammalian neuromuscular junction. J Neurosci 14, 2809–2817

Christopherson, K. S., Ullian, E. M., Stokes, C. C., Mullowney, C. E., Hell, J. W., Agah, A., Lawler, J., Mosher, D. F., Bornstein, P., and Barres, B. A. (2005). Thrombospondins are astrocyte-secreted proteins that promote CNS synaptogenesis. Cell 120, 421–433

Cohen, M. W., and Godfrey, E. W. (1992). Early appearance of and neuronal contribution to agrin-like molecules at embryonic frog nerve-muscle synapses formed in culture. J Neurosci 12, 2982–2992

Cohen, M. W., Hoffstrom, B. G., and DeSimone, D. W. (2000). Active zones on motor nerve terminals contain alpha 3beta 1 integrin. J Neurosci 20, 4912–4921

Colognato, H., and Yurchenco, P. D. (2000). Form and function: the laminin family of heterotrimers. Dev Dyn 218, 213–234

Colomar, A., and Robitaille, R. (2004). Glial modulation of synaptic transmission at the neuromuscular junction. Glia 47, 284–289

Connold, A. L., Evers, J. V., and Vrbova, G. (1986). Effect of low calcium and protease inhibitors on synapse elimination during postnatal development in the rat soleus muscle. Brain Res 393, 99–107

Cote, P. D., Moukhles, H., Lindenbaum, M., and Carbonetto, S. (1999). Chimaeric mice deficient in dystroglycans develop muscular dystrophy and have disrupted myoneural synapses. Nat Genet 23, 338–342

Couteaux, R. (1944). Nouvelles observations sur la structure de la plaque motrice et interprétation des rapports myo-neuraux. CR Soc Biol 138, 976–979

Couteaux, R. (1946). Sur les gouttières synaptiques du muscle strié. CR Soc Biol 140, 270–273

Couteaux, R., and Pecot-Dechavassine, M. (1970). Synaptic vesicles and pouches at the level of "active zones" of the neuromuscular junction. C R Acad Sci Hebd Seances Acad Sci D 271, 2346–2349

Covault, J., and Sanes, J. R. (1985). Neural cell adhesion molecule (N-CAM) accumulates in denervated and paralyzed skeletal muscles. Proc Natl Acad Sci U S A 82, 4544–4548

Covault, J., and Sanes, J. R. (1986). Distribution of N-CAM in synaptic and extrasynaptic portions of developing and adult skeletal muscle. J Cell Biol 102, 716–730

Dai, Z., and Peng, H. B. (1995). Presynaptic differentiation induced in cultured neurons by local application of basic fibroblast growth factor. J Neurosci 15, 5466–5475

Dale, H. H., Feldberg, W., and Vogt, M. (1936). Release of acetylcholine at voluntary motor nerve endings. J Physiol 86, 353–380
DeChiara, T. M., Bowen, D. C., Valenzuela, D. M., Simmons, M. V., Poueymirou, W. T., Thomas, S., Kinetz, E., Compton, D. L., Rojas, E., Park, J. S., et al. (1996). The receptor tyrosine kinase MuSK is required for neuromuscular junction formation in vivo. Cell 85, 501–512
del Castillo, J., and Katz, B. (1956). Localization of active spots within the neuromuscular junction of the frog. J Physiol 132, 630–649
Demestre, M., Orth, M., Wells, G. M., Gearing, A. J., Hughes, R. A., and Gregson, N. A. (2005). Characterization of matrix metalloproteinases in denervated muscle. Neuropathol Appl Neurobiol 31, 545–555
Dimitropoulou, A., and Bixby, J. L. (2005). Motor neurite outgrowth is selectively inhibited by cell surface MuSK and agrin. Mol Cell Neurosci 28, 292–302
Douville, P. J., Harvey, W. J., and Carbonetto, S. (1988). Isolation and partial characterization of high affinity laminin receptors in neural cells. J Biol Chem 263, 14964–14969
Duclert, A., and Changeux, J. P. (1995). Acetylcholine receptor gene expression at the developing neuromuscular junction. Physiol Rev 75, 339–368
Edmonds, B., Gibb, A. J., and Colquhoun, D. (1995a). Mechanisms of activation of glutamate receptors and the time course of excitatory synaptic currents. Annu Rev Physiol 57, 495–519
Edmonds, B., Gibb, A. J., and Colquhoun, D. (1995b). Mechanisms of activation of muscle nicotinic acetylcholine receptors and the time course of endplate currents. Annu Rev Physiol 57, 469–493
Egles, C., Claudepierre, T., Manglapus, M. K., Champliaud, M. F., Brunken, W. J., and Hunter, D. D. (2007). Laminins containing the beta2 chain modulate the precise organization of CNS synapses. Mol Cell Neurosci 34, 288–298
English, A. W.(2003). Cytokines, growth factors and sprouting at the neuromuscular junction. J Neurocytol 32, 943–960
Ervasti, J. M., and Campbell, K. P. (1991). Membrane organization of the dystrophin-glycoprotein complex. Cell 66, 1121–1131
Escher, P., Lacazette, E., Courtet, M., Blindenbacher, A., Landmann, L., Bezakova, G., Lloyd, K. C., Mueller, U., and Brenner, H. R. (2005). Synapses form in skeletal muscles lacking neuregulin receptors. Science 308, 1920–1923
Fallon, J. R., Nitkin, R. M., Reist, N. E., Wallace, B. G., and McMahan, U. J. (1985). Acetylcholine receptor-aggregating factor is similar to molecules concentrated at neuromuscular junctions. Nature 315, 571–574
Fallon, J. R., and Gelfman, C. E. (1989). Agrin-related molecules are concentrated at acetylcholine receptor clusters in normal and aneural developing muscle. J Cell Biol 108, 1527–1535
Fallon, J. R., and Hall, Z. W. (1994). Building synapses: agrin and dystroglycan stick together. Trends Neurosci 17, 469–473
Falls, D. L., Rosen, K. M., Corfas, G., Lane, W. S., and Fischbach, G. D. (1993). ARIA, a protein that stimulates acetylcholine receptor synthesis, is a member of the neu ligand family. Cell 72, 801–815
Falls, D. L. (2003). Neuregulins and the neuromuscular system: 10 years of answers and questions. J Neurocytol 32, 619–647
Fatt, P., and Katz, B. (1950). Some observations on biological noise. Nature 166, 597–598
Fatt, P., and Katz, B. (1952). Spontaneous subthreshold activity at motor nerve endings. J Physiol 117, 109–128
Feng, G., Krejci, E., Molgo, J., Cunningham, J. M., Massoulie, J., and Sanes, J. R. (1999). Genetic analysis of collagen Q: roles in acetylcholinesterase and butyrylcholinesterase assembly and in synaptic structure and function. J Cell Biol 144, 1349–1360
Feng, Z., Koirala, S., and Ko, C. P. (2005). Synapse-glia interactions at the vertebrate neuromuscular junction. Neuroscientist 11, 503–513

Ferns, M. J., Campanelli, J. T., Hoch, W., Scheller, R. H., and Hall, Z. (1993). The ability of agrin to cluster AChRs depends on alternative splicing and on cell surface proteoglycans. Neuron 11, 491–502

Ferns, M., and Carbonetto, S. (2001). Challenging the neurocentric view of neuromuscular synapse formation. Neuron 30, 311–314

Fertuck, H. C., and Salpeter, M. M. (1974). Localization of acetylcholine receptor by 125I-labeled alpha-bungarotoxin binding at mouse motor endplates. Proc Natl Acad Sci U S A 71, 1376–1378

Fertuck, H. C., and Salpeter, M. M. (1976). Quantitation of junctional and extrajunctional acetylcholine receptors by electron microscope autoradiography after 125I-alpha-bungarotoxin binding at mouse neuromuscular junctions. J Cell Biol 69, 144–158

Flanagan-Steet, H., Fox, M. A., Meyer, D., and Sanes, J. R. (2005). Neuromuscular synapses can form in vivo by incorporation of initially aneural postsynaptic specializations. Development 132, 4471–4481

Flucher, B. E., and Daniels, M. P. (1989). Distribution of Na+ channels and ankyrin in neuromuscular junctions is complementary to that of acetylcholine receptors and the 43 kd protein. Neuron 3, 163–175

Fontaine, B., Klarsfeld, A., Hokfelt, T., and Changeux, J. P. (1986). Calcitonin gene-related peptide, a peptide present in spinal cord motoneurons, increases the number of acetylcholine receptors in primary cultures of chick embryo myotubes. Neurosci Lett 71, 59–65

Fox, M. A., and Umemori, H. (2006). Seeking long-term relationship: axon and target communicate to organize synaptic differentiation. J Neurochem 97, 1215–1231.

Fox, M. A., Latvanlehto, A., Pihlajaniemi, T., and Sanes, J. R. (2006). Collagen XIII is critical for postsynaptic differentiation and maturation at the NMJ. Paper presented at: Society for Neuroscience. (Atlanta, GA.)

Fox, M. A., Sanes, J. R., Borza, D. B., Eswarakumar, V. P., Fassler, R., Hudson, B. G., John, S. W., Ninomiya, Y., Pedchenko, V., Pfaff, S. L., et al. (2007a). Distinct target-derived signals organize formation, maturation, and maintenance of motor nerve terminals. Cell 129, 179–193

Fox, M. A., Smyth, N., and Sanes, J. R. (2007b). Nidogen at the neuromuscular junction. Paper presented at: Society for Neuroscience. (San Diego, CA)

Frank, E., and Fischbach, G. D. (1979). Early events in neuromuscular junction formation in vitro: induction of acetylcholine receptor clusters in the postsynaptic membrane and morphology of newly formed synapses. J Cell Biol 83, 143–158

Gautam, M., Noakes, P. G., Moscoso, L., Rupp, F., Scheller, R. H., Merlie, J. P., and Sanes, J. R. (1996). Defective neuromuscular synaptogenesis in agrin-deficient mutant mice. Cell 85, 525–535

Gawlak, M., Górkiewicz, T., Gorlewicz, A., Konopacki, F. A., Kaczmarek, L., and Wilczynski, G. M. (2009). High resolution in situ zymography reveals matrix metalloproteinase activity at glutamatergic synapses. Neuroscience 158(1), 167-176

Gee, S. H., Montanaro, F., Lindenbaum, M. H., and Carbonetto, S. (1994). Dystroglycan-alpha, a dystrophin-associated glycoprotein, is a functional agrin receptor. Cell 77, 675–686

Gilbert, J. J., Steinberg, M. C., and Banker, B. Q. (1973). Ultrastructural alterations of the motor end plate in myotonic dystrophy of the mouse (dy2J dy2J). J Neuropathol Exp Neurol 32, 345–364

Glass, D. J., DeChiara, T. M., Stitt, T. N., DiStefano, P. S., Valenzuela, D. M., and Yancopoulos, G. D. (1996). The receptor tyrosine kinase MuSK is required for neuromuscular junction formation and is a functional receptor for agrin. Cold Spring Harb Symp Quant Biol 61, 435–444

Glass, D. J., Apel, E. D., Shah, S., Bowen, D. C., DeChiara, T. M., Stitt, T. N., Sanes, J. R., and Yancopoulos, G. D. (1997). Kinase domain of the muscle-specific receptor tyrosine

kinase (MuSK) is sufficient for phosphorylation but not clustering of acetylcholine receptors: required role for the MuSK ectodomain? Proc Natl Acad Sci U S A 94, 8848–8853
Glicksman, M. A., and Sanes, J. R. (1983). Differentiation of motor nerve terminals formed in the absence of muscle fibres. J Neurocytol 12, 661–671
Godfrey, E. W., Nitkin, R. M., Wallace, B. G., Rubin, L. L., and McMahan, U. J. (1984). Components of Torpedo electric organ and muscle that cause aggregation of acetylcholine receptors on cultured muscle cells. J Cell Biol 99, 615–627
Goridis, C., and Brunet, J. F. (1992). NCAM: structural diversity, function and regulation of expression. Semin Cell Biol 3, 189–197
Gould, D. B., Phalan, F. C., Breedveld, G. J., van Mil, S. E., Smith, R. S., Schimenti, J. C., Aguglia, U., van der Knaap, M. S., Heutink, P., and John, S. W. (2005). Mutations in Col4a1 cause perinatal cerebral hemorrhage and porencephaly. Science 308, 1167–1171
Green, T. L., Hunter, D. D., Chan, W., Merlie, J. P., and Sanes, J. R. (1992). Synthesis and assembly of the synaptic cleft protein S-laminin by cultured cells. J Biol Chem 267, 2014–2022
Heino, J. (2007). The collagen family members as cell adhesion proteins. Bioessays 29, 1001–1010
Ho, M. S., Bose, K., Mokkapati, S., Nischt, R., and Smyth, N. (2008). Nidogens-Extracellular matrix linker molecules. Microsc Res Tech 71, 387–395
Hoch, W., Ferns, M., Campanelli, J. T., Hall, Z. W., and Scheller, R. H. (1993). Developmental regulation of highly active alternatively spliced forms of agrin. Neuron 11, 479–490
Hodges, S. H., Anderson, A. L., and Connor, N. P. (2004). Remodeling of neuromuscular junctions in aged rat genioglossus muscle. Ann Otol Rhinol Laryngol 113, 175–179
Hudson, B. G., Tryggvason, K., Sundaramoorthy, M., and Neilson, E. G. (2003). Alport's syndrome, Goodpasture's syndrome, and type IV collagen. N Engl J Med 348, 2543–2556
Hunter, D. D., Shah, V., Merlie, J. P., and Sanes, J. R. (1989). A laminin-like adhesive protein concentrated in the synaptic cleft of the neuromuscular junction. Nature 338, 229–234
Hunter, D. D., Cashman, N., Morris-Valero, R., Bulock, J. W., Adams, S. P., and Sanes, J. R. (1991). An LRE (leucine-arginine-glutamate)-dependent mechanism for adhesion of neurons to S-laminin. J Neurosci 11, 3960–3971
Hynes, R. O. (2002). Integrins: bidirectional, allosteric signaling machines. Cell 110, 673–687
Ibraghimov-Beskrovnaya, O., Ervasti, J. M., Leveille, C. J., Slaughter, C. A., Sernett, S. W., and Campbell, K. P. (1992). Primary structure of dystrophin-associated glycoproteins linking dystrophin to the extracellular matrix. Nature 355, 696–702
Israel, M., Manaranche, R., Mastour-Frachon, P., and Morel, N. (1976). Isolation of pure cholinergic nerve endings from the electric organ of Torpedo marmorata. Biochem J 160, 113–115
Jacobson, C., Cote, P. D., Rossi, S. G., Rotundo, R. L., and Carbonetto, S. (2001). The dystroglycan complex is necessary for stabilization of acetylcholine receptor clusters at neuromuscular junctions and formation of the synaptic basement membrane. J Cell Biol 152, 435–450
Jaworski, A., and Burden, S. J. (2006). Neuromuscular synapse formation in mice lacking motor neuron- and skeletal muscle-derived Neuregulin-1. J Neurosci 26, 655–661
Jennings, C. G., Dyer, S. M., and Burden, S. J. (1993). Muscle-specific trk-related receptor with a kringle domain defines a distinct class of receptor tyrosine kinases. Proc Natl Acad Sci U S A 90, 2895–2899
Jessell, T. M., Siegel, R. E., and Fischbach, G. D. (1979). Induction of acetylcholine receptors on cultured skeletal muscle by a factor extracted from brain and spinal cord. Proc Natl Acad Sci U S A 76, 5397–5401
Jin, Y., and Garner, C. C. (2008). Molecular mechanisms of presynaptic differentiation. Annu Rev Cell Dev Biol 24, 237–262. Review
Juranek, J., Mukherjee, K., Rickmann, M., Martens, H., Calka, J., Südhof, T. C., and Jahn, R. (2006). Differential expression of active zone proteins in neuromuscular junctions suggests functional diversification. Eur J Neurosci 24, 3043–3052

Kadler, K. E., Baldock, C., Bella, J., and Boot-Handford, R. P. (2007). Collagens at a glance. J Cell Sci 120, 1955–1958
Kalcheim, C., Duksin, D., and Vogel, Z. (1982a). Involvement of collagen in the aggregation of acetylcholine receptors on cultured muscle cells. J Biol Chem 257, 12722–12727
Kalcheim, C., Duksin, D., and Vogel, Z. (1982b). Aggregation of acetylcholine receptors in nerve-muscle cocultures is decreased by inhibitors of collagen production. Neurosci Lett 31, 265–270
Kalluri, R. (2003). Basement membranes: structure, assembly and role in tumour angiogenesis. Nat Rev Cancer 3, 422–433
Kang, H., Tian, L., and Thompson, W. (2003). Terminal Schwann cells guide the reinnervation of muscle after nerve injury. J Neurocytol 32, 975–985
Kawashima, S., Imamura, Y., Chandana, E. P., Noda, T., Takahashi, R., Adachi, E., Takahashi, C., and Noda, M. (2008). Localization of the membrane-anchored MMP-regulator RECK at the neuromuscular junctions. J Neurochem 104, 376–385
Keller-Peck, C. R., Walsh, M. K., Gan, W. B., Feng, G., Sanes, J. R., and Lichtman, J. W. (2001). Asynchronous synapse elimination in neonatal motor units: studies using GFP transgenic mice. Neuron 31, 381–394
Kherif, S., Dehaupas, M., Lafuma, C., Fardeau, M., and Alameddine, H. S. (1998). Matrix metalloproteinases MMP-2 and MMP-9 in denervated muscle and injured nerve. Neuropathol Appl Neurobiol 24, 309–319
Kim, N., and Burden, S. J. (2008). MuSK controls where motor axons grow and form synapses. Nat Neurosci 11, 19–27
Knight, D., Tolley, L. K., Kim, D. K., Lavidis, N. A., and Noakes, P. G. (2003). Functional analysis of neurotransmission at beta2-laminin deficient terminals. J Physiol 546, 789–800
Koirala, S., Reddy, L. V., and Ko, C. P. (2003). Roles of glial cells in the formation, function, and maintenance of the neuromuscular junction. J Neurocytol 32, 987–1002
Ksiazek, I., Burkhardt, C., Lin, S., Seddik, R., Maj, M., Bezakova, G., Jucker, M., Arber, S., Caroni, P., Sanes, J. R., et al. (2007). Synapse loss in cortex of agrin-deficient mice after genetic rescue of perinatal death. J Neurosci 27, 7183–7195
Kuffler, S. W., and Yoshikami, D. (1975). The number of transmitter molecules in a quantum: an estimate from iontophoretic application of acetylcholine at the neuromuscular synapse. J Physiol 251, 465–482
Kummer, T. T., Misgeld, T., Lichtman, J. W., and Sanes, J. R. (2004). Nerve-independent formation of a topologically complex postsynaptic apparatus. J Cell Biol 164, 1077–1087
Kummer, T. T., Misgeld, T., and Sanes, J. R. (2006). Assembly of the postsynaptic membrane at the neuromuscular junction: paradigm lost. Curr Opin Neurobiol 16, 74–82
Kühne, W. (1862). Über die peripherischen Endorgane der motorischen Nerven (Leipzig, W. Engelmann)
Kühne, W. (1887). Neue Untersuchungen über motorische Nervenendigungen. Z Biol 23, 1–148
Kühne, W. (1888). On the origin and the causation of vital movement. Proc R Soc Lond B 4, 427–447
Lai, K. O., and Ip, N. Y. (2003). Postsynaptic signaling of new players at the neuromuscular junction. J Neurocytol 32, 727–741
Land, B. R., Harris, W. V., Salpeter, E. E., and Salpeter, M. M. (1984). Diffusion and binding constants for acetylcholine derived from the falling phase of miniature endplate currents. Proc Natl Acad Sci U S A 81, 1594–1598
Lee, C. Y., Tseng, L. F., and Chiu, T. H. (1967). Influence of denervation on localization of neurotoxins from clapid venoms in rat diaphragm. Nature 215, 1177–1178
Libby, R. T., Lavallee, C. R., Balkema, G. W., Brunken, W. J., and Hunter, D. D. (1999). Disruption of laminin beta2 chain production causes alterations in morphology and function in the CNS. J Neurosci 19, 9399–9411

Lin, W., Burgess, R. W., Dominguez, B., Pfaff, S. L., Sanes, J. R., and Lee, K. F. (2001). Distinct roles of nerve and muscle in postsynaptic differentiation of the neuromuscular synapse. Nature 410, 1057–1064
Lin, W., Dominguez, B., Yang, J., Aryal, P., Brandon, E. P., Gage, F. H., and Lee, K. F. (2005). Neurotransmitter acetylcholine negatively regulates neuromuscular synapse formation by a Cdk5-dependent mechanism. Neuron 46, 569–579
Lin, S., Landmann, L., Ruegg, M. A., and Brenner, H. R. (2008). The role of nerve- versus muscle-derived factors in mammalian neuromuscular junction formation. J Neurosci 28, 3333–3340
Liu, Y., Fields, R. D., Festoff, B. W., and Nelson, P. G. (1994). Proteolytic action of thrombin is required for electrical activity-dependent synapse reduction. Proc Natl Acad Sci U S A 91, 10300–10304
Lluri, G., Langlois, G. D., McClellan, B., Soloway, P. D., and Jaworski, D. M. (2006). Tissue inhibitor of metalloproteinase-2 (TIMP-2) regulates neuromuscular junction development via a beta1 integrin-mediated mechanism. J Neurobiol 66, 1365–1377
Loeb, J. A. (2003). Neuregulin: an activity-dependent synaptic modulator at the neuromuscular junction. J Neurocytol 32, 649–664
Loewi, O. (1921). Über humorale Übertragbarkeit der Herznervenwirkung. Pflügers Archiv, 239–242
Luo, Z., Wang, Q., Dobbins, G. C., Levy, S., Xiong, W. C., and Mei, L. (2003). Signaling complexes for postsynaptic differentiation. J Neurocytol 32, 697–708
Lupa, M. T., and Hall, Z. W. (1989). Progressive restriction of synaptic vesicle protein to the nerve terminal during development of the neuromuscular junction. J Neurosci 9, 3937–3945
Lwebuga-Mukasa, J. S., Lappi, S., and Taylor, P. (1976). Molecular forms of acetylcholinesterase from Torpedo californica: their relationship to synaptic membranes. Biochemistry 15, 1425–1434
Marnay, A., and Nachmansohn, D. (1937). Cholinesterase in voluntary frog's muscle. J Physiol 89, 359–367
Marques, M. J., Conchello, J. A., and Lichtman, J. W. (2000). From plaque to pretzel: fold formation and acetylcholine receptor loss at the developing neuromuscular junction. J Neurosci 20, 3663–3675
Martin, P. T., Ettinger, A. J., and Sanes, J. R. (1995). A synaptic localization domain in the synaptic cleft protein laminin beta 2 (s-laminin). Science 269, 413–416
Martin, P. T., and Sanes, J. R. (1995). Role for a synapse-specific carbohydrate in agrin-induced clustering of acetylcholine receptors. Neuron 14, 743–754
Martin, P. T., and Sanes, J. R. (1997). Integrins mediate adhesion to agrin and modulate agrin signaling. Development 124, 3909–3917
Martin, P. T., Scott, L. J., Porter, B. E., and Sanes, J. R. (1999). Distinct structures and functions of related pre- and postsynaptic carbohydrates at the mammalian neuromuscular junction. Mol Cell Neurosci 13, 105–118
Martin, P. T. (2003). Glycobiology of the neuromuscular junction. J Neurocytol 32, 915–929
McMahan, U. J. (1990). The agrin hypothesis. Cold Spring Harb Symp Quant Biol 55, 407–418
Miledi, R., Molinoff, P., and Potter, L. T. (1971). Isolation of the cholinergic receptor protein of Torpedo electric tissue. Nature 229(5286), 554–557
Miner, J. H., and Sanes, J. R. (1994). Collagen IV alpha 3, alpha 4, and alpha 5 chains in rodent basal laminae: sequence, distribution, association with laminins, and developmental switches. J Cell Biol 127, 879–891
Miner, J. H., and Sanes, J. R. (1996). Molecular and functional defects in kidneys of mice lacking collagen alpha 3(IV): implications for Alport syndrome. J Cell Biol 135, 1403–1413.
Miner, J. H., Patton, B. L., Lentz, S. I., Gilbert, D. J., Snider, W. D., Jenkins, N. A., Copeland, N. G., and Sanes, J. R. (1997). The laminin alpha chains: expression,

developmental transitions, and chromosomal locations of alpha1-5, identification of heterotrimeric laminins 8–11, and cloning of a novel alpha3 isoform. J Cell Biol 137, 685–701.

Miner, J. H., Cunningham, J., and Sanes, J. R. (1998). Roles for laminin in embryogenesis: exencephaly, syndactyly, and placentopathy in mice lacking the laminin alpha5 chain. J Cell Biol 143, 1713–1723

Misgeld, T., Kummer, T. T., Lichtman, J. W., and Sanes, J. R. (2005). Agrin promotes synaptic differentiation by counteracting an inhibitory effect of neurotransmitter. Proc Natl Acad Sci U S A 102, 11088–11093

Moscoso, L. M., Cremer, H., and Sanes, J. R. (1998). Organization and reorganization of neuromuscular junctions in mice lacking neural cell adhesion molecule, tenascin-C, or fibroblast growth factor-5. J Neurosci 18, 1465–1477

Mott, J. D., and Werb, Z. (2004). Regulation of matrix biology by matrix metalloproteinases. Curr Opin Cell Biol 16, 558–564

Myllyharju, J., and Kivirikko, K. I. (2004). Collagens, modifying enzymes and their mutations in humans, flies and worms. Trends Genet 20, 33–43

Nelson, P. G., Lanuza, M. A., Jia, M., Li, M. X., and Tomas, J. (2003). Phosphorylation reactions in activity-dependent synapse modification at the neuromuscular junction during development. J Neurocytol 32, 803–816

Nishimune, H., Sanes, J. R., and Carlson, S. S. (2004). A synaptic laminin-calcium channel interaction organizes active zones in motor nerve terminals. Nature 432, 580–587

Nishimune, H., Miner, J. H., and Sanes, J. R. (2005). Roles of individual synaptic laminin chains in presynaptic differentiation at the neuromuscular junction. Paper presented at: Society for Neuroscience. (Washington, DC)

Nitkin, R. M., Smith, M. A., Magill, C., Fallon, J. R., Yao, Y. M., Wallace, B. G., and McMahan, U. J. (1987). Identification of agrin, a synaptic organizing protein from Torpedo electric organ. J Cell Biol 105, 2471–2478

Noakes, P. G., Gautam, M., Mudd, J., Sanes, J. R., and Merlie, J. P. (1995). Aberrant differentiation of neuromuscular junctions in mice lacking s-laminin/laminin beta 2. Nature 374, 258–262

O'Brien, R. A., Ostberg, A. J., and Vrbova, G. (1978). Observations on the elimination of polyneuronal innervation in developing mammalian skeletal muscle. J Physiol 282, 571–582

O'Brien, R. A., Ostberg, A. J., and Vrbova, G. (1984). Protease inhibitors reduce the loss of nerve terminals induced by activity and calcium in developing rat soleus muscles in vitro. Neuroscience 12, 637–646

Ornitz, D. M., and Itoh, N. (2001). Fibroblast growth factors. Genome Biol 2, REVIEWS3005

Ortega, N., and Werb, Z. (2002). New functional roles for non-collagenous domains of basement membrane collagens. J Cell Sci 115, 4201–4214

Panzer, J. A., Gibbs, S. M., Dosch, R., Wagner, D., Mullins, M. C., Granato, M., and Balice-Gordon, R. J. (2005). Neuromuscular synaptogenesis in wild-type and mutant zebrafish. Dev Biol 285, 340–357

Panzer, J. A., Song, Y., and Balice-Gordon, R. J. (2006). In vivo imaging of preferential motor axon outgrowth to and synaptogenesis at prepatterned acetylcholine receptor clusters in embryonic zebrafish skeletal muscle. J Neurosci 26, 934–947

Patton, B. L., Miner, J. H., Chiu, A. Y., and Sanes, J. R. (1997). Distribution and function of laminins in the neuromuscular system of developing, adult, and mutant mice. J Cell Biol 139, 1507–1521

Patton, B. L., Cunningham, J. M., Thyboll, J., Kortesmaa, J., Westerblad, H., Edstrom, L., Tryggvason, K., and Sanes, J. R. (2001). Properly formed but improperly localized synaptic specializations in the absence of laminin alpha4. Nat Neurosci 4, 597–604

Patton, B. L. (2003). Basal lamina and the organization of neuromuscular synapses. J Neurocytol 32, 883–903

Peng, H. B., Xie, H., Rossi, S. G., and Rotundo, R. L. (1999). Acetylcholinesterase clustering at the neuromuscular junction involves perlecan and dystroglycan. J Cell Biol 145, 911–921

Poberai, M., Savay, G., and Csillik, B. (1972). Function-dependent proteinase activity in the neuromuscular synapse. Neurobiology 2, 1–7

Poberai, M., and Savay, G. (1976). Time course of proteolytic enzyme alterations in the motor end-plates after stimulation. Acta Histochem 57, 44–48

Porter, S., and Froehner, S. C. (1983). Characterization and localization of the Mr = 43,000 proteins associated with acetylcholine receptor-rich membranes. J Biol Chem 258, 10034–10040

Porter, B. E., Weis, J., and Sanes, J. R. (1995). A motoneuron-selective stop signal in the synaptic protein S-laminin. Neuron 14, 549–559

Poschl, E., Schlotzer-Schrehardt, U., Brachvogel, B., Saito, K., Ninomiya, Y., and Mayer, U. (2004). Collagen IV is essential for basement membrane stability but dispensable for initiation of its assembly during early development. Development 131, 1619–1628

Prakash, Y. S., and Sieck, G. C. (1998). Age-related remodeling of neuromuscular junctions on type-identified diaphragm fibers. Muscle Nerve 21, 887–895

Pun, S., Sigrist, M., Santos, A. F., Ruegg, M. A., Sanes, J. R., Jessell, T. M., Arber, S., and Caroni, P. (2002). An intrinsic distinction in neuromuscular junction assembly and maintenance in different skeletal muscles. Neuron 34, 357–370

Rafuse, V. F., Polo-Parada, L., and Landmesser, L. T. (2000). Structural and functional alterations of neuromuscular junctions in NCAM-deficient mice. J Neurosci 20, 6529–6539

Reddy, L. V., Koirala, S., Sugiura, Y., Herrera, A. A., and Ko, C. P. (2003). Glial cells maintain synaptic structure and function and promote development of the neuromuscular junction in vivo. Neuron 40, 563–580

Redfern, P. A. (1970). Neuromuscular transmission in new-born rats. J Physiol 209, 701–709

Rimer, M. (2003). Neuregulins: primary or secondary signals for the control of synapse-specific gene expression. J Neurocytol 32, 665–675

Rimer, M. (2007). Neuregulins at the neuromuscular synapse: past, present, and future. J Neurosci Res 85, 1827–1833

Robertson, J. D. (1956a). Some features of the ultrastructure of reptilian skeletal muscle. J Biophys Biochem Cytol 2, 369–380

Robertson, J. D. (1956b). The ultrastructure of a reptilian myoneural junction. J Biophys Biochem Cytol 2, 381–394

Robitaille, R., Adler, E. M., and Charlton, M. P. (1990). Strategic location of calcium channels at transmitter release sites of frog neuromuscular synapses. Neuron 5, 773–779

Robitaille, R., Adler, E. M., and Charlton, M. P. (1993a). Calcium channels and calcium-gated potassium channels at the frog neuromuscular junction. J Physiol Paris 87, 15–24

Robitaille, R., Garcia, M. L., Kaczorowski, G. J., and Charlton, M. P. (1993b). Functional colocalization of calcium and calcium-gated potassium channels in control of transmitter release. Neuron 11, 645–655

Rosenheimer, J. L., and Smith, D. O. (1985). Differential changes in the end-plate architecture of functionally diverse muscles during aging. J Neurophysiol 53, 1567–1581

Rosenthal, J. L., and Taraskevich, P. S. (1977). Reduction of multiaxonal innervation at the neuromuscular junction of the rat during development. J Physiol 270, 299–310

Rotundo, R. L. (2003). Expression and localization of acetylcholinesterase at the neuromuscular junction. J Neurocytol 32, 743–766

Rotundo, R. L., Ruiz, C. A., Marrero, E., Kimbell, L. M., Rossi, S. G., Rosenberry, T., Darr, A., and Tsoulfas, P. (2008). Assembly and regulation of acetylcholinesterase at the vertebrate neuromuscular junction. Chem Biol Interact

Ruegg, M. A., Tsim, K. W., Horton, S. E., Kroger, S., Escher, G., Gensch, E. M., and McMahan, U. J. (1992). The agrin gene codes for a family of basal lamina proteins that differ in function and distribution. Neuron 8, 691–699

Salpeter, M. M., and Loring, R. H. (1985). Nicotinic acetylcholine receptors in vertebrate muscle: properties, distribution and neural control. Prog Neurobiol 25, 297–325
Sanes, J. R., Marshall, L. M., and McMahan, U. J. (1978). Reinnervation of muscle fiber basal lamina after removal of myofibers. Differentiation of regenerating axons at original synaptic sites. J Cell Biol 78, 176–198
Sanes, J. R., and Hall, Z. W. (1979). Antibodies that bind specifically to synaptic sites on muscle fiber basal lamina. J Cell Biol 83, 357–370
Sanes, J. R., and Cheney, J. M. (1982). Lectin binding reveals a synapse-specific carbohydrate in skeletal muscle. Nature 300, 646–647
Sanes, J. R., Schachner, M., and Covault, J. (1986). Expression of several adhesive macromolecules (N-CAM, L1, J1, NILE, uvomorulin, laminin, fibronectin, and a heparan sulfate proteoglycan) in embryonic, adult, and denervated adult skeletal muscle. J Cell Biol 102, 420–431
Sanes, J. R., Engvall, E., Butkowski, R., and Hunter, D. D. (1990). Molecular heterogeneity of basal laminae: isoforms of laminin and collagen IV at the neuromuscular junction and elsewhere. J Cell Biol 111, 1685–1699
Sanes, J. R., Apel, E. D., Gautam, M., Glass, D., Grady, R. M., Martin, P. T., Nichol, M. C., and Yancopoulos, G. D. (1998). Agrin receptors at the skeletal neuromuscular junction. Ann N Y Acad Sci 841, 1–13
Sanes, J. R., and Lichtman, J. W. (1999). Development of the vertebrate neuromuscular junction. Annu Rev Neurosci 22, 389–442
Sanes, J. R., and Lichtman, J. W. (2001). Induction, assembly, maturation and maintenance of a postsynaptic apparatus. Nat Rev Neurosci 2, 791–805
Sanes, J. R. (2003). The basement membrane/basal lamina of skeletal muscle. J Biol Chem 278, 12601–12604
Santos, A. F., and Caroni, P. (2003). Assembly, plasticity and selective vulnerability to disease of mouse neuromuscular junctions. J Neurocytol 32, 849–862
Schoch, S., and Gundelfinger, E. D. (2006). Molecular organization of the presynaptic active zone. Cell Tissue Res 326, 379–391
Schumacher, M., Camp, S., Maulet, Y., Newton, M., MacPhee-Quigley, K., Taylor, S. S., Friedmann, T., and Taylor, P. (1986). Primary structure of Torpedo californica acetylcholinesterase deduced from its cDNA sequence. Nature 319, 407–409
Schwander, M., Shirasaki, R., Pfaff, S. L., and Muller, U. (2004). Beta1 integrins in muscle, but not in motor neurons, are required for skeletal muscle innervation. J Neurosci 24, 8181–8191
Scott, L. J., Bacou, F., and Sanes, J. R. (1988). A synapse-specific carbohydrate at the neuromuscular junction: association with both acetylcholinesterase and a glycolipid. J Neurosci 8, 932–944
Scott, L. J., Balsamo, J., Sanes, J. R., and Lilien, J. (1990). Synaptic localization and neural regulation of an N-acetylgalactosaminyl transferase in skeletal muscle. J Neurosci 10, 346–350
Sealock, R., Wray, B. E., and Froehner, S. C. (1984). Ultrastructural localization of the Mr 43,000 protein and the acetylcholine receptor in Torpedo postsynaptic membranes using monoclonal antibodies. J Cell Biol 98, 2239–2244
Sheridan, M. N. (1965). The Fine Structure of the Electric Organ of Torpedo Marmorata. J Cell Biol 24, 129–141
Sigoillot, M., Lambergeon, M., Bourgeois, F., and Legay, C. (2007). The role of ColQ, a specific collagen in the postsynaptic organization of the neuromuscular junction. Paper presented at: Society for Neuroscience. (San Diego, CA)
Slater, C. R. (1982). Postnatal maturation of nerve-muscle junctions in hindlimb muscles of the mouse. Dev Biol 94, 11–22
Slater, C. R., Lyons, P. R., Walls, T. J., Fawcett, P. R., and Young, C. (1992). Structure and function of neuromuscular junctions in the vastus lateralis of man. A motor point biopsy study of two groups of patients. Brain 115 (Pt 2), 451–478

Slater, C. R. (2003). Structural determinants of the reliability of synaptic transmission at the vertebrate neuromuscular junction. J Neurocytol 32, 505–522
Smirnov, S. P., Barzaghi, P., McKee, K. K., Ruegg, M. A., and Yurchenco, P. D. (2005). Conjugation of LG domains of agrins and perlecan to polymerizing laminin-2 promotes acetylcholine receptor clustering. J Biol Chem 280, 41449–41457
Steinbach, J. H. (1981). Neuromuscular junctions and alpha-bungarotoxin-binding sites in denervated and contralateral cat skeletal muscles. J Physiol 313, 513–528
Sternlicht, M. D., and Werb, Z. (2001). How matrix metalloproteinases regulate cell behavior. Annu Rev Cell Dev Biol 17, 463–516
Stevens, B., Allen, N. J., Vazquez, L. E., Howell, G. R., Christopherson, K. S., Nouri, N., Micheva, K. D., Mehalow, A. K., Huberman, A. D., Stafford, B., et al. (2007). The classical complement cascade mediates CNS synapse elimination. Cell 131, 1164–1178
Storms, S. D., Kim, A. C., Tran, B. H., Cole, G. J., and Murray, B. A. (1996). NCAM-mediated adhesion of transfected cells to agrin. Cell Adhes Commun 3, 497–509
Strochlic, L., Cartaud, A., and Cartaud, J. (2005). The synaptic muscle-specific kinase (MuSK) complex: new partners, new functions. Bioessays 27, 1129–1135
Südhof, T. C. (2004). The synaptic vesicle cycle. Annu Rev Neurosci 27, 509–547
Sunesen, M., and Changeux, J. P. (2003). Transcription in neuromuscular junction formation: who turns on whom? J Neurocytol 32, 677–684
Tai, K., Bond, S. D., MacMillan, H. R., Baker, N. A., Holst, M. J., and McCammon, J. A. (2003). Finite element simulations of acetylcholine diffusion in neuromuscular junctions. Biophys J 84, 2234–2241
Takahashi, T., Nakajima, Y., Hirosawa, K., Nakajima, S., and Onodera, K. (1987). Structure and physiology of developing neuromuscular synapses in culture. J Neurosci 7, 473–481
Tello, J.F. (1907). Dègeneration et règeneration des plaques motrices après la section des nerfs. Trav Lab Recherches Biol 5, 117–149
Todd, K. J., Auld, D. S., and Robitaille, R. (2007). Neurotrophins modulate neuron-glia interactions at a vertebrate synapse. Eur J Neurosci 25, 1287–1296
Trimble, W. S., Cowan, D. M., and Scheller, R. H. (1988). VAMP-1: a synaptic vesicle-associated integral membrane protein. Proc Natl Acad Sci U S A 85, 4538–4542
Umemori, H., Linhoff, M. W., Ornitz, D. M., and Sanes, J. R. (2004). FGF22 and its close relatives are presynaptic organizing molecules in the mammalian brain. Cell 118, 257–270
Urbano, F. J., Rosato-Siri, M. D., and Uchitel, O. D. (2002). Calcium channels involved in neurotransmitter release at adult, neonatal and P/Q-type deficient neuromuscular junctions (Review). Mol Membr Biol 19, 293–300
Usdin, T. B., and Fischbach, G. D. (1986). Purification and characterization of a polypeptide from chick brain that promotes the accumulation of acetylcholine receptors in chick myotubes. J Cell Biol 103, 493–507
Vaisanen, M. R., Vaisanen, T., and Pihlajaniemi, T. (2004). The shed ectodomain of type XIII collagen affects cell behaviour in a matrix-dependent manner. Biochem J 380, 685–693
Valenstein, E. S. (2002). The discovery of chemical neurotransmitters. Brain Cogn 49, 73–95
Valenzuela, D. M., Stitt, T. N., DiStefano, P. S., Rojas, E., Mattsson, K., Compton, D. L., Nunez, L., Park, J. S., Stark, J. L., Gies, D. R., and et al. (1995). Receptor tyrosine kinase specific for the skeletal muscle lineage: expression in embryonic muscle, at the neuromuscular junction, and after injury. Neuron 15, 573–584
VanSaun, M., and Werle, M. J. (2000). Matrix metalloproteinase-3 removes agrin from synaptic basal lamina. J Neurobiol 43, 140–149
VanSaun, M., Herrera, A. A., and Werle, M. J. (2003). Structural alterations at the neuromuscular junctions of matrix metalloproteinase 3 null mutant mice. J Neurocytol 32, 1129–1142
Vogel, Z., Christian, C. N., Vigny, M., Bauer, H. C., Sonderegger, P., and Daniels, M. P. (1983). Laminin induces acetylcholine receptor aggregation on cultured myotubes and enhances the receptor aggregation activity of a neuronal factor. J Neurosci 3, 1058–1068

Wallace, B. G., Nitkin, R. M., Reist, N. E., Fallon, J. R., Moayeri, N. N., and McMahan, U. J. (1985). Aggregates of acetylcholinesterase induced by acetylcholine receptor-aggregating factor. Nature 315, 574–577
Werle, M. J., and VanSaun, M. (2003). Activity dependent removal of agrin from synaptic basal lamina by matrix metalloproteinase 3. J Neurocytol 32, 905–913
Wilczynski, G. M., Konopacki, F. A., Wilczek, E., Lasiecka, Z., Gorlewicz, A., Michaluk, P., Wawrzyniak, M., Malinowska, M., Okulski, P., Kolodziej, L. R., et al. (2008). Important role of matrix metalloproteinase 9 in epileptogenesis. J Cell Biol 180, 1021–1035
Witzemann, V. (2006). Development of the neuromuscular junction. Cell Tissue Res 326, 263–271
Wokke, J. H., Jennekens, F. G., van den Oord, C. J., Veldman, H., Smit, L. M., and Leppink, G. J. (1990). Morphological changes in the human end plate with age. J Neurol Sci 95, 291–310
Wood, S. J., and Slater, C. R. (2001). Safety factor at the neuromuscular junction. Prog Neurobiol 64, 393–429
Wyatt, R. M., and Balice-Gordon, R. J. (2003). Activity-dependent elimination of neuromuscular synapses. J Neurocytol 32, 777–794
Yang, X., Li, W., Prescott, E. D., Burden, S. J., and Wang, J. C. (2000). DNA topoisomerase IIbeta and neural development. Science 287, 131–134
Yang, X., Arber, S., William, C., Li, L., Tanabe, Y., Jessell, T. M., Birchmeier, C., and Burden, S. J. (2001). Patterning of muscle acetylcholine receptor gene expression in the absence of motor innervation. Neuron 30, 399–410
Young, S. H., and Poo, M. M. (1983). Spontaneous release of transmitter from growth cones of embryonic neurones. Nature 305, 634–637
Yurchenco, P. D., Amenta, P. S., and Patton, B. L. (2004). Basement membrane assembly, stability and activities observed through a developmental lens. Matrix Biol 22, 521–538
Zhang, J., Lefebvre, J. L., Zhao, S., and Granato, M. (2004). Zebrafish unplugged reveals a role for muscle-specific kinase homologs in axonal pathway choice. Nat Neurosci 7, 1303–1309
Zoubine, M. N., Ma, J. Y., Smirnova, I. V., Citron, B. A., and Festoff, B. W. (1996). A molecular mechanism for synapse elimination: novel inhibition of locally generated thrombin delays synapse loss in neonatal mouse muscle. Dev Biol 179, 447–457

Chapter 4
Synapse Formation in the Mammalian Central Nervous System

Masahiro Yasuda and Hisashi Umemori

Abstract Synapses are highly organized molecular complexes at which neurons communicate with each other in the brain. To form a functional synapse, hundreds of molecules need to be organized at the contact site between the axon and its target in the developing brain. Converging evidence now suggests that several families of cell adhesion molecules (CAMs) play important roles in differentiation, maturation, and maintenance of synapses. In this chapter, we will describe the structure of synapses, pre- and postsynaptic scaffolding molecules, steps of synapse formation, and synaptogenic molecules, including CAMs, in the mammalian central nervous system.

Keywords EphrinB · FGF22 · Narp · Neurexin · Neuroligin · SynCAM · Thrombospondin · Wnt7a

4.1 Introduction

Synapses are highly specialized apparatuses for neurotransmission. In the central nervous system (CNS), they mediate synaptic transmission from the presynaptic terminal to the postsynaptic membrane of the target neuron. Synapses are composed of a highly organized multimolecular complex of proteins, including synapse-specific proteins. This suggests that synaptogenesis involves a cascade of molecular interactions between synaptic proteins, which are induced by signals from the contact between axons and their targets. In this chapter we will describe molecular components of pre- and postsynaptic apparatuses and provide an overview of the molecules that are involved in the organization of CNS synapses.

H. Umemori (✉)
Molecular and Behavioral Neuroscience Institute, University of Michigan, Ann Arbor, MI 48109, USA
e-mail: umemoh@umich.edu

M. Hortsch, H. Umemori (eds.), *The Sticky Synapse*,
DOI 10.1007/978-0-387-92708-4_4, © Springer Science+Business Media, LLC 2009

4.2 Structures and Molecules of CNS Synapses

4.2.1 Ultrastructure of CNS Synapses

A CNS neuron has a highly polarized cell shape, which can be morphologically divided into dendrites with spines, a cell body, an axon, and presynaptic terminals. The specialized neuronal contact sites between presynaptic terminals and postsynaptic dendrites are called "synapses", the term introduced by Sir Michael Foster and Sir Charles Sherrington in 1897. At that time it was described as a contact without the continuity of substances between two neuronal cells. In 1959, with the advent of electron microscopy, George Gary provided the first clear evidence of synaptic contacts between neurons in the CNS (Gray 1959). He also described that there are two different types of synapses in the CNS, asymmetric and symmetric; the former was found onto spines, the latter mostly onto cell bodies. Now these two types of synapses are called Gray type I and Gray type II synapses, corresponding to excitatory and inhibitory synapses, respectively. Ultrastructurally, the organization of a synapse can be divided into several domains (Fig. 4.1). The presynaptic terminal consists of the mitochondria, synaptic vesicle (SV), and active zone domain. Presynaptic terminals are packed with clear SVs (50nm in diameter), which are filled with neurotransmitters. The membrane of active zones is characterized as an electron-dense area with docked SVs. Active zones are sites of neurotransmitter release. The gap between the presynaptic and postsynaptic membrane is called the synaptic cleft into which neurotransmitters are released. At the postsynaptic membrane, electron-dense structures are found, which are referred to as postsynaptic densities (PSDs). They represent the location of neurotransmitter-activated ion channels anchored by scaffolding proteins (Palay 1956, Scannevin and Huganir 2000, Dresbach et al. 2001, Sheng 2001). PSDs are formed directly apposed to active zones.

Excitatory synapses (type I synapses), which use glutamate as a neurotransmitter in the CNS, have a wider synaptic cleft than inhibitory synapses (type II

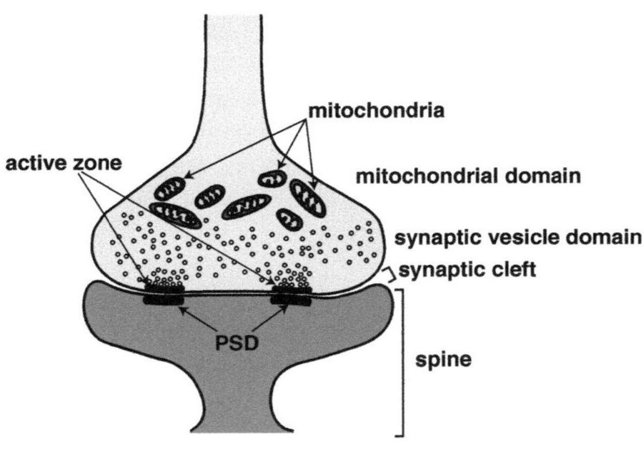

Fig. 4.1 Schematic illustration of the structural organization of synapses in the CNS. See the text for details. PSD, postsynaptic density

synapses). The clefts of excitatory synapses are around 30 nm wide and contain electron-dense extracellular materials. The active zone of excitatory synapses is 1–2 μm^2 in area. The most prominent feature of an excitatory synapse is found at the postsynaptic site. The excitatory synapse has PSDs of a 30–40 nm thick meshwork, which is located just below the postsynaptic membrane (Harris and Stevens 1989). These PSDs are large protein complexes, which include glutamate-responding ion channels. In contrast, at type II inhibitory synapses, the synaptic cleft is 20 nm wide, and there is little or no electron-dense extracellular materials within the cleft. The active zone is smaller (less than 1 μm^2), and the presynaptic membrane specializations and dense regions are less obvious. In addition, PSDs at the postsynaptic site are less obvious compared to those at excitatory synapses (Gray 1959).

4.2.2 Molecules at CNS Synapses

In mature neurons, SVs accumulate in the presynaptic terminal, and thus SV-associated proteins are abundant in the presynaptic terminal. Most protein components of SVs are basically the same in all neurons, including motoneurons (see Chapter 3). However, the neurotransmitter transporters, which load neurotransmitters into the vesicles, are different: in general in CNS neurons, vesicles at the excitatory presynaptic terminal contain vesicular glutamate transporter (VGLUT) to transport glutamate into the vesicle, and vesicles at the inhibitory presynaptic terminal have vesicular GABA transporter (VGAT) to load the vesicle with GABA.

During the past decade, a large number of synaptic components have been identified and cloned. Recent mass spectrometry and immunoblotting of a purified PSD fraction supported an estimate that approximately 600 proteins, including various postsynaptic scaffolding proteins, are components of PSDs (Collins et al. 2006). Compared to postsynaptic scaffolding proteins, the identification of presynaptic scaffolding proteins in the CNS is lagging behind. However, several presynaptic scaffolding proteins were identified by screening for mammalian homologs of genes, which are mutated in neurotransmitter release-deficit insects, and by using antibodies that recognize presynaptic terminals. The existence of these synapse-specific scaffolding molecules suggests that, viewed from a molecular perspective, synapse formation is a process during which hundreds of pre- and postsynaptic molecules are assembled into a highly organized synaptic structure (McAllister 2007, Scheiffele 2003, Waites et al. 2005, Craig et al. 2006, Garner et al. 2006). Since scaffolding proteins anchor many membrane proteins at synapses, specifically neurotransmitter receptors and CAMs, we will first briefly review this group of pre- and postsynaptic proteins.

4.2.2.1 Presynaptic Scaffold Molecules in the CNS

Proteins in the active zone form a multimolecular complex for Ca^{2+}-dependent neurotransmitter release. Several classes of cytoplasmic scaffold proteins have been identified as components of active zones. These proteins interact with each other, with Ca^{2+} channels and with CAMs to form a complex protein network at the active zone (Dresbach et al. 2001, Lonart 2002). Here, we describe presynaptic scaffold molecules with known functions in the CNS that have been analyzed in mutant mice.

Munc13-1

Munc13-1 was isolated as a mammalian homolog of a phorbol ester/diacylglycerol-binding protein, which had been originally identified in a *C. elegans* mutant with a severe uncoordinated movement phenotype (Maruyama and Brenner 1991, Brose et al. 1995, 2000). Munc13-1 is an ~200 kDa protein that contains a C1 domain and three C2 domains. C domains are functional domains found in protein kinase C. The C1 domain is a known diacylglycerol/phorbol ester-binding region, and the C2 domain has phospholipid-dependent Ca^{2+} binding activity, although the ability of the Munc13-1 C2 domain to bind Ca^{2+} has not yet been demonstrated (Perin et al. 1990, Betz et al. 1998, Brose et al. 2000). Munc13-1 is localized in the active zone. However, not all active zones contain Munc13-1: for example, 96% of synapses in the stratum lucidum, 56% in the stratum radiatum, and 45% in the parietal cortex are Munc13-1 positive (Betz et al. 1998). Munc 13-1 knockout mice form ultrastructurally normal synapses, but show a reduction in evoked glutamate release due to a reduction in the size of the readily releasable vesicle pool (Augustin et al. 1999). In contrast, GABA release is not changed by a loss of Munc13-1. These results suggest that Munc13-1 plays a role in the formation of a readily releasable SV pool at excitatory presynaptic sites.

RIM1

RIM1 was isolated as a presynaptic protein that interacts with the small GTPase Rab3 (Wang et al. 1997). RIM 1 is a multiple domain protein that has two Zn^{2+} finger domains at its N-terminal, a PDZ domain in its central region, and two C2 domains, one is downstream of the PDZ domain and the other is at the C-terminus. The N-terminal Zn^{2+} finger domains interact with the SV protein Rab3 and with Munc13-1 (Wang et al. 1997, 2000). Since the genetic disruption of RIM1 drastically reduces the efficacy of neurotransmitter release without obvious morphological deficits of synapses, it has been postulated that RIM1 plays a role in the regulation of vesicle fusion rather than the maintenance of synapse structures (Schoch et al. 2002, Takao-Rikitsu et al. 2004, Südhof 2004).

Bassoon and Piccolo

Bassoon and Piccolo were isolated from rat brain by a screening for synaptic junction components (Cases-Langhoff et al. 1996, Langnaese et al. 1996, tom Dieck et al. 1998). Bassoon and Piccolo are large proteins with sizes of 420 and 530 kDa, respectively. They are closely related to each other and with the exception of the C-terminus region share a common protein domain structure (Fenster et al. 2000). They are localized near active zones and interact with other proteins, such as the Rab3 acceptor protein PRA1 and the cytomatrix at the active zone (CAZ)-associated protein CAST. This suggests that they are scaffolding proteins, which assemble molecules involved in neurotransmitter release (Fenster et al. 2000, Takao-Rikitsu et al. 2004). Interestingly, Bassoon and Piccolo are present at excitatory and at inhibitory CNS synapses, but appear to be absent from cholinergic synapses such as the NMJ (Cases-Langhoff et al. 1996, Fenster et al. 2000, tom Dieck et al. 1998, Richter et al. 1999). Although Bassoon and Piccolo are closely related, their functions in the presynaptic terminal are distinct. Mutant mice lacking the central part of Bassoon have ultrastructurally normal brain synapses, with normal numbers of SVs clustered and docked. Electrophysiological experiments with hippocampal CA1 synapses in slices and in culture showed a 30–50% reduction in overall synaptic strength, but no change in mEPSC amplitude and postsynaptic responsiveness. These results suggest that Bassoon plays an important role for SVs to become fusion-competent (Dick et al. 2003, Altrock et al. 2003). In addition, the formation of retinal photoreceptor ribbon synapse is impaired in Bassoon mutant mice. RNAi knockdown of Piccolo in cultured hippocampal neurons indicates that Piccolo negatively regulates SV exocytosis (Leal-Ortiz et al. 2008).

CASK

CASK is a MAGuKs superfamily protein, which consists of a PDZ domain, an SH3 domain and a guanylate kinase (GK)-like domain. MAGuK proteins are involved in the assembly and organization of a variety of cell junctions. The MAGuKs family members that are present at the presynaptic terminal of neurons (including NMJs) are SAP90, SAP97, and CASK. Unlike other MAGuKs, CASK has a Ca^{2+}/calmodulin-dependent kinase-like domain at its N-terminus region (Hata et al. 1996). CASK interacts with many other presynaptic proteins, including cytoplasmic and transmembrane proteins. Examples for CASK-interacting proteins are cell adhesion molecules β-neurexins (Hata et al. 1996), syndecan 2 (Hsueh et al. 1998), voltage-gated Ca^{2+} channels (Maximov et al. 1999), and cytoplasmic proteins Veli/Lin-7 and Mint1 (Butz et al. 1998). CASK knockout mice die right after birth of a severe postnatal respiratory failure. However, they have normal evoked synaptic transmission and normal synaptic structures with a reduced level of CASK-interacting proteins, Mints, Veli/Mals, and neurexins. CASK knockout mice have an increased spontaneous glutamatergic synaptic release (mEPSC) and a reduced spontaneous GABAergic synaptic release (mIPSC), suggesting that CASK has a role during neurotransmitter release at the CNS presynaptic active zone.

4.2.2.2 Postsynaptic Scaffold Molecules in the CNS

For proper synaptic transmission, postsynaptic sites need to form a cluster of neurotransmitter receptors, which matches the neurotransmitter that is released from the apposed presynaptic terminal. In the CNS postsynaptic sites must accumulate glutamate receptors for excitatory synapses, and GABA or glycine receptors for inhibitory synapses. Scaffolding proteins that anchor these receptors depend on the type of synapses, i.e., excitatory synapses have excitatory synapse-specific scaffolding proteins, and inhibitory synapses have inhibitory synapse-specific scaffolding proteins. This indicates that scaffold proteins play important roles in the organization of synapse-type-specific postsynaptic apparatuses (Kennedy 2000, Sheng 2001, Kim and Sheng 2004).

Postsynaptic Scaffold Proteins at the Excitatory Synapse

The PSD95/SAP90 Family

The PSD consists of a large collection of structural and scaffold proteins that serve to cluster neurotransmitter receptors and their signaling machinery. PSD95 is abundant in biochemically prepared PSD fractions and it was one of the first proteins identified as a PSD component (Cho et al. 1992). The family of PSD95 proteins includes PSD95/SAP90, PSD93/chapsin-110, SAP102 and SAP97, which are also members of the MAGuK-protein family at the postsynaptic site. PSD95 family proteins have three PDZ domains, an SH3 domain, and a GK domain. PSD95 forms multimers, which may facilitate the clustering of multimolecular assemblies at specific sites, such as the PSD. PSD95 family proteins interact with many proteins, including glutamate channel subunits NR2A, NR2B, KA2, and GluR6 (Kornau et al. 1995, Roche et al. 1999, Garcia et al. 1998).

ProSAP/Shank Family Proteins

Proteins belonging to the ProSAP/Shank family constitute another type of scaffolding proteins at the PSD. ProSAP/Shank was isolated as a binding protein of GKAP (a PSD95-associated protein) in a yeast two-hybrid screen (Naisbitt et al. 1999, Sheng and Kim 2000, Boeckers et al. 2002). ProSAP/Shank family proteins consist of multiple ankyrin repeats, an SH3 domain, a PDZ domain, a proline-rich region, which makes up for half of the entire protein sequence, and a sterile alpha domain. All these domains are known to be involved in protein–protein interactions (Boeckers et al. 1999, Naisbitt et al. 1999, Tu et al. 1999, Yao et al. 1999, Du et al. 1998). An immunoelectron microscopic study showed that, when compared with SAP90/PSD-95, ProSAP/Shank proteins are concentrated in deeper layers of the PSD. This suggests that

ProSAP/Shank proteins function as master scaffold molecules to hold components of excitatory synapses through protein–protein interactions (Sheng and Kim 2000, Valtschanoff and Weinberg 2001, Boeckers et al. 2002, Boeckers 2006). Shank may form a polymeric structure with Homer (Hayashi et al. 2009).

Postsynaptic Scaffold Protein at the Inhibitory Synapse

Gephyrin

When compared with postsynaptic scaffold proteins located at the excitatory synapse, much less is known about equivalent molecules at inhibitory synapses. Gephyrin is the only known scaffold protein that is specifically located at the inhibitory postsynaptic site. Gephyrin plays a role in clustering GABAA and glycine receptors at inhibitory synapses in hippocampal neurons (Kneussel et al. 1999, Fischer et al. 2000, Levi et al. 2004).

4.3 Synaptogenesis in the CNS

Synapses are induced as the tip of axons (growth cone) interacts with the dendrites (or cell somas) of target cells. During the early stage of synaptogenesis, both axons and dendrites extend filopodia. Because of the small size of synapses, electron microscopy is most effective to examine the sequence of structural changes in developing CNS, from the initial contact to the mature synapse. Presynaptically, SV aggregates at the contact site, active zone are formed, mitochondria are recruited, and presynaptic buttons enlarge during synapse formation. Postsynaptically, PSDs are formed, and the shape of spines changes from filopodia to stubby spines to mushroom-like spines with a thin neck (Fiala et al. 1998, Harris 1999) (Fig. 4.2). These observations suggest that the process of synapse formation involves dramatic structural changes. Although synaptogenesis is a consecutive process from the initial axo-dendritic contact to the maturation of a fully functional synaptic junction, it can be divided into several distinct steps: initial contact, differentiation, and maturation (Fig. 4.2) (Waites et al. 2005, Garner et al. 2006).

4.3.1 Initial Contact of the Axon with its Target and Differentiation of CNS Synapses

After an axon reaches the target neuron, both partners need to recognize each other to form a synapse. This initial step is believed to be regulated by cell adhesion molecules (CAMs). As single CAM deletions in knockout mice have exhibited no deficits in the initial contact, multiple CAMs may function together during this first step of synaptogenesis. The following phases of synapse formation include the assembly of pre- and postsynaptic molecules at

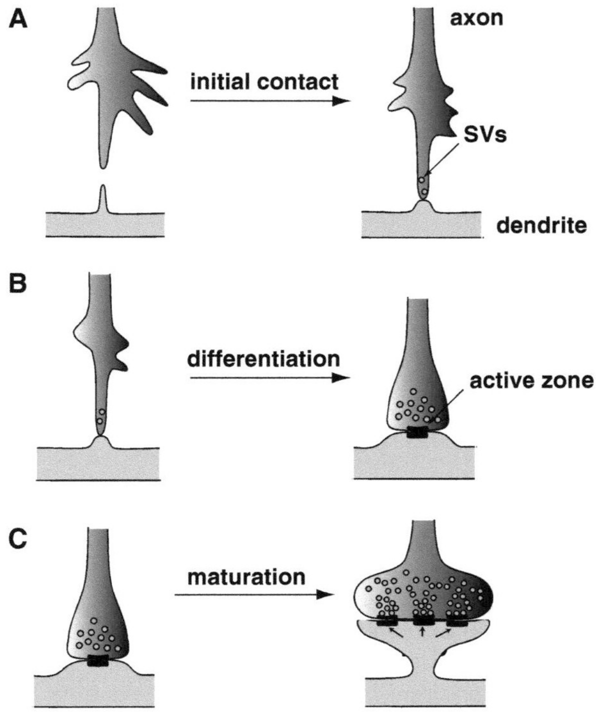

Fig. 4.2 The three steps of CNS synapse formation. Synapse formation is a multi-step process that centers around the organization and reorganization of synaptic proteins. (**A**) Initial contact. Both axons and dendrites extend filopodia to identify their appropriate partner. (**B**) Differentiation of synapses. After the initial contact between appropriate presynaptic axons and postsynaptic dendrites, synaptogenic molecules that are exchanged between axons and dendrites promote the accumulation of synaptic proteins for the development of functional synapses. On the presynaptic side, SVs are accumulated and the apparatus for vesicle fusion is formed (active zone). At the apposed postsynaptic site, neurotransmitter receptors and scaffolding proteins are clustered to form PSDs. (**C**) Maturation. As synapses mature, they increase in size and in their number of vesicles, active zones, and PSDs. In addition, the spine of excitatory synapses becomes mushroom shaped

the contact site. Electron microscopic studies have shown that the axonal site forms active zones with a few SVs, while the dendritic site accumulates electron-dense material to form PSDs apposed to active zones in the developing rat hippocampus CA1 (postnatal day 1–12) (Fiala et al. 1998). Even before initial contact, neurons already express pre- and postsynaptic proteins. There are two possible mechanisms for the recruitment of synaptic proteins to synaptic sites. One is the local recruitment of the individual proteins. In this case, signals that are generated at the initial contact site between an axon and its target induce the accumulation of synaptic components. The other possible mechanism involves the transport of synaptic components as preassembled "packets" to the initial contact site. The identification of packets containing active zone components in the cultured neurons argues in support of the second mechanism (Ahmari et al. 2000, Zhai et al. 2001, Shapira et al. 2003).

Immunohistochemical studies during excitatory synapse formation in vitro suggest that the differentiation of presynaptic sites precedes postsynaptic differentiation (Friedman et al. 2000, Zhai et al. 2000). This finding was subsequently confirmed by an elegant study using fusion molecules consisting of synaptic proteins with fluorescent proteins. In vitro live imaging of GFP fused to PSD95, which serves as a marker for postsynaptic differentiation, and of CFP fused to synaptophysin, which is used as a presynaptic differentiation marker, in the hippocampal cultured neuron (12–18 days in vitro) showed that the contact of an axonal filopodium to a dendrite induces the accumulation of PSD95 (Okabe et al. 2001). The accumulation of PSD95 at the contact site is always associated with clustering of synaptophysin molecules. Furthermore, the clustering of SVs, which are labeled by CFP-tagged synaptophysin, precedes the formation of PSD95 puncta at new synaptic sites (Okabe et al. 2001).

However, when comparing the above experiments with relatively young neuronal cultures (5–7 days in vitro), this may not always be the case. A recent report suggests that extrasynaptic clusters of PSD proteins, which contain the CAM neuroligin, form a stationary complex of PSD proteins in the dendrite, while others without neuroligin were motile in the dendrite (Gerrow et al. 2006). The time-lapse imaging of fluorescent-tagged synaptic markers suggested that stationary extrasynaptic clusters of PSD proteins, but not motile aggregates, induce the accumulation of SVs that are juxtaposed to the PSD clusters (Gerrow et al. 2006).

It has been shown that neurotransmitter release is not essential for the initial synapse formation to occur. Cultured neurons, in which synaptic release or action potential was pharmacologically inhibited, formed normal synapses (Craig 1998, Bacci et al. 2001). In vivo, Munc 13-1 or Munc 18-1 knockout mice, both molecules being essential for neurotransmitter release, also form structurally normal synapses in the brain (Augustin et al. 1999, Verhage et al. 2000). These results give strong support to the notion that initial formation of synapses solely depends on the interactions of synaptic molecules and not on neural activity.

4.3.2 Maturation and Maintenance of CNS Synapses

After nascent synapses have formed, they differentiate into stable synapses. During this maturation step, maturing synapses increase the number of vesicles at the presynaptic site and widen the contact site between the axon and the dendrite (Vaughn 1989, Fiala et al. 1998, Harris et al. 1992). This does not involve any dramatic change of their molecular composition, suggesting that maturation of synapses rather involves a reorganization of synaptic proteins to become more structurally stable synapses. During this process excitatory postsynaptic sites change their shape from stubby spines to mushroom spines (Fiala et al. 1998, Yuste and Bonhoeffer 2004). This morphological change involves the stabilization of cytoskeletal elements, because the actin cytoskeleton of immature synapses is affected by actin depolymerization drugs, while that of mature synapses is resistant to these drugs (Zhang and Benson 2001).

Some synaptic components change during the maturation process. At the presynaptic site, N-type voltage-dependent calcium channels are replaced by P/Q type channels. At the postsynaptic site, the NR2B subunit of the NMDA receptors is exchanged with the NR2A subunit (Pravettoni et al. 2000, Sheng et al. 1994, Carmignoto and Vicini 1992, Okabe et al. 1998, Petralia et al. 2005).

At excitatory synapses, glutamate release activates two types of ion channels, NMDA and AMPA receptors. Electrophysiological approaches suggest that synapses in young brains show NMDA receptor-mediated responses, but not AMPA receptor-meditated responses. This type of synapse is referred to as "silent synapse". Silent synapses are often found in developing brains, but not often in adult brains. The recruitment of AMPA receptors into silent synapses can be induced by a high-frequency stimulation of presynaptic axons, suggesting that this is an activity-dependent mechanism of synaptic maturation (Liao et al. 1995, Isaac et al. 1995, Durand and Konnerth 1996, Wu et al. 1996, Isaac et al. 1997, Feldman and Knudsen 1998, Liao et al. 1999). In cultured neurons, a chronic change of neural activity alters the rate of synaptic protein turnovers (Ehlers 2003). Thus, synaptic activity is also involved in the maintenance of mature synapses.

4.4 Synaptogenic Molecules in the CNS

Although the pre- and postsynaptic structures of neuromuscular and central synapses are very similar, the width of their synaptic clefts differs. The width of a CNS synaptic cleft is narrower than that of a neuromuscular junction (NMJ) ($\sim$20 nm for CNS and $\sim$50 nm for NMJ). This difference may be due to the lack of basal lamina at CNS synapses (Sanes and Lichtman 1999). Instead, CNS synapses contain various cell adhesion molecules (CAMs), which are expressed by both pre- and postsynaptic neurons. The width of a synaptic cleft at a CNS synapse may therefore be limited to allow these CAMs to interact with each other.

CNS neurons express a number of different CAMs. The majority of these CAMs belong to either the cadherin (Tepass et al. 2000) or the immunoglobulin superfamily (Rougon and Hobert 2003). Synaptic CAMs may have two different functions at CNS synapses. First, to provide mechanical stability to synapses, and second, to function as signal transduction molecules to facilitate the communication between pre- and postsynaptic neurons (Yamagata et al. 2003). The importance of CAMs during synapse formation in the CNS is well supported by the synaptic deficits that have been demonstrated in CAM knock-out mice (Table 4.1; see corresponding chapters). In the rest of this chapter we will focus on molecules that are shown to have synaptogenic ability in the CNS. Several molecules other than CAMs also have synaptogenic activity and these molecules might function cooperatively with synaptogenic CAMs.

Table 4.1 Summary of loss-of-function phenotypes of vertebrate CAMs involved in synapse formation

Molecules	Loss-of-function phenotypes	References	Chapters
N-cadherin	Normal synapse formation in hippocampal culture, although brain structures were randomized.	Kadowaki et al. (2007)	7
Cadherin 11, 13	RNA*i*: reduced synapse density in hippocampal culture	Paradis et al. (2007)	7
γProtocadherins	Reduced synapse number in the ventral spinal cord	Weiner et al. (2007)	7
Nectin	Reduced number of puncta adherence between mossy fiber and CA3 synapses; abnormal mossy fiber trajectories	Honda et al. (2006)	8
DSCAM	Defects in self-avoidance of neurite arbors in two subpopulations of retina amacrine cells	Fuerst et al. (2008)	9
Sidekick	Loss of laminar-specific targeting of processes in the inner plexiform layer of the chicken retina	Yamagata and Sanes (2008)	10
NEPH2 or 3 (SYG)	GOF: Generation of duplicated target glomeruli in the olfactory bulb. Olfactory axon sorting and targeting deficits	Serizawa et al. (2006)	11
Ll-CAM	Axon guidance error in various brain regions. Dendritic misorientation of cortical neurons.	Demyanenko et al. (1999) Cohen et al. (1998)	12
Neurofascin			12
NF186	Targeting deficit of climbing fibers to Purkinje cells	Ango et al. (2004)	
NF166	RNA*i*: Decreased clustering of gephyrin in hippocampal cultured neurons	Burkarth et al. (2007)	
NCAM	Abnormal mossy fiber projections to CA3. Reduced LTP at mossy fiber-CA3 synapses	Cremer et al. (1997) Cremer et al. (1998)	13
MHC	Deficits in selective retraction of redundant synaptic terminals during pruning in the visual systems and in synaptic stripping of motoneurons	Huh et al. (2000) Oliveira et al. (2004) Goddard et al. (2007)	14

Table 4.1 (continued)

Molecules	Loss-of-function phenotypes	References	Chapters
Sema3A	Decreased spines on layerV pyramidal neurons	Morita et al. (2006)	15
EphB1, 2, 3	Small postsynaptic densities and abnormal dendritic spines	Dalva et al. (2000)	16
Neuroligin			17
1	Decreased NMDA receptor mediated synaptic transmission.	Chubykin et al. (2007)	
2	Reduced inhibitory postsynaptic current amplitude	Chubykin et al. (2007)	
1,2,3 triple knockout	Normal synapse formation	Varoqueaux et al. (2006)	
SALM	RNA*i*: Decrease in outgrowth and process length in hippocampal culture. Loss of excitatory synapses and dendritic spines	Ko et al. (2006) Wang et al. (2008)	18
Integrin (α5)	RNA*i*: Defects in maturation of excitatory synapses, spine elongation, and new protrusions	Webb et al. (2007)	19
Thrombospondin 1,2	Reduced number of synapses in the cortex in the double knockout.	Christopherson et al. (2005)	
Agrin	Smaller number of synapses in the brain.	Ksiazek et al. (2007)	20

Synaptic phenotypes of CAM knockout mice are listed. For molecules that knockout mice are not yet published, results from RNAi or GOF (gain-of-function) are listed. See each chapter for more details.

4.4.1 WNT7a

WNT proteins comprise a large family of secreted polypeptides that are expressed in a tissue-specific pattern during the vertebrate embryonic development. Mutational analysis in mice has shown that WNTs play roles in controlling diverse developmental processes such as the patterning of body axes, CNS, and limbs (Logan and Nusse 2004, Ciani and Salinas 2005). The expression pattern of Wnt7a in cerebellar granule cells during neurite extension and synaptogenesis pointed to a function in synapse formation (Lucas and Salinas 1997). First evidence that WNT7a induces synaptogenesis was provided by experiments using pontine nuclei explants culture (Hall et al. 2000). Axons from pontine nuclei explants form synapses onto cerebellar granule neurons. The addition of soluble WNT7a into the pontine nuclei explants culture medium induced the differentiation of the pontine nuclei growth cone. This effect was blocked when the WNT7a antagonist sFRP-1 was added to the culture (Hall et al. 2000). In addition, co-culture of pontine nuclei explants with WNT7a-transfected, non-neuronal cells increased the clustering of a SV-associated protein synapsin I when compared to mock-transfected cells. Furthermore, Wnt7a knockout mice show a deficit in synapsin I accumulation in their pontine axon terminals. However this deficit is subtle and transient, which may be due to a compensatory mechanism that is induced by the closely related molecule Wnt7b, which is also expressed in cerebellar granule cells. The possibility of compensation is supported by the observation that a double knockout of Wnt7a and its receptor Dishevelled 1 shows more severe deficits than each single mutant (Ahmad-Annuar et al. 2006).

4.4.2 Neurexin/Neuroligin

Neurexins and neuroligins (see also Chapter 17) are ligands and receptors that are located at the presynaptic terminal and the postsynaptic site of the synapse, respectively. The role of their interaction during synaptogenesis has been studied extensively by using in vitro culture systems. In a co-culture system with pontine nuclei explants and neuroligin-transfected non-neuronal cells, SV clustering in the pontine neurite was induced at sites of contact with neuroligin-expressing cells. As shown by an EM analysis, the contacting neurite formed a structure resembling that of a presynaptic terminal. In addition, this structure was functional: it recycled SVs spontaneously and in an activity-dependent mode (Ichtchenko et al. 1995, Nguyen and Südhof 1997, Song et al. 1999, Scheiffele et al. 2000, Boucard et al. 2005, Sara et al. 2005). Conversely, the expression of β-neurexin in the non-neural cells induced the aggregation of the excitatory postsynaptic components PSD95 and NMDA receptors, as well as the inhibitory postsynaptic components Gephyrin and GABAA receptors at contact sites with co-cultured hippocampal neurons (Graf et al. 2004, Nam and Chen 2005). However, although the pups die within 24 hrs after birth, triple knockout mice, in which all neuroligin isoforms are eliminated, still form ultrastructurally normal synapses (Varoqueaux et al. 2006). Even in vitro cultured neurons that are prepared from these triple knockout mice form normal synapses. These results suggest that neurexin/neuroligin may be mainly involved in synapse maturation and maintenance. It appears that neuroligin1 and neuroligin2 control the balance between excitatory and inhibitory synapses (Graf et al. 2004, Chih et al. 2005, Chubykin et al. 2007).

4.4.3 SynCAM (or Nectin-Like Molecules)

SynCAMs (see also Chapter 8) are immunoglobulin-superfamily members with a PDZ-domain-binding motif in their intracellular domain. Comparisons of SynCAM sequences from different species show that SynCAMs are evolutionally well conserved (Biederer et al. 2002). These molecules have a homophilic-binding ability and their expression at synapses suggests that they function as adhesive bridges between the presynaptic and the postsynaptic membrane (Biederer et al. 2002). In fact, contact sites of hippocampal neurons with SynCAM-transfected non-neuronal cells formed functional, presynaptic sites in vitro (Sara et al. 2005). These in vitro molecular experiments indicate that SynCAMs are transsynaptic adhesion molecules, which may be involed in synapse formation.

4.4.4 FGF22

FGF22, a member of fibroblast growth factor family, was identified as a protein that induces SV clustering in the neurites of cultured neurons. Its close homologs, FGF7 and FGF10, have a similar vesicle clustering ability and all of these growth factors share the same receptor, FGFR2b (Ornitz et al. 1996, Umemori et al. 2004). In vivo experiments confirmed that FGF22 is a presynaptic differentiation factor. This was demonstrated by two different approaches: first, by the injection of a recombinant soluble fragment of FGFR2b, which binds to the intrinsic FGF22 and neutralizes it, into mouse brains, and secondly, by using FGFR2 knockout mice. Both experiments showed a significant reduction in SV accumulation at synaptic sites and a decrease in the number of active zones in the cerebellum (Umemori et al. 2004). Taken together, these findings confirm that FGF22 is a molecule to induce presynaptic differentiation thorough its interaction with FGFR2b.

4.4.5 Narp

Narp was first identified as an immediate early response gene, which is activated by seizures (Tsui et al. 1996). It is structurally similar to a secreted protein, Pentraxin, and localizes to synapses. Narp binds to the extracellular domain of AMPA receptors. Overexpression of Narp in cultured spinal cord neurons increased the clustering of AMPA receptors at synapses (O'Brien et al. 1999, 2002). The AMPA receptor clustering was blocked by a dominant-negative Narp mutant expressed in spinal cord neurons, suggesting that endogenous Narp, which is secreted at the synapse, induces the clustering of AMPA receptors in vitro (O'Brien et al. 2002). This Narp function of clustering AMPA receptors at contact sites is specific for certain types of neurons. Narp expressed by spinal neurons can induce the clustering of AMPA receptors at the contact sites with dendritic shafts of co-cultured hippocampal inhibitory neurons, but not with excitatory neurons (Mi et al. 2002). These findings suggest that synapse formation with excitatory neurons and with inhibitory neurons depends on different synaptogenic molecules.

4.4.6 EphrinB

Ephrins (see also Chapter 16) are transmembrane proteins, which were isolated as ligands for Eph tyrosine kinase receptors. Both Ephs and Ephrins are expressed in the CNS, and their developmental functions as axon guidance molecules during the establishment of topographic maps in the CNS is well known (Flanagan and Vanderhaeghen 1998, Yamaguchi and Pasquale 2004, McLaughlin and O'Leary 2005). The role of Eph and Ephrins in synapse formation was first suggested from results, which showed that EphrinB induces the clustering of NMDA receptors by binding to EphB. EphB interacts with NMDA receptors through its extracellular domain (Dalva et al. 2000). Triple knockout mice of EphB1, 2, and 3 show smaller postsynaptic sites and deficits in spine morphology at excitatory synapses, but not at inhibitory synapses (Henkemeyer et al. 2003).

4.4.7 Thrombospondins

The molecules described above are all expressed by neurons. However, recent work shows that proteins secreted from astrocytes also induce synapse formation. Thrombospondins (TSPs, see also Chapter 20) were isolated from glia-conditioned culture media as factors that increase the number of synapses when applied to cultured neurons. The synaptogenic effect on retinal ganglion cells was dose dependent and these TSP-induced synapses are positive for both presynaptic and postsynaptic marker proteins. Moreover, double knockout mice of TSP-1 and -2 show a reduced number of synapses in the cortex (Christopherson et al. 2005). Two possibilities have been put forward on how TSP induces synapse formation. The first is that TSPs directly induce the differentiation of presynaptic and postsynaptic specializations. In this case, TSPs activate their receptors (e.g., integrins and integrin-associated proteins) and thereby induce the differentiation of both sides of the synapse. The second possibility assumes that TSPs enhance the adhesion between the presynaptic and the postsynaptic membrane, which subsequently results in synapse formation by other synaptogenic molecules (Christopherson et al. 2005).

4.4.8 NGLs, SIRPs, and LRRTMs

Recently, three additional families of CAMs, NGLs, (Kim et al. 2006, Woo et al. 2009), SIRPs (Umemori et al. 2008) and LRRTMs (Linhoff et al. 2009), are shown to have synaptogenic ability. When non-neuronal cells expressing these CAMs are cocultured with primary neurons, SV clustering is induced in the axons at contact sites with non-neuronal cells. NGLs: netrin G ligands; SIRPs: signal regulatory proteins; LRRTMs: leucine rich repeat transmembrane proteins.

4.5 Conclusions

It has been of great interest to neuroscientists as to how neurons form precise synaptic connections in the brain. The search for synaptogenic molecules in the mammalian CNS has identified key molecules that are involved in synaptogenesis. The initial identification of these molecules relied either on the isolation of mammalian homologs, which were first identified in mutant flies and worms, or on a candidate approach, which was based on their developmental expression patterns in the brain. However, these types of approaches have proven to be of very limited use for the identification of "real" synaptogenic molecules. In fact, recent non-biased synaptogenic molecule screenings have identified several new molecules, including several non-neuronal proteins. Still, no knockout mice have been reported that completely fail to form synapses, suggesting that more important synaptogenic molecules await their discovery.

The successful development of neuronal culture systems, and the fluorescent protein tagging of synaptic component proteins combined with confocal time-lapse imaging techniques, has enabled us to monitor the dynamics of synapse formation. Genetically engineered mice, which express fluorescent proteins in a subset of neurons (Feng et al. 2000, Umemori et al. 2004, Livet et al. 2007), and the development of two-photon microscopes provide new promising avenues to monitor synapse formation in living animals (Trachtenberg et al. 2002, Grutzendler et al. 2002). These techniques will help us better understand the dynamics of synapse formation. Future questions, which need to be addressed, include the role of synaptogenic molecules in activity-dependent events, such as learning and memory formation, and in neurodevelopmental disorders, like Down syndrome and autism.

References

Ahmad-Annuar A, Ciani L, Simeonidis I et al. (2006) Signaling across the synapse: a role for Wnt and Dishevelled in presynaptic assembly and neurotransmitter release. J Cell Biol 174:127–139

Ahmari SE, Buchanan J and Smith SJ (2000) Assembly of presynaptic active zones from cytoplasmic transport packets. Nat Neurosci 3:445–451

Altrock WD, tom Dieck S, Sokolov M et al. (2003) Functional inactivation of a fraction of excitatory synapses in mice deficient for the active zone protein bassoon. Neuron 37:787–800

Ango F, di Cristo G, Higashiyama H et al. (2004) Ankyrin-based subcellular gradient of neurofascin, an immunoglobulin family protein, directs GABAergic innervation at purkinje axon initial segment. Cell 119:257–272

Augustin I, Rosenmund C, Südhof TC et al. (1999) Munc13-1 is essential for fusion competence of glutamatergic synaptic vesicles. Nature 400:457–461

Bacci A, Coco S, Pravettoni E et al. (2001) Chronic blockade of glutamate receptors enhances presynaptic release and downregulates the interaction between synaptophysin-synaptobrevin-vesicle-associated membrane protein 2. J Neurosci 21:6588–6596

Betz A, Ashery U, Rickmann M et al. (1998) Munc13-1 is a presynaptic phorbol ester receptor that enhances neurotransmitter release. Neuron 21:123–136

Biederer T, Sara Y, Mozhayeva M et al. (2002) SynCAM, a synaptic adhesion molecule that drives synapse assembly. Science 297:1525–1531

Boeckers TM (2006) The postsynaptic density. Cell Tissue Res 326:409–422

Boeckers TM, Bockmann J, Kreutz MR et al. (2002) ProSAP/Shank proteins – a family of higher order organizing molecules of the postsynaptic density with an emerging role in human neurological disease. J Neurochem 81:903–910
Boeckers TM, Kreutz MR, Winter C et al. (1999) Proline-rich synapse-associated protein-1/ cortactin binding protein 1 (ProSAP1/CortBP1) is a PDZ-domain protein highly enriched in the postsynaptic density. J Neurosci 19:6506–6518
Boucard AA, Chubykin AA, Comoletti D et al. (2005) A splice code for trans-synaptic cell adhesion mediated by binding of neuroligin 1 to alpha- and beta-neurexins. Neuron 48:229–236
Brose N, Hofmann K, Hata Y et al. (1995) Mammalian homologues of Caenorhabditis elegans unc-13 gene define novel family of C2-domain proteins. J Biol Chem 270:25273–25280
Brose N, Rosenmund C and Rettig J (2000) Regulation of transmitter release by Unc-13 and its homologues. Curr Opin Neurobiol 10:303–311
Burkarth N, Kriebel M, Kranz EU et al. (2007) Neurofascin regulates the formation of gephyrin clusters and their subsequent translocation to the axon hillock of hippocampal neurons. Mol Cell Neurosci 36:59–70
Butz S, Okamoto M and Südhof TC (1998) A tripartite protein complex with the potential to couple synaptic vesicle exocytosis to cell adhesion in brain. Cell 94:773–782
Carmignoto G and Vicini S (1992) Activity-dependent decrease in NMDA receptor responses during development of the visual cortex. Science 258:1007–1011
Cases-Langhoff C, Voss B, Garner AM et al. (1996) Piccolo, a novel 420 kDa protein associated with the presynaptic cytomatrix. Eur J Cell Biol 69:214–223
Chih B, Engelman H and Scheiffele P (2005) Control of excitatory and inhibitory synapse formation by neuroligins. Science 307:1324–1328
Cho KO, Hunt CA and Kennedy MB (1992) The rat brain postsynaptic density fraction contains a homolog of the Drosophila discs-large tumor suppressor protein. Neuron 9:929–942
Christopherson KS, Ullian EM, Stokes CC et al. (2005) Thrombospondins are astrocyte-secreted proteins that promote CNS synaptogenesis. Cell 120:421–433
Chubykin AA, Atasoy D, Etherton MR et al. (2007) Activity-dependent validation of excitatory versus inhibitory synapses by neuroligin-1 versus neuroligin-2. Neuron 54:919–931
Ciani L and Salinas PC (2005) WNTs in the vertebrate nervous system: from patterning to neuronal connectivity. Nat Rev Neurosci 6:351–362
Cohen NR, Taylor JS, Scott LB et al. (1998) Errors in corticospinal axon guidance in mice lacking the neural cell adhesion molecule L1. Curr Biol 8:26–33
Collins MO, Husi H, Yu L et al. (2006) Molecular characterization and comparison of the components and multiprotein complexes in the postsynaptic proteome. J Neurochem 97 Suppl 1:16–23
Craig AM (1998) Activity and synaptic receptor targeting: the long view. Neuron 21:459–462
Craig AM, Graf ER and Linhoff MW (2006) How to build a central synapse: clues from cell culture. Trends Neurosci 29:8–20
Cremer H, Chazal G, Carleton A et al. (1998) Long-term but not short-term plasticity at mossy fiber synapses is impaired in neural cell adhesion molecule-deficient mice. Proc Natl Acad Sci U S A 95:13242–13247
Cremer H, Chazal G, Goridis C et al. (1997) NCAM is essential for axonal growth and fasciculation in the hippocampus. Mol Cell Neurosci 8:323–335
Dalva MB, Takasu MA, Lin MZ et al. (2000) EphB receptors interact with NMDA receptors and regulate excitatory synapse formation. Cell 103:945–956
Demyanenko GP, Tsai AY and Maness PF (1999) Abnormalities in neuronal process extension, hippocampal development, and the ventricular system of L1 knockout mice. J Neurosci 19:4907–4920
Dick O, tom Dieck S, Altrock WD et al. (2003) The presynaptic active zone protein bassoon is essential for photoreceptor ribbon synapse formation in the retina. Neuron 37:775–786
Dresbach T, Qualmann B, Kessels MM et al. (2001) The presynaptic cytomatrix of brain synapses. Cell Mol Life Sci 58:94–116

Du Y, Weed SA, Xiong WC et al. (1998) Identification of a novel cortactin SH3 domain-binding protein and its localization to growth cones of cultured neurons. Mol Cell Biol 18:5838–5851

Durand GM and Konnerth A (1996) Long-term potentiation as a mechanism of functional synapse induction in the developing hippocampus. J Physiol Paris 90:313–315

Ehlers MD (2003) Activity level controls postsynaptic composition and signaling via the ubiquitin-proteasome system. Nat Neurosci 6:231–242

Feldman DE and Knudsen EI (1998) Experience-dependent plasticity and the maturation of glutamatergic synapses. Neuron 20:1067–1071

Feng G, Mellor RH, Bernstein M et al. (2000) Imaging neuronal subsets in transgenic mice expressing multiple spectral variants of GFP. Neuron 28:41–51

Fenster SD, Chung WJ, Zhai R et al. (2000) Piccolo, a presynaptic zinc finger protein structurally related to bassoon. Neuron 25:203–214

Fiala JC, Feinberg M, Popov V et al. (1998) Synaptogenesis via dendritic filopodia in developing hippocampal area CA1. J Neurosci 18:8900–8911

Fischer F, Kneussel M, Tintrup H et al. (2000) Reduced synaptic clustering of GABA and glycine receptors in the retina of the gephyrin null mutant mouse. J Comp Neurol 427:634–648

Flanagan JG and Vanderhaeghen P (1998) The ephrins and Eph receptors in neural development. Annu Rev Neurosci 21:309–345

Friedman HV, Bresler T, Garner CC et al. (2000) Assembly of new individual excitatory synapses: time course and temporal order of synaptic molecule recruitment. Neuron 27:57–69

Fuerst PG, Koizumi A, Masland RH et al. (2008) Neurite arborization and mosaic spacing in the mouse retina require DSCAM. Nature 451:470–474

Garcia EP, Mehta S, Blair LA et al. (1998) SAP90 binds and clusters kainate receptors causing incomplete desensitization. Neuron 21:727–739

Garner CC, Waites CL and Ziv NE (2006) Synapse development: still looking for the forest, still lost in the trees. Cell Tissue Res 326:249–262

Gerrow K, Romorini S, Nabi SM et al. (2006) A preformed complex of postsynaptic proteins is involved in excitatory synapse development. Neuron 49:547–562

Goddard CA, Butts DA and Shatz CJ (2007) Regulation of CNS synapses by neuronal MHC class I. Proc Natl Acad Sci U S A 104:6828–6833

Graf ER, Zhang X, Jin SX et al. (2004) Neurexins induce differentiation of GABA and glutamate postsynaptic specializations via neuroligins. Cell 119:1013–1026

Gray EG (1959) Axo-somatic and axo-dendritic synapses of the cerebral cortex: an electron microscope study. J Anat 93:420–433

Grutzendler J, Kasthuri N and Gan WB (2002) Long-term dendritic spine stability in the adult cortex. Nature 420:812–816

Hall AC, Lucas FR and Salinas PC (2000) Axonal remodeling and synaptic differentiation in the cerebellum is regulated by WNT-7a signaling. Cell 100:525–535

Harris KM (1999) Structure, development, and plasticity of dendritic spines. Curr Opin Neurobiol 9:343–348

Harris KM, Jensen FE and Tsao B (1992) Three-dimensional structure of dendritic spines and synapses in rat hippocampus (CA1) at postnatal day 15 and adult ages: implications for the maturation of synaptic physiology and long-term potentiation. J Neurosci 12:2685–2705

Harris KM and Stevens JK (1989) Dendritic spines of CA 1 pyramidal cells in the rat hippocampus: serial electron microscopy with reference to their biophysical characteristics. J Neurosci 9:2982–2997

Hata Y, Butz S and Südhof TC (1996) CASK: a novel dlg/PSD95 homolog with an N-terminal calmodulin-dependent protein kinase domain identified by interaction with neurexins. J Neurosci 16:2488–2494

Hayashi MK, Tang C, Verpelli C et al. (2009) The postsynaptic density proteins Homer and Shank form a polymeric network structure. Cell 137:159–171
Henkemeyer M, Itkis OS, Ngo M et al. (2003) Multiple EphB receptor tyrosine kinases shape dendritic spines in the hippocampus. J Cell Biol 163:1313–1326
Honda T, Sakisaka T, Yamada T et al. (2006) Involvement of nectins in the formation of puncta adherentia junctions and the mossy fiber trajectory in the mouse hippocampus. Mol Cell Neurosci 31:315–325
Hsueh YP, Yang FC, Kharazia V et al. (1998) Direct interaction of CASK/LIN-2 and syndecan heparan sulfate proteoglycan and their overlapping distribution in neuronal synapses. J Cell Biol 142:139–151
Huh GS, Boulanger LM, Du H et al. (2000) Functional requirement for class I MHC in CNS development and plasticity. Science 290:2155–2159
Ichtchenko K, Hata Y, Nguyen T et al. (1995) Neuroligin 1: a splice site-specific ligand for beta-neurexins. Cell 81:435–443
Isaac JT, Crair MC, Nicoll RA et al. (1997) Silent synapses during development of thalamocortical inputs. Neuron 18:269–280
Isaac JT, Nicoll RA and Malenka RC (1995) Evidence for silent synapses: implications for the expression of LTP. Neuron 15:427–434
Kadowaki M, Nakamura S, Machon O et al. (2007) N-cadherin mediates cortical organization in the mouse brain. Dev Biol 304:22–33
Kennedy MB (2000) Signal-processing machines at the postsynaptic density. Science 290:750–754
Kim E and Sheng M (2004) PDZ domain proteins of synapses. Nat Rev Neurosci 5:771–781
Kim S, Burette A, Chung HS et al. (2006) NGL family PSD-95-interacting adhesion molecules regulate excitatory synapse formation. Nat Neurosci 9:1294–1301
Kneussel M, Brandstatter JH, Laube B et al. (1999) Loss of postsynaptic GABA(A) receptor clustering in gephyrin-deficient mice. J Neurosci 19:9289–9297
Ko J, Kim S, Chung HS et al. (2006) SALM synaptic cell adhesion-like molecules regulate the differentiation of excitatory synapses. Neuron 50:233–245
Kornau HC, Schenker LT, Kennedy MB et al. (1995) Domain interaction between NMDA receptor subunits and the postsynaptic density protein PSD-95. Science 269:1737–1740
Ksiazek I, Burkhardt C, Lin S et al. (2007) Synapse loss in cortex of agrin-deficient mice after genetic rescue of perinatal death. J Neurosci 27:7183–7195
Langnaese K, Seidenbecher C, Wex H et al. (1996) Protein components of a rat brain synaptic junctional protein preparation. Brain Res Mol Brain Res 42:118–122
Leal-Ortiz S, Waites CL, Terry-Lorenzo R et al. (2008) Piccolo modulation of Synapsin1a dynamics regulates synaptic vesicle exocytosis. J Cell Biol 181:831–846
Levi S, Logan SM, Tovar KR et al. (2004) Gephyrin is critical for glycine receptor clustering but not for the formation of functional GABAergic synapses in hippocampal neurons. J Neurosci 24:207–217
Liao D, Hessler NA and Malinow R (1995) Activation of postsynaptically silent synapses during pairing-induced LTP in CA1 region of hippocampal slice. Nature 375:400–404
Liao D, Zhang X, O'Brien R et al. (1999) Regulation of morphological postsynaptic silent synapses in developing hippocampal neurons. Nat Neurosci 2:37–43
Linhoff MW, Lauren J, Cassidy RM et al. (2009) An unbiased expression screen for synaptogenic proteins identifies the LRRTM protein family as synaptic organizers. Neuron 61:734–749
Livet J, Weissman TA, Kang H et al. (2007) Transgenic strategies for combinatorial expression of fluorescent proteins in the nervous system. Nature 450:56–62
Logan CY and Nusse R (2004) The Wnt signaling pathway in development and disease. Annu Rev Cell Dev Biol 20:781–810
Lonart G (2002) RIM1: an edge for presynaptic plasticity. Trends Neurosci 25:329–332
Lucas FR and Salinas PC (1997) WNT-7a induces axonal remodeling and increases synapsin I levels in cerebellar neurons. Dev Biol 192:31–44

Maruyama IN and Brenner S (1991) A phorbol ester/diacylglycerol-binding protein encoded by the unc-13 gene of Caenorhabditis elegans. Proc Natl Acad Sci U S A 88:5729–5733
Maximov A, Südhof TC and Bezprozvanny I (1999) Association of neuronal calcium channels with modular adaptor proteins. J Biol Chem 274:24453–24456
McAllister AK (2007) Dynamic aspects of CNS synapse formation. Annu Rev Neurosci 30:425–450
McLaughlin T and O'Leary DD (2005) Molecular gradients and development of retinotopic maps. Annu Rev Neurosci 28:327–355
Mi R, Tang X, Sutter R et al. (2002) Differing mechanisms for glutamate receptor aggregation on dendritic spines and shafts in cultured hippocampal neurons. J Neurosci 22:7606–7616
Morita A, Yamashita N, Sasaki Y et al. (2006) Regulation of dendritic branching and spine maturation by semaphorin3A-Fyn signaling. J Neurosci 26:2971–2980
Naisbitt S, Kim E, Tu JC et al. (1999) Shank, a novel family of postsynaptic density proteins that binds to the NMDA receptor/PSD-95/GKAP complex and cortactin. Neuron 23:569–582
Nam CI and Chen L (2005) Postsynaptic assembly induced by neurexin-neuroligin interaction and neurotransmitter. Proc Natl Acad Sci U S A 102:6137–6142
Nguyen T and Südhof TC (1997) Binding properties of neuroligin 1 and neurexin 1beta reveal function as heterophilic cell adhesion molecules. J Biol Chem 272:26032–26039
O'Brien R, Xu D, Mi R et al. (2002) Synaptically targeted narp plays an essential role in the aggregation of AMPA receptors at excitatory synapses in cultured spinal neurons. J Neurosci 22:4487–4498
O'Brien RJ, Xu D, Petralia RS et al. (1999) Synaptic clustering of AMPA receptors by the extracellular immediate-early gene product Narp. Neuron 23:309–323
Okabe S, Collin C, Auerbach JM et al. (1998) Hippocampal synaptic plasticity in mice overexpressing an embryonic subunit of the NMDA receptor. J Neurosci 18:4177–4188
Okabe S, Miwa A and Okado H (2001) Spine formation and correlated assembly of presynaptic and postsynaptic molecules. J Neurosci 21:6105–6114
Oliveira AL, Thams S, Lidman O et al. (2004) A role for MHC class I molecules in synaptic plasticity and regeneration of neurons after axotomy. Proc Natl Acad Sci USA 101: 17843–17848
Ornitz DM, Xu J, Colvin JS et al. (1996) Receptor specificity of the fibroblast growth factor family. J Biol Chem 271:15292–15297
Palay SL (1956) Synapses in the central nervous system. J Biophys Biochem Cytol 2:193–202
Paradis S, Harrar DB, Lin Y et al. (2007) An RNAi-based approach identifies molecules required for glutamatergic and GABAergic synapse development. Neuron 53:217–232
Perin MS, Fried VA, Mignery GA et al. (1990) Phospholipid binding by a synaptic vesicle protein homologous to the regulatory region of protein kinase C. Nature 345:260–263
Petralia RS, Sans N, Wang YX et al. (2005) Ontogeny of postsynaptic density proteins at glutamatergic synapses. Mol Cell Neurosci 29:436–452
Pravettoni E, Bacci A, Coco S et al. (2000) Different localizations and functions of L-type and N-type calcium channels during development of hippocampal neurons. Dev Biol 227:581–594
Richter K, Langnaese K, Kreutz MR et al. (1999) Presynaptic cytomatrix protein bassoon is localized at both excitatory and inhibitory synapses of rat brain. J Comp Neurol 408:437–448
Roche KW, Ly CD, Petralia RS et al. (1999) Postsynaptic density-93 interacts with the delta2 glutamate receptor subunit at parallel fiber synapses. J Neurosci 19:3926–3934
Rougon G and Hobert O (2003) New insights into the diversity and function of neuronal immunoglobulin superfamily molecules. Annu Rev Neurosci 26:207–238
Sanes JR and Lichtman JW (1999) Development of the vertebrate neuromuscular junction. Annu Rev Neurosci 22:389–442

Sara Y, Biederer T, Atasoy D et al. (2005) Selective capability of SynCAM and neuroligin for functional synapse assembly. J Neurosci 25:260–270
Scannevin RH and Huganir RL (2000) Postsynaptic organization and regulation of excitatory synapses. Nat Rev Neurosci 1:133–141
Scheiffele P (2003) Cell-cell signaling during synapse formation in the CNS. Annu Rev Neurosci 26:485–508
Scheiffele P, Fan J, Choih J et al. (2000) Neuroligin expressed in nonneuronal cells triggers presynaptic development in contacting axons. Cell 101:657–669
Schoch S, Castillo PE, Jo T et al. (2002) RIM1alpha forms a protein scaffold for regulating neurotransmitter release at the active zone. Nature 415:321–326
Serizawa S, Miyamichi K, Takeuchi H et al. (2006) A neuronal identity code for the odorant receptor-specific and activity-dependent axon sorting. Cell 127:1057–1069
Shapira M, Zhai RG, Dresbach T et al. (2003) Unitary assembly of presynaptic active zones from Piccolo-Bassoon transport vesicles. Neuron 38:237–252
Sheng M (2001) Molecular organization of the postsynaptic specialization. Proc Natl Acad Sci U S A 98:7058–7061
Sheng M, Cummings J, Roldan LA et al. (1994) Changing subunit composition of heteromeric NMDA receptors during development of rat cortex. Nature 368:144–147
Sheng M and Kim E (2000) The Shank family of scaffold proteins. J Cell Sci 113 (Pt 11): 1851–1856
Sherrington CS (1897) The central nervous system, Vol. III. In: M. Foster (Ed.) A Textbook of Physiology, 7th Edition. London: Macmillan. p. 60
Song JY, Ichtchenko K, Südhof TC et al. (1999) Neuroligin 1 is a postsynaptic cell-adhesion molecule of excitatory synapses. Proc Natl Acad Sci U S A 96:1100–1105
Südhof TC (2004) The synaptic vesicle cycle. Annu Rev Neurosci 27:509–547
Takao-Rikitsu E, Mochida S, Inoue E et al. (2004) Physical and functional interaction of the active zone proteins, CAST, RIM1, and Bassoon, in neurotransmitter release. J Cell Biol 164:301–311
Tepass U, Truong K, Godt D et al. (2000) Cadherins in embryonic and neural morphogenesis. Nat Rev Mol Cell Biol 1:91–100
tom Dieck S, Sanmarti-Vila L, Langnaese K et al. (1998) Bassoon, a novel zinc-finger CAG/glutamine-repeat protein selectively localized at the active zone of presynaptic nerve terminals. J Cell Biol 142:499–509
Trachtenberg JT, Chen BE, Knott GW et al. (2002) Long-term in vivo imaging of experience-dependent synaptic plasticity in adult cortex. Nature 420:788–794
Tsui CC, Copeland NG, Gilbert DJ et al. (1996) Narp, a novel member of the pentraxin family, promotes neurite outgrowth and is dynamically regulated by neuronal activity. J Neurosci 16:2463–2478
Tu JC, Xiao B, Naisbitt S et al. (1999) Coupling of mGluR/Homer and PSD-95 complexes by the Shank family of postsynaptic density proteins. Neuron 23:583–592
Umemori H, Linhoff MW, Ornitz DM et al. (2004) FGF22 and its close relatives are presynaptic organizing molecules in the mammalian brain. Cell 118:257–270
Umemori H and Sanes JR (2008) Signal regulatory proteins (SIRPS) are secreted presynaptic organizing molecules. J Biol Chem 283:34053–34061
Valtschanoff JG and Weinberg RJ (2001) Laminar organization of the NMDA receptor complex within the postsynaptic density. J Neurosci 21:1211–1217
Varoqueaux F, Aramuni G, Rawson RL et al. (2006) Neuroligins determine synapse maturation and function. Neuron 51:741–754
Vaughn JE (1989) Fine structure of synaptogenesis in the vertebrate central nervous system. Synapse 3:255–285
Verhage M, Maia AS, Plomp JJ et al. (2000) Synaptic assembly of the brain in the absence of neurotransmitter secretion. Science 287:864–869
Waites CL, Craig AM and Garner CC (2005) Mechanisms of vertebrate synaptogenesis. Annu Rev Neurosci 28:251–274

Wang PY, Seabold GK and Wenthold RJ (2008) Synaptic adhesion-like molecules (SALMs) promote neurite outgrowth. Mol Cell Neurosci 39:83–94

Wang Y, Okamoto M, Schmitz F et al. (1997) Rim is a putative Rab3 effector in regulating synaptic-vesicle fusion. Nature 388:593–598

Wang Y, Sugita S and Südhof TC (2000) The RIM/NIM family of neuronal C2 domain proteins. Interactions with Rab3 and a new class of Src homology 3 domain proteins. J Biol Chem 275:20033–20044

Webb DJ, Zhang H, Majumdar D et al. (2007) alpha5 integrin signaling regulates the formation of spines and synapses in hippocampal neurons. J Biol Chem 282:6929–6935

Weiner JA, Wang X, Tapia JC et al. (2005) Gamma protocadherins are required for synaptic development in the spinal cord. Proc Natl Acad Sci U S A 102:8–14

Woo J, Kwon SK, Choi S et al. (2009) Trans-synaptic adhesion between NGL-3 and LAR regulates the formation of excitatory synapses. Nat Neurosci 12:428–437

Wu G, Malinow R and Cline HT (1996) Maturation of a central glutamatergic synapse. Science 274:972–976

Yamagata M and Sanes JR (2008) Dscam and Sidekick proteins direct lamina-specific synaptic connections in vertebrate retina. Nature 451:465–469

Yamagata M, Sanes JR and Weiner JA (2003) Synaptic adhesion molecules. Curr Opin Cell Biol 15:621–632

Yamaguchi Y and Pasquale EB (2004) Eph receptors in the adult brain. Curr Opin Neurobiol 14:288–296

Yao I, Hata Y, Hirao K et al. (1999) Synamon, a novel neuronal protein interacting with synapse-associated protein 90/postsynaptic density-95-associated protein. J Biol Chem 274:27463–27466

Yuste R and Bonhoeffer T (2004) Genesis of dendritic spines: insights from ultrastructural and imaging studies. Nat Rev Neurosci 5:24–34

Zhai R, Olias G, Chung WJ et al. (2000) Temporal appearance of the presynaptic cytomatrix protein bassoon during synaptogenesis. Mol Cell Neurosci 15:417–428

Zhai RG, Vardinon-Friedman H, Cases-Langhoff C et al. (2001) Assembling the presynaptic active zone: a characterization of an active one precursor vesicle. Neuron 29:131–143

Zhang W and Benson DL (2001) Stages of synapse development defined by dependence on F-actin. J Neurosci 21:5169–5181

Chapter 5
Developmental Axonal Pruning and Synaptic Plasticity

Bibiana Scelfo and Mario Rosario Buffelli

Abstract The functioning of the nervous system depends upon the underlying specific and highly ordered patterns of neuronal connections. The embryonic pattern of neuronal connectivity is essentially established by specific molecular cues and is refined by synapse elimination and axonal pruning of exuberant or inaccurate connections during later developmental stages. In this chapter we provide a general description of the phenomena of synapse elimination, axonal pruning, and synaptic plasticity in the central and the peripheral nervous system. Also, we briefly describe the role of the adhesion molecules in these processes.

Keywords Synapse elimination · Polyneuronal innervation · Axonal pruning · Neuronal connectivity · Synaptic plasticity

5.1 Introduction

Experiences throughout lifetime sculpt the synaptic connections in the nervous system. These modifications are particularly pronounced during the developmental period called the *critical period* that leads to a highly refined degree of neuronal connections, which characterize the mature stage. However, also in the adult nervous system, synaptic strength and connections maintain a certain degree of plasticity in order to ensure experience-dependent adaptation of the nervous system to environmental stimuli. The word plasticity has been applied to a wide variety of nervous system changes. It is used for changes in synaptic strength and in synaptic connectivity. However, it is also improperly used to describe the changes of the neural circuitry in pathologic disorders. Synaptic plasticity specifically refers to activity-dependent modifications of the strength or efficacy of synaptic transmission at preexisting synapses. It is thought to play a central role in learning and memory processes.

M.R. Buffelli (✉)
Dipartimento di Scienze Neurologiche e della Visione, Sezione di Fisiologia, Università di Verona, Strada Le Grazie n.8, 37134, Verona, Italia
e-mail: mario.buffelli@univr.it

M. Hortsch, H. Umemori (eds.), *The Sticky Synapse*,
DOI 10.1007/978-0-387-92708-4_5,

In this chapter, we focus on regressive processes that occur during the development, such as synapse elimination and axonal pruning. We describe some general features of these phenomena and examine a few examples in the peripheral and central nervous system. Furthermore, we are considering various aspects that are implicated in these processes, such as cellular mechanisms, the role of activity, and the involvement of adhesion/repulsion molecules in synapse remodeling. A complete description of these molecules is presented in the cell adhesion molecule (CAM)-specific chapters of this book.

5.2 Stereotyped and Stochastic Axonal Pruning

The complete development of the nervous system of higher vertebrates is achieved through three transient stages: (i) growth cones travel for long distances to find the proper target region (the so-called pathway selection); (ii) they contact their target neurons establishing an initial pattern of connections which often involves the formation of redundant synapses (target selection); and (iii) through axonal competitions, neural connections are refined by pruning or elimination to form a highly specific pattern of innervation (address selection) (Goodman and Shatz 1993, Katz and Shatz 1996, Nguyen and Lichtman 1996, Sanes and Lichtman 1999, Buffelli et al. 2004).

Synapse elimination is a widespread phenomenon, which occurs in rodents between the time of birth and postnatal day (P) 15. It was observed for the first time in the neuromuscular junction (NMJ) by Redfern (reviewed in Jansen and Fladby 1990). Later, it was observed and confirmed by other researchers in many other locations of the peripheral nervous system (PNS) and central nervous system (CNS), e.g., in autonomic ganglia, the cerebellum, the thalamus, and the cerebral cortex (Purves and Lichtman 1980, Lohof et al. 1996, Kantor and Kolodkin 2003, Liu et al. 2005). In each of these cases, target cells lose most neuronal inputs and ultimately become strongly innervated by relatively few axons and the final pattern of connections is shaped by branch retraction.

Developmental axonal pruning can take place in different ways, from the molecular to the functional level. This depends on the structure where it occurs and on several other parameters that regulate this process. In the last years a distinction between local *stochastic* pruning and long-range *stereotyped* pruning has been drawn (Bagri et al. 2003, Kantor and Kolodkin 2003, Luo and O'Leary 2005). The characterization of the two processes has gained interest as it identified various mechanisms, which distinguish this process in a number of involved structures in both the peripheral and central nervous system (Bagri et al. 2003, Luo and O'Leary 2005). In stereotyped pruning the branches that are retracted form a stereotypic predictable pattern and are therefore identifiable prior to the onset of pruning (Bagri et al. 2003, Kantor and Kolodkin 2003). In contrast, local pruning affects subsets of branches and occurs within target regions to regulate terminal arborization. The pruned branches are selected stochastically, perhaps reflecting competitive forces (Bernstein and Lichtman 1999). Moreover, as in the case of visual cortical efferents that project to the spinal cord, during stereotyped axonal pruning the length of the axon branches that is retracted can be of many millimeters (O'Leary and Koester 1993). On the contrary, stochastic terminal arbor pruning usually involves a few micrometer long branches. The two processes differ by 3 orders of magnitude in the length of axon that is retracted (Bagri et al. 2003).

Another aspect, which differentiates the two kinds of pruning, is related to their molecular control mechanisms. One prevalent hypothesis suggests that stereotyped pruning is entirely preprogrammed within neurons and is perhaps triggered by intrinsic transcriptional switches like the expression of the Otx1 gene (Weimann et al. 1999). Therefore, it is not dependent on any external signals acting locally on terminal axon branches. Alternatively, stereotyped pruning might be regulated by extrinsic signals, e.g., by retraction inducers like members of the semaphorin family and their receptors (Bagri et al. 2003). In contrast, stochastic elimination of redundant motoneuron terminals in the developing NMJ and redundant climbing fibers in the cerebellar cortex is dependent on neuronal activity or maintenance factors (Lohof et al. 1996, Bernstein and Lichtman 1999, Buffelli et al. 2004).

The functional elimination of inappropriate projections usually occurs in two main mechanisms: retraction and degeneration. Recently, a third mechanism has been proposed, called axosome shedding. It incorporates features of both degeneration and retraction. This mechanism seems to be responsible for synaptic pruning in different systems. It has been described in the developing NMJ where a retraction bulb was noted at the distal tip of retreating axons and pieces of the axon, called axosomes, shed from the tip and are engulfed by Schwann cells. These axosomes still contained vesicles and mitochondria. These findings suggest that this mechanism is very different from degeneration (Bishop et al. 2004).

In both the peripheral and central nervous system several neuronal structures offer suitable models for investigating the physiological and molecular mechanisms underlying synapse and branch elimination during development. In the following paragraphs we will describe a few classic examples for synapse elimination in the peripheral and the central nervous system.

5.3 Synapse Elimination in the Peripheral Nervous System

Most of our knowledge about the molecular mechanisms underlying synapse formation and elimination are based on the work on the vertebrate neuromuscular junction. In neonatal animals each muscle fiber is innervated by several axons that are collaterals of different motoneurons (Fig. 5.1A), while in the adult only one axonal branch makes contact with them (Fig. 5.1B). In rodents synapse elimination at the NMJ occurs between the time of birth and P15–P20.

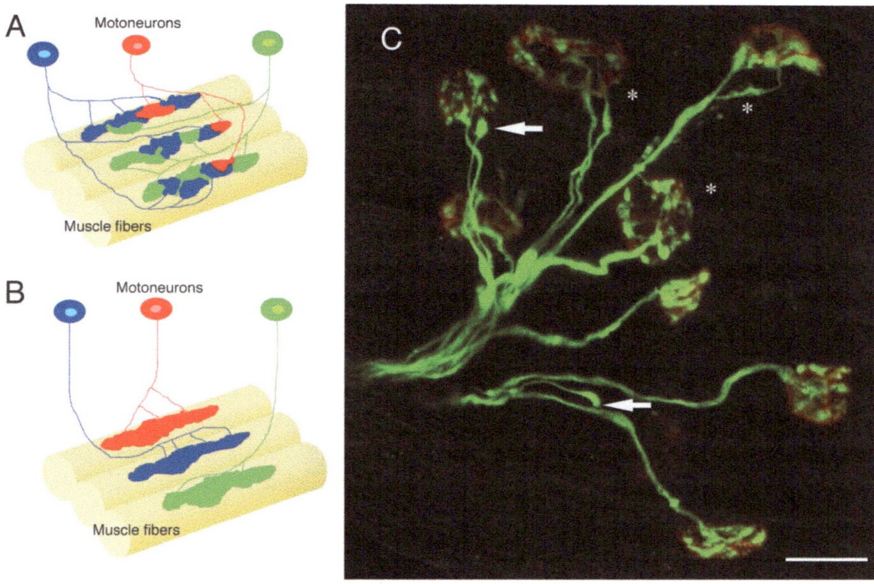

Fig. 5.1 Synapse elimination in the developing neuromuscular junction. (**A** and **B**) Schematic representation of the changes in innervation of individual NMJs during the first 2 weeks of postnatal life. Each NMJ changes from multiple converging axons innervations to only one axon. (**A**) Polyneuronal innervation in neonatal muscle fibers where the axonal terminals of different motor neurons are intermingled at each cluster of acetylcholine receptors. (**B**) Mononeuronal innervation in the adult NMJ. (**C**) Poly- and mononeuronal innervations in a soleus muscle of a newborn rat (P15) are shown by confocal fluorescence microscopy. NMJs are labeled with rhodamine-α-bungarotoxin (AChRs, *red*), anti-neurofilament (axons, *green*), and anti-synaptophysin (terminals, *green*) fluorescent antibodies. *Asterisks* indicate polyneuronally innervated NMJs. *Arrow*: retraction bulb with atrophic axon. Scale bar 20 μm

Synapse elimination is a peripheral process that only involves motor nerve endings and their axonal branch. The number of motoneurons in the spinal cord stays constant and programmed cell death is already over. Each motor unit undergoes a structural change from a large size, which overlaps with other motor units, to a much smaller non-overlapping one. The neonatal/adult motor unit size ratio varies for different muscles and is estimated to reach a value of about 5 in soleus muscle.

The presence of multiple innervations in neonatal muscle fibers can be assessed by electrophysiology and morphology techniques. By intracellular recording from muscle fibers in vitro, the graded single shock electrical stimulation of the nerve evokes endplate potentials (EPPs) of amplitudes increasing in discrete steps in the neonatal (polyneuronal innervation) and only one step of the EPP (mononeuronal innervation) in adult muscle fibers (see reviews Jansen and Fladby 1990, Buffelli et al. 2004). Multiple innervation in neonatal muscle fibers can also be assessed by morphological studies, either on fixed tissues or in in vivo preparations (Gan et al. 1999, Kasthuri and Lichtman 2003, Walsh and Lichtman 2003). Imaging experiments revealed a peculiar feature of multiple innervation in muscle fibers. Multiple axons do not reach separate AChR aggregates, but rather converge on the same aggregate and the areas occupied by the different terminals initially intermingle with one another in a complex fashion. During the following days the terminals, which converge on each AChR aggregate and initially remain in close contact at their boundaries, become segregated and finally withdraw with the exception of one synapse that takes over the relinquished AChR territory. Only occasionally this take-over does not occur and in such cases the vacated AChR region disappears altogether (Walsh and Lichtman 2003). Moreover, in vivo imaging studies have also demonstrated that (a) an input can initiate the process of elimination and then subsequently reverse this process by growing to become the single input that is permanently maintained. This suggests that the process is reversible and continually driven until an entire input is ultimately eliminated (Walsh and Lichtman 2003); (b) input elimination occurs asynchronously among branches of a single motoneuron without any apparent regional bias. This argues for a local control at each muscle fiber (Keller-Peck et al. 2001); (c) intrinsic properties of motoneurons establish a hierarchy of competitive dominance, so that the outcome, in terms of winner and loser, is the same when terminals belonging to two identified neurons converge on the same endplate (Kasthuri and Lichtman 2003). The retraction bulb (Fig. 5.1C) is the frequent morphological indicator of a terminal undergoing elimination. Serial electron microscopy and time lapse of fluorescent-labeled axons have shown that the mechanism underlying elimination of nerve ending and atrophic branches differs from that of wallerian degeneration. The retracting axon shed membrane-enclosed, called axosomes, into the surrounding glia cells (Bishop et al. 2004).

Synapse elimination is not a random process by which each motoneuron loses a percentage of its terminals, but it clearly involves a competition between terminals from different motor axons. This was shown by experiments in which the fate of a particular axonal input was influenced by the presence of other inputs (Brown et al. 1976, Kuffler et al. 1977, Thompson and Jansen 1977, Betz et al. 1980, Fladby 1987). Impulse activity is an important candidate in mediating this competitive process, albeit probably not the only one. The in vivo application of tetrodotoxin (TTX), tubocurarine, or bungarotoxin (BTX) during the critical period inhibits neuromuscular activity and prevents neuromuscular synapse elimination (Thompson et al. 1979, Duxson 1982, Jansen and Fladby 1990). On the contrary, the chronic stimulation of motor nerves accelerates input synapse elimination (Thompson 1983, Busetto et al. 2000). However, the role of neuronal activity on synapse elimination is complex. We recently studied the physiological role of synchronous–asynchronous firing of motoneurons in synaptic competition and elimination at the developing NMJ (Busetto et al. 2000, Buffelli et al. 2002, 2004, Favero et al. 2007). In these studies we showed, in different paradigms of adult muscle reinnervation, that the complete substitution of the physiological asynchronous firing of motor axons with a synchronous activity profoundly inhibits the competition and elimination of the multiple inputs (Busetto et al. 2000, Favero et al. 2007). These findings indicate that synapse elimination is strongly dependent upon asynchronous firing of motor axons. Similar observations were made in the visual cortex, where synchronous activity also prevents synapse elimination (reviewed by Katz and Crowley 2002, Hua and Smith 2004). In different sets of experiments, we found that motoneuronal spontaneous activity is synchronized at perinatal age, thus most likely favoring the formation of multiple inputs on myofibers in the embryonic life. It desynchronizes soon after birth, leading to the elimination of redundant inputs (Buffelli et al. 2002).

As predicted by Hebb's hypothesis (Hebb 1949) it is now well established that competition among neurons innervating the same target cell profoundly influences the strength and structure of synapses. Inputs that fire coordinately with postsynaptic cells are strengthened, while inputs that are not are weakened. Thus, we investigated the role of differences in the amount of activity between competing inputs on synapse elimination. We compared two different types of inputs on the same myofiber. Whereas one was able to evoke postsynaptic responses and normally released acetylcholine (ACh), the other did not because of a conditional genetic manipulation of choline acetyltransferase (ChAT), the synthetic enzyme for ACh. Therefore, in a given pool some motoneurons were unable to synthesize ACh due to lack of ChAT. Our results indicate that the competent terminals always overcome the incompetent ones (Buffelli et al. 2003).

Thus, while competition among presynaptic inputs influences their targets, postsynaptic target cells also modulate competition, in part through retrograde interactions (Fitzsimonds and Poo 1998). Jennings (1994) proposed a model to explain how activity may drive synapse elimination. In this model strong synapses, which are effective in driving postsynaptic responses, actively punish and eliminate nearby weaker synapses by inducing two postsynaptic signals. First, a short-range "protective" signal and second a longer range elimination or "punishment" signal. A similar, activity-based competition process was postulated to be responsible for the elimination of terminal branches of retinal axons (Hua and Smith 2004). Interestingly, it was also shown that synapse elimination can occur in reinnervated adult muscles in the complete absence of activity. This suggests that activity may modulate, rather than directly cause synapse elimination at the regenerating NMJ (Costanzo et al. 2000).

Although the underlying molecular mechanisms of synapse elimination are not understood, the role of postsynaptic ACh receptors (AChRs) and calcium influx into postsynaptic cells is considered crucial (Srihari and Vrbova 1978, O'Brien et al. 1980, Cash et al. 1996, Sanes and Lichtman 1999). It was shown that focal blockade of receptor activation by α-bungarotoxin within a small region of the NMJ induces elimination of the portion of motor terminal overlying this site (Balice-Gordon and Lichtman 1994). Moreover, electrophysiological recordings at the mammalian NMJ, correlate with AChR staining, demonstrated that removal of postsynaptic receptors can precede the retraction of the presynaptic elements (Colman et al. 1997, Akaaboune et al. 1999). However, live imaging experiments at the NMJ also provide clear evidence of retracting presynaptic elements at sites where postsynaptic receptors persist (Walsh and Lichtman 2003). Recently, many retrograde signals, such as neurotrophins (neurotrophins 3 and 4/5, brain-derived neurotrophic factor, etc.), have been identified both at the NMJ and at central synapses (Fitzsimonds and Poo 1998, Sanes and Lichtman 1999, Tao and Poo 2001). However, we know little about how such signals are triggered, maintained, or regulated by synaptic activity. At least two mechanisms are commonly thought to underlay axon competition. In the first, neuromuscular activity and muscle contraction control the release of trophic factors from muscles and the uptake by the nerve (Snider and Lichtman 1996). Several factors, such as glia-derived neurotrophic factor, which are produced by muscles are able to retard synapse elimination when applied to muscle (English and Schwartz 1995, Kwon et al. 1995, Kwon and Gurney 1996) or their overexpression in transgenic animal (Nguyen et al. 1998). However, these effects may be relatively non-specific. For example, excess neurotrophins or cytokines also induce the production of supernumerary axonal branches and thus indirectly affect the time course of synapse elimination (Sanes and Lichtman 1999). In the second mechanism, inactive axons are actively pruned or retracted in response to negative signals, e.g., calcium-activated protease (Connold et al. 1986) or serine protease (Liu et al. 1994).

Recently an extracellular matrix protein called "reelin", also possessing a serine protease activity, has been claimed to be important for the physiology of synapse elimination. Quattrocchi et al. (2003) have reported that synapse elimination did not take place in the *reeler* mice, in which reelin is lacking, since innervation of muscle fibers remains polyneuronal into adulthood. A theoretical interpretation of how this suggested physiological function of reelin could be linked to the well-established activity dependence of neuromuscular synapse elimination was offered by others (Chih and Scheiffele 2003). However, a subsequent thorough re-examination of the findings of Quattrocchi et al., which was conducted in parallel with two other laboratories, has not confirmed these original findings. The new studies concluded that in *reeler* mice synapse elimination proceeds normally and occurs at the right developmental time points (Bidoia et al. 2004).

We previously showed that local activity enhances local axonal growth. However, at the same time it actively inhibits the growth of inactive axon collaterals, which are derived from neighboring motoneurons (Buffelli et al. 2003). This finding suggests a mechanism for the selection of one axon collateral over another and raises the possibility that activity-dependent neurotrophin secretion may allow active axons to inhibit competing axons from growing into the same target area. Recently a novel mechanism was described, in which active axons can eliminate less-active, competing axons by directly promoting p75 neurotrophin receptor (p75NTR)-mediated axonal degeneration (Singh and Miller 2005, Singh et al. 2008). The winning sympathetic axon secretes brain-derived neurotrophic factor (BDNF) in an activity-dependent fashion. This BDNF binds to the p75NTR on the losing axons to cause their degeneration. Thus, synapse pruning requires both activity-dependent BDNF secretion by the dominant neuron and p75NTR expression by the losing neurons (Singh and Miller 2005, Singh et al. 2008). Moreover, a mathematical model was proposed to explain the relation between innervation-dependent expression of protection signals (NGF/TrkA) and neuron protection/apoptotic cues (BDNF/NT-4) in the sensory system (Deppmann et al. 2008). In summary, the evidence described above strongly suggests that both presynaptic activity and postsynaptic factors participate in PNS synapse elimination during development.

5.4 Axonal Pruning in the Central Nervous System

5.4.1 Axon Pruning in the Hippocampus: The Development of the Infrapyramidal Bundle

The hippocampus plays an extremely important role in the consolidation of new memories, emotions, and spatial orientation (Knight 1996, Maguire et al. 1998, 1999, Mainen et al. 1999). Imaging studies in humans and pharmacological and surgical manipulations in animal models suggest that different regions of the hippocampus may subserve different functions (Mizumori et al. 1999, LeDoux 2000, Maren 2001). Moreover, anatomical alterations in the hippocampus were found in many neuropsychiatric disorders (Geuze et al. 2005b, a), thus knowing the mechanisms that underlie normal hippocampal development may help to understand the pathogenesis of many psychiatric disorders.

In the hippocampus, granule cells of the dentate gyrus extend their axons, the mossy fibers, to the adjacent region to innervate pyramidal cells in the CA3 region. In early postnatal life, the dentate granule cells send two axon bundles: the main bundle (MB) and a transient set of axons, the infrapyramidal bundle (IPB). Projections of the MB lies adjacent to the apical dendrites (AD) of CA3 pyramidal cells and the IPB extends through the *stratum oriens* to project to the CA3 basal dendrites (BD) (see Fig. 5.2A). Later on, between P20 and P30 in mice, IPB axons are pruned back until attaining their final length by P45, when they are confined largely to the hilus (Fig. 5.2B) (Bagri et al. 2003). Axons from the main bundle form stable "synaptic complexes" in which a presynaptic mossy fiber bouton establishes multiple excitatory synapses with CA3 apical dendritic spines (Amaral and Dent 1981). Similar synaptic complexes are formed between boutons of the mossy fibers from the IBP and the basal dendrite of pyramidal cells in CA3 prior to pruning. Between P20 and P30, first the elimination of synaptic complexes occurs and then axons are pruned (Liu et al. 2005).

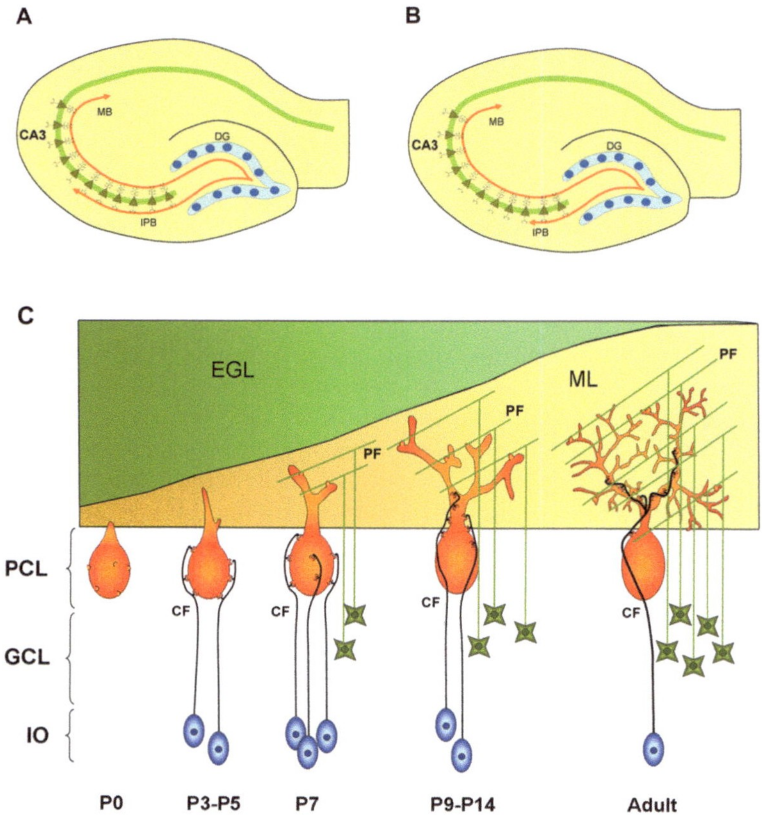

Fig. 5.2 Hippocampal and cerebellar axon pruning. (**A**) During the early postnatal period a main bundle of mossy fibers (MB) and a developing infrapyramidal bundle (IPB) both originate from the dentate gyrus (DG) elongate and form synaptic complexes with the apical and basal dendrites of the CA3 pyramidal cells, respectively. (**B**) Within the first month of postnatal life synaptic complexes formed by the MB are maintained while those formed by the IPB are eliminated and the IPB is pruned back. (**C**) Diagram of a developing cerebellar cortex. In the early postnatal life immature Purkinje cells (PCs) are multiply innervated by climbing fibers (CFs), which establish their synaptic contacts mainly on the somatic domain of PCs (P7). As the dendrite of the PC develop, multiple CFs translocate and innervation with parallel fibers (PFs) is established (P9–P14). At this stage the process of pruning takes place until CF monoinnervation is reached by the third postnatal week and the cerebellar cortex reaches the mature stage (adult). PCL: Purkinje cell layer; GCL: granule cell layer; IO: inferior olive; EGL: external granule cell layer; ML: molecular layer

The molecular mechanism of mossy fiber pruning in the hippocampus is still unclear. There is evidence that ephrins and Eph receptors modulate the refinement and synaptogenesis of mossy fibers (Martinez et al. 2005). Ephrin receptors form a family of 14 receptor tyrosines, which based on their affinity for ephrin-A or ephrin-B ligands is divided into subclasses A and B (Gale et al. 1996, Aoto and Chen 2007; see Chapter 16). Ephrin ligands play a crucial role in the formation of diverse projections in the developing brain (Kullander and Klein 2002). As both ligands and receptors are transmembrane proteins ephrin/Eph signaling is mediated by cell-to-cell contacts (Davis et al. 1994) and the effects of their interaction will depend on the cell types involved. Several ephrins and Eph receptors are expressed in the principal neurons of the developing hippocampal formation (Mori et al. 1995, Mackarehtschian et al. 1999, Yue et al. 2002). In vitro studies have indicated that ephrin-A2, ephrin-A3, and ephrin-A5 and the receptor EphA5 are involved in the formation of the lamina-specific pattern of hippocampal connections (Gao et al. 1996, Brownlee et al. 2000). Granule cells and CA3 pyramidal neurons express ephrin-A and EphA (Gao et al. 1996, Mackarehtschian et al. 1999, Yue et al. 2002). In a transgenic mouse model, in which ephrin-A signaling is disrupted by the overexpression of a soluble form of the EphA5 receptor, the mossy fiber projections were analyzed during postnatal development (Martinez et al. 2005). In both control and mutant mice the suprapyramidal and the infrapyramidal bundles were present and the IBP reached the distal part of CA3 at an early postnatal age. While postnatal development proceeded, the length of the infrapyramidal bundle became progressively reduced in control animals and was restricted to the proximal portion of the CA3 region, near the dentate gyrus. In contrast, in transgenic mice, the IPB did not retract but persisted throughout the CA3 region. These data indicate that ephrin-A/EphA signaling is required for the normal targeting of the IBP and thus is involved in patterning of the hippocampal mossy fiber (Martinez et al. 2005).

In addition to ephrin-A/EphA signaling, semaphorin family and their neuropilin and plexin receptors are involved in the pruning of IPB in the hippocampus (Bagri et al. 2003). Semaphorins are a large family of secreted and transmembrane molecules with important roles in axon guidance (Kolodkin 1998, Huber et al. 2003). One semaphorin subfamily, semaphorin 3F (Sema3F), acting through receptor complexes consisting of a neuropilin protein, which functions in ligand binding, and a plexin protein, which functions in signal transduction, has a role in triggering the pruning in the hippocampus. Plexins are present on the dentate granule cells, while the semaphorins are present in the path of the IPB. Sema3F expression is downregulated early in postnatal development, but resumes at the time when the axons are being pruned (Bagri et al. 2003). In plexin-A3, neuropilin-2, and Sema3F mutants knockout mice the IPB fails to prune and the adult retains projections throughout the field of CA3 as it does in the earliest postnatal stage (Chen et al. 2000, Cheng et al. 2001, Bagri et al. 2003, Liu et al. 2005). In plexin-A3 mutants, the synaptic complexes that normally are eliminated during pruning continue to mature and reach a form that is very similar to that of the synaptic complexes of mature MB axons (Liu et al. 2005). There is some evidence suggesting that there is a relationship between semaphorins and activity in the hippocampus. The application of Sema3F to hippocampal slices increased the amplitude and frequency of miniature EPSCs in granule cells of the dentate gyrus and pyramidal cells of CA1 in a dose-dependent manner (Sahay et al. 2005). However, more studies are needed to understand the mechanism by which semaphorins cause synapse elimination and their role in the synaptic function.

Concerning the cellular mechanism of IPB pruning, axon retraction seems to be the most likely responsible mechanism. This would involve the recycling via retrograde transport of components in the distal regions of the axon (Riley 1981). Although retraction bulbs have not been observed during IPB pruning, several types of evidence exclude both degeneration and axosome shedding. In fact, silver stains specific for degeneration (Bagri et al. 2003) and ultrastructural analysis (Liu et al. 2005) have demonstrated the absence of blebbing on the retracting axons and no signs of glial engulfment, which is typical of Wallerian degeneration.

Finally, another important issue, which should be considered, is the functional significance of developmental pruning in the IBP. Several studies provide evidence that there is a direct correlation between the size of the IBP and the degree of performance in hippocampal tasks (Crusio and Schwegler 1987, Schopke et al. 1991). In a study of different mice strains with heritable variations in the length of the IPB, increased length was associated with fewer errors in the radial maze task and a better performance in the water maze task (Crusio and Schwegler 1987, Schopke et al. 1991). In addition to these genetic differences, changes in the IPB length have been artificially induced using chemical manipulations in animals (Lauder and Mugnaini 1980, Schwegler et al. 1991). Daily injections of thyroxine induce hyperthyroidism in P0–P11 mice and result in an increased IPB area (Schwegler et al. 1991). These animals performed significantly better than control animals in the radial maze task, supporting the findings for mice strains with a genetically extended IPB. Ultrastructural studies demonstrated that synapses are formed between the IPB mossy fibers and the basal dendrites of pyramidal cells in CA3 in thyroxine-treated, but not in control, animals (Lauder and Mugnaini 1980).

On the other hand, there is evidence that genetic mutations leading to an impairment of IBP pruning also affect hippocampal tasks. The fragile X syndrome is the main inheritable genetic cause of mental retardation and is triggered by a mutation of gene FMR1 that encodes the fragile X mental retardation protein (FMRP). Morphological studies suggest that this protein plays a role in normal synapse maturation and neuronal plasticity (Greenough et al. 1985). Postmortem studies on brains from patients suffering from fragile X syndrome have shown that mossy fiber maturation and pruning is altered. Both the density and morphology of dendritic spines remain at immature stages in the adult hippocampus of affected individuals (Hinton et al. 1991). This phenotype is also present in *fmr1* knockout mice, which show deficits in the water maze navigation task (Comery et al. 1997).

Two papers describe conflicting IPB phenotype in *fmr1* knockout mice: one showed a pruning defect in the IPB (Ivanco and Greenough 2002) and the other showed a decrease in the length of the IPB compared to control (Mineur et al. 2002). According to Faulkner et al. (2007) this discrepancy in the results so far obtained might be due to the use of different strains of mice where the size of

IPB in normal conditions can be very different and a comparison among the obtained observations is very hard to do. This may be explained with the possibility that the mechanism engaged for pruning is related to the length of the axon being pruned, as described above (Chapter 5.1; Luo and O'Leary 2005), and there can be different mechanisms in mossy fiber pruning depending on the considered model. Still the hippocampal formation constitutes an interesting model for the study of developmental axonal pruning. The fact that stereotyped mechanisms of synapse elimination are engaged by this formation allows to get additional insights on this process which cannot be evinced by the study of the classical stochastic pruning.

5.4.2 Axon Pruning in the Cerebellum: The Regression of Redundant Climbing Fibers

The cerebellum, a very distinct region of the brain, occupies most of the posterior cranial fossa and derives its name as a diminutive of the word *cerebrum*. The classical view is that the cerebellum plays a fundamental role in motor coordination (Flourens 1824) and learning (Ito 1984, Thompson 1990). More recently, several experiments have demonstrated that the cerebellum is also involved in complex functions like cognition and emotions (see, for review, Thach 2007). The cerebellar cortex consists of three layers (Ramon y Cajal 1911, Palay, Billings-Gagliardi and Chan-Palay 1974). The outermost is the *molecular layer*, which consists of the dendrites of the Purkinje cells and a high number of parallel fibers that run along the axis of the folium. The parallel fibers make contact with the Purkinje cell dendrites and constitute one of their excitatory inputs (see Fig. 5.2C). Below the molecular layer is the *Purkinje cell layer*, which is made up of a single row of Purkinje cell somata. Finally, the innermost layer is the *granule cell layer* that is mainly constituted by a high number of granule cells, which give rise to the parallel fibers. The climbing fibers originate from the inferior olive in the brain stem and represent the second excitatory input by impinging on the Purkinje cell dendritic proximal domain. At the adult stage each Purkinje cell is innervated by only one climbing fiber, but receives hundreds of thousands of synaptic inputs from parallel fibers (Eccles et al. 1966, Napper and Harvey 1988).

The contacts between climbing fibers and Purkinje cells are a powerful model to study the principles underlying developmental synapse elimination. During early postnatal development each Purkinje cell is innervated by multiple climbing fibers. Elimination of synapses formed by supernumerary fibers occurs through a stochastic process of pruning until monoinnervation is reached (see Fig. 5.2C). Already by the end of the second postnatal week most Purkinje cells are innervated by a single climbing fiber and the process is completed 3 weeks after birth (Crepel et al. 1981, Kano et al. 1995). Morphological and electrophysiological data suggest that in rats functional synapses between climbing fibers and Purkinje cells appear at P2 (Crepel et al. 1981). At P5, virtually all Purkinje cells have multiple innervation. On average, 3.5 climbing fibers contact each Purkinje cell (Crepel et al. 1981). At this developmental stage, when Purkinje cells are morphologically immature with only rudimentary dendrites, the initial contacts between climbing fibers and Purkinje cells are made at the perisomatic level (Altman 1972). It has been argued that confining inputs to the limited space of the cell soma may trigger the competitive process leading to synaptic pruning (Hume and Purves 1981, Purves and Hume 1981). However, recent studies in mice have demonstrated that the strength of multiple climbing fibers impinging on the same target cell increases or at least is stable up to P7 with no signs of regression (Scelfo and Strata 2005, Bosman et al. 2008). Only when functional synapses with the parallel fibers appear, the process of competition among multiple climbing fibers and the pruning of the redundant contacts starts. Moreover, it has been demonstrated that the time course of their elimination directly correlates with the maturation of the parallel fiber innervation (Scelfo and Strata 2005). By P13 Purkinje cell morphology is reminiscent of that of adult animals (Mason et al. 1990). During the time window P7–P13 climbing fiber branches gradually move to the growing dendrite and complete their translocation to attain their final innervation at the proximal portion of the Purkinje cell dendrites.

The role of parallel fibers in climbing fiber regression has previously been recognized (Mariani and Changeux 1980). In fact, when granule cells are deleted (Mariani and Changeux 1980, Crepel et al. 1981), are degenerate (Rabacchi et al. 1992b), are impaired in their function (Rabacchi et al. 1992a), or if their transduction pathways are deficient (Conquet et al. 1994, Kano et al. 1995, Hashimoto et al. 2001a, b), the regression of the redundant climbing fibers is hampered and multiple innervation persists into adulthood. All conditions, which lead to multiple innervation by climbing fibers in adulthood, are pathological and are characterized by severe impairment in motor performances and by deficiencies in learning and cognitive functions such as fear memory (Conquet et al. 1994, Chen et al. 1995, Sacchetti et al. 2004).

Imaging experiments, which were performed on polyinnervated Purkinje cells, provided insights into the functional mechanism of stochastic pruning in the developing cerebellum. A recent study has shown that more than one climbing fiber invades the dendritic tree and that part of the competition occurs on the dendritic territory where climbing fibers termination overlap. This close vicinity is probably necessary to allow local mechanisms of synapse selection (Scelfo et al. 2003). Morphological studies involving in situ imaging of poly-innervated muscle fiber and culture systems containing motoneurons and myocytes revealed that the degree of space separation between different innervation sites can be a determining factor in the process of synapse elimination (Lo and Poo 1991, Gan and Lichtman 1998). It has been postulated that lack of climbing fiber segregation in normal cerebellar development facilitates the competition between developing climbing fiber terminal arbors (Scelfo et al. 2003). Such a scheme would explain experimental findings that in several cerebellar models, in which synapses between parallel fibers and Purkinje cells are deficient, polyinnervation by climbing fibers persists and their innervation territories are segregated (Bravin et al. 1995, Hashimoto et al. 2001a). This demonstrates that the interaction between climbing fibers and parallel fibers is an enduring process, which starts in the postnatal period and persists throughout adulthood. Whenever one of the two afferent systems is affected, the other one takes over. In the same way as an impairment of parallel fiber synapses leads to redundant climbing fiber innervations, an impairment of climbing fiber innervation, which is induced in mice by a mutation in a P/Q-type Ca^{2+} channel $\alpha1$ subunit, does lead to an overextension of the parallel fiber territory of innervation, which occupies also the proximal portion of the Purkinje cell dendritic domain. The regression from multiple to monoinnervation of climbing fibers remains incomplete and located on the basal dendrites of the Purkinje (Miyazaki et al. 2004).

In contrast to the NMJ, where synapse pruning occurs through a process of homosynaptic competition among motoneurons, the regression and perhaps the choice of the surviving climbing fiber arbor in the cerebellar cortex are under the dominant control of the parallel fibers. Therefore, heterosynaptic interactions appear to be involved in climbing fiber selection. It is likely that the climbing fiber terminal arbor, which remains connected with the Purkinje cell, is selected by the concerted patterns of activity of both inputs. The mechanism how the parallel fibers influence such a selection is unknown. The fact that altering the temporal pattern of CF activity during development impairs regression (Andjus et al. 2003) suggests that a time correlation of the firing of the surviving climbing fiber terminal arbor with that of the parallel fibers might be important as well as spatial integration of the two inputs. In fact, experimental evidence from Purkinje cells indicates that different synaptic responses are found along the somato-dendritic membrane. The postsynaptic response may thus differ in the kinetics and the intracellular pathway, which eventually is activated. Therefore, inputs from different parts of the dendrite may be accompanied by different receptor properties and intracellular biochemical machinery (Midtgaard 1995). This has led to the hypothesis that the molecular pathways involved in the activation of parallel fibers and climbing fibers during development and throughout adulthood may involve calcium signals, which are regulated by their interaction. In fact the climbing fiber-evoked Ca^{2+} influx into the spiny dendrites is finely graded by parallel fiber-induced depolarization (Midtgaard 1995). This conclusion is supported by the finding that mutant mice bearing a deficiency in protein kinases and phospholipases (both triggered by calcium influx) display multiple climbing fiber innervations and impairments in motor performances and synaptic plasticity (Chen et al. 1995, Miyata et al. 2001).

The cellular mechanism of synaptic pruning in the cerebellum differs from the mechanism at work during hippocampal formation. In the hippocampus axon degeneration and glial engulfment seem to be the processes involved in elimination of exuberant synapses. Although during peak periods of climbing fiber elimination no clear sign of degeneration is observed, an increase in double membrane-bound terminals surrounded and taken up by the Bergmann glia has been reported (Eckenhoff and Pysh 1979). This could be the equivalent of axosome shedding, which has been observed in other developing structures of the central nervous system.

To date the role of CAMs in cerebellar synaptic pruning still has to be elucidated. Evidence from studies directed at describing the expression and distribution of different classes of CAMs in the cerebellum strongly point toward a role of the neurexin/neuroligin complex in the process of synapse formation and selection (Patrizi et al. 2008). Neurexins are neuronal cell surface proteins that are expressed in thousands of different protein isoforms, which fall into two main protein classes: α- and β-neurexins. Neurexins participate in intercellular junctions by binding to members of a second class of neuronal cell surface receptors: neuroligins (see, for review, Shapiro et al. 2007, Südhof 2008 and Chapter 17). Neuroligin-1 localizes to the postsynaptic density of glutamatergic synapses (Song et al. 1999) and likely forms a complex with β-neurexins, which are present at the presynaptic terminals. The association between β-neurexins and neuroligins is Ca^{2+} dependent (Ichtchenko et al. 1996) and the underlying cytoplasmic complexes associated with the two proteins involve CASK, a scaffolding protein acting as a kinase, on the neurexin side and PSD95 on the neuroligin side (Hata et al. 1996).

PSD95 is a PDZ-domain protein that is thought to be involved in the assembly and organization of signal transduction complexes at postsynaptic densities (Kornau and Seeburg 1997). The effect of neuroligin binding to PSD95 is not well understood, but it seems that neuroligins serve as synaptic CAMs that organize the postsynaptic assembly of protein complexes, which are involved in signal transduction (Hunt et al. 1996). Since the formation of the neurexin–neuroligin complex is calcium dependent and neuroligins specify synapses in an activity-dependent manner (Araç et al. 2007) precise patterns of neuronal activity may induce the proper calcium signals, which either favor the stabilization or the disruption of active synapses.

It is widely accepted that the timing of parallel and climbing fiber activation is a basic mechanism in the cerebellar function (Ito 1984). This feature is present already during development where precisely timed heterosynaptic interactions between parallel and climbing fibers are needed to properly prune redundant climbing fibers and achieve the correct cerebellar architecture.

5.4.3 Axon Pruning in the Visual System

The visual cortex is a highly organized structure in which neurons that have similar response properties are grouped together to form orderly represented maps of primary visual cortex features, such as ocular dominance, retinotopic maps, ocularity, orientation, direction, and spatial frequency (Crowley and Katz 2002). In the visual cortex of many mammalian species, the thalamocortical axons arising from the distinct eye-specific layers of the lateral geniculate nucleus (LGN) form a series of alternating input-specific stripes within cortical layer 4. These stripes form the structural basis for the functionally defined ocular dominance columns. Early in development, ocular dominance stripes in layer 4 are absent and axon terminations are overlapped (see Fig. 5.3). However, by selective pruning of arbors within the appropriate regions LGN axon gradually form dense eye-specific patches (Katz and Shatz 1996, Crowley and Katz 2002). This pattern of innervation in the visual cortex can be demonstrated either electrophysiologically (Hubel and Wiesel 1963) or anatomically by injecting a radioactive tracer into one eye that is anterogradely transported to the cortex (Hubel et al. 1977). The anatomical rearrangement of the thalamocortical axons is accompanied by a corresponding change in the synaptic physiology of layer 4 neurons. The majority of these changes are initially activated by stimuli, which are presented to either eye but finally come to respond to visual stimulation through one eye only (LeVay et al. 1978).

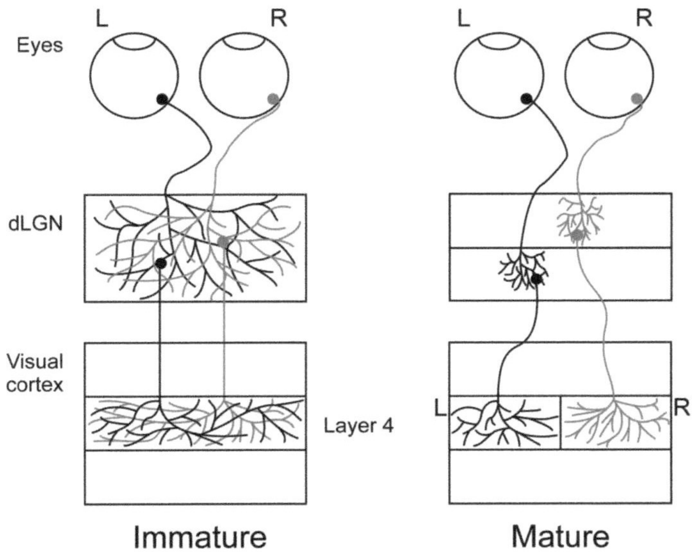

Fig. 5.3 Axon pruning in the mammalian visual system. In the mature mammalian visual system, the axons of retinal ganglion cells (RGCs) from the nasal portion of each retina cross the optic chiasm and project to the same dorsal lateral geniculate nucleus (dLGN) as the axons from the temporal portion of the other eye. These projections form eye-specific layers in dLGN. The projection from the dLGN to layer 4 of the primary visual cortex maintains this eye-specific segregation by terminating in eye-specific patches that are the anatomical basis for ocular dominance columns. During development, inputs representing the two eyes segregate first in the dLGN and then in the cortex. In each projection system, the two monocular inputs are initially overlapped. As development proceeds, the selective axon pruning of overlapping parts of axonal arbors form the eye-specific layers in the dLGN and the ocular dominance columns in the visual cortex using activity-dependent mechanisms

Separation of inputs from the two eyes occurs prior to the emergence of ocular dominance columns in the dorsal lateral geniculate nucleus (dLGN), where input from different eyes segregate into separate layers (see Fig. 5.3). In embryonic cats, axonal terminals of retinal ganglion cells (RGCs) from the two eyes overlap within the dLGN before forming the specific eye layers. This refinement process involves the retraction of axon branches from inappropriate regions of the geniculate nucleus and the formation of synapses in the correct layer (Shatz 1990, Wong 1999). In the absence of retinal activity, RGC inputs to dLGN do not segregate into eye-specific layers (Penn et al. 1998). Altering spontaneous activity in one retina relative to the other alters the normal pattern of eye-specific connections (Penn et al. 1998, Stellwagen and Shatz 2002). This form of axonal pruning is complete prior to the onset of visually evoked activity and it is coincident with retinal waves that spontaneously propagate across the retina (Wong 1999). These findings indicate that RGC activities are required to drive the axon elimination and eye-specific segregation of retino-geniculate arborizations.

The molecular links between neuronal activity and segregation of retino-geniculate arborizations are not yet precisely defined. Some evidence suggests that class I major histocompatibility complex (MHC class I) proteins are involved in activity-dependent plasticity in the developing visual system (Boulanger and Shatz 2004). MHC class I are present in select subsets of neurons within the CNS. The expression of MHC class I proteins becomes reduced upon blockade of neural activity with tetrodotoxin in vivo during the period when spontaneous retinal activity is needed for synaptic refinement of overlapping eye-specific input to dLGN neurons (Corriveau et al. 1998). In mice with deficient expression of MHC class I proteins, RGCs fails to segregate into eye-specific layers within the dLGN despite the presence of normal retinal activity (Huh et al. 2000). These findings suggest a novel non-immune function for MHC class I molecules in the CNS.

A recent study by Stevens et al. (2007) identified an unexpected and novel role for astrocytes and C1q, the initiating protein of the classical complement cascade, in the eye-specific segregation of connections in dLGN. They found that (a) a signal released by immature astrocytes induced the neuronal expression of C1q; (b) C1q is highly expressed in the immature brain and retina during the period of synapse elimination; and (c) mice deficient in complement protein C1q or the downstream complement protein C3 exhibit large defects in the eye-specific segregation of RGC input in the dLGN. These new findings also suggest an important role of astrocytes and classical complement proteins in the synapse elimination and refinement in the CNS.

There is evidence suggesting that the initial organization of the visual cortex is regulated by intrinsic factors and is therefore independent of any thalamic influence (see, for review, Sur and Leamey 2001, Majewska and Sur 2006). For example, the EphA receptor tyrosine kinases and their ligands show differential gradients of expression in the cortex before thalamic innervation (Donoghue and Rakic 1999a, b, Mackarehtschian et al. 1999). Alteration of these expression gradients modifies the location and the size of visual cortical areas (Cang et al. 2005). The subsequent thalamo innervation is fundamental for the correct development and maintenance of cortical structures. Early thalamic ablation results in a thinner cortex and a decrease in the size of cortical areas (Windrem and Finlay 1991).

After eye-opening, visual-driven activity significantly influences cortical connections. The pioneering work of Hubel and Wiesel on the effect of monocular deprivation on the development of visual systems clearly shows the role of experience-driven electrical activities for the refinement of neural circuits in the primary visual cortex over a well-defined critical period (Wiesel and Hubel 1963, Hubel and Wiesel 1970, Hubel et al. 1977). If one eye is deprived of any visual input/stimulus, even temporally, during this critical period, most mature visual cortical neurons are only responsive to stimuli from the eye that remained open. Within layer 4, early eye closure greatly enlarges the patches of input from LGN axons representing the open eye, whereas those representing the closed eye are relegated to very small regions (Hubel et al. 1977). Remarkably, the same manipulation had no effect on the visual cortex in the adult animals. Furthermore, binocular eye closure during the critical period has a smaller effect than monocular deprivation. These results suggest that (a) changes in the synapses and connectivity are rapid and dramatic during the critical period and (b) synapse elimination in the visual system is due to an activity-mediated competition between the connections driven by the two eyes.

In the visual system, the refinement process is not confined to the separation of eye input, but may also occur during the formation of retinotopic maps in the superior colliculus, in the lateral geniculate nucleus, and in the visual cortex in mammals. These maps convey positional information of objects in the visual field to different brain structures. During early development, when retinal ganglion cells (RGCs) overextend their axons beyond their final targets, the retinotopic map is less reined. Following branching and pruning of axon collaterals to the proper targets can only partially refine the map. This first step is regulated by a signaling gradient that is triggered by the Eph–ephrin family of tyrosine kinases (Cheng et al. 1995, Nakamoto et al. 1996, Yates et al. 2001). However, a pattern of correlated electrophysiological activity is required to complete the retinotopic map (McLaughlin et al. 2003). Despite the evidence that the pattern of neuronal activity influences the organization of circuits throughout the lateral geniculate nucleus and the visual cortex, the mechanism by which this activity induces long-term structural changes in connections remains unknown.

There is evidence that neurotrophins might have an important role for ocular dominance column formation and plasticity. For example, the decrease of retinal activity results in a decrease in the expression of brain-derived neurotrophic factor (BDNF) (Rossi et al. 1999, Lein and Shatz 2000). The blockade of BDNF receptors prevents the formation of ocular dominance columns (Cabelli et al. 1997). A link between BDNF, the maturation of inhibitory synapses, and the critical period for ocular dominance plasticity has also been established. In transgenic mice, which overexpress BDNF, an accelerated rate of maturation of the GABAergic system and an earlier than normal end to the critical period have been observed (Huang et al. 1999). Conversely, in transgenic mice with reduced GABAergic function the critical period onset is delayed (Hensch et al. 1998). However, a critical period can be induced at any age, in these transgenic mice, by artificially enhancing GABA-mediated inhibition or by increasing the expression of BDNF in the visual cortex (Fagiolini and Hensch 2000, Maya Vetencourt et al. 2008). These observations strongly suggest that (a) inhibitory thresholds control the onset, duration, and closure of the critical period (Huang et al. 1999, Feldman 2000) and (b) BDNF controls the time course of the critical period by accelerating the maturation of GABA-mediated inhibition.

The extracellular matrix (ECM) also plays an important role in the control of experience-dependent plasticity in both the developing and adult visual cortex (Berardi et al. 2004). During postnatal development, chondroitin sulfate proteoglycans (CSPGs) progressively accumulate around somatic and dendritic synapses of certain neurons and contribute to the formation of the so-called perineuronal nets (for review, see Celio et al. 1998). These ECM structures appear to restrict plasticity in the adult brain. In the visual cortex, the appearance of these perineuronal nets corresponds with the end of the critical period for plasticity. Dark rearing from birth delays not only the end of the visual cortex critical period but also the formation of perineuronal nets (Guimaraes et al. 1990). Interestingly, the degradation of CSPGs from the adult visual cortex with the enzyme chondroitinase ABC restores ocular dominance plasticity in adult animals. This alteration of the inhibitory properties of the perineuronal net or the removal of physical barriers induces functional and structural plasticity and suggests that the adult ECM exerts a powerful inhibitory control on plasticity in the visual cortex (Pizzorusso et al. 2002).

In addition, the serine protease tissue-type plasminogen activator (tPA), which is able to degrade extracellular matrix, is also involved in ocular dominance plasticity. The proteolytic activity of tPA in the binocular zone increases 2 days after monocular deprivation during the critical period (Mataga et al. 2002). Moreover, mice with normal neuronal activity, but deficient in tPA do not exhibit a loss of neuronal responsiveness to the deprived eye and no shift toward the open eye after monocular deprivation (Mataga et al. 2002). Rapid changes of the dendritic spines in the visual cortex occur after brief monocular deprivation and the degradation of ECM with tPA induces the same changes in the spines motility (Oray et al. 2004) and in the spines loss (Mataga et al. 2004). The effects of tPA are occluded by prior monocular deprivation, suggesting that tPA might act in a lamina-specific manner to enable structural synaptic plasticity following brief monocular deprivation (Oray et al. 2004). Moreover, there is evidence that tPA is released after depolarization (Gualandris et al. 1996) or after an induction of long-lasting potentiation (Baranes et al. 1998). This suggests that tPA could be responsible for activity-dependent signals and may provide a flexible and permissive extracellular matrix environment for competitive structural plasticity during the critical period. In conclusion, it appears likely that several factors work in concert with neurotrophins and CAMs to regulate plasticity in the visual cortex.

5.5 Synaptic Plasticity

Synaptic plasticity is thought to be the cellular mechanism for learning and memory processes. At the physiological level specific patterns of presynaptic and postsynaptic activities are believed to result in changes of synaptic efficacy and neural excitability. This activity-dependent modification of synapses is a powerful mechanism for shaping and modifying the response properties of neurons and subsequently of neural networks by allowing the nervous system to adapt to environmental stimuli and to sensory input from developmental through the adult stage.

During development activity-dependent synaptic plasticity has been hypothesized to participate in network refinement and to result in a precise map of synaptic contacts, which form a functional brain. Spontaneous patterns of synchronized activities would help to determine such synaptic maps. Connections between cells that fire together would be reinforced, while the contacts between cells that fire asynchronous would be depressed or even eliminated (Lowel and Singer 1992). The concept that cells, which fire together, are also wired together was postulated by Donald Hebb (1949). His original theory describes a basic mechanism for synaptic plasticity wherein an increase in synaptic efficacy arises when the presynaptic cell repeatedly and persistently activates the postsynaptic cell. This process is also called *Hebb's rule*. In this way the two cells or systems of cells that are concomitantly active at the same time will tend to become "associated", so that activity in one facilitates activity in the other. The theory is commonly evoked to explain some types of associative learning in which simultaneous activation of cells leads to pronounced increases in synaptic strength. Such a learning is known as *Hebbian learning*.

Neural activity has an important role in the synaptic plasticity. In 1973 Bliss and Lomo published the first demonstration that neuronal activity can produce a persistent change in synaptic strength. They observed that brief trains of stimuli resulted in an increased efficiency of transmission at the perforant path–granule cell synapses in the rabbit hippocampus and found that this effect could last for hours (Bliss and Lomo 1973). This increase in synaptic strength is called long-term potentiation (LTP) (Alger and Teyler 1976) and the reciprocal form of plasticity, where a stimulus produces a reduction in synaptic strength, is called long-term depression (LTD). LTD was first discovered in the cerebellum at the parallel fiber–Purkinje cell synapse (Ito and Kano 1982; reviewed in Ito 2001). A great deal of effort has focused on understanding the molecular mechanism how synaptic activity patterns are converted into long-lasting changes of synaptic efficacy and how they are related to behavior and cognitive functions.

Besides long-term synaptic changes, which can last up to several hours, novel mechanisms of short-term synaptic plasticity have been identified. Neurons fire repetitively at frequencies that range from less than one to hundreds of hertz during the course of their activity (Zucker 1989). Repetitive firing can temporarily change synaptic strength, resulting in various forms of short-term plasticity, such as facilitation (which lasts for less than a few seconds), depression (which lasts from a few to tens of seconds), and post-tetanic potentiation (which can last for several minutes). These forms of short-term plasticity are crucial for neuronal network computations (Abbott and Regehr 2004). Several lines of evidence demonstrate that different molecular pathways underlie short- and long-term synaptic changes and that short-term plasticity often constitutes the early phase of long-lasting synaptic potentiation or depression (Lauri et al. 2007).

Usually short-term synaptic plasticity involves local phenomena like the phosphorylation of postsynaptic receptors, which is induced by calcium influx or retrograde signaling. One such signaling process is provided by the endocannabinoid pathway, which modulates presynaptic transmitter release (Kreitzer and Regehr 2001, Raymond et al. 2003). On the other hand, long-term modifications of synaptic function can also involve the persistent activation of protein kinases (like the CaMKII). This results in changes of gene expression and protein synthesis, which have been postulated to provide a mechanism for memory formation (see, for review, Colbran and Brown 2004).

In addition to the distinction between short- and long-term forms of synaptic plasticity also the presynaptic and the postsynaptic loci, where the modification occurs, confer different characteristics to this process. Plasticity-induced changes can occur on both sides of the synapse by altering neurotransmitter release from the presynaptic nerve terminal and/or by modifying the response to neurotransmitter in the postsynaptic neuron. Plasticity changes occurring at synapses often utilize both presynaptic and postsynaptic expression mechanisms. The ability to study the efficacy of transmission at single synapses by using a combination of traditional electrophysiological approaches and imaging techniques has recently allowed to experimentally distinguish the two mechanisms. The dynamics of presynaptic transmitter release are often matched to the physiological properties and function of the postsynaptic cell. Through quantal analysis it is possible to assess the presynaptic nature of the plastic change. Structural alterations at presynaptic release sites exert their effect by changing release probability and not the amount of neurotransmitter released per synaptic vesicle or the proportion of presynaptically active sites (Zakharenko et al. 2001, Krueger and Fitzsimonds 2006).

Postsynaptic plasticity is initiated by modifications of the postsynaptic response without altering the presynaptic release machinery. The mechanisms underlying this phenomenon often involve the activation of second messenger systems. This leads to the phosphorylation of postsynaptic receptors, which can either enhance (Barria et al. 1997) or reduce (Ito 1989) their function by altering the number of receptors expressed on the postsynaptic membrane or by changing their sensitivity or even their subunit composition (Nicoll and Malenka 1999, Hirai 2001, Shi et al. 2001).

Recently a learning-related form of LTP has been demonstrated by our group in the cerebellar cortex following associative fear conditioning. This form of plasticity involves both excitatory and inhibitory synapses onto

Purkinje cells (Sacchetti et al. 2004, Scelfo et al. 2008). The excitatory synapses are postsynaptically potentiated through augmented AMPA receptor function (Sacchetti et al. 2004). In contrast, the inhibitory GABAergic transmission is enhanced by means of a presynaptic mechanism, which implies an increase of the release probability (see Fig. 5.4) (Scelfo et al. 2008). We speculate that the potentiation of both excitation and inhibition allows for the better detection of coincident activation of parallel fibers and ensures that the temporal fidelity of the neuronal circuit is maintained.

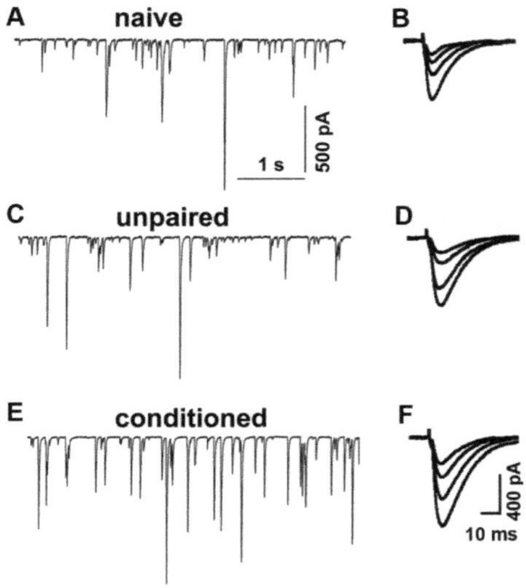

Fig. 5.4 Presynaptic and postsynaptic LTPs in the cerebellar cortex related to associative learning. In this experiment young rats underwent associative fear conditioning. Three groups of animals were used: conditioned rats, which were submitted to the associative protocol; unpaired rats, which received the two conditioning stimuli in an uncorrelated manner; and naïve rats, which were left in their home cage. Twenty-four hours after the conditioning session cerebellar slices from the three groups were obtained and both excitatory and inhibitory activities were tested. (**A**, **C**, **E**) Spontaneous GABAergic inhibitory events recorded by voltage clamping from Purkinje cells in the naïve, unpaired, and conditioned test groups. The GABAergic activity was presynaptically potentiated as a result of the associative learning. The frequency, but not the amplitude of these events, was significantly greater in the conditioned group (**E**). (**B**, **D**, **F**) Excitatory postsynaptic currents evoked in Purkinje cells following parallel fiber activation. Twenty-four hours after the conditioning session excitatory transmission on the Purkinje cells was postsynaptically potentiated. In this case the amplitude of the postsynaptic response was significantly greater in conditioned animals (**F**)

Finally in the last decade another form of synaptic plasticity, called "homeostatic synaptic plasticity" has been described (Turrigiano 1999, Abbott and Nelson 2000, Turrigiano and Nelson 2004). This form of plasticity implies that neurons and neuronal circuits are able to maintain stable function in the face of perturbations such as developmental or activity-dependent changes in synaptic strength, neuronal excitability, and structural changes (Davis and

Goodman 1998) by means of negative feedback signaling (Marder and Prinz 2003, Zhang and Linden 2003). This mechanism allows to compare the level of synaptic activity and strength with an internal set point and then adjust synaptic properties to keep activity close to this set point. Homeostatic synaptic plasticity can be considered as an important complement to Hebbian forms of plasticity in the activity-dependent refinement of synaptic connectivity, hence multiple and probably redundant mechanisms are employed to confine neuronal excitability in the vicinity of the set equilibrium point. These mechanisms include (a) activity-dependent regulation of intrinsic neuronal firing properties (Marder and Prinz 2003, Zhang and Linden 2003); (b) presynaptic and postsynaptic forms of excitatory synaptic plasticity, such as synaptic scaling, that adjust the strength of all excitatory synapses of a neuron to stabilize firing (Turrigiano and Nelson 2004, Davis 2006); (c) balancing of excitation and inhibition within recurrently connected neuronal networks (Maffei et al. 2004, Gonzalez-Islas and Wenner 2006); and (d) compensatory changes in synapse number (Kirov et al. 1999, Wierenga et al. 2006).

Synaptic homeostasis can be expressed in different forms depending on the physiological and molecular mechanisms engaged; however, the complete signaling pathways underlying this form of plasticity are far to be fully understood. Various examples of homeostatic plasticity have been reported in the peripheral and central nervous system. One of the best described occurs at the NMJ, where perturbations in presynaptic function lead to compensatory changes in postsynaptic excitability, and vice versa (Davis and Bezprozvanny 2001). This has led to the idea that during development nerve excitation keeps up with muscle growth through homeostatic compensatory changes in synaptic transmission (Pulver et al. 2005). It has been shown in *Drosophila* that genetic reductions in glutamate receptor function or chronic hyperpolarization of the muscle lead to compensatory and long-lasting (several days) increases in transmitter release (Paradis et al. 2001).

In vitro studies on central neurons have shown that manipulations of neuronal activity were followed by modifications of synapse strength due to a simple change in the accumulation of postsynaptic glutamate receptors, with no perceptible changes in presynaptic function (Lissin et al. 1998, O'Brien et al. 1998). For example, treatment of neocortical cultures with tetrodotoxin (TTX) increased the amplitude but not the frequency of miniature excitatory postsynaptic currents, and subsequent studies found no changes in release probability, synapse number, or short-term plasticity (Turrigiano et al. 1998, Wierenga et al. 2005). By contrast, several other in vitro studies (using hippocampal preparations) have reported that activity deprivation induces changes in both presynaptic release probability and number of release sites (Murthy et al. 2001, Thiagarajan et al. 2005). Although the time of deprivation was the same in these studies the difference in the results seems to be due to the duration of the time in vitro. Before 3 weeks in vitro the adaptation response was purely postsynaptic but when the period in vitro was longer than 3 weeks the adaptation response had a presynaptic and a postsynaptic component (Wierenga et al. 2006).

The molecular identity of the signaling components of homeostatic plasticity has not been well established. Recent work suggests that multiple forms of synaptic homeostasis are mediated by distinct signaling pathways with distinct expression mechanisms. These include presynaptic forms that depend on retrograde signaling to presynaptic Ca^{2+} channels (Leslie et al. 2001) and postsynaptic forms influenced by BDNF (Desai et al. 1999, Swanwick et al. 2006), TNF-α (Beattie et al. 2002), and Arc signaling (Shepherd et al. 2006).

Although neuronal activity plays a fundamental role in the establishment of synaptic plasticity, recent data also support the involvement of several CAMs in this process. The suspected underlying mechanism involves interactions between different classes of adhesion molecules with cell surface receptors, such as ion channels and neurotransmitter receptors. Because synaptic plasticity mechanisms are very diverse, it is reasonable to imagine that more than one class of CAMs regulates synaptic changes both in the developing and in the adult nervous system. The first evidence of an involvement of a CAM in synaptic plasticity was reported by Xiao et al. (1991). Their experiments demonstrated that an incubation of hippocampal slices with a blocker of the integrins leads to an impairment in the expression of LTP.

Subsequently several other classes of CAMs have been implicated in the expression of synaptic plasticity. Among them are PSA-CAM (Luthl et al. 1994), cadherins (Tang et al. 1998), and syndecans (Lauri et al. 1999). The common mechanism seems to involve the remodeling of synaptic structures in response to neuronal activity (Segal and Andersen 2000, Bonhoeffer and Yuste 2002). Besides a role in structural remodeling, several ECM components and CAMs directly participate in the regulation of synaptic transmission. Narp (neuronal activity-regulated pentraxin) and tenascin-C can directly affect functional properties of AMPA and GABA-B neurotransmitter receptors, respectively (O'Brien et al. 1999, Saghatelyan et al. 2000). Furthermore, both tenascin-C and tenascin-R bind voltage-gated sodium channels and have been suggested to play an important role in the modulation of their activity and localization (Srinivasan et al. 1998, Xiao et al. 1999). Tenascin-C has also been implicated in L-type voltage-dependent calcium channel signaling (Evers et al. 2002), which is involved in several forms of synaptic plasticity (Perrier et al. 2002, Haas et al. 2006).

In conclusion, the proposed mechanisms by which CAMs affect neuronal activity and synaptic plasticity can be classified into two main types. One involves structural alterations that are associated with synaptic plasticity. The second involves the coordination of transsynaptic signaling processes during neuronal activity. These mechanisms may act synergistically and provide a powerful means for structural and functional regulation of the nervous system. The understanding of their dynamics is fundamental for elucidating the neuronal mechanisms underlying cognitive processes, especially of learning and memory, as well as for pathological conditions, which relate to neuronal degeneration and regeneration.

5.6 Conclusions

The evidence, which has been reviewed in this chapter, clearly indicates that the selective pruning of neuronal connections is a regressive phenomenon that refines the territory of neuronal innervation during development and an important mechanism for plasticity during both the developing and the adult nervous system. Dendritic and axonal pruning occurs by retraction and by degenerative processes. Both retraction and degeneration also contribute to neurological and neurodegenerative disease phenotypes, suggesting the possibility of common mechanisms. Thus, the knowledge of the mechanisms that underlie developmental pruning could also yield insights into degenerative disorders. In recent years, huge advances have been made in identifying the molecules involved in synapse formation. However, significantly less progress has been made in the understanding of the mechanisms that underlie axon growth, pruning, degeneration/regeneration, and synaptic plasticity. It is possible that the same adhesion molecules that are involved in axon guidance and synapse formation are also implicated in the pruning and plasticity of synaptic contacts and in linking lifetime experiences and critical periods with behavior and cognition.

Acknowledgments We thank Telethon-Italy for the Grant No. GGP030164, MIUR-Italy for the Grant PRIN-2006, Fondazione Cariverona, University of Verona, the National Institute of Neuroscience-Italy and the Regione Piemonte, grant for Ricerca Sanitaria Finalizzata 2006.

References

Abbott LF and Nelson SB (2000) Synaptic plasticity: taming the beast. Nat Neurosci 3 Suppl:1178–1183

Abbott LF and Regehr WG (2004) Synaptic computation. Nature 431:796–803

Akaaboune M, Culican SM, Turney SG et al. (1999) Rapid and reversible effects of activity on acetylcholine receptor density at the neuromuscular junction in vivo [see comments]. Science 286:503–507

Alger BE and Teyler TJ (1976) Long-term and short-term plasticity in the CA1, CA3, and dentate regions of the rat hippocampal slice. Brain Res 110:463–480

Altman J (1972) Postnatal development of the cerebellar cortex in the rat. II. Phases in the maturation of Purkinje cells and of the molecular layer. J Comp Neurol 145:399–463

Amaral DG and Dent JA (1981) Development of the mossy fibers of the dentate gyrus: I. A light and electron microscopic study of the mossy fibers and their expansions. J Comp Neurol 195:51–86

Andjus PR, Zhu L, Cesa R et al. (2003) A change in the pattern of activity affects the developmental regression of the Purkinje cell polyinnervation by climbing fibers in the rat cerebellum. Neuroscience 121:563–572

Aoto J and Chen L (2007) Bidirectional ephrin/Eph signaling in synaptic functions. Brain Res 1184:72–80

Araç D, Boucard AA, Ozkan E et al. (2007) Structures of neuroligin-1 and the neuroligin-1/neurexin-1 beta complex reveal specific protein-protein and protein-Ca2+ interactions. Neuron 56:992–1003

Bagri A, Cheng HJ, Yaron A et al. (2003) Stereotyped pruning of long hippocampal axon branches triggered by retraction inducers of the semaphorin family. Cell 113:285–299
Balice-Gordon RJ and Lichtman JW (1994) Long-term synapse loss induced by focal blockade of postsynaptic receptors. Nature 372:519–524
Baranes D, Lederfein D, Huang YY et al. (1998) Tissue plasminogen activator contributes to the late phase of LTP and to synaptic growth in the hippocampal mossy fiber pathway. Neuron 21:813–825
Barria A, Muller D, Derkach V et al. (1997) Regulatory phosphorylation of AMPA-type glutamate receptors by CaM-KII during long-term potentiation. Science 276:2042–2045
Beattie EC, Stellwagen D, Morishita W et al. (2002) Control of synaptic strength by glial TNFalpha. Science 295:2282–2285
Berardi N, Pizzorusso T and Maffei L (2004) Extracellular matrix and visual cortical plasticity: freeing the synapse. Neuron 44:905–908
Bernstein M and Lichtman JW (1999) Axonal atrophy: the retraction reaction. Curr Opin Neurobiol 9:364–370
Betz WJ, Caldwell JH and Ribchester RR (1980) Sprouting of active nerve terminals in partially inactive muscles of the rat. J Physiol 303:281–297
Bidoia C, Misgeld T, Weinzierl E et al. (2004) Comment on "Reelin promotes peripheral synapse elimination and maturation". Science 303:1977; author reply 1977
Bishop DL, Misgeld T, Walsh MK et al. (2004) Axon branch removal at developing synapses by axosome shedding. Neuron 44:651–661
Bliss TV and Lomo T (1973) Long-lasting potentiation of synaptic transmission in the dentate area of the anaesthetized rabbit following stimulation of the perforant path. J Physiol 232:331–356
Bonhoeffer T and Yuste R (2002) Spine motility. Phenomenology, mechanisms, and function. Neuron 35:1019–1027
Bosman LW, Takechi H, Hartmann J et al. (2008) Homosynaptic long-term synaptic potentiation of the "winner" climbing fiber synapse in developing Purkinje cells. J Neurosci 28:798–807
Boulanger LM and Shatz CJ (2004) Immune signalling in neural development, synaptic plasticity and disease. Nat Rev Neurosci 5:521–531
Bravin M, Rossi F and Strata P (1995) Different climbing fibres innervate separate dendritic regions of the same Purkinje cell in hypogranular cerebellum. J Comp Neurol 357:395–407
Brown MC, Jansen JKS and Essen DV (1976) Polyneuronal innervation of skeletal muscle in new-born rats and its elimination during maturation. J Physiol 261:387–424
Brownlee H, Gao PP, Frisen J et al. (2000) Multiple ephrins regulate hippocampal neurite outgrowth. J Comp Neurol 425:315–322
Buffelli M, Burgess RW, Feng G et al. (2003) Genetic evidence that relative synaptic efficacy biases the outcome of synaptic competition. Nature 424:430–434
Buffelli M, Busetto G, Bidoia C et al. (2004) Activity-dependent synaptic competition at mammalian neuromuscular junctions. News Physiol Sci 19:85–91
Buffelli M, Busetto G, Cangiano L et al. (2002) Perinatal switch from synchronous to asynchronous activity of motoneurons: link with synapse elimination. Proc Natl Acad Sci USA 99:13200–13205
Busetto G, Buffelli M, Tognana E et al. (2000) Hebbian mechanisms revealed by electrical stimulation at developing rat neuromuscular junctions. J Neurosci 20:685–695
Cabelli RJ, Shelton DL, Segal RA et al. (1997) Blockade of endogenous ligands of trkB inhibits formation of ocular dominance columns. Neuron 19:63–76
Cang J, Kaneko M, Yamada J et al. (2005) Ephrin-as guide the formation of functional maps in the visual cortex. Neuron 48:577–589
Cash S, Dan Y, Poo MM et al. (1996) Postsynaptic elevation of calcium induces persistent depression of developing neuromuscular synapses. Neuron 16:745–754
Celio MR, Spreafico R, De Biasi S et al. (1998) Perineuronal nets: past and present. Trends Neurosci 21:510–515
Chen C, Kano M, Abeliovich A et al. (1995) Impaired motor coordination correlates with persistent multiple climbing fiber innervation in PKC gamma mutant mice. Cell 83:1233–1242

Chen H, Bagri A, Zupicich JA et al. (2000) Neuropilin-2 regulates the development of selective cranial and sensory nerves and hippocampal mossy fiber projections. Neuron 25:43–56
Cheng HJ, Bagri A, Yaron A et al. (2001) Plexin-A3 mediates semaphorin signaling and regulates the development of hippocampal axonal projections. Neuron 32:249–263
Cheng HJ, Nakamoto M, Bergemann AD et al. (1995) Complementary gradients in expression and binding of ELF-1 and Mek4 in development of the topographic retinotectal projection map. Cell 82:371–381
Chih B and Scheiffele P (2003) Is reelin the answer to synapse elimination at the neuromuscular junction? Sci STKE 2003:pe45
Colbran RJ and Brown AM (2004) Calcium/calmodulin-dependent protein kinase II and synaptic plasticity. Curr Opin Neurobiol 14:318–327
Colman H, Nabekura J and Lichtman JW (1997) Alterations in synaptic strength preceding axon withdrawal [see comments]. Science 275:356–361
Comery TA, Harris JB, Willems PJ et al. (1997) Abnormal dendritic spines in fragile X knockout mice: maturation and pruning deficits. Proc Natl Acad Sci USA 94:5401–5404
Connold AL, Evers JV and Vrbova G (1986) Effect of low calcium and protease inhibitors on synapse elimination during postnatal development in the rat soleus muscle. Brain Res 393:99–107
Conquet F, Bashir ZI, Davies CH et al. (1994) Motor deficit and impairment of synaptic plasticity in mice lacking mGluR1. Nature 372:237–243
Corriveau RA, Huh GS and Shatz CJ (1998) Regulation of class I MHC gene expression in the developing and mature CNS by neural activity. Neuron 21:505–520
Costanzo EM, Barry JA and Ribchester RR (2000) Competition at silent synapses in reinnervated skeletal muscle. Nat Neurosci 3:694–700
Crepel F, Dhanjal SS and Garthwaite J (1981) Morphological and electrophysiological characteristics of rat cerebellar slices maintained in vitro. J Physiol 316:127–138
Crowley JC and Katz LC (2002) Ocular dominance development revisited. Curr Opin Neurobiol 12:104–109
Crusio WE and Schwegler H (1987) Hippocampal mossy fiber distribution covaries with open-field habituation in the mouse. Behav Brain Res 26:153–158
Davis GW (2006) Homeostatic control of neural activity: from phenomenology to molecular design. Annu Rev Neurosci 29:307–323
Davis GW and Bezprozvanny I (2001) Maintaining the stability of neural function: a homeostatic hypothesis. Annu Rev Physiol 63:847–869
Davis GW and Goodman CS (1998) Genetic analysis of synaptic development and plasticity: homeostatic regulation of synaptic efficacy. Curr Opin Neurobiol 8:149–156
Davis S, Gale NW, Aldrich TH et al. (1994) Ligands for EPH-related receptor tyrosine kinases that require membrane attachment or clustering for activity. Science 266:816–819
Deppmann CD, Mihalas S, Sharma N et al. (2008) A model for neuronal competition during development. Science 320:369–373
Desai NS, Rutherford LC and Turrigiano GG (1999) BDNF regulates the intrinsic excitability of cortical neurons. Learn Mem 6:284–291
Donoghue MJ and Rakic P (1999a) Molecular evidence for the early specification of presumptive functional domains in the embryonic primate cerebral cortex. J Neurosci 19:5967–5979
Donoghue MJ and Rakic P (1999b) Molecular gradients and compartments in the embryonic primate cerebral cortex. Cereb Cortex 9:586–600
Duxson MJ (1982) The effect of postsynaptic block on development of the neuromuscular junction in postnatal rats. J Neurocytol 11:395–408
Eccles JC, Llinas R and Sasaki K (1966) The excitatory synaptic action of climbing fibres on the Purkinje cells of the cerebellum. J Physiol 182:268–296
Eckenhoff MF and Pysh JJ (1979) Double-walled coated vesicle formation: evidence for massive and transient conjugate internalization of plasma membranes during cerebellar development. J Neurocytol 8:623–638

English AW and Schwartz G (1995) Both basic fibroblast growth factor and ciliary neurotrophic factor promote the retention of polyneuronal innervation of developing skeletal muscle fibers. Dev Biol 169:57–64
Evers MR, Salmen B, Bukalo O et al. (2002) Impairment of L-type Ca2+ channel-dependent forms of hippocampal synaptic plasticity in mice deficient in the extracellular matrix glycoprotein tenascin-C. J Neurosci 22:7177–7194
Fagiolini M and Hensch TK (2000) Inhibitory threshold for critical-period activation in primary visual cortex. Nature 404:183–186
Faulkner RL, Low LK and Cheng HJ (2007) Axon pruning in the developing vertebrate hippocampus. Dev Neurosci 29:6–13
Favero M, Lorenzetto E, Bidoia C et al. (2007) Synapse formation and elimination: role of activity studied in different models of adult muscle reinnervation. J Neurosci Res 85:2610–2619
Feldman DE (2000) Inhibition and plasticity. Nat Neurosci 3:303–304
Fitzsimonds RM and Poo MM (1998) Retrograde signaling in the development and modification of synapses. Physiol Rev 78:143–170
Fladby T (1987) Postnatal loss of synaptic terminals in the normal mouse soleus muscle. Acta Physiol Scand 129:229–238
Flourens P (1824) Recherches expérimentales sur les propriétés et les fonctions du système nerveux dans les animaux vertébrés. In: Clarke EaOM, C. (eds) The Human Brain and Spinal Cord. University of California Press, Berkeley and Los Angeles
Gale NW, Holland SJ, Valenzuela DM et al. (1996) Eph receptors and ligands comprise two major specificity subclasses and are reciprocally compartmentalized during embryogenesis. Neuron 17:9–19
Gan WB, Bishop DL, Turney SG et al. (1999) Vital imaging and ultrastructural analysis of individual axon terminals labeled by iontophoretic application of lipophilic dye. J Neurosci Methods 93:13–20
Gan WB and Lichtman JW (1998) Synaptic segregation at the developing neuromuscular junction. Science 282:1508–1511
Gao PP, Zhang JH, Yokoyama M et al. (1996) Regulation of topographic projection in the brain: Elf-1 in the hippocamposeptal system. Proc Natl Acad Sci USA 93:11161–11166
Geuze E, Vermetten E and Bremner JD (2005a) MR-based in vivo hippocampal volumetrics: 1. Review of methodologies currently employed. Mol Psychiatry 10:147–159
Geuze E, Vermetten E and Bremner JD (2005b) MR-based in vivo hippocampal volumetrics: 2. Findings in neuropsychiatric disorders. Mol Psychiatry 10:160–184
Gonzalez-Islas C and Wenner P (2006) Spontaneous network activity in the embryonic spinal cord regulates AMPAergic and GABAergic synaptic strength. Neuron 49:563–575
Goodman CS and Shatz CJ (1993) Developmental mechanisms that generate precise patterns of neuronal connectivity. Cell 72 Suppl:77–98
Greenough WT, Hwang HM and Gorman C (1985) Evidence for active synapse formation or altered postsynaptic metabolism in visual cortex of rats reared in complex environments. Proc Natl Acad Sci USA 82:4549–4552
Gualandris A, Jones TE, Strickland S et al. (1996) Membrane depolarization induces calcium-dependent secretion of tissue plasminogen activator. J Neurosci 16:2220–2225
Guimaraes A, Zaremba S and Hockfield S (1990) Molecular and morphological changes in the cat lateral geniculate nucleus and visual cortex induced by visual deprivation are revealed by monoclonal antibodies Cat-304 and Cat-301. J Neurosci 10:3014–3024
Haas JS, Nowotny T and Abarbanel HD (2006) Spike-timing-dependent plasticity of inhibitory synapses in the entorhinal cortex. J Neurophysiol 96:3305–3313
Hashimoto K, Ichikawa R, Takechi H et al. (2001a) Roles of glutamate receptor delta 2 subunit (GluRdelta 2) and metabotropic glutamate receptor subtype 1 (mGluR1) in climbing fiber synapse elimination during postnatal cerebellar development. J Neurosci 21:9701–9712
Hashimoto K, Miyata M, Watanabe M et al. (2001b) Roles of phospholipase Cbeta4 in synapse elimination and plasticity in developing and mature cerebellum. Mol Neurobiol 23:69–82

Hata Y, Butz S and Südhof TC (1996) CASK: a novel dlg/PSD95 homolog with an N-terminal calmodulin-dependent protein kinase domain identified by interaction with neurexins. J Neurosci 16:2488–2494
Hebb D (1949) The Organization of Behavior. Wiley, New York
Hensch TK, Fagiolini M, Mataga N et al. (1998) Local GABA circuit control of experience-dependent plasticity in developing visual cortex. Science 282:1504–1508
Hinton VJ, Brown WT, Wisniewski K et al. (1991) Analysis of neocortex in three males with the fragile X syndrome. Am J Med Genet 41:289–294
Hirai H (2001) Modification of AMPA receptor clustering regulates cerebellar synaptic plasticity. Neurosci Res 39:261–267
Hua JY and Smith SJ (2004) Neural activity and the dynamics of central nervous system development. Nat Neurosci 7:327–332
Huang ZJ, Kirkwood A, Pizzorusso T et al. (1999) BDNF regulates the maturation of inhibition and the critical period of plasticity in mouse visual cortex. Cell 98:739–755
Hubel DH and Wiesel TN (1963) Receptive Fields of Cells in Striate Cortex of Very Young, Visually Inexperienced Kittens. J Neurophysiol 26:994–1002
Hubel DH and Wiesel TN (1970) The period of susceptibility to the physiological effects of unilateral eye closure in kittens. J Physiol 206:419–436
Hubel DH, Wiesel TN and LeVay S (1977) Plasticity of ocular dominance columns in monkey striate cortex. Philos Trans R Soc Lond B Biol Sci 278:377–409
Huber AB, Kolodkin AL, Ginty DD et al. (2003) Signaling at the growth cone: ligand-receptor complexes and the control of axon growth and guidance. Annu Rev Neurosci 26:509–563
Huh GS, Boulanger LM, Du H et al. (2000) Functional requirement for class I MHC in CNS development and plasticity. Science 290:2155–2159
Hume RI and Purves D (1981) Geometry of neonatal neurones and the regulation of synapse elimination. Nature 293:469–471
Hunt CA, Schenker LJ and Kennedy MB (1996) PSD-95 is associated with the postsynaptic density and not with the presynaptic membrane at forebrain synapses. J Neurosci 16:1380–1388
Ichtchenko K, Nguyen T and Südhof TC (1996) Structures, alternative splicing, and neurexin binding of multiple neuroligins. J Biol Chem 271:2676–2682
Ito M (1984) The Cerebellum and Neural Control. Raven Press, New York
Ito M (1989) Long-term depression. Annu Rev Neurosci 12:85–102
Ito M (2001) Cerebellar long-term depression: characterization, signal transduction, and functional roles. Physiol Rev 81:1143–1195
Ito M and Kano M (1982) Long-lasting depression of parallel fiber-Purkinje cell transmission induced by conjunctive stimulation of parallel fibers and climbing fibers in the cerebellar cortex. Neurosci Lett 33:253–258
Ivanco TL and Greenough WT (2002) Altered mossy fiber distributions in adult Fmr1 (FVB) knockout mice. Hippocampus 12:47–54
Jansen JK and Fladby T (1990) The perinatal reorganization of the innervation of skeletal muscle in mammals. Prog Neurobiol 34:39–90
Jennings C (1994) Developmental neurobiology. Death of a synapse. Nature 372:498–499
Kano M, Hashimoto K, Chen C et al. (1995) Impaired synapse elimination during cerebellar development in PKC gamma mutant mice. Cell 83:1223–1231
Kantor DB and Kolodkin AL (2003) Curbing the excesses of youth: molecular insights into axonal pruning. Neuron 38:849–852
Kasthuri N and Lichtman JW (2003) The role of neuronal identity in synaptic competition. Nature 424:426–430
Katz LC and Crowley JC (2002) Development of cortical circuits: lessons from ocular dominance columns. Nat Rev Neurosci 3:34–42

Katz LC and Shatz CJ (1996) Synaptic activity and the construction of cortical circuits. Science 274:1133–1138
Keller-Peck CR, Walsh MK, Gan WB et al. (2001) Asynchronous synapse elimination in neonatal motor units: studies using GFP transgenic mice. Neuron 31:381–394.
Kirov SA, Sorra KE and Harris KM (1999) Slices have more synapses than perfusion-fixed hippocampus from both young and mature rats. J Neurosci 19:2876–2886
Knight R (1996) Contribution of human hippocampal region to novelty detection. Nature 383:256–259
Kolodkin AL (1998) Semaphorin-mediated neuronal growth cone guidance. Prog Brain Res 117:115–132
Kornau HC and Seeburg PH (1997) Partner selection by PDZ domains. Nat Biotechnol 15:319
Kreitzer AC and Regehr WG (2001) Retrograde inhibition of presynaptic calcium influx by endogenous cannabinoids at excitatory synapses onto Purkinje cells. Neuron 29:717–727
Krueger S and Fitzsimonds RM (2006) Remodeling the plasticity debate: the presynaptic locus revisited. Physiology (Bethesda) 21:346–351
Kuffler D, Thompson W and Jansen J (1977) The elimination of synapses in multiply-innervated skeletal muscle fibres of the rat: dependence on the distance between end-plates. Brain Res 138:353–358
Kullander K and Klein R (2002) Mechanisms and functions of Eph and ephrin signalling. Nat Rev Mol Cell Biol 3:475–486
Kwon YW, Abbondanzo SJ, Stewart CL et al. (1995) Leukemia inhibitory factor influences the timing of programmed synapses withdrawal from neonatal muscles. J Neurobiol 28:35–50
Kwon YW and Gurney ME (1996) Brain-derived neurotrophic factor transiently stabilizes silent synapses on developing neuromuscular junctions. J Neurobiol 29:503–516
Lauder JM and Mugnaini E (1980) Infrapyramidal mossy fibers in the hippocampus of the hyperthyroid rat. A light and electron microscopic study. Dev Neurosci 3:248–265
Lauri SE, Kaukinen S, Kinnunen T et al. (1999) Regulatory role and molecular interactions of a cell-surface heparan sulfate proteoglycan (N-syndecan) in hippocampal long-term potentiation. J Neurosci 19:1226–1235
Lauri SE, Palmer M, Segerstrale M et al. (2007) Presynaptic mechanisms involved in the expression of STP and LTP at CA1 synapses in the hippocampus. Neuropharmacology 52:1–11
LeDoux JE (2000) Emotion circuits in the brain. Annu Rev Neurosci 23:155–184
Lein ES and Shatz CJ (2000) Rapid regulation of brain-derived neurotrophic factor mRNA within eye-specific circuits during ocular dominance column formation. J Neurosci 20:1470–1483
Leslie KR, Nelson SB and Turrigiano GG (2001) Postsynaptic depolarization scales quantal amplitude in cortical pyramidal neurons. J Neurosci 21:RC170
LeVay S, Stryker MP and Shatz CJ (1978) Ocular dominance columns and their development in layer IV of the cat's visual cortex: a quantitative study. J Comp Neurol 179:223–244
Lissin DV, Gomperts SN, Carroll RC et al. (1998) Activity differentially regulates the surface expression of synaptic AMPA and NMDA glutamate receptors. Proc Natl Acad Sci USA 95:7097–7102
Liu XB, Low LK, Jones EG et al. (2005) Stereotyped axon pruning via plexin signaling is associated with synaptic complex elimination in the hippocampus. J Neurosci 25:9124–9134
Liu Y, Fields RD, Fitzgerald S et al. (1994) Proteolytic activity, synapse elimination, and the Hebb synapse. J Neurobiol 25:325–335
Lo YJ and Poo MM (1991) Activity-dependent synaptic competition in vitro: heterosynaptic suppression of developing synapses. Science 254:1019–1022
Lohof AM, Delhaye-Bouchaud N and Mariani J (1996) Synapse elimination in the central nervous system: functional significance and cellular mechanisms. Rev Neurosci 7:85–101

Lowel S and Singer W (1992) Selection of intrinsic horizontal connections in the visual cortex by correlated neuronal activity. Science 255:209–212
Luo L and O'Leary DD (2005) Axon retraction and degeneration in development and disease. Annu Rev Neurosci 28:127–156
Luthl A, Laurent JP, Figurov A et al. (1994) Hippocampal long-term potentiation and neural cell adhesion molecules L1 and NCAM. Nature 372:777–779
Mackarehtschian K, Lau CK, Caras I et al. (1999) Regional differences in the developing cerebral cortex revealed by ephrin-A5 expression. Cereb Cortex 9:601–610
Maffei A, Nelson SB and Turrigiano GG (2004) Selective reconfiguration of layer 4 visual cortical circuitry by visual deprivation. Nat Neurosci 7:1353–1359
Maguire EA, Burgess N and O'Keefe J (1999) Human spatial navigation: cognitive maps, sexual dimorphism, and neural substrates. Curr Opin Neurobiol 9:171–177
Maguire EA, Frith CD, Burgess N et al. (1998) Knowing where things are parahippocampal involvement in encoding object locations in virtual large-scale space. J Cogn Neurosci 10:61–76
Mainen ZF, Maletic-Savatic M, Shi SH et al. (1999) Two-photon imaging in living brain slices. Methods 18:231–239, 181
Majewska AK and Sur M (2006) Plasticity and specificity of cortical processing networks. Trends Neurosci 29:323–329
Marder E and Prinz AA (2003) Current compensation in neuronal homeostasis. Neuron 37:2–4
Maren S (2001) Neurobiology of Pavlovian fear conditioning. Annu Rev Neurosci 24:897–931
Mariani J and Changeux JP (1980) Multiple innervation of Purkinje cells by climbing fibers in the cerebellum of the adult staggerer mutant mouse. J Neurobiol 11:41–50
Martinez A, Otal R, Sieber BA et al. (2005) Disruption of ephrin-A/EphA binding alters synaptogenesis and neural connectivity in the hippocampus. Neuroscience 135:451–461
Mason CA, Christakos S and Catalano SM (1990) Early climbing fiber interactions with Purkinje cells in the postnatal mouse cerebellum. J Comp Neurol 297:77–90
Mataga N, Mizuguchi Y and Hensch TK (2004) Experience-dependent pruning of dendritic spines in visual cortex by tissue plasminogen activator. Neuron 44:1031–1041
Mataga N, Nagai N and Hensch TK (2002) Permissive proteolytic activity for visual cortical plasticity. Proc Natl Acad Sci USA 99:7717–7721
Maya Vetencourt JF, Sale A, Viegi A et al. (2008) The antidepressant fluoxetine restores plasticity in the adult visual cortex. Science 320:385–388
McLaughlin T, Hindges R and O'Leary DD (2003) Regulation of axial patterning of the retina and its topographic mapping in the brain. Curr Opin Neurobiol 13:57–69
Midtgaard J (1995) Spatial synaptic integration in Purkinje cell dendrites. J Physiol Paris 89:23–32
Mineur YS, Sluyter F, de Wit S et al. (2002) Behavioral and neuroanatomical characterization of the Fmr1 knockout mouse. Hippocampus 12:39–46
Miyata M, Kim HT, Hashimoto K et al. (2001) Deficient long-term synaptic depression in the rostral cerebellum correlated with impaired motor learning in phospholipase C beta4 mutant mice. Eur J Neurosci 13:1945–1954
Miyazaki T, Hashimoto K, Shin HS et al. (2004) P/Q-type Ca2+ channel alpha1A regulates synaptic competition on developing cerebellar Purkinje cells. J Neurosci 24:1734–1743
Mizumori SJ, Ragozzino KE, Cooper BG et al. (1999) Hippocampal representational organization and spatial context. Hippocampus 9:444–451
Mori T, Wanaka A, Taguchi A et al. (1995) Differential expressions of the eph family of receptor tyrosine kinase genes (sek, elk, eck) in the developing nervous system of the mouse. Brain Res Mol Brain Res 29:325–335
Murthy VN, Schikorski T, Stevens CF et al. (2001) Inactivity produces increases in neurotransmitter release and synapse size. Neuron 32:673–682

Nakamoto M, Cheng HJ, Friedman GC et al. (1996) Topographically specific effects of ELF-1 on retinal axon guidance in vitro and retinal axon mapping in vivo. Cell 86:755–766
Napper RM and Harvey RJ (1988) Number of parallel fiber synapses on an individual Purkinje cell in the cerebellum of the rat. J Comp Neurol 274:168–177
Nguyen QT and Lichtman JW (1996) Mechanism of synapse disassembly at the developing neuromuscular junction. Curr Opin Neurobiol 6:104–112
Nguyen QT, Parsadanian AS, Snider WD et al. (1998) Hyperinnervation of neuromuscular junctions caused by GDNF overexpression in muscle. Science 279:1725–1729
Nicoll RA and Malenka RC (1999) Expression mechanisms underlying NMDA receptor-dependent long-term potentiation. Ann N Y Acad Sci 868:515–525
O'Brien RA, Ostberg AJ and Vrbova G (1980) The effect of acetylcholine on the function and structure of the developing mammalian neuromuscular junction. Neuroscience 5:1367–1379
O'Brien RJ, Lau LF and Huganir RL (1998) Molecular mechanisms of glutamate receptor clustering at excitatory synapses. Curr Opin Neurobiol 8:364–369
O'Brien RJ, Xu D, Petralia RS et al. (1999) Synaptic clustering of AMPA receptors by the extracellular immediate-early gene product Narp. Neuron 23:309–323
O'Leary DD and Koester SE (1993) Development of projection neuron types, axon pathways, and patterned connections of the mammalian cortex. Neuron 10:991–1006
Oray S, Majewska A and Sur M (2004) Dendritic spine dynamics are regulated by monocular deprivation and extracellular matrix degradation. Neuron 44:1021–1030
Palay SL, Billings-Gagliardi S and Chan-Palay V (1974) Neuronal perikarya with dispersed, single ribosomes in the visual cortex of *Macaca mulatta*. J Cell Biol 63:1074–1089
Paradis S, Sweeney ST and Davis GW (2001) Homeostatic control of presynaptic release is triggered by postsynaptic membrane depolarization. Neuron 30:737–749
Patrizi A, Scelfo B, Viltono L et al. (2008) Synapse formation and clustering of neuroligin-2 in the absence of GABAA receptors. Proc Natl Acad Sci USA 105:13151–13156
Penn AA, Riquelme PA, Feller MB et al. (1998) Competition in retinogeniculate patterning driven by spontaneous activity. Science 279:2108–2112
Perrier JF, Alaburda A and Hounsgaard J (2002) Spinal plasticity mediated by postsynaptic L-type Ca2+ channels. Brain Res Brain Res Rev 40:223–229
Pizzorusso T, Medini P, Berardi N et al. (2002) Reactivation of ocular dominance plasticity in the adult visual cortex. Science 298:1248–1251
Pulver SR, Bucher D, Simon DJ et al. (2005) Constant amplitude of postsynaptic responses for single presynaptic action potentials but not bursting input during growth of an identified neuromuscular junction in the lobster, *Homarus americanus*. J Neurobiol 62:47–61
Purves D and Hume RI (1981) The relation of postsynaptic geometry to the number of presynaptic axons that innervate autonomic ganglion cells. J Neurosci 1:441–452
Purves D and Lichtman JW (1980) Elimination of synapses in the developing nervous system. Science 210:153–157
Quattrocchi CC, Huang C, Niu S et al. (2003) Reelin promotes peripheral synapse elimination and maturation. Science 301:649–653
Rabacchi S, Bailly Y, Delhaye-Bouchaud N et al. (1992a) Involvement of the N-methyl D-aspartate (NMDA) receptor in synapse elimination during cerebellar development. Science 256:1823–1825
Rabacchi SA, Bailly Y, Delhaye-Bouchaud N et al. (1992b) Role of the target in synapse elimination: studies in cerebellum of developing lurcher mutants and adult chimeric mice. J Neurosci 12:4712–4720
Ramon y Cajal S (1911) Histologie du Systeme Nerveux de l'Homme et des Vertebres. Maloine, Paris
Raymond CR, Ireland DR and Abraham WC (2003) NMDA receptor regulation by amyloid-beta does not account for its inhibition of LTP in rat hippocampus. Brain Res 968:263–272

Riley DA (1981) Ultrastructural evidence for axon retraction during the spontaneous elimination of polyneuronal innervation of the rat soleus muscle. J Neurocytol 10:425–440
Rossi FM, Bozzi Y, Pizzorusso T et al. (1999) Monocular deprivation decreases brain-derived neurotrophic factor immunoreactivity in the rat visual cortex. Neuroscience 90:363–368
Sacchetti B, Scelfo B, Tempia F et al. (2004) Long-term synaptic changes induced in the cerebellar cortex by fear conditioning. Neuron 42:973–982
Saghatelyan AK, Gorissen S, Albert M et al. (2000) The extracellular matrix molecule tenascin-R and its HNK-1 carbohydrate modulate perisomatic inhibition and long-term potentiation in the CA1 region of the hippocampus. Eur J Neurosci 12:3331–3342
Sahay A, Kim CH, Sepkuty JP et al. (2005) Secreted semaphorins modulate synaptic transmission in the adult hippocampus. J Neurosci 25:3613–3620
Sanes JR and Lichtman JW (1999) Development of the vertebrate neuromuscular junction. Annu Rev Neurosci 22:389–442
Scelfo B, Sacchetti B and Strata P (2008) Learning-related long-term potentiation of inhibitory synapses in the cerebellar cortex. Proc Natl Acad Sci USA 105:769–774
Scelfo B and Strata P (2005) Correlation between multiple climbing fibre regression and parallel fibre response development in the postnatal mouse cerebellum. Eur J Neurosci 21:971–978
Scelfo B, Strata P and Knöpfel T (2003) Sodium imaging of climbing fiber innervation fields in developing mouse Purkinje cells. J Neurophysiol 89:2555–2563
Schopke R, Wolfer DP, Lipp HP et al. (1991) Swimming navigation and structural variations of the infrapyramidal mossy fibers in the hippocampus of the mouse. Hippocampus 1:315–328
Schwegler H, Crusio WE, Lipp HP et al. (1991) Early postnatal hyperthyroidism alters hippocampal circuitry and improves radial-maze learning in adult mice. J Neurosci 11:2102–2106
Segal M and Andersen P (2000) Dendritic spines shaped by synaptic activity. Curr Opin Neurobiol 10:582–586
Shapiro L, Love J and Colman DR (2007) Adhesion molecules in the nervous system: structural insights into function and diversity. Annu Rev Neurosci 30:451–474
Shatz CJ (1990) Competitive interactions between retinal ganglion cells during prenatal development. J Neurobiol 21:197–211
Shepherd JD, Rumbaugh G, Wu J et al. (2006) Arc/Arg3.1 mediates homeostatic synaptic scaling of AMPA receptors. Neuron 52:475–484
Shi S, Hayashi Y, Esteban JA et al. (2001) Subunit-specific rules governing AMPA receptor trafficking to synapses in hippocampal pyramidal neurons. Cell 105:331–343
Singh KK and Miller FD (2005) Activity regulates positive and negative neurotrophin-derived signals to determine axon competition. Neuron 45:837–845
Singh KK, Park KJ, Hong EJ et al. (2008) Developmental axon pruning mediated by BDNF-p75NTR-dependent axon degeneration. Nat Neurosci 11:649–658
Snider WD and Lichtman JW (1996) Are neurotrophins synaptotrophins? Mol Cell Neurosci 7:433–442
Song JY, Ichtchenko K, Südhof TC et al. (1999) Neuroligin 1 is a postsynaptic cell-adhesion molecule of excitatory synapses. Proc Natl Acad Sci USA 96:1100–1105
Srihari T and Vrbova G (1978) The role of muscle activity in the differentiation of neuromuscular junctions in slow and fast chick muscles. J Neurocytol 7:529–540
Srinivasan J, Schachner M and Catterall WA (1998) Interaction of voltage-gated sodium channels with the extracellular matrix molecules tenascin-C and tenascin-R. Proc Natl Acad Sci USA 95:15753–15757
Stellwagen D and Shatz CJ (2002) An instructive role for retinal waves in the development of retinogeniculate connectivity. Neuron 33:357–367
Stevens B, Allen NJ, Vazquez LE et al. (2007) The classical complement cascade mediates CNS synapse elimination. Cell 131:1164–1178
Südhof TC (2008) Neuroligins and neurexins link synaptic function to cognitive disease. Nature 455:903–911

Sur M and Leamey CA (2001) Development and plasticity of cortical areas and networks. Nat Rev Neurosci 2:251–262
Swanwick CC, Murthy NR and Kapur J (2006) Activity-dependent scaling of GABAergic synapse strength is regulated by brain-derived neurotrophic factor. Mol Cell Neurosci 31:481–492
Tang L, Hung CP and Schuman EM (1998) A role for the cadherin family of cell adhesion molecules in hippocampal long-term potentiation. Neuron 20:1165–1175
Tao HW and Poo M (2001) Retrograde signaling at central synapses. Proc Natl Acad Sci USA 98:11009–11015
Thach WT (2007) On the mechanism of cerebellar contributions to cognition. Cerebellum 6:163–167
Thiagarajan TC, Lindskog M and Tsien RW (2005) Adaptation to synaptic inactivity in hippocampal neurons. Neuron 47:725–737
Thompson RF (1990) Neural mechanisms of classical conditioning in mammals. Philos Trans R Soc Lond B Biol Sci 329:161–170
Thompson W and Jansen JK (1977) The extent of sprouting of remaining motor units in partly denervated immature and adult rat soleus muscle. Neuroscience 2:523–535
Thompson WJ (1983) Synapse elimination in neonatal rat muscle is sensitive to the pattern of muscle use. Nature 302:614–616
Thompson WJ, Kuffler DP and Jansen JKS (1979) The effect of prolonged reversible block of nerve impulses on the elimination of polyneuronal innervation of new-born rat skeletal muscle. Neuroscience 4:271–281
Turrigiano GG (1999) Homeostatic plasticity in neuronal networks: the more things change, the more they stay the same. Trends Neurosci 22:221–227
Turrigiano GG, Leslie KR, Desai NS et al. (1998) Activity-dependent scaling of quantal amplitude in neocortical neurons. Nature 391:892–896
Turrigiano GG and Nelson SB (2004) Homeostatic plasticity in the developing nervous system. Nat Rev Neurosci 5:97–107
Walsh MK and Lichtman JW (2003) In vivo time-lapse imaging of synaptic takeover associated with naturally occurring synapse elimination. Neuron 37:67–73
Weimann JM, Zhang YA, Levin ME et al. (1999) Cortical neurons require Otx1 for the refinement of exuberant axonal projections to subcortical targets. Neuron 24:819–831
Wierenga CJ, Ibata K and Turrigiano GG (2005) Postsynaptic expression of homeostatic plasticity at neocortical synapses. J Neurosci 25:2895–2905
Wierenga CJ, Walsh MF and Turrigiano GG (2006) Temporal regulation of the expression locus of homeostatic plasticity. J Neurophysiol 96:2127–2133
Wiesel TN and Hubel DH (1963) Effects of Visual Deprivation on Morphology and Physiology of Cells in the Cats Lateral Geniculate Body. J Neurophysiol 26:978–993
Windrem MS and Finlay BL (1991) Thalamic ablations and neocortical development: alterations of cortical cytoarchitecture and cell number. Cereb Cortex 1:230–240
Wong RO (1999) Retinal waves and visual system development. Annu Rev Neurosci 22:29–47
Xiao P, Bahr BA, Staubli U et al. (1991) Evidence that matrix recognition contributes to stabilization but not induction of LTP. Neuroreport 2:461–464
Xiao ZC, Ragsdale DS, Malhotra JD et al. (1999) Tenascin-R is a functional modulator of sodium channel beta subunits. J Biol Chem 274:26511–26517
Yates PA, Roskies AL, McLaughlin T et al. (2001) Topographic-specific axon branching controlled by ephrin-As is the critical event in retinotectal map development. J Neurosci 21:8548–8563
Yue Y, Chen ZY, Gale NW et al. (2002) Mistargeting hippocampal axons by expression of a truncated Eph receptor. Proc Natl Acad Sci USA 99:10777–10782
Zakharenko SS, Zablow L and Siegelbaum SA (2001) Visualization of changes in presynaptic function during long-term synaptic plasticity. Nat Neurosci 4:711–717
Zhang W and Linden DJ (2003) The other side of the engram: experience-driven changes in neuronal intrinsic excitability. Nat Rev Neurosci 4:885–900
Zucker RS (1989) Short-term synaptic plasticity. Annu Rev Neurosci 12:13–31

Chapter 6
Cell Adhesion Molecules in Synaptopathies

Thomas Bourgeron

Abstract Synaptopathies are human disorders caused by a defect in synapse formation or function. In the first 3 years of life, the ability of children for learning is impressive and correlates with an intense phase of synaptogenesis in their brains. During this critical period, cell adhesion molecules (CAMs) are crucial factors for the identification of the appropriate partner cell and the formation of a functional synapse. Consistent with their key roles in brain development, mutations in brain CAMs can lead to a variety of neurological disorders such as deafness, epilepsy, mental retardation, and autism spectrum conditions (ASC). Furthermore, polymorphisms of brain CAMs within the human population may also play a role in the susceptibility to milder cognitive disorders. This chapter reports several examples of CAM mutations that are associated with human brain disorders and highlights the emerging key roles of these molecules in the susceptibility to neurologic and psychiatric conditions.

Keywords Autism · Psychiatry · Synapse · Synaptogenesis

6.1 Introduction

The wiring of the brain is the result of a mixture of information coming from genetic, epigenetic, and environmental factors. The actual contribution of these factors in the susceptibility to psychiatric conditions remains a matter of considerable debate. Recently, an emerging category of disorders caused by a defect in synapse formation or function, the so-called synaptopathies, focuses attention on several synaptic genes that cause neurological/psychiatric disorders. Among the syndromes influenced by CAMs, autism affects about 0.7% of children and is characterized by deficits in social communication, absence or

T. Bourgeron (✉)
Human Genetics and Cognitive Functions Units, Institut Pasteur, 25 rue du Docteur Roux 75724 Paris Cedex 15, Paris, France
e-mail: thomasb@pasteur.fr

M. Hortsch, H. Umemori (eds.), *The Sticky Synapse*,
DOI 10.1007/978-0-387-92708-4_6,

delay in language, and stereotyped and repetitive behaviors (Freitag 2007). Beyond this unifying definition lies an extreme degree of clinical heterogeneity, ranging from debilitating impairments to mild personality traits. Hence autism is not a single entity, but rather a complex phenotype encompassing either multiple "autistic disorders" or a continuum of autistic-like traits and behaviors. To take into account this heterogeneity, the term autism spectrum condition/disorder (ASC/ASD) is now used.

ASC is usually diagnosed before 3 years of age, a period characterized by intense synaptogenesis in the human brain (Huttenlocher and Dabholkar 1997). In that perspective, CAMs represent excellent candidates for acting as susceptibility factors for this syndrome. Indeed, brain CAMs are crucial factors for synaptic contact initiation, recruitment of presynaptic and postsynaptic proteins, synapse maturation/stabilization or elimination, and synaptic plasticity (Dalva et al. 2007). Besides their importance in synaptogenesis, two lines of evidence make CAMs compelling candidates for modulating inter-individual susceptibility to psychiatric conditions.

First, their functions may be significant, but is often not vital for the organism. In contrast to genes involved in the early stage of the human brain development, some CAMs may only play a role in the specificity of the wiring (Mattson and van Praag 2008). Thus, a slight atypical wiring by mutations in CAMs may cause very specific cognitive disorders and/or susceptibility to personality dimensions without causing a severe neurological condition. Consistent with this hypothesis, several mutant mice for CAMs do not always display obvious phenotypes (Tabuchi et al. 2007, Jamain et al. 2008b).

Second, CAM-encoding genes represent a significant portion of the human genome (Li et al. 2008) and each gene usually has several paralogous genes forming an extended gene family (Figs. 6.1 and 6.2). Therefore, due to molecular redundancy, a mutation in a single gene could lead to very subtle or specific cognitive alterations rather than to a severe neurological disorder. Furthermore, several CAM possess a large combinatory of alternative promoters and splicing exons. This ability to create more than 100 different proteins from a single gene makes CAMs among the best candidates for encoding neuronal identity (Shapiro et al. 2007). Along this line, it is very likely that an anomaly in neuronal identity may increase the susceptibility to a psychiatric condition. In this chapter, I will present several case reports that illustrate the diversity of CAM mutations in humans and discuss how dysfunction of these adhesive systems may contribute to these disorders.

6.2 Neuroligins and Neurexins

Neuroligins and neurexins are postsynaptic and presynaptic proteins involved in synapse formation and maintenance (Craig and Kang 2007, see also Chapter 17). Mutations affecting these proteins, as well as SHANK3, a scaffolding protein of the postsynaptic density (PSD) that binds to neurologins, are

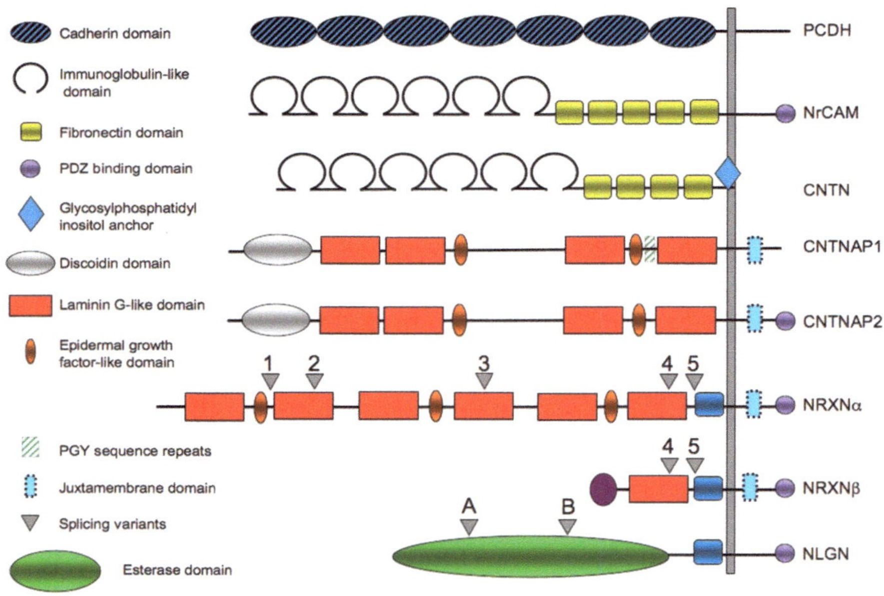

Fig. 6.1 Structure of CAMs associated with synaptopathies in humans

associated with ASC (Jamain et al. 2003, Durand et al. 2007). There are five neuroligin genes *NLGN1*, *NLGN2*, *NLGN3*, *NLGN4X*, and *NLGN4Y* in the human genome, but only the X-linked genes *NLGN3* and *NLGN4X* are strongly associated with human disorders. The first report concerned a frame-shift mutation in the *NLGN4X* gene that was identified in two brothers, one with autism and the other with Asperger syndrome (a milder form of autism without language alteration) (Jamain et al. 2003). In the same study, a non-synonymous mutation (R451C) of the X-linked *NLGN3* gene, which affects a highly conserved amino acid in the esterase domain, was identified in a second family with two brothers, one with autism and the other with Asperger syndrome. These mutations were studied at the functional level and were found to alter the property of their neuroligin protein products to trigger synapse formation in cultured neuronal cells (Chih et al. 2004, Comoletti et al. 2004).

Several laboratories replicated the original finding and have identified independent *NLGN* point mutations (Laumonnier et al. 2004, Yan et al. 2004) or deletions (Chocholska et al. 2006, Macarov et al. 2007, Kent et al. 2008, Lawson-Yuen et al. 2008). Among these variations, one concerns the Y-linked gene *NLGN4Y*, an interesting member of the neuroligin family that is specific to primates and is only present in males (Yan et al. 2008a). In addition, a de novo mutation in the promoter of *NLGN4X* was shown to increase the transcript level of *NLGN4X* in one boy with ASC (Daoud et al. 2008). Finally, in a recent study focused on X-linked ichthyosis (XLI) (steroid sulfatase [STS] deficiency),

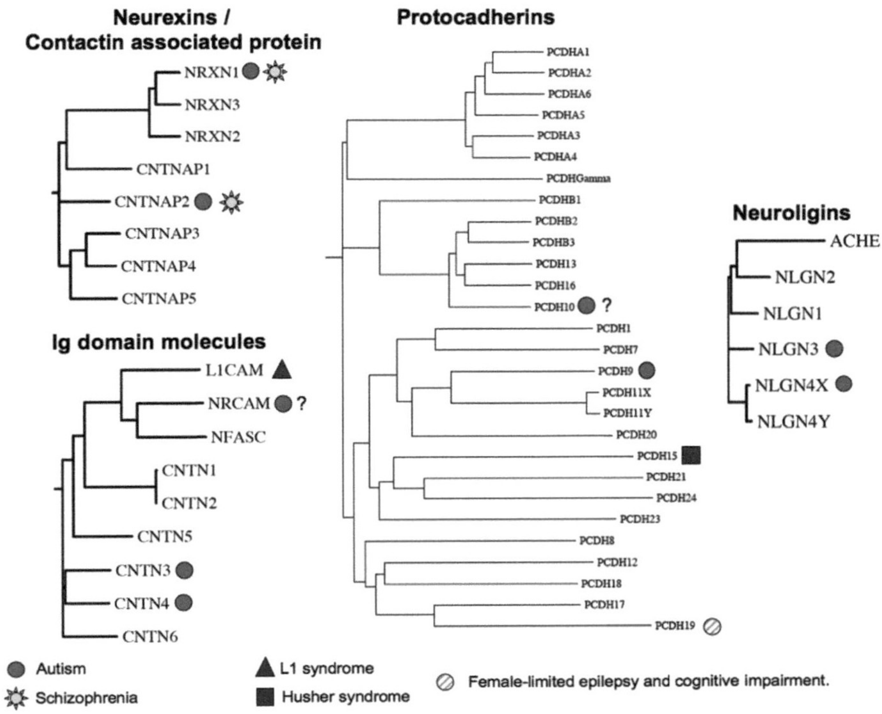

Fig. 6.2 Phylogeny of the human CAMs with their respective associated disorders. The question marks indicate genes proposed as susceptibility genes for autism, but with no formal proof for association. The phylogeny trees were obtained using human protein sequences and the neighbor-joining method implemented in the CLUTALW2 program (http://www.ebi.ac.uk/Tools/clustalw2/index.html)

five affected boys from three unrelated families had an unusually large deletion that encompass STS, the causative gene for XLI, and the *NLGN4X* gene (Kent et al. 2008). Remarkably, all these patients fulfilled criteria for an autism or a related language/communication difficulty. In contrast, none of the boys with a deletion or presumed point mutations of STS only demonstrated autistic difficulties. Besides mutations in the coding sequence, abnormal *NLGN3* and *NLGN4X* spliced isoforms were detected in blood cells from individuals with ASC (Talebizadeh et al. 2006). If these abnormal transcripts are actually present in the brain of the affected individuals, this finding may represent a new type of NLGN alteration in ASC patients.

Although independent mutations in *NLGN3* and *NLGN4X* were identified and functionally validated as susceptibility factors for ASC, these mutations may only concern a limited number of cases (<1% of the individuals) (Vincent et al. 2004, Gauthier et al. 2005, Ylisaukko-oja et al. 2005, Blasi et al. 2006). In addition, the specificity of the disorder associated with *NLGN4X* mutations can greatly differ from one individual to another, even when they belong to the same family and carry the same mutation. Indeed, mutations within *NLGN4X* were associated with mental retardation (Laumonnier et al. 2004), typical autism (Jamain et al. 2003, Yan et al. 2004), Asperger syndrome (Jamain et al. 2003), and more recently with Gilles de la Tourette syndrome (Macarov et al. 2007, Lawson-Yuen et al. 2008), a neurological disorder that is characterized by motor and vocal tics and behavioral anomalies. To date, only a single case of an *NLGN4X* deletion in a male with normal intelligence and apparently no autistic features has been reported (Macarov et al. 2007).

Following these initial findings, *Nlgn3* R451C knockin and *Nlgn4* knockout mice were generated. Both mutant mice display reduced social interactions when compared with wild-type mice (Tabuchi et al. 2007, Jamain et al. 2008a). The Nlgn3 R451C knockin mice also show increased GABAergic synapses and inhibitory postsynaptic currents (IPSC). Interestingly, *Nlgn4* KO mice exhibit reduced ultrasonic vocalizations (USV). This seems to be the consequence of a reduced motivation to "communicate" rather than a problem in the vocalization process per se. Thus, together with mutant mice for *Fmr1* (fragile X syndrome) and *Mecp2* (Rett syndrome), the neuroligin mutant mice represent the first animal models for ASC based on results obtained directly from human genetics studies.

A third gene in the neuroligin pathway, *SHANK3*, is associated with ASC, but does not code for a CAM. SHANK3 is a scaffolding protein of the postsynaptic density (PSD), which binds to NLGN, and is known to regulate the structural organization of dendritic spines (Boeckers et al. 2002, Meyer et al. 2004). The *SHANK3* gene is located on chromosome 22q13, a region deleted in several individuals with ASC (Manning et al. 2004, Durand and Bourgeron 2008). Mutations in or a loss of one copy of *SHANK3* are associated with autism, whereas the presence of an extra copy appears to be associated with Asperger syndrome or attention-deficit hyperactivity disorder (ADHD) (Durand et al. 2007, Moessner et al. 2007). Among the variations identified, a de novo frame-shift mutation that originated in a mother with germinal mosaicism has been found in two brothers with autism. Expression experiments of the rat *Shank3* cDNA carrying the frame-shift mutation in cultured neurons indicate that the truncated protein, in contrast to the full-length protein, is absent from the dendritic spines. Recently, several independent mutations and deletions of SHANK3 were identified in individuals with ASC (Moessner et al. 2007, Sebat et al. 2007, Gauthier et al. 2008), providing further evidences that the synaptic pathway, which includes *NLGN* and *SHANK3*, is associated with ASC.

Finally, the fourth ASC susceptibility protein which was identified within the neuroligin pathway is NRXN1, the presynaptic binding partner of NLGNs. *NRXN1* is located on chromosome 2p16 and codes for the long neurexin α and the short neurexin β form (Fig. 6.1). Furthermore, NRXN transcripts are extensively differentially spliced, leading in theory to more than 100 different isoforms (Tabuchi and Südhof 2002). In humans, the first *NRXN1* alteration was identified in a 7-year-old boy with an IQ of 74 and a complex psychiatric phenotype including attention-deficit disorder and Asperger syndrome (Friedman et al. 2006, Zahir et al. 2008). The patient carried a de novo 320 kb deletion of *NRXN1* that removes the promoter and first exon of *NRXN1a*, but apparently does not affect *NRXN1b*. The patient also had dismorphic features, such as frontal bossing, anomalies in the anterior and posterior hairlines, and dorsal scoliosis with 13 ribs on left, bifid right second rib, hemivertebrae and fusions of T2, T3, T4, and fusion of L4 and L5. The second *NRXN1* genetic alteration was detected using a whole genome approach performed by the Autism Genome Project Consortium that investigated 1168 multiplex families for the presence of small genomic alterations called copy number variants (CNV) (Szatmari et al. 2007). This analysis detected a de novo deletion in the *NRXN1* gene, removing several exons from both the *NRXN1α* and *NRXN1β*, in two sisters with typical ASC. More recently, two cases with ASC harboring translocations within or near *NRXN1* were identified (Kim et al. 2008). The first one is a female with ASC and carrying a paternally inherited translocation that directly disrupts *NRXN1α* within intron 5. The second case is a male patient carrying a de novo translocation with a breakpoint at ~750 kb of the *NRXN1* gene.

Following these reports, a mutation screening of the *NRXN1* coding sequence was performed in two cohorts of ASD subjects (Kim et al. 2008, Yan et al. 2008b). A number of rare variants altering evolutionary conserved residues were identified. However, as for some translocations, these point mutations were inherited from healthy parents, indicating that other factors are also involved in producing ASC. In addition, like *NLGN4X*, the phenotypes associated with *NRXN1* mutations are not specific to ASC. Indeed, one deletion, which partially overlaps with the first deletion described in an ASC patient, was identified in two siblings with schizophrenia and in their asymptomatic mother (Kirov et al. 2008).

Only limited data are available for understanding the role of the NRXN–NLGN–SHANK pathway at synapses in the human brain, but studies using neuronal cell culture and animal models have provided crucial information. First, NLGNs and NRXNs enhance synapse formation in vitro (Scheiffele et al. 2000), but are not required for the generation of synapses in vivo (Varoqueaux et al. 2006). Indeed, results from knockout mice demonstrate that neither NLGNs nor NRXNs are required for the initial formation of synapses, but both are essential for synaptic function and the survival of the mouse. Therefore, NLGNs may not establish synapses, but may specify and validate them via an activity-dependent mechanism, with different neuroligins acting on distinct types of synapses. This model, proposed by Chubykin et al. (2007), reconciles the overexpression and KO phenotypes and suggests that NLGNs contribute to the activity-dependent formation of neural circuits (Chubykin et al. 2007) (see Chapter 17).

Second, NLGNs and NRXNs are also emerging as central organizing molecules for excitatory glutamatergic and inhibitory GABAergic synapses in the mammalian brain (Graf et al. 2004, Prange et al. 2004). NLGN1 and NLGN3 are specific to glutamatergic synapses, whereas NLGN2 is restricted to

GABAergic synapses. Whether NLGN4 and NLGN4Y are localized to specific synapses remains unknown. The selectivity for glutamatergic vs. GABAergic synapses is also conferred by alternative splicing of NRXN/NLGN (Chih et al. 2006, Comoletti et al. 2006). Consistent with this role in the establishment of a specific synapse, the mutant mice carrying the R451C *Nlgn3* mutation show an increased number of GABAergic synapses and inhibitory currents, suggesting that the R451C mutation is a gain-of-function mutation (Tabuchi et al. 2007). Although *NLGN2* mutations were not reported in ASC, mutant mice with enhanced expression of Nlgn2 display enlarged synaptic contact size and vesicle reserve pool in frontal cortex synapses and an overall reduction in the excitation/inhibition ratio (Hines et al. 2008). These animals also manifest a stereotypic jumping behavior, anxiety, impaired social interactions, and enhanced incidence of spike-wave discharges. This role of NRXN/NLGN for synaptic specificity is highly relevant to ASC since an imbalance between excitation and inhibition can lead to epilepsy, a disease observed in almost 25% of individuals with ASC.

Third, the scaffold formed by SHANK3 proteins at the postsynaptic density (PSD), which binds to the NLGN, is known to regulate the structural organization of dendritic spines (Roussignol et al. 2005). Shank proteins are a family of three members composed of Shank1, Shank2, and Shank3, which are crucial components of the postsynaptic density. Shank proteins link ionotropic and metabotropic glutamate receptor complexes to the cytoskeleton. Shank proteins and their binding partners are involved, in vitro, in regulating the size and shape of dendritic spines (Roussignol et al. 2005). Remarkably, mice carrying a null mutation of *Shank1* exhibit smaller dendritic spines, weaker synaptic transmission, increased anxiety-related behavior, and impaired contextual fear memory, but show enhanced spatial learning (Hung et al. 2008).

Taken together, these results strongly suggest that synapse specificity and maintenance have an important role in the susceptibility to ASC (Fig. 6.2). However, the complex genotype–phenotype relationships indicate that other factors modulate the expression and the severity of the syndrome. Among other synaptic genes that could play such a modifier role, CAMs belonging to the contactin (CNTN) and the contactin-associated protein (CNTNAP) families represent very promising candidates.

6.3 Contactin and Contactin-Associated Proteins

The human contactin family is a part of the immunoglobulin superfamily and consists of six members: CNTN1 (contactin or F3), CNTN2 (TAG-1), CNTN3 (BIG-1), CNTN4 (BIG-2), CNTN5 (NB-2), and CNTN6 (NB-3). CNTNs are characterized by six Ig-like and four fibronectin type III (FN III)-like domains followed by a glycosylphosphatidyl inositol (GPI) moiety at the COOH terminal (Fig. 6.1). In mice, contactins act in different processes, such as axonal and dendritic interactions as well as in the organization of the nodes of Ranvier.

CNTN1 and CNTN2 are essential for the molecular organization of paranodes and juxtaparanodes of myelinated fibers, respectively (Boyle et al. 2001, Traka et al. 2003). Mice deficient for CNTN1 display a severe ataxic phenotype, which is consistent with defects in the cerebellum and survive only until postnatal day 18 (Boyle et al. 2001). CNTN2 is required for proper neuronal migration of the precerebellar nuclei (Denaxa et al. 2005). CNTN4-deficient mice are less affected and display aberrant projection of axons from olfactory sensory neurons to multiple glomeruli (Kaneko-Goto et al. 2008), whereas CNTN5-deficient mice show aberrant responses to acoustic stimuli (Li et al. 2003). Finally, CNTN6 deficiency leads to impaired motor coordination (Takeda et al. 2003).

In humans, so far only alterations of *CNTN3* and *CNTN4* have been reported. *CNTN4* is located in the region for 3p deletion syndrome, a rare contiguous-gene disorder characterized by developmental delay, growth retardation, and dysmorphic features. In 2004, Fernandez et al. identified a child with the characteristic physical features of 3p deletion syndrome, who carried a de novo balanced translocation that disrupts *CNTN4* (Fernandez et al. 2004). Interestingly, this patient, who exhibited both verbal and nonverbal developmental delays, was reevaluated in 2008 and was found to fulfill all criteria for ASC (Fernandez et al. 2008). Two additional families were reported (Roohi et al. 2008). One family included three affected siblings with ASC: two of them had a paternally inherited *CNTN4* deletion, but one affected brother did not have the deletion. In the second family, the affected boy carried a paternally inherited duplication. Finally, a homozygous deletion of approximately 500 kb from the *CNTN3* gene was recently identified in one child with ASC born from consanguineous parents (Morrow et al. 2008).

Although genetic data point at CNTN3/CNTN4 as a susceptibility gene for ASC, the lack of straight segregation between the genetic alterations and the disorder indicates that a single CNTN mutation is not enough to produce ASC. The strongest support for a role of the CNTN pathway in the susceptibility to ASC came from the study of CNTNAP2 (Caspr2), a binding partner of CNTN that possesses strong homology to NRXN (Fig. 6.1) and is associated with ASC in independent samples of patients (Strauss et al. 2006, Bakkaloglu et al. 2008, Alarcon et al. 2008, Arking et al. 2008).

CNTNAP2 mutations were originally identified in Amish individuals, who were diagnosed with recessive cortical dysplasia-focal epilepsy syndrome (CDFE) and language regression (Strauss et al. 2006). Notably, two-thirds of the affected individuals also met criteria for ASC. Especially when the carboxy-terminal of the CTNAP2 protein is truncated, *CTNAP2* mutations are associated with severe autism with medication-insensitive temporal lobe seizures, language regression, and low IQ. Following this initial report, several independent studies showed that chromosomal alterations and rare single base pair mutations, as well as common variation in *CNTNAP2*, can contribute to ASC risk. First, Bakkaloglu et al. reported a de novo 7q35 inversion that disrupts *CNTNAP2* in a child with autistic features (Bakkaloglu et al. 2008). This finding led them to sequence all 24 exons of the gene in 635 affected and

942 control individuals. Thirteen rare variants were identified in affected cases and of these eight were predicted to be deleterious. Although these variants were also found in the control cohort, they were roughly twice more frequent in affected than in control individuals. One variant (I869T) was found to be present in four affected individuals from three different families, but was not present in >4000 chromosomes from unaffected individuals.

Alarcón et al. genotyped 172 parent–child trios at 2758 single nucleotide polymorphisms (SNPs) across a 10 cM linkage peak at 7q35 (Alarcon et al. 2008). Only SNP rs2710102, which is located in *CNTNAP2*, showed a significant association with ASC, but only in male probands. This association was specifically related to the delay in the "age of first word". In a child with ASC the authors also reported a paternally inherited microdeletion within an intron of *CNTNAP2* that was not seen in 1000 control chromosomes. Finally, another common SNP variant (rs7794745) within the *CNTNAP2* gene was identified by association mapping with strictly defined autism cases. It was also confirmed in an independent replication population (with broader diagnostic inclusion criteria) (Arking et al. 2008).

Similar to *NLGN4X* and *NRXN1*, *CNTNAP2* alterations are not specific to ASC, but are also associated with other disorders such as epilepsy and schizophrenia (Friedman et al. 2008) and Gilles de la Tourette syndrome (Verkerk et al. 2003). Moreover, a translocation disrupting *CNTNAP2* was reported for several members of a family with multiple congenital malformations (scoliosis, single kidney, hearing loss, ptosis, and vision loss because of cataract and affected optic nerve), severe mental retardation, and an absence of language development (Belloso et al. 2007). Remarkably, in the same family, several individuals carrying the *CNTNAP2* translocation were phenotypically normal.

The association between the *CNTN/CNTNAP* pathway and ASC is especially interesting as it provides new information on the possible mechanisms leading to ASC. CNTN4 had a neurite outgrowth-promoting activity when used as a substrate for mouse neurons in vitro. In vivo, *Cntn4* expression is regulated by neural activity and its product acts as an axon guidance molecule that mediates proper neuronal wiring in the mouse olfactory system (Kaneko-Goto et al. 2008). It is also highly expressed in the cerebellum and in the CA1 pyramidal cells in the hippocampus (Yoshihara et al. 1995). Thus, it is likely that CNTN4 plays functional roles in the formation and maintenance of neural circuits in these regions. Cntap2 (Caspr2), at least in mice, is differentially expressed in distinct neuronal structures, including the soma and dendrites, and in specific short-segmented pairs along myelinated axons. Its expression in myelinated nerves is mostly confined to the axon at the juxtaparanodal region and to some isolated paranodal loops. In the juxtaparanodal region, Cntnap2 is precisely colocalized with Shaker-like potassium channels. The complete loss of Cntnap2 eliminates spatial clustering of axonal inwardly rectifying potassium channels, but does not result in overt cortical dysplasia or spontaneous seizures (Poliak et al. 2003). Therefore, patients carrying mutations in *CNTNAP2* might have slight to severe alterations of the attachment of the axon to the glia cell and mislocalization of ion channels at the juxtaparanodal junction leading to an alteration of neurotransmission velocity. However, in addition to their scaffolding roles at the nodes of Ranvier, CNTN/CNTNAP could also be involved in cortical histogenesis and may mediate intercellular interactions during latter phases of neuroblast migration, laminar organization, or axonal pathfinding. Finally, the possibility cannot be excluded that some members of the CNTN/CNTNAP family directly participate in synapse formation/validation in a similar way as the NLGN–NRXN–SHANK pathway. Interestingly, the expression pattern of human *CNTNAP2* greatly differs from its mouse ortholog (Abrahams et al. 2007). Indeed, human *CNTNAP2* is consistently expressed at high levels in the prefrontal and anterior temporal cortex, as well as in the dorsal thalamus, caudate, putamen, and amygdala in midgestation fetal brains. In contrast to the findings in humans, mouse Cntnap2 is broadly expressed in the developing rodent brain. Therefore, it was hypothesized that human *CNTNAP2* has acquired a new specific role in circuits that are involved in higher cortical functions, including language (Abrahams et al. 2007).

6.4 Cadherins and Protocadherins

In the human genome, the most compelling candidate genes for determining a part of neuronal identity are those belonging to the cadherin family (see Chapter 7). Cadherins (CDH) are single-pass transmembrane proteins, which are characterized by the presence of cadherin repeat protein domains in their extracellular segments. Each of these 110 amino acid repeats forms a β-sandwich and the presence of calcium is essential for cadherin-adhesive function. The human genome encodes more than 100 CDH, most of which are expressed in the nervous system. Cadherins have been classified into several subfamilies, including the classical cadherins, which are linked to the actin cytoskeleton through catenin, and the protocadherin family, which has a complex genomic organization with multiple variable exons and a set of constant exons resembling immunoglobulin (Ig) and T-cell receptor (TCR) genes (Hirayama and Yagi 2006). Several human disorders are caused by mutations in CDH/PCDH. These include Usher syndrome, EFMR (epilepsy and mental retardation limited to females) syndrome, and most likely ASC.

Usher syndrome (USH) is the most frequent cause of combined deafblindness in man (Roux et al. 2006). It is a clinically and genetically heterogeneous syndrome caused by at least eight identified USH genes. The two most prevalent mutated genes encode CAMs: the Cadherin 23 (CDH23) and the protocadherin 15 (PCDH15). Mutations in *CDH23* are linked to the recessive USH1D and to the recessive non-syndromic deafness DFN12. Mutations in *PCDH15* are responsible for the recessive USH1F and recessive deafness DFNB23. Analyses of the mice bearing mutations in the pcdh15 and Cdh23 gene revealed that these CAMs play important roles in membrane adhesion at specialized synapses of photoreceptor and inner hair cells. In the inner ear, CDH23 and PCDH15 interact to form tip-link filaments in sensory hair cells (Kazmierczak et al. 2007). In the retina, they are localized in the connecting cilium and the basal body complex, as well as the ribbon synapses of rod and cone photoreceptor cells. Most interestingly, by 5 months of age mice double heterozygous for *Cdh23* and *Pcdh15* mutations exhibit deafness and abnormal stereocilia in the outer and inner hair cells of the organ of corti. Single heterozygotes lack this pathology (Zheng et al. 2005). This digenic inheritance of a USH1 phenotype was also reported in three unrelated families with affected individuals carrying heterozygous mutations in both *CDH23* and *PCDH15*.

Another remarkable example of a human disorder that is caused by a CDH mutation is EFMR, an X-linked disorder, which is characterized by epilepsy and mental retardation and has a unique inheritance pattern (Dibbens et al. 2008). In general, disorders arising from mutations on the X chromosome typically affect males and have unaffected carrier females. In contrast, EFMR spares transmitting males and affects only carrier females. Recently, Dibbens et al. identified several independent *protocadherin 19* (*PCDH19*) gene mutations in seven families. Five mutations result in the introduction of a premature termination codon. A study of two of these demonstrated a nonsense codon-mediated decay of *PCDH19* mRNA. Two missense mutations are predicted to affect the adhesiveness of the PCDH19 protein by impairing calcium binding. *PCDH19* is expressed in human and mouse developing brain, but its actual role remains unknown. To explain the sex-limited mode of inheritance the authors proposed a mechanism of rescue in males by a Y chromosome gene, *PCDH11Y* (Durand et al. 2006). *PCDH11Y* originates from a translocation of the *PCDH11X* gene after the divergence between chimpanzees and humans; therefore, it is one of the few genes that are specific to the hominoid lineage and are absent from other primates.

However, an alternative hypothesis could explain the female-limited phenotype (Fig. 6.2). First, it should be noted that *PCDH19* is subjected to X inactivation and therefore a single neuron most likely do not express both alleles. The alternative hypothesis proposes that the disorder is the consequence of an "allelic incompatibility" between two different sets of neurons, one expressing the wild-type sequence and the other expressing the mutant sequence. Indeed, heterozygous-affected females have two sets of neurons, which either express the wild-type or the mutated PCDH19 gene. In contrast, unaffected males carrying a *PCDH19* mutation only express the mutated forms of PCDH19, and control males and females only express the wt sequence. Therefore, the presence of two distinct populations of neurons on the basis of PCDH19 expression may scramble specific neuronal identity during development. If this hypothesis is correct, female homozygous for the mutation should not be affected and males with an X chromosome mosaicism should be affected. However, as *PCDH19* mutations are very rare no such case has so far been described (Fig. 6.3).

Allelic incompatibility is a well-known cause of maternal–fetal incompatibility reactions and this concept may also apply to brain development/function at the neuronal level. This hypothesis is strengthened by the finding that even autosomal

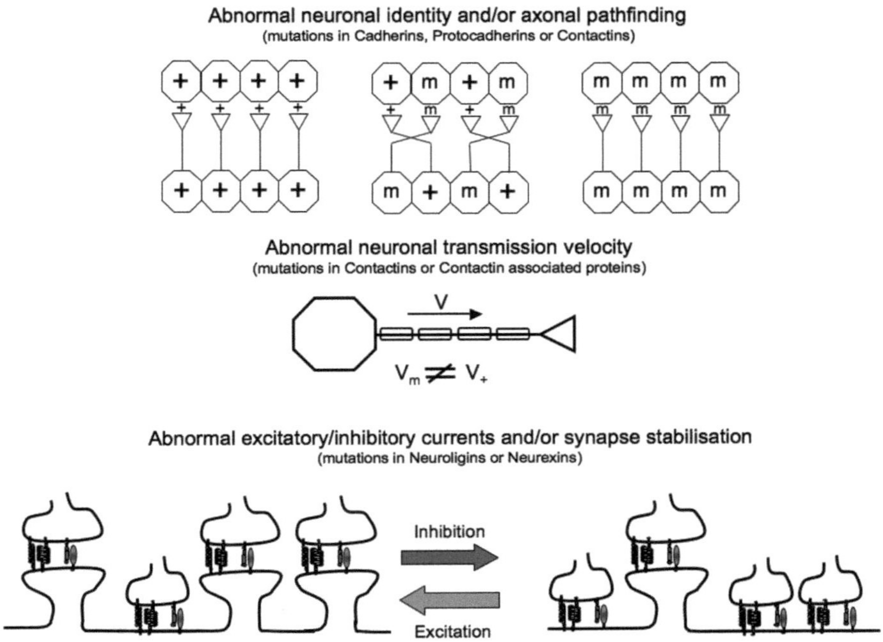

Fig. 6.3 Possible brain wiring alterations caused by CAM mutations in synaptopathies. Mutations in cadherins, protocadherins, or contactins could lead to abnormal brain wiring by scrambling neuronal identity and/or axonal pathfinding cues. In heterozygote individuals (m/+), the presence of two different sets of neurons could have a more drastic effect than in homozygote individuals who harbor in both cases (+/+ or m/m) a homogenous population of neurons. Mutations in the contactin and contactin-associated proteins could lead to uncoordinated information processing by modifying neuronal transmission velocity (V) in different parts of the brain. Mutations in neuroligins and neurexins could lead to abnormal information processing by modifying the excitation/inhibition ratio and/or synapse stabilization. m, mutated allele; +, control allele

PCDH genes are monoallelically expressed in single neurons (Esumi et al. 2005). Therefore, this huge diversity of PCDH expression (Hilschmann et al. 2001) combined with a modulation between different allelic interactions could increase the number of specific individual neuronal cell identities and could play a crucial role in establishing the blueprint for the neuronal networks in each individual human.

6.5 CAMs Polymorphisms and the Susceptibility to Psychiatric Conditions

In the previous sections, we mainly focused on CAM mutations, which cause monogenic inheritable conditions. However, the sequence of the human genome revealed that a relatively large number of genetic variations exist between individuals. The most frequent and studied variations concern SNPs and

CNVs. The Hapmap consortium explored human genome diversity and has identified more than 3 million SNPs that are distributed among the human population (Frazer et al. 2007). The actual number of reported CNVs (>1500) is less clear as such small structural variants are sometimes difficult to detect (Redon et al. 2006).

It is still difficult to evaluate the contribution of these polymorphisms to human disorders. Concerning CAMs, their different allelic forms among individuals may play a key role in the susceptibility to psychiatric disorders since they increase the number of distinct CAM interactions during axonal pathfinding and synapse formation. Thus, in theory, CAM polymorphisms may cause slight differences in the wiring of each individual brain. Several CAM polymorphisms were tested for correlating with a susceptibility to psychiatric conditions. For example, *NrCAM*, *PCDH8*, and *PCDH11Y* were studied in patients with ASC and schizophrenia (Bray et al. 2002, Hutcheson et al. 2004, Bonora et al. 2005, Durand et al. 2006, Sakurai et al. 2006). However, none of these studies has provided clear evidence to implicate any of these genes in these disorders. In parallel, whole genome analyses (WGA) represent powerful methodologies to detect susceptibility genes for human disorders. These unbiased approaches sometimes detect signals within CAMs. Interestingly, two independent studies detected SNPs within the *NRXN1* gene in strong association with nicotine dependence (Nussbaum et al. 2008) and within *NRXN3* associated with susceptibility for polysubstance addiction (Hishimoto et al. 2007). Although these WGA studies need to be replicated and validated at the functional levels, they will most likely shed light on new pathways that contribute to the susceptibility for specific human brain disorders.

6.6 Conclusions and Perspectives

Several considerations can be made from the examples presented in this chapter. First, mutations in CAMs can obviously affect the wiring and/or the plasticity of synapses in the human nervous system. However, it is not clear whether the disorder is the consequence of an abnormal wiring or due to an alteration in synaptic plasticity or to both processes. The second consideration concerns the very complex genotype–phenotype relationship, which has been observed for CAM mutations. Indeed, the specificity and the severity of the disease conditions that are associated with brain CAM mutations remain very difficult to predict. An additional example of this diversity is illustrated by the mutations in the X-linked *L1CAM* gene that are associated with at least four different syndromes: HSAS (X-linked hydrocephalus with stenosis of the aqueduct of Sylvius), MASA syndrome (mental retardation, aphasia (delayed speech), spastic paraplegia, adducted thumbs), SPG1 (X-linked complicated hereditary spastic paraplegia type 1), and X-linked complicated corpus callosum agenesis. These disorders are now collectively referred to as L1 syndrome.

This complex genotype–phenotype relationship is most likely due to several intrinsic properties of the CAMs: (1) their huge diversity within the human genome; (2) their close relationship with neuronal activities and plasticity; and (3) the modifying roles of CAM polymorphisms in the establishment of the neuronal networks. Therefore, the severity and the specificity of the disorders may be the consequence of a combination of mutations in different CAMs, as illustrated by the CDH23 and PCDH15 digenic inheritance of Usher syndrome (Zheng et al. 2005).

Only a global genetic and functional analysis of all CAMs may enable us to better predict the consequence of CAM mutations in humans. These investigations are warranted since mutations in brain CAMs are obviously associated with an increasing number of human disorders. Indeed, several high-throughput sequencing projects are currently in progress to discover new genes that are associated with a susceptibility to psychiatric disorders, such as ASC, schizophrenia, or bipolar disorders. A number of synaptic CAMs, such as SynCAMs, DsCAMs, Sidekicks, have not been associated with a human disorder. However, the discovery of such associations is probably just a question of time.

Acknowledgments I would like to thank Michael Hortsch and Hisashi Umemori for their critical reading of this chapter. This work was supported by the Pasteur Institute, INSERM, Assistance Publique-Hôpitaux de Paris, FP6 TISM MOLGEN, FP6 ENI-NET, Fondation Orange, Fondation de France, Fondation biomédicale de la Mairie de Paris, Fondation pour la Recherche Médicale and FondaMental.

References

Abrahams BS, Tentler D, Perederiy JV et al. (2007) Genome-wide analyses of human perisylvian cerebral cortical patterning. Proc Natl Acad Sci USA 104:17849–17854

Alarcon M, Abrahams BS, Stone JL et al. (2008) Linkage, association, and gene-expression analyses identify CNTNAP2 as an autism-susceptibility gene. Am J Hum Genet 82:150–159

Arking DE, Cutler DJ, Brune CW et al. (2008) A common genetic variant in the neurexin superfamily member CNTNAP2 increases familial risk of autism. Am J Hum Genet 82:160–164

Bakkaloglu B, O'oak BJ, Louvi A et al. (2008) Molecular cytogenetic analysis and resequencing of contactin associated protein-like 2 in autism spectrum disorders. Am J Hum Genet 82:165–173

Belloso JM, Bache I, Guitart M et al. (2007) Disruption of the CNTNAP2 gene in a t(7;15) translocation family without symptoms of Gilles de la Tourette syndrome. Eur J Hum Genet 15:711–713

Blasi F, Bacchelli E, Pesaresi G et al. (2006) Absence of coding mutations in the X-linked genes neuroligin 3 and neuroligin 4 in individuals with autism from the IMGSAC collection. Am J Med Genet B Neuropsychiatr Genet 141:220–221

Boeckers TM, Bockmann J, Kreutz MR et al. (2002) ProSAP/Shank proteins – a family of higher order organizing molecules of the postsynaptic density with an emerging role in human neurological disease. J Neurochem 81:903–910

Bonora E, Lamb JA, Barnby G et al. (2005) Mutation screening and association analysis of six candidate genes for autism on chromosome 7q. Eur J Hum Genet 13:198–207

Boyle ME, Berglund EO, Murai KK et al. (2001) Contactin orchestrates assembly of the septate-like junctions at the paranode in myelinated peripheral nerve. Neuron 30:385–397
Bray NJ, Kirov G, Owen RJ et al. (2002) Screening the human protocadherin 8 (PCDH8) gene in schizophrenia. Genes Brain Behav 1:187–191
Chih B, Afridi SK, Clark L et al. (2004) Disorder-associated mutations lead to functional inactivation of neuroligins. Hum Mol Genet 13:1471–1477
Chih B, Gollan L and Scheiffele P (2006) Alternative splicing controls selective trans-synaptic interactions of the neuroligin-neurexin complex. Neuron 51:171–178
Chocholska S, Rossier E, Barbi G et al. (2006) Molecular cytogenetic analysis of a familial interstitial deletion Xp22.2–22.3 with a highly variable phenotype in female carriers. Am J Med Genet A 140:604–610
Chubykin AA, Atasoy D, Etherton MR et al. (2007) Activity-Dependent Validation of Excitatory versus Inhibitory Synapses by Neuroligin-1 versus Neuroligin-2. Neuron 54:919–931
Comoletti D, De Jaco A, Jennings LL et al. (2004) The Arg451Cys-neuroligin-3 mutation associated with autism reveals a defect in protein processing. J Neurosci 24:4889–4893
Comoletti D, Flynn RE, Boucard AA et al. (2006) Gene selection, alternative splicing, and post-translational processing regulate neuroligin selectivity for beta-neurexins. Biochemistry 45:12816–12827
Craig AM and Kang Y (2007) Neurexin-neuroligin signaling in synapse development. Curr Opin Neurobiol 17:43–52
Dalva MB, McClelland AC and Kayser MS (2007) Cell adhesion molecules: signalling functions at the synapse. Nature Rev 8:206–220
Daoud H, Bonnet-Brilhault F, Védrine S et al. (2008) NLGN4X Gene overexpression is associated with autism and profound mental retardation. Poster at the International Meeting for Autism Research (IMFAR), London, 2008
Denaxa M, Kyriakopoulou K, Theodorakis K et al. (2005) The adhesion molecule TAG-1 is required for proper migration of the superficial migratory stream in the medulla but not of cortical interneurons. Dev Biol 288:87–99
Dibbens LM, Tarpey PS, Hynes K et al. (2008) X-linked protocadherin 19 mutations cause female-limited epilepsy and cognitive impairment. Nature Genet 40:776–781
Durand CM, Betancur C, Boeckers TM et al. (2007) Mutations in the gene encoding the synaptic scaffolding protein SHANK3 are associated with autism spectrum disorders. Nature Genet 39:25–27
Durand CM and Bourgeron T (2008) Genetic, neurobiological and clinical findings related to SHANK3 mutations and 22q13 chromosomal rearrangements in autism spectrum disorders. Eur J Psychiatry 1:58–61
Durand CM, Kappeler C, Betancur C et al. (2006) Expression and genetic variability of PCDH11Y, a gene specific to Homo sapiens and candidate for susceptibility to psychiatric disorders. Am J Med Genet B Neuropsychiatr Genet 141:67–70
Esumi S, Kakazu N, Taguchi Y et al. (2005) Monoallelic yet combinatorial expression of variable exons of the protocadherin-alpha gene cluster in single neurons. Nature Genet 37:171–176
Fernandez T, Morgan T, Davis N et al. (2004) Disruption of contactin 4 (CNTN4) results in developmental delay and other features of 3p deletion syndrome. Am J Hum Genet 74:1286–1293
Fernandez T, Morgan T, Davis N et al. (2008) Disruption of Contactin 4 (CNTN4) results in developmental delay and other features of 3p deletion syndrome. Am J Hum Genet 82:1385
Frazer KA, Ballinger DG, Cox DR et al. (2007) A second generation human haplotype map of over 3.1 million SNPs. Nature 449:851–861
Freitag CM (2007) The genetics of autistic disorders and its clinical relevance: a review of the literature. Mol Psychiatry 12:2–22

Friedman JI, Vrijenhoek T, Markx S et al. (2008) CNTNAP2 gene dosage variation is associated with schizophrenia and epilepsy. Mol Psychiatry 13:261–266
Friedman JM, Baross A, Delaney AD et al. (2006) Oligonucleotide microarray analysis of genomic imbalance in children with mental retardation. Am J Hum Genet 79:500–513
Gauthier J, Bonnel A, St-Onge J et al. (2005) NLGN3/NLGN4 gene mutations are not responsible for autism in the Quebec population. Am J Med Genet B Neuropsychiatr Genet 132:74–75
Gauthier J, Spiegelman D, Piton A et al. (2008) Novel de novo SHANK3 mutation in autistic patients. Am J Med Genet B Neuropsychiatr Genet 150B(3):421–424
Graf ER, Zhang X, Jin SX et al. (2004) Neurexins induce differentiation of GABA and glutamate postsynaptic specializations via neuroligins. Cell 119:1013–1026
Hilschmann N, Barnikol HU, Barnikol-Watanabe S et al. (2001) The immunoglobulin-like genetic predetermination of the brain: the protocadherins, blueprint of the neuronal network. Die Naturwissenschaften 88:2–12
Hines RM, Wu L, Hines DJ et al. (2008) Synaptic imbalance, stereotypies, and impaired social interactions in mice with altered neuroligin 2 expression. J Neurosci 28:6055–6067
Hirayama T and Yagi T (2006) The role and expression of the protocadherin-alpha clusters in the CNS. Curr Opin Neurobiol 16:336–342
Hishimoto A, Liu QR, Drgon T et al. (2007) Neurexin 3 polymorphisms are associated with alcohol dependence and altered expression of specific isoforms. Hum Mol Genet 16:2880–2891
Hung AY, Futai K, Sala C et al. (2008) Smaller dendritic spines, weaker synaptic transmission, but enhanced spatial learning in mice lacking Shank1. J Neurosci 28:1697–1708
Hutcheson HB, Olson LM, Bradford Y et al. (2004) Examination of NRCAM, LRRN3, KIAA0716, and LAMB1 as autism candidate genes. BMC Med Genet 5:12
Huttenlocher PR and Dabholkar AS (1997) Regional differences in synaptogenesis in human cerebral cortex. J Comp Neurol 387:167–178
Jamain S, Quach H, Betancur C et al. (2003) Mutations of the X-linked genes encoding neuroligins NLGN3 and NLGN4 are associated with autism. Nature Genet 34:27–29
Jamain S, Radyushkin K, Hammerschmidt K et al. (2008a) Reduced social interaction and ultrasonic communication in a mouse model of monogenic heritable autism. Proc Natl Acad Sci USA 105:1710–1715
Jamain S, Radyushkin K, Hammerschmidt K et al. (2008b) Reduced social interaction and ultrasonic communication in a mouse model of monogenic heritable autism. Proc Natl Acad Sci USA 105:1710–1715
Kaneko-Goto T, Yoshihara S, Miyazaki H et al. (2008) BIG-2 mediates olfactory axon convergence to target glomeruli. Neuron 57:834–846
Kazmierczak P, Sakaguchi H, Tokita J et al. (2007) Cadherin 23 and protocadherin 15 interact to form tip-link filaments in sensory hair cells. Nature 449:87–91
Kent L, Emerton J, Bhadravathi V et al. (2008) X linked ichthyosis (steroid sulphatase deficiency) is associated with increased risk of attention deficit hyperactivity disorder, autism and social communication deficits. J Med Genet 45:519–524
Kim HG, Kishikawa S, Higgins AW et al. (2008) Disruption of neurexin 1 associated with autism spectrum disorder. Am J Hum Genet 82:199–207
Kirov G, Gumus D, Chen W et al. (2008) Comparative genome hybridization suggests a role for NRXN1 and APBA2 in schizophrenia. Hum Mol Genet 17:458–465
Laumonnier F, Bonnet-Brilhault F, Gomot M et al. (2004) X-linked mental retardation and autism are associated with a mutation in the NLGN4 gene, a member of the neuroligin family. Am J Hum Genet 74:552–557
Lawson-Yuen A, Saldivar JS, Sommer S et al. (2008) Familial deletion within NLGN4 associated with autism and Tourette syndrome. Eur J Hum Genet 16:614–618
Li CY, Liu QR, Zhang PW et al. (2008) OKCAM: an ontology-based, human-centered knowledgebase for cell adhesion molecules. Nucleic Acids Res 37 (Database issue) :D251–60

Li H, Takeda Y, Niki H et al. (2003) Aberrant responses to acoustic stimuli in mice deficient for neural recognition molecule NB-2. Euro J Neurosci 17:929–936
Macarov M, Zeigler M, Newman JP et al. (2007) Deletions of VCX-A and NLGN4: a variable phenotype including normal intellect. J Intellect Disabil Res 51:329–333
Manning MA, Cassidy SB, Clericuzio C et al. (2004) Terminal 22q deletion syndrome: a newly recognized cause of speech and language disability in the autism spectrum. Pediatrics 114:451–457
Mattson MP and van Praag H (2008) TAGing APP constrains neurogenesis. Nature Cell Biol 10:249–250
Meyer G, Varoqueaux F, Neeb A et al. (2004) The complexity of PDZ domain-mediated interactions at glutamatergic synapses: a case study on neuroligin. Neuropharmacology 47:724–733
Moessner R, Marshall CR, Sutcliffe JS et al. (2007) Contribution of SHANK3 mutations to autism spectrum disorder. Am J Hum Genet 81:1289–1297
Morrow EM, Yoo SY, Flavell SW et al. (2008) Identifying autism loci and genes by tracing recent shared ancestry. Science (New York, NY) 321:218–223
Nussbaum J, Xu Q, Payne TJ et al. (2008) Significant association of the neurexin-1 gene (NRXN1) with nicotine dependence in European- and African-American smokers. Hum Mol Genet 17:1569–1577
Poliak S, Salomon D, Elhanany H et al. (2003) Juxtaparanodal clustering of Shaker-like K + channels in myelinated axons depends on Caspr2 and TAG-1. J Cell Biol 162:1149–1160
Prange O, Wong TP, Gerrow K et al. (2004) A balance between excitatory and inhibitory synapses is controlled by PSD-95 and neuroligin. Proc Natl Acad Sci USA 101:13915–13920
Redon R, Ishikawa S, Fitch KR et al. (2006) Global variation in copy number in the human genome. Nature 444:444–454
Roohi J, Montagna C, Tegay DH et al. (2008) Disruption of Contactin 4 in 3 Subjects with Autism Spectrum Disorder. J Med Genet 46:176–182
Roussignol G, Ango F, Romorini S et al. (2005) Shank expression is sufficient to induce functional dendritic spine synapses in aspiny neurons. J Neurosci 25:3560–3570
Roux AF, Faugere V, Le Guedard S et al. (2006) Survey of the frequency of USH1 gene mutations in a cohort of Usher patients shows the importance of cadherin 23 and protocadherin 15 genes and establishes a detection rate of above 90%. J Med Genet 43:763–768
Sakurai T, Ramoz N, Reichert JG et al. (2006) Association analysis of the NrCAM gene in autism and in subsets of families with severe obsessive-compulsive or self-stimulatory behaviors. Psychiatric Genet 16:251–257
Scheiffele P, Fan J, Choih J et al. (2000) Neuroligin expressed in nonneuronal cells triggers presynaptic development in contacting axons. Cell 101:657–669
Sebat J, Lakshmi B, Malhotra D et al. (2007) Strong association of de novo copy number mutations with autism. Science (New York, NY) 316:445–449
Shapiro L, Love J and Colman DR (2007) Adhesion molecules in the nervous system: structural insights into function and diversity. Ann Rev Neurosci 30:451–474
Strauss KA, Puffenberger EG, Huentelman MJ et al. (2006) Recessive symptomatic focal epilepsy and mutant contactin-associated protein-like 2. N Engl J Med 354:1370–1377
Szatmari P, Paterson AD, Zwaigenbaum L et al. (2007) Mapping autism risk loci using genetic linkage and chromosomal rearrangements. Nature Genet 39:319–328
Tabuchi K, Blundell J, Etherton MR et al. (2007) A neuroligin-3 mutation implicated in autism increases inhibitory synaptic transmission in mice. Science (New York, NY) 318:71–76
Tabuchi K and Südhof TC (2002) Structure and evolution of neurexin genes: insight into the mechanism of alternative splicing. Genomics 79:849–859
Takeda Y, Akasaka K, Lee S et al. (2003) Impaired motor coordination in mice lacking neural recognition molecule NB-3 of the contactin/F3 subgroup. J Neurobiol 56:252–265
Talebizadeh Z, Lam DY, Theodoro MF et al. (2006) Novel splice isoforms for NLGN3 and NLGN4 with possible implications in autism. J Med Genet 43:e21

Traka M, Goutebroze L, Denisenko N et al. (2003) Association of TAG-1 with Caspr2 is essential for the molecular organization of juxtaparanodal regions of myelinated fibers. J Cell Biol 162:1161–1172

Varoqueaux F, Aramuni G, Rawson RL et al. (2006) Neuroligins determine synapse maturation and function. Neuron 51:741–754

Verkerk AJ, Mathews CA, Joosse M et al. (2003) CNTNAP2 is disrupted in a family with Gilles de la Tourette syndrome and obsessive compulsive disorder. Genomics 82:1–9

Vincent JB, Kolozsvari D, Roberts WS et al. (2004) Mutation screening of X-chromosomal neuroligin genes: no mutations in 196 autism probands. Am J Med Genet B Neuropsychiatr Genet 129B:82–84

Yan J, Feng J, Schroer R et al. (2008a) Analysis of the neuroligin 4Y gene in patients with autism. Psychiatric Genet 18:204–207

Yan J, Noltner K, Feng J et al. (2008b) Neurexin 1alpha structural variants associated with autism. Neurosci Lett 438:368–370

Yan J, Oliveira G, Coutinho A et al. (2004) Analysis of the neuroligin 3 and 4 genes in autism and other neuropsychiatric patients. Mol Psychiatry 10:329–332

Ylisaukko-oja T, Rehnstrom K, Auranen M et al. (2005) Analysis of four neuroligin genes as candidates for autism. Eur J Hum Genet 13:1285–1292

Yoshihara Y, Kawasaki M, Tamada A et al. (1995) Overlapping and differential expression of BIG-2, BIG-1, TAG-1, and F3: four members of an axon-associated cell adhesion molecule subgroup of the immunoglobulin superfamily. J Neurobiol 28:51–69

Zahir FR, Baross A, Delaney AD et al. (2008) A patient with vertebral, cognitive and behavioural abnormalities and a de novo deletion of NRXN1alpha. J Med Genet 45:239–243

Zheng QY, Yan D, Ouyang XM et al. (2005) Digenic inheritance of deafness caused by mutations in genes encoding cadherin 23 and protocadherin 15 in mice and humans. Hum Mol Genet 14:103–111

Chapter 7
The Cadherin Superfamily in Synapse Formation and Function

Andrew M. Garrett, Dietmar Schreiner, and Joshua A. Weiner

Abstract The cadherin superfamily of adhesion molecules encompasses over 100 members, which can be subdivided into classical cadherins, desmosomal cadherins, and protocadherins. Many classical cadherins and protocadherins are expressed in the central nervous system (CNS), where they have been implicated in a wide variety of processes. The ~20 classical cadherins have 5 characteristic cadherin repeats and an intracellular domain that contains binding sites for members of the catenin family. The classical cadherins are important for a number of processes integral to the synapse, including axon targeting, dendrite and dendritic spine maturation, and synapse function and plasticity. The more than 70 protocadherins (Pcdhs) make up the largest subgroup of the cadherin superfamily, which includes the clustered α-, β-, and γ-Pcdhs, the δ-Pcdhs, the seven-transmembrane (7-TM) Pcdhs, and the giant fat Pcdhs. These diverse molecules, all of which contain a varying number of cadherin repeats, play a multitude of roles in synapse development and function. Of the clustered Pcdhs, the γ-Pcdhs are required for synapse development and interneuron survival in the spinal cord, while the α-Pcdhs influence axon targeting, as do the 7-TM Pcdhs. Several of the δ-Pcdhs and fat Pcdhs have demonstrated and/or suggested functions at the synapse. This chapter reviews the vast extant literature on the cadherin superfamily, summarizing key findings regarding its diverse members and their functions at central synapses. Particular attention is paid to the most recent results, including the fascinating demonstration of cooperation between synaptic classical cadherins and Pcdhs.

Keywords Cadherin · Catenin · Protocadherin · Synapse · Adhesion

---

J.A. Weiner (✉)
Department of Biology, Graduate Program in Neuroscience, The University of Iowa, Iowa City, IA 52242, USA
e-mail: joshua-weiner@uiowa.edu

M. Hortsch, H. Umemori (eds.), *The Sticky Synapse*,
DOI 10.1007/978-0-387-92708-4_7, © Springer Science+Business Media, LLC 2009

7.1 Introduction

The cadherin superfamily of adhesion molecules encompasses over 100 members, which can be subdivided into classical cadherins, desmosomal cadherins, and protocadherins (Morishita and Yagi 2007, Takeichi 2007). Many classical cadherins and protocadherins are expressed in the central nervous system (CNS), where they have been implicated in a wide variety of processes, including morphogenesis; axon pathfinding; and the formation, maturation, and plasticity of synapses. As befits the importance of the cadherin superfamily to CNS function, there have been many excellent recent reviews of the vast literature on these molecules (Salinas and Price 2005, Anderson and Benson 2006, Bruses 2006, Morishita and Yagi 2007, Takeichi 2007, Arikkath and Reichardt 2008, Suzuki and Takeichi 2008, Tai et al. 2008, Yagi 2008). Here, we focus primarily on the roles played by classical cadherins and protocadherins in synapse formation and function, paying particular attention to the most recent and significant advances in our understanding. Most of the studies discussed below are summarized for convenience in Tables 7.1, 7.2, and 7.3.

The sine qua non of the cadherin superfamily is the presence of a variable number of repetitive subdomains, termed cadherin repeats, in the extracellular protein domain (Fig. 7.1). These cadherin repeats contain sites that mediate calcium-dependent cell–cell adhesion. The ~20 classical cadherins have 5

Table 7.1 N-cadherin

Process	Summary	Model system	Source
LTP/LTD	Function-blocking antibodies decrease LTP	Rat hippocampal slices	Tang et al. (1998)
	Expression of β-catenin mutants that cannot regulate NCad endocytosis inhibits NMDA-dependent LTD	Hippocampal cultures	Tai et al. (2007)
Targeting	Function-blocking antibodies against NCad disrupt laminar axon targeting	Chick Slice culture	Inoue and Sanes (1997) and Poskanzer et al. (2003)
	Genetic mutations of CadN disrupt targeting of photoreceptor projections to lamina and the medulla and the targeting of lamina projections to the medulla	*Drosophila*	Lee et al. (2001), Iwai et al. (2002) and Nern et al. (2008)
	Dendrites from CadN knockout olfactory projection neurons do not innervate a single glomerulus	*Drosophila*	Zhu and Luo (2004)
Spines and dendrites	Overexpression of NCad increased dendritic arborization	Hippocampal cultures	Yu and Malenka (2003)
	siRNA directed against NCad caused a reduction in spine density if introduced early in culture development (6 DIV) but not if introduced late (14 DIV)	Hippocampal cultures	Saglietti et al. (2007)

Table 7.1 (continued)

Process	Summary	Model system	Source
Synapse development	NCad is at early synaptic sites, and synapse development coincides with increased NCad clustering	Hippocampal cultures Hippocampal slices *Drosophila* Chick Rat	Benson and Tanaka (1998), Bozdagi et al. (2000), Prakash et al. (2005), Rubio et al. (2005) and Huntley and Benson (1999)
	In telencephalon-specific NCad knockout mice, in NCad mutant zebrafish retina, and in cultures made from NCad knockout tissue, synapse development occurs	Mouse, hippocampal cultures Zebrafish ES cell-derived neuron cultures	Kadowaki et al. (2007), Erdmann et al. (2003) and Jüngling et al. (2006)
Vesicle release	Neurons lacking NCad (genetic deletion) have reduced presynaptic activity under high stimulation and a fast decrease in vesicles available for release, as well as reduced frequency of mEPSCs (siRNA knockdown)	ES cell-derived neuron cultures siRNA in Hippocampal cultures	Jüngling et al. (2006) and Saglietti et al. (2007)
Response to activity	Activity stimulates NCad accumulation at synapses	Hippocampal slices	Bozdagi et al. (2000)
	NMDAR activity triggers β-catenin enrichment at spines, increasing association with NCad to inhibit NCad internalization	Hippocampal cultures	Tai et al. (2007)
	NMDAR activity stimulates NCad dimerization	Hippocampal cultures	Tanaka et al. (2000)
	NCad dimerizes with arcadlin, triggering internalization	Hippocampal cultures	Yasuda et al. (2007)

Table 7.2 Other classical cadherins/catenins

Process	Summary	Model system	Source
LTP/LTD	There is increased CA1 LTP in Cad11 knockout mice	Mouse	Manabe (2000)
Spines and dendrites	Disrupting signaling between αN- or β-catenin and cadherins caused spines to become more motile and immature and reduced synaptic activity. Catenin overexpression increased spine density and dendritic field size	Hippocampal cultures • αN-catenin deletion • Floxed β-catenin mutant • Dominant-negative NCad • Catenin overexpression	Abe et al. (2004), Okuda et al. (2007), Togashi et al. (2002) and Yu and Malenka (2003)
	Dominant-negative NCad expression in horizontal cells in the retina resulted in smaller dendritic fields with fewer synaptic puncta	Chick	Tanabe et al. (2006)
	Deleting p120-catenin using a floxed allele crossed to EMX1-Cre results in reduced dendritic tree complexity and spine and synapse number in vivo and in vitro	Mouse, hippocampal neurons	Elia et al. (2006)

Table 7.2 (continued)

Process	Summary	Model system	Source
Synapse formation	Transfection of a dominant-negative NCad or RNAi knockdown of Cad11 or Cad13 significantly reduced the density of synaptic puncta	Hippocampal cultures • Dominant-negative • RNAi	Togashi et al. (2002) and Paradis et al. (2007)
Synapse function	Transfection of a dominant-negative NCad resulted in smaller, less differentiated synaptic contacts with reduced synaptic vesicles cycling	Hippocampal cultures	Bozdagi et al. (2004)
	When β-catenin was deleted from the hippocampus by crossing a mouse harboring a floxed β-catenin allele to a CamKII Cre line, there was an increase in synapse number identified by TEM, but a decrease in the number of undocked vesicles per synapse and a faster decline of the reserve pool of synaptic vesicles upon stimulation	Mouse	Bamji et al. (2003)
	Cad8 interactions between DRG sensory neurons and interneurons in spinal cord laminae I and IIo are not required for synapse formation, but are required for the normal function of the circuit (Cad8 genetic deletion mice)	Mouse	Suzuki et al. (2007)
	Deletion of β-catenin in postsynaptic neurons reduced mEPSC amplitude and caused defects in quantal scaling	Hippocampal cultures	Okuda et al. (2007)

characteristic cadherin repeats and an intracellular domain that contains binding sites for members of the catenin family (Kemler 1993, Aberle et al. 1994, Lilien and Balsamo 2005, Kwiatkowski et al. 2007). Both β-catenin and δ-catenin/p120 catenin bind directly to distinct sites on the cadherin cytoplasmic domain and engage in a wide variety of protein–protein interactions to transduce the adhesive signal. β-Catenin interacts with α-catenin, a major regulator of the actin cytoskeleton; both β- and δ-catenin also interact with a number of synaptic scaffolding proteins (Kwiatkowski et al. 2007, Arikkath and Reichardt 2008). Classical cadherins can be divided further into type I cadherins (e.g., NCad and ECad), which are primarily homophilic, and type II cadherins (e.g., Cad8, Cad11, and Cad13), which are more likely to engage in heterophilic

Table 7.3 Protocadherins

Process	Summary	Model system	Source
Axonal targeting	Flamingo-mediated adhesion between growth cones in drosophila photoreceptor cells is required for appropriate targeting on lamina neurons	*Drosophila*	Lee et al. (2003) and Chen and Clandinin (2008)
	Celsr3 is required in axons and guidepost cells for appropriate tract formation	Mouse	Tissir et al. (2005) and Zhou et al. (2008)
	Pcdh-α is required for olfactory sensory neurons to target to their appropriate glomerulus	Mouse	Hasegawa et al. (2008)
Response to activity	Arcadlin/Pcdh8/PAPC (a δ2Pcdh) is expressed and transported to the synapse in response to activity where it acts to internalize NCad, a process that regulates dendritic spine density	Hippocampal cultures	Yasuda et al. (2007)
Synapse development and function	Deletion of the *Pcdh-γ* locus causes a severe reduction of synapses in the ventral spinal cord leading to interneuron cell death	Mouse	Wang et al. (2002), Weiner et al. (2005), Prasad et al. (2008)
	Several *Pcdh-α*, *Pcdh-β*, and *Pcdh-γ* isoforms localize to synapses	Mouse	Kohmura et al. (1998), Junghans et al. (2008),and Phillips et al. (2003)
	Pcdh19 mutations in humans are linked to epilepsy and mental retardation limited to females, implicating *Pcdh19* (a δ2-protocadherin) in synapse development or function	Human	Dibbens et al. (2008)
	Fat1 interacts with Ena/VASP and Homer, proteins involved in the organization of the synapse	Mouse	Moeller et al. (2004), Tanoue and Takeichi (2004) and Schreiner et al. (2006)
	Flamingo mutant motor axons form ectopic synapses while other neurons display axonal degeneration and synapse loss	*Drosophila*	Bao et al. (2007)

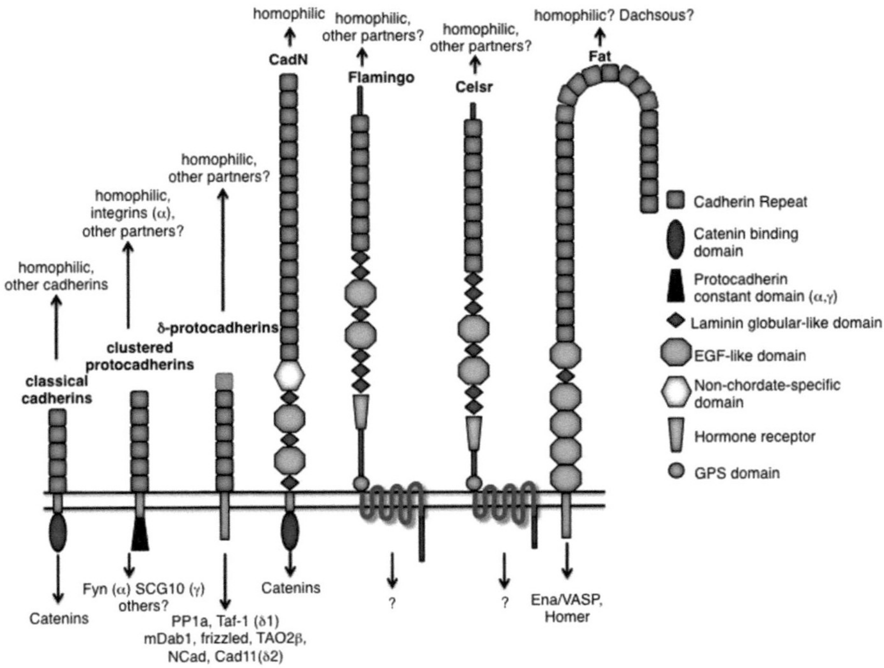

Fig. 7.1 Schematic structures of neuronally expressed cadherin superfamily members. Key to domain structure is presented at the *right*. Reported intracellular signaling partners are presented at the *bottom* and extracellular binding interactions noted above

adhesion with other classical cadherins in addition to homophilic interactions (Shimoyama et al. 2000).

The more than 70 protocadherins (Pcdhs) make up the largest and most diverse subgroup of the cadherin superfamily, including the α-, β-, and γ-protocadherins, the δ-protocadherins, the seven-transmembrane protocadherins, and the giant fat protocadherins (Frank and Kemler 2002, Junghans et al. 2005, Redies et al. 2005, Weiner 2006, Morishita and Yagi 2007). Over 50 protocadherin genes are clustered in three groups, termed *Pcdh-α*, *Pcdh-β*, and *Pcdh-γ*, that form a tandem array on a single chromosome in mammals and, with considerable variation, in other vertebrates (Fig. 7.2; Wu and Maniatis 1999, Wu et al. 2001, Noonan et al. 2004, Tada et al. 2004, Wu 2005, Zou et al. 2007). Members of all three clusters have been localized to synapses (Kohmura et al. 1998, Wang et al. 2002, Phillips et al. 2003, Junghans et al. 2008) as has the δ-Pcdh arcadlin/Pcdh8 (Yasuda et al. 2007). Protocadherins do not interact with catenins, and in most cases their signaling partners are entirely unknown. Although several protocadherins have been shown to interact homophilically, they generally mediate adhesion that is weaker than that of

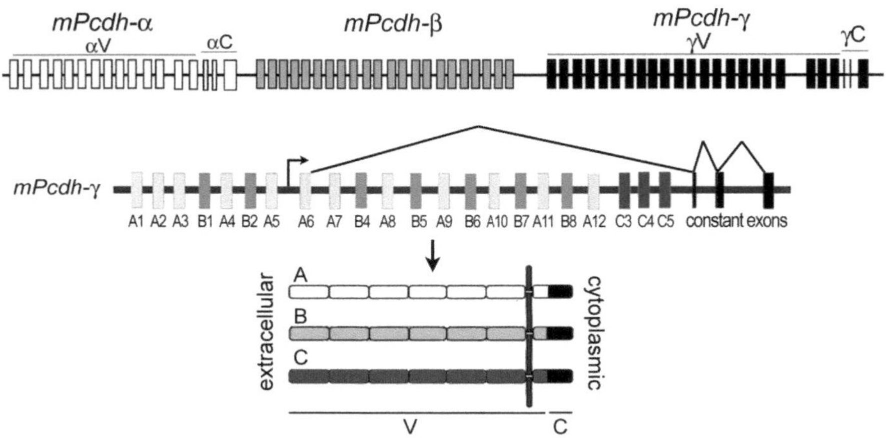

Fig. 7.2 The murine *Pcdh-α, Pcdh-β,* and *Pcdh-γ* gene clusters. *Top*: A schematic representation of the three clustered *Pcdh* loci on ~900 kb of mouse chromosome 18. V, variable exon; C, constant exon. *Middle*: An expanded view of the *Pcdh-γ* locus, showing the three subfamilies (**A**–**C**) of V exons. Splicing pattern of an example transcript, initiating at V exon A6, is indicated. *Bottom*: Each V exon, which encodes most of a single γ-Pcdh isoform, is expressed from its own promoter and subsequently spliced to the C exons, which encode a shared C-terminal cytoplasmic domain

the classical cadherins (Sano et al. 1993, Obata et al. 1995, Sago et al. 1995, Frank et al. 2005), and heterophilic ligands are likely (Mutoh et al. 2004).

7.2 Classical Cadherins and Catenins

N-cadherin (NCad), the canonical neuronal adhesion molecule, was the first member of the cadherin superfamily to be identified in the CNS. Its presence at the synapse was first demonstrated by the laboratories of Josh Sanes (Yamagata et al. 1995) and Masatoshi Takeichi (Uchida et al. 1996) using immunoelectron microscopy. Since then, it has been found that NCad is present at nascent synaptic sites in a number of model systems, and accumulates and becomes clustered as synapses mature (Benson and Tanaka 1998, Huntley and Benson 1999, Bozdagi et al. 2000, Jontes et al. 2004, Prakash et al. 2005, Rubio et al. 2005). Many studies have taken advantage of *Drosophila* genetics, focusing on the fly homologue to NCad, which is termed CadN (alternatively, DNCad). While CadN differs from its vertebrate counterpart in a number of ways, the most prominent being the presence of 15 cadherin repeats, its actions have much in common with mammalian NCad. While NCad continues to be the classical cadherin that has been most studied at the synapse, several recent publications have provided strong support implicating a number of type II classical cadherins in the formation and function of synapses (Manabe et al. 2000, Paradis et al.

2007, Suzuki et al. 2007). It should be noted that many studies have broadly demonstrated synaptic roles for the classical cadherins by expressing dominant-negative NCad molecules lacking the extracellular domain, which should generally disrupt cadherin signaling through the catenins or by direct depletion of catenins themselves. In these cases, it is usually unclear which of the classical cadherins are the most relevant for the synaptic process under investigation.

Aside from their purely structural role in holding presynaptic and postsynaptic membranes in close apposition, many researchers have hypothesized that the ~20 classical cadherins might form an adhesive "code" that mediates specific recognition between subsets of synaptic partners. The data available thus far, however, do not particularly support this hypothesis; it is primarily members of the immunoglobulin superfamily, rather than cadherins, which have been implicated in synaptic specificity (Yamagata et al. 2002, Shen and Bargmann 2003, Shen et al. 2004, Yamagata and Sanes 2008). Nevertheless, the classical cadherins are clearly important for several necessary steps leading to synaptogenesis, including axon targeting and dendritic spine morphogenesis, as well as for synaptic plasticity. In the following sections we review recent progress in defining functions for NCad, the catenins, and other classical cadherins in processes intrinsic to the development of neuronal circuitry.

7.2.1 Roles in Axon Targeting

There is evidence from several different model systems that NCad is important for axon targeting. Experiments using function-blocking antibodies to NCad in chick tectum in vivo (Inoue and Sanes 1997) or in thalamus-cortex slice co-cultures in vitro (Poskanzer et al. 2003) suggest that NCad may act as an adhesive "stop signal" to specify laminar targeting of particular axons. Additionally, Cad7 and Cad11 mark distinct retinal ganglion cell populations as well as specific retinorecipient layers of the chick tectum, suggesting that homophilic interactions between these type II cadherins may be involved in laminar targeting (Yamagata et al. 2006). Consistent with these data are experiments utilizing the *Drosophila* visual system, in which simple circuit formation allows for reproducible investigation of the targeting of individual axons. The *Drosophila* compound eye is made up of 800 ommatidia, each of which contains eight photoreceptor neurons, termed R1–R8. R1–R6 photoreceptor axons from each ommatidium fasciculate and project to the lamina, where they defasciculate and reassociate to project to cartridges such that each cartridge receives input from the same area of the visual field. The postsynaptic lamina neurons then project to specific layers in the deeper medulla. R7 and R8 photoreceptor axons project directly to distinct layers within the medulla (reviewed by Mast et al. 2006).

Flies harboring CadN mutations that are homozygous only in the retina exhibit targeting defects in R1–R6 projections: The axons reach the lamina in mutants, but the fibers fail to defasiculate and innervate their appropriate cartridges (Lee et al. 2001). In a different mutant line in which CadN expression is hypomorphic throughout the nervous system, R1–R6 axons defasiculate, but the terminals do not stay correctly positioned during synaptogenesis. Furthermore, glial processes invade areas that normally would be occupied by dendritic spines (Iwai et al. 2002). When CadN mutations are limited to R7 cells, their projections stop short of their targets in medulla layer M6, innervating instead layer M3, the normal target of R8 projections (Lee et al. 2001). Nern and colleagues (2008) recently investigated the role of CadN in targeting of projections from lamina neurons, which are postsynaptic to R1–R6, to the medulla. They show that CadN on growth cones interacts homophilically with CadN on the branches of adjacent axons (Nern et al. 2008). Using the mosaic analysis with a repressible cell marker (MARCM) technique to analyze individual neurons expressing a truncated form of CadN on a background of wild-type cells, Nern et al. (2008) found that the mutant neurons responded aberrantly to specific targeting choice points, which turn out to correlate with high CadN expression.

7.2.2 Roles in Dendrite and Dendritic Spine Morphogenesis

NCad has been shown to be an important regulator of dendrite morphogenesis and of dendritic spine maturity. In *Drosophila*, olfactory receptor neurons and olfactory projection neurons form synapses together at glomeruli. Without CadN, projection neurons fail to restrict their dendritic arbors to a single glomerulus, as they do in controls (Zhuo et al. 2001). Consistent with the data discussed above on axon targeting in the visual system, CadN on olfactory projection neuron dendrites does not seem to mediate this function by signaling across the synapse, but rather by signaling between dendrites (Zhu et al. 2004). In contrast, Marrs et al. (2006) found that a cell-permeable peptide that blocks NCad signaling to β1-integrin via its juxtamembrane domain resulted in the collapse of retinal ganglion cell dendritic arbors in retinal explant cultures. Similarly, expression of a dominant-negative form of NCad, which lacks the extracellular domain, in chick retina results in smaller dendritic fields with fewer GluR4-positive synaptic puncta, but does not affect axon elongation (Tanabe et al. 2006). Conversely, overexpression of NCad, αN-catenin, or β-catenin increases dendritic arborization in hippocampal cultures (Yu and Malenka 2003). Interestingly, neural activity induced by high K+ media also increases arborization in a β-catenin- and Wnt-dependent mechanism that does not involve transcription (Yu and Malenka 2003).

Studies investigating the role of cadherin/catenin signaling in dendritic spine morphology have primarily been carried out in hippocampal cultures. Togashi et al. (2002) took two approaches to block cadherin function: transfection of a dominant-negative NCad construct lacking the extracellular domain and targeted mutation of the neuronal form of α-catenin (αN). Both manipulations resulted in a more immature, elongated spine structure, though only the former had a severe effect on the accumulation of synaptic proteins (Togashi et al. 2002). The immature dendritic spines in αN-catenin mutant hippocampal neurons also exhibit increased motility and decreased stability. The converse manipulation, overexpression of αN-catenin, causes an increase in spine density and a decrease in spine turnover (Abe et al. 2004). Okuda et al. (2007) cultured hippocampal neurons from mice harboring a floxed β-catenin conditional mutant allele. Once the cultures had matured, these neurons were transfected with Cre recombinase to determine the postsynaptic role of β-catenin signaling. In Cre-transfected mutant neurons, dendritic spines are elongated, and the amplitude of mEPSCs is lower, again indicating that the spines have reverted to a more immature state (Okuda et al. 2007). In contrast to the effect of disrupting αN- and β-catenin, deletion of the gene encoding p120 catenin leads not to elongated spines, but rather a severe decrease in spine density both in vivo and in vitro, as well as a reduction in the number of terminal dendritic branches (Elia et al. 2006).

7.2.3 Roles in Synapse Formation and Maturation

Many studies utilizing a wide array of model systems have demonstrated that NCad localizes to nascent synaptic sites, and that synapse maturation coincides with increased NCad clustering (Benson and Tanaka 1998, Huntley and Benson 1999, Bozdagi et al. 2000, Jontes et al. 2004, Prakash et al. 2005, Rubio et al. 2005). Any role of NCad in promoting synapse maturation may be restricted to excitatory synapses, as in hippocampal cultures, NCad is present at all nascent synapses but becomes excluded from GABAergic synapses as development proceeds (Benson and Tanaka 1998). Functional studies testing the role of classical cadherins in synapse formation have produced some conflicting results, and thus it remains somewhat unclear whether synapse formation per se is a primary function of NCad and other classical cadherins. Unlike neuroligin (Scheiffele et al. 2000) and SynCAM (Biederer et al. 2002), NCad is incapable of inducing presynaptic assembly on contacting axons when exogenously expressed in HEK293 cells. One difficulty in interpreting the results of cadherin disruption is in isolating effects on synapse formation from those on the requisite steps preceding this, such as dendritic arborization or axon outgrowth. For example, disrupting cadherin function in the chick retina causes a reduction in synapse number on horizontal cells (Tanabe et al. 2006). However, it is unclear whether this is secondary to the reduction in dendritic field size. Similarly, the loss of synaptic density in the hippocampus of p120 catenin null mice is difficult to dissociate from the many effects of this mutation on dendrite morphology and dendritic spines (Elia et al. 2006). Below, we briefly summarize several studies that have examined synapse development in the absence of classical cadherin function.

In both mouse hippocampal neurons in vivo and in vitro (Kadowaki et al. 2007) and in zebrafish retinal neurons in vivo (Erdmann et al. 2003), genetic disruption of the NCad gene does not prevent synapse formation, although it remains unclear whether quantitative differences exist in these mutants. Similarly, synapses can still form in cultured hippocampal neurons transfected with a dominant-negative NCad construct lacking the extracellular domain (Bozdagi et al. 2004). This manipulation, which should disrupt multiple classical cadherins, does, however, affect the synapses that form, resulting in smaller, less differentiated synaptic contacts with reduced recycling of synaptic vesicles (Bozdagi et al. 2004). Similarly, synapse formation is not grossly affected in neurons derived from NCad knockout ES cells (Jungling et al. 2006). On the other hand, Togashi and colleagues (2002) reported a significant reduction in both presynaptic and postsynaptic puncta in hippocampal cultures transfected with a dominant-negative construct similar to that used by Bozdagi et al. (2004). Additionally, a recent study that used an RNAi screen in cultured hippocampal neurons identified roles for two type II cadherins, in the formation of synapses (Paradis et al. 2007). Knockdown of either Cad11, which had previously been shown to localize to synapses in hippocampal neurons (Manabe et al. 2000), or the GPI-linked Cad13/TCad led to significant reductions in the density of co-localized excitatory and inhibitory presynaptic and postsynaptic puncta, and in the amplitude and frequency of recorded mEPSCs. Conversely, overexpression of Cad11, but not Cad13, increased the number of excitatory synapses (Paradis et al. 2007).

Because synapse formation is a highly dynamic and asynchronous process, with many steps that may be controlled by varying mechanisms in different neuronal types, it is difficult to disentangle the role of cadherins in the formation of synapses from those in their maturation or stabilization. Indeed, there is evidence from hippocampal cultures that spines and synapses in young neurons (<10 days in vitro) may be more susceptible to cadherin disruption than are those in more mature neurons (Bozdagi et al. 2004, Saglietti et al. 2007). Thus, some of the disparity in the results summarized above may be due simply to the use of different model systems, different means of disrupting cadherin function, or analysis at different time points. That said, it seems likely that NCad, at least, is more critical for the stabilization and maturation of nascent synaptic contacts than it is for the initial interaction of presynaptic and postsynaptic terminals.

7.2.4 Roles in Synapse Function and Plasticity

Several studies have shown that the localization, clustering, and stability of NCad at the membrane are regulated by synaptic activity. When late-phase long-term potentiation (L-LTP) is induced in hippocampal slices, NCad accumulation increases at synaptic sites (Bozdagi et al. 2000). NMDA treatment in dissociated hippocampal cultures also causes NCad to dimerize and become protease resistant (Tanaka et al. 2000). Consistent with this, a recent study found that activity mediated by the NMDA receptor (NMDAR) regulates the endocytosis of NCad via β-catenin (Tai et al. 2007). The same group had previously shown that stimulation of hippocampal neurons with KCl causes β-catenin to move from dendritic shafts into spines and that this synaptic redistribution is negatively regulated by phosphorylation of a single tyrosine residue (Y654, Murase et al. 2002). NMDAR activation also induces β-catenin enrichment at spines, increasing its association with NCad. This stabilizes NCad at the membrane by decreasing its endocytosis (Tai et al. 2007). Furthermore, in neurons expressing a mutant β-catenin construct that exhibits enhanced association with NCad (Y654F, which prevents phosphorylation), NMDA-dependent long-term depression (LTD) is inhibited (Tai et al. 2007). A role for classical cadherins in the regulation of synaptic plasticity is further supported by studies of LTP in hippocampal slices and knockout mice. Treatment of slices with function-blocking antibodies against the extracellular domain of NCad causes a significant reduction in CA1 LTP (Tang et al. 1998, Bozdagi et al. 2000). In contrast, slices cultured from mice lacking Cad11 exhibit a modest *increase* in CA1 LTP (Manabe et al. 2000).

Cadherin/catenin signaling has been implicated in the control of both presynaptic and postsynaptic functions. Neurons derived from NCad knockout ES cells exhibit a variety of presynaptic defects, including impaired synaptic vesicle exocytosis during high-frequency stimulation, and a greater susceptibility to synaptic depression (Jungling et al. 2006). Consistent with this, mice harboring a hippocampus-restricted β-catenin mutation exhibit a marked decrease in the number of undocked vesicles per synapse identified by electron microscopy (Bamji et al. 2003). Recordings in hippocampal slices from these mutant mice found that the response from prolonged low-frequency stimulation declined much more quickly than in controls, indicating that β-catenin is required to maintain the reserve pool of synaptic vesicles (Bamji et al. 2003). Okuda et al. took a similar but modified approach to examine the role of β-catenin in postsynaptic function. Hippocampal neurons cultured from β-catenin conditional floxed mutant mice were transfected at low efficiency with a plasmid encoding a Cre-IRES-GFP cassette. In this culture, the authors identified fluorescent mutant neurons postsynaptic to primarily wild-type partners (Okuda et al. 2007). In mutant neurons, recorded mEPSCs were significantly reduced in amplitude, which could be rescued by reintroduction of wild-type β-catenin, but not by β-catenin lacking the cadherin-binding armadillo (ARM) domain. Structure–function analyses indicate that the role of β-catenin in regulating postsynaptic quantal AMPA responses requires both the ARM and PDZ-binding domains. Consistent with this, the reduction in mEPSC amplitude was mimicked by transfection of neurons with a dominant-negative NCad construct lacking the extracellular domain. β-Catenin's modulation of AMPA responses contributes to homeostatic synaptic plasticity, as mutant neurons were deficient in quantal scaling in response to changes in activity levels (Okuda et al. 2007).

Two recent papers go beyond the model system of hippocampal neurons to elucidate roles for several classical cadherins in the control of spinal cord circuitry. Suzuki et al. (2007) found that Cad8, a type II cadherin, is expressed by dorsal root ganglia (DRG) sensory neurons that express TRPM8, an ion channel sensitive to menthol and cold stimuli, and project their axons to laminae I and IIo in the dorsal horn. Cad8 is also expressed by neurons situated in these laminae, and the protein localizes to synapses between them and Cad8-positive sensory terminals. Null mutation of Cad8 in mice did not appear to cause a reduction of synapses in the dorsal horn, but the function of these synapses was impaired (Suzuki et al. 2007). As measured by whole-cell patch clamp in spinal cord slice cultures, menthol application increased synaptic frequency in controls, but had no effect in Cad8 knockouts. Furthermore, mutant animals were less responsive to cold stimuli than were controls, together indicating that the role of Cad8, like that demonstrated for NCad in many studies, is not primarily to promote synapse formation, but rather to regulate synaptic function (Suzuki et al. 2007). The fact that two other classical cadherins, NCad and ECad, also localize to specific subsets of dorsal horn synapses and respond differentially to sciatic nerve axotomy (NCad expression persists while ECad expression is completely lost from the sciatic nerve termination zone, Brock et al. 2004) suggests that the functions uncovered for Cad8 by Suzuki et al. (2007) may be carried out by multiple classical cadherins at various types of spinal cord synapses.

7.3 Protocadherins

The term "protocadherin" was first used by Sano et al. (1993) to describe a group of novel cDNAs identified using a degenerate PCR screen in a wide range of species, including humans, rodents, *Xenopus*, *Drosophila*, and *Caenorhabditis elegans,* to search for cadherin-like extracellular domains. Because sequences encoding similar domains were found in all species examined, the authors believed that they had found a primordial type of cadherin, thus the prefix "proto". It has since been realized that the worm and fly genes encode proteins that signal through catenin-like molecules, and that are now considered to be divergent classical cadherins. The term "protocadherin" is now taken to refer to those proteins that contain a varying number of cadherin-like extracellular domains but lack the classical catenin-binding sites in their cytoplasmic domains. More than 70 such protocadherins have now been identified, with the majority exhibiting expression in the nervous system (Morishita and Yagi 2007). Below, we describe several protocadherin subfamilies and discuss evidence for their functional roles at the synapse.

7.3.1 Clustered Protocadherins

From the beginning, it was speculated that protocadherins (Pcdhs) might act as diverse cell adhesion cues that help specify neuronal network formation (Sano et al. 1993), an idea that was bolstered by the discovery of the clustered Pcdh gene families by Wu and Maniatis (1999). The clustered Pcdhs comprise over 50 genes, divided into 3 subfamilies termed *Pcdh-α, Pcdh-β,* and *Pcdh-γ*, tandemly arrayed across ~900 kb on human chromosome 5q31. Similar clusters varying in gene number and organization have been identified in rodents (Wu et al. 2001, Yanase et al. 2004, Wu 2005), chicken (Sugino et al. 2004), and fish (Noonan et al. 2004, Tada et al. 2004) (Fig. 7.2). The organization of the *Pcdh-α, Pcdh-β,* and *Pcdh-γ* gene clusters is somewhat reminiscent of the immunoglobulin and T-cell receptor genomic loci, in that there are multiple "variable" (V) exons encoding variant isoforms, which, for the *Pcdh-α* and *Pcdh-γ* clusters, are expressed from their own individual upstream promoters and spliced to three short, invariant "constant" (C) exons (Fig. 7.2; the *Pcdh-β* locus is similar but lacks any constant exons; Wu and Maniatis 1999, Wu et al. 2001, Tasic et al. 2002, Wang et al. 2002, Wu 2005).

Each large V exon encodes six cadherin extracellular domains, a transmembrane domain, and a proximal cytoplasmic domain of a single Pcdh isoform, with the C-terminal ~125 amino acids of the α- and γ-Pcdhs encoded by their respective C exons (Fig. 7.2). The *Pcdh-γ* genes can be further grouped into 3 subfamilies: A (12 genes), B (7 genes), and C (3 genes), based on sequence similarity (Fig. 7.2); 2 of the *Pcdh-α* genes are referred to as *C1* and *C2* because their similarity to the *Pcdh-γ* C subfamily (*C3, C4,* and *C5*) is greater than their similarity to the other 12 *Pcdh-α* genes. The resulting α- and γ-Pcdh proteins are thus exactly the type of molecules that one might predict would underlie a "chemoaffinity" mechanism such as that predicted by the classic studies of Roger Sperry (1963): their divergent extracellular domains could engage in a variety of homo- or heterophilic interactions, while their shared cytoplasmic domain could be involved in conserved signaling mechanisms leading to the assembly or the stabilization of the synaptic apparatus. Consistent with this hypothesis, the clustered Pcdh families are highly expressed in the developing nervous system, with each neuron apparently expressing a distinct subset of these genes (Esumi et al. 2005, Kaneko et al. 2006, Zou et al. 2007).

Thus far, the best evidence in favor of a role for the clustered Pcdhs at the synapse has been obtained for the *Pcdh-γ* family (Wang et al. 2002, Weiner et al. 2005). *Pcdh-γ* genes are expressed widely in the CNS, and γ-Pcdh proteins are found immunohistochemically at some, though far from all, synapses (Wang et al. 2002, Phillips et al. 2003). Mice in which all 22 *Pcdh-γ* genes have been deleted via homologous recombination lack voluntary movements, exhibit deficient reflexes, do not feed, and die within hours of birth. This phenotype is apparently due to severe apoptosis and neurodegeneration of spinal interneurons and a concomitant loss of synapses in the late embryonic period (Wang et al. 2002). This reduction in synaptic density is a primary function of the γ-Pcdhs and not merely secondary to the neurodegeneration observed: When apoptosis is blocked in *Pcdh-γ* null mutant mice by the additional loss of the pro-apoptotic gene *Bax* (Deckwerth et al. 1996), the neurodegenerative phenotype is rescued, but both excitatory and inhibitory synaptic puncta remain reduced by 40–50% and the double-mutant mice still die within hours of birth (Weiner et al. 2005). A similar phenotype of primary synaptic defects in the absence of significant apoptosis is observed in a mutant mouse that expresses hypomorphic levels of the γ-Pcdhs. Spinal interneurons cultured from this mouse line exhibit reduced synaptic current amplitudes (Weiner et al. 2005).

The mechanisms by which the γ-Pcdhs influence synapse development are far from being understood, and little is known about their adhesive proclivities and intracellular signaling partners. Although γ-Pcdh proteins are present in synaptic membrane fractions, much of the protein is clearly extrasynaptic (Wang et al. 2002). Using immunogold electron microscopy, Phillips et al. (2003) showed that the γ-Pcdhs are localized not only to some synaptic membranes but also to "tubulovesicular" structures in both axons and dendrites that are likely to be part of an endosomal system. This suggests that γ-Pcdhs are actively cycled between these tubulovesicular structures and synaptic membranes in a dynamic process that may regulate synaptic specificity or plasticity via differential stability (Phillips et al. 2003, Jontes and Phillips 2006). While it has been shown that some of the γ-Pcdhs can interact homophilically (Sano et al. 1993, Obata et al. 1995, Frank et al. 2005), they appear to generate only a weak adhesive strength and the possibility of heterophilic interactions has not yet been examined. Several studies have demonstrated that the γ-Pcdhs can be cleaved extracellularly by metalloproteinases such as ADAM10, which may modulate their adhesive properties (Haas et al. 2005, Hambsch et al. 2005, Reiss et al. 2006). Similarly, the proteins undergo intracellular cleavage by presenilin, which appears to allow a fragment of the cytoplasmic domain to translocate to the nucleus (Haas et al. 2005). The variable cytoplasmic domain of particular γ-Pcdh members can interact in vitro with SCG10, a stathmin homologue that regulates microtubule dynamics (Gayet et al. 2004). However, it is unclear whether this interaction occurs in neurons and whether it is relevant to synaptic function.

Despite being the first clustered Pcdhs to be identified as synaptic molecules (Kohmura et al. 1998), the role of the α-Pcdhs (originally termed cadherin-related neuronal receptors or CNRs) in synapse formation and function remains unclear. α-Pcdhs localize to developing axons (Blank et al. 2004, Morishita et al. 2004), where their expression at the membrane is apparently negatively regulated by myelination (Morishita et al. 2004). It has been reported that α-Pcdhs require *cis* interactions with γ-Pcdhs for efficient cell surface localization (Murata et al. 2004). This suggests that some of the phenotypes observed in *Pcdh-γ* null mice could be due to loss of membrane α-Pcdh. A recent report describing the first genetic analysis of the *Pcdh-α* gene cluster, however, suggests otherwise (Hasegawa et al. 2008). Mice in which the *Pcdh-α* constant exons have been deleted develop normally, survive to adulthood, and are fertile, but exhibit an axonal targeting defect in the olfactory system. Olfactory sensory axons expressing a given odorant receptor normally coalesce on a single glomerulus in the olfactory bulb, but in *Pcdh-α* null mutants, these axons are scattered among several glomeruli. These disorganized axons appear to be able to form nerve terminals and synapses at the glomeruli they contact. Thus α-Pcdhs may be more important for axon guidance than they are for synaptogenesis (Hasegawa et al. 2008). Since α-Pcdh proteins are highly expressed in both olfactory axons and glomeruli, the phenotype in these mutant mice may reflect the loss of a homophilic adhesion mechanism. It should be noted, however, that at least some α-Pcdhs appear to engage preferentially in heterophilic interactions with β1-integrins (Mutoh et al. 2004).

The *Pcdh-β* cluster remains the least studied of the three. The lack of a shared constant domain among its members makes it much more difficult to study, as the expression, localization, and function of each family member must be addressed on an individual basis. A recent study raised antibodies specific for two β-Pcdh proteins, β16 and β22, and examined their localization patterns in the rodent nervous system (Junghans et al. 2008). Both β16 and β22 are present in the synaptic zones of the retina, the inner and outer plexiform layers, though only β16 was tightly localized to synapses, primarily the postsynaptic compartment (Junghans et al. 2008). To date, no genetic analysis of the *Pcdh-β* cluster has been reported. However, mice with *loxP* sequences targeted to various sites within the clustered Pcdh loci, designed for studies using trans-allelic somatic recombination to create large-scale genomic deletions, have been reported (Wu et al. 2007). Forthcoming analyses of such mice should yield the first functional information on the role of the β-Pcdhs in the nervous system.

7.3.2 Fat-Type and 7-Transmembrane Protocadherins

There are two families of large atypical Pcdhs: those typified by *Drosophila* fat, a giant protein with 34 cadherin repeats in its extracellular domain, and those typified by *Drosophila* flamingo, a 7-transmembrane (7TM) domain protein with 8 cadherin repeats, along with their many vertebrate homologues (Halbleib and Nelson 2006) (Fig. 7.1). Of these, the 7TM Pcdhs have been best studied in the nervous system. In *Drosophila* flamingo mutants, motor axons form ectopic synapses on inappropriate muscles and ectopic accumulations of presynaptic proteins along axons within the nerve. In addition, flamingo mutants exhibit axonal degeneration and concomitant synapse loss in an age-dependent manner (Bao et al. 2007). Several studies have also demonstrated important roles for flamingo and its vertebrate homologue Celsr3 in axon guidance as well as in the control of planar cell polarity (Lu et al. 1999, Usui et al. 1999, Curtin et al. 2003). Flamingo mediates adhesion between growth cones of *Drosophila* photoreceptor axons, which is required for appropriate targeting (Lee et al. 2003). Chen and Clandinin (2008) used a reverse MARCM technique to fluorescently label single wild-type axons in flies that also contained flamingo mutant axons. These wild-type axons exhibited targeting defects in the lamina only when cells adjacent to them in the same ommatidium were flamingo mutant. Interestingly, however, single mutant axons themselves targeted normally, confirming that flamingo is critical for axon–axon interactions that guide axons to their appropriate synaptic targets

(Chen and Clandinin 2008). Whether flamingo also plays a later, direct synaptic role remains to be determined. That a primary role of the 7TM Pcdhs is in the control of axon targeting is further supported by genetic analyses of one of flamingo's three mammalian homologues, Celsr3. Celsr3 is expressed in post-migratory neurons during development, but is downregulated as maturation occurs (Tissir et al. 2002). In mice in which the *Celsr3* gene has been inactivated, animals die neonatally with respiratory failure and exhibit major defects in the formation of several axonal tracts. The anterior commissure and internal capsule are completely absent, while longitudinal tracts such as the corticospinal and spinocerebellar tracts display marked anomalies (Tissir et al. 2005). By using a *Celsr3* conditional mutant and a variety of telencephalic cell-type-specific Cre lines, Zhou et al. (2008) showed that Celsr3 is required in both axons and the guidepost cells along their trajectory in order for normal wiring to occur (Zhou et al. 2008).

Fat-like protocadherins are single-pass transmembrane proteins characterized by an unusually large number (27–34) of cadherin repeats in their extracellular domain (Fig. 7.1). In *Drosophila* there are three fat-like Pcdhs (fat, fat-like, and dachsous), which have been intensively studied in recent years and are known to be involved in the establishment of planar cell polarity (Saburi and McNeill 2005), as well as in the Hippo/Warts pathway that controls organ size (Bennett and Harvey 2006, Silva et al. 2006, Willecke et al. 2006). All of the fat-like Pcdhs are also expressed in the *Drosophila* nervous system (Fung et al. 2008). In mammals the fat-like subfamily consists of six members: Fat1, Fat2, Fat3, Fat-J (Fat4), and two dachsous homologues, Dchs1 and Dchs2, all of which are highly expressed in the brain (Nollet et al. 2000, Rock et al. 2005, Halbleib and Nelson 2006).

Fat2 is expressed almost exclusively in the cerebellum, where it localizes to parallel fibers, the axons of cerebellar granule neurons (Nakayama et al. 2002). Parallel fibers form synapses primarily on the dendritic spines of Purkinje cells. Fat2, however, does not appear to localize to these synaptic junctions (Nakayama et al. 2002). This suggests that like the 7TM Pcdhs, Fat2 may primarily be involved in axon development rather than in synapse formation. This is supported by the fact that CDH-4, a *C. elegans* homologue of *Drosophila* fat-like protein, was recently shown to be involved in the control of axon fasciculation (Schmitz et al. 2008). Furthermore, Fat3 was shown to localize to axon fascicles in a variety of regions (Nagae et al. 2007), indicating that it may also play a role in organization of fiber tracts.

Like other members of the fat-like Pcdh subfamily, Fat1 is expressed in the brain, with strongest expression in the cerebellum and the olfactory bulb (Rock et al. 2005). A pivotal role of Fat1 in mammalian development was demonstrated through gene targeting. Fat1 knockout mice die within hours of birth, apparently from severe renal dysfunction (Ciani et al. 2003). However, severe but incompletely penetrant defects in forebrain and eye development were also observed, indicating a role for Fat1 in brain development that awaits elucidation by analysis of conditional mutant mice. The cytosolic domain of Fat1 has been demonstrated to interact with both Ena/VASP proteins (Moeller et al. 2004, Tanoue and Takeichi 2004) and with Homer proteins (Schreiner et al. 2006). Both Ena/VASP and Homer proteins play important roles in the organization of synaptic protein complexes, indicating a possible synaptic role for Fat1; indeed, Fat1 protein is highly enriched in the postsynaptic density fraction (D.S., unpublished observation). A synaptic role for Fat1 is also suggested by a report that certain variations of the human *Fat1* gene are associated with bipolar disorder (Abou Jamra et al. 2008).

7.3.3 δ-Protocadherins

Nine unclustered Pcdh genes have been grouped together as "δ-Pcdhs" based on their sequence homology. The δ-Pcdhs can be divided into δ1 and δ2 subgroups based on overall homology, the presence or absence of a conserved cytoplasmic motif, and the number of cadherin repeats, seven for the former subgroup and six for the latter (Fig. 7.1; Vanhalst et al. 2001). Several Pcdhs that had previously been extensively studied during vertebrate development (albeit not at the synapse) fall into the δ-Pcdh family, including paraxial Pcdh (PAPC/Pcdh8), axial Pcdh (AXPC/Pcdh1), and NF-Pcdh (Pcdh7) (Redies et al. 2005). One δ2-Pcdh, Pcdh19, recently has been linked to a human disorder that presumably involves synaptic dysfunctions. The *Pcdh19* locus is on human chromosome Xq22, a region linked to the disorder EFMR (*e*pilepsy and *m*ental *r*etardation limited to *f*emales). This is an atypical X-linked disorder in which the phenotype is only present in *females*. Dibbens et al. (2008) found mutations in *Pcdh19* in all seven of the affected families they tested. Five of the mutations resulted in truncations of the *Pcdh19* transcript (likely leading to nonsense-mediated decay of the protein) while the other two are likely to disrupt the adhesive properties of the protein (Dibbens et al. 2008). The mechanism of Pcdh19 action remains unknown, as does the explanation for the unusual female-only phenotypic inheritance pattern, but this study clearly implies that Pcdh19 plays a role in synapse development or function.

Intriguingly, one of the δ2-Pcdhs, arcadlin/PAPC/Pcdh8, has been shown recently to interact with NCad in a way that controls NCad's endocytosis in response to activity (Yasuda et al. 2007). Arcadlin was identified in 1999 as a cadherin-like molecule that is rapidly upregulated by synaptic activity and is localized to synapses. Function-blocking antibodies raised against arcadlin can reduce EPSP amplitudes and block the induction of LTP in hippocampal slices (Yamagata et al. 1999). Yasuda et al. (2007) found that when at the synapse, arcadlin binds in *cis* to NCad and in *trans* homophilically to arcadlin molecules across the cleft. This *trans* interaction activates a splice form of TAO2 kinase (TAO2β), a MAPKKK, which binds to the intracellular domain of arcadlin. TAO2β activates p38 MAPK, which then phosphorylates TAO2β in a feedback loop, triggering the endocytosis of the arcadlin/NCad complex. The arcadlin-mediated endocytosis of NCad allows for activity-dependent remodeling of spine synapses, and consistent with this, cultured hippocampal neurons deficient in arcadlin have a higher spine density (Yasuda et al. 2007). These exciting results demonstrate that arcadlin is a Pcdh that interacts with a classical cadherin to link synaptic activity to synapse remodeling, which is likely to be important for normal plasticity.

7.4 Concluding Remarks

From the foregoing brief summary of the vast extant literature on the cadherin superfamily, it should be clear that its diverse members play many critical roles in the formation, dynamic modulation, and mature function of synapses. And yet, in a sense, it can be said that the full measure of their importance is far from being grasped. Among the $\sim$20 classical cadherins, synaptic functions have been analyzed for only a few, namely NCad, Cad8, Cad11, and Cad13. Synapse-centric studies of the much more diverse Pcdhs have been similarly limited to a few members, primarily the γ-Pcdhs and arcadlin. Thus, despite the vast amount of progress that has been made over the past decade, clearly much remains to be elucidated. Individual classical cadherins are expressed by overlapping but distinct populations of neurons (Redies et al. 2003), suggesting that each may be critical for synapse formation and function in subsets of circuits. Many of these molecules play roles in other organ systems that are crucial for early embryonic development, thus precluding, in many cases, the use of knockout mice to study synaptic functions. With the increasing utilization of techniques for selective gene disruption in mice, including RNAi and the Cre/loxP system, functional analyses of the remaining classical cadherins in specific brain regions should be forthcoming.

The Pcdhs remain enigmatic in comparison to their better studied classical cousins. They are generally considered to be adhesion molecules because they contain cadherin repeats similar to those in the classical cadherins. However, whenever their adhesive properties have been tested, the Pcdhs have generally proven to be weakly adhesive at best (Sano et al. 1993, Obata et al. 1995, Sago et al. 1995, Frank et al. 2005, Yasuda et al. 2007). It should be noted that this could reflect either the difficulty of obtaining sufficient surface expression in heterologous cells (Murata et al. 2004, Mutoh et al. 2004, A.M.G., D.S. and J.A.W. unpublished observations) or a tight regulation of their membrane localization by proteases (Haas et al. 2005, Hambsch et al. 2005, Reiss et al. 2006). Nevertheless, this raises the question of whether Pcdhs act primarily as adhesion molecules that ensure close apposition of adjacent membranes or whether they are better characterized as ligands and receptors that convey extracellular signals to the cytoplasm. Consistent with the latter model, a number of Pcdhs have been shown to localize to apical membranes of cells at which there is ostensibly no adjacent membrane to which to adhere. For instance, mu-Pcdh localizes to the apical surface along the brush border in the kidney (Goldberg et al. 2002). *Drosophila* Cad99 (the orthologue of the human Usher syndrome gene PCDH15) is restricted to apical microvilli in ovarian follicle cells (D'Alterio et al. 2005, Schlichting et al. 2006), and the γ-Pcdhs localize to the apical surface of the choroid plexus epithelium in the brain (L. Helsper, D.S. and J.A.W. in preparation). The observation that at least one Pcdh, Arcadlin/PAPC, can interact in *cis* to modulate classical cadherin-mediated adhesion (Chen and Gumbiner 2006, Yasuda et al. 2007) suggests the exciting possibility that widespread cooperation between different branches of the cadherin superfamily tree may serve to greatly increase the complexity and diversity of cell–cell interactions between neurons. Elucidation of Pcdh interaction partners and signaling mechanisms, which remain largely unknown, should lead to a much better understanding of their role in nervous system function.

References

Abe K, Chisaka O, Van Roy F et al. (2004) Stability of dendritic spines and synaptic contacts is controlled by alpha N-catenin. Nat Neurosci 7:357–363

Aberle H, Butz S, Stappert J et al. (1994) Assembly of the cadherin-catenin complex in vitro with recombinant proteins. J Cell Sci 107 (Pt 12):3655–3663

Abou Jamra R, Becker T, Georgi A et al. (2008) Genetic variation of the FAT gene at 4q35 is associated with bipolar affective disorder. Mol Psychiatry 13:277–284

Anderson TR and Benson DL (2006) Cadherin-mediated adhesion and signaling during vertebrate central synapse formation. In: Dityatev A and El-Husseini A. (eds) Molecular Mechanisms of Synaptogenesis, Springer Science, New York

Arikkath J and Reichardt LF (2008) Cadherins and catenins at synapses: roles in synaptogenesis and synaptic plasticity. Trends Neurosci 31:487–494

Bamji SX, Shimazu K, Kimes N et al. (2003) Role of beta-catenin in synaptic vesicle localization and presynaptic assembly. Neuron 40:719–731

Bao H, Berlanga ML, Xue M et al. (2007) The atypical cadherin flamingo regulates synaptogenesis and helps prevent axonal and synaptic degeneration in Drosophila. Mol Cell Neurosci 34:662–678

Bennett FC and Harvey KF (2006) Fat cadherin modulates organ size in Drosophila via the Salvador/Warts/Hippo signaling pathway. Curr Biol 16:2101–2110

Benson DL and Tanaka H (1998) N-cadherin redistribution during synaptogenesis in hippocampal neurons. J Neurosci 18:6892–6904

Biederer T, Sara Y, Mozhayeva M et al. (2002) SynCAM, a synaptic adhesion molecule that drives synapse assembly. Science 297:1525–1531

Blank M, Triana-Baltzer GB, Richards CS et al. (2004) Alpha-protocadherins are presynaptic and axonal in nicotinic pathways. Mol Cell Neurosci 26:530–543

Bozdagi O, Shan W, Tanaka H et al. (2000) Increasing numbers of synaptic puncta during late-phase LTP: N-cadherin is synthesized, recruited to synaptic sites, and required for potentiation. Neuron 28:245–259

Bozdagi O, Valcin M, Poskanzer K et al. (2004) Temporally distinct demands for classic cadherins in synapse formation and maturation. Mol Cell Neurosci 27:509–521

Brock JH, Elste A and Huntley GW (2004) Distribution and injury-induced plasticity of cadherins in relationship to identified synaptic circuitry in adult rat spinal cord. J Neurosci 24:8806–8817

Bruses JL (2006) N-cadherin signaling in synapse formation and neuronal physiology. Mol Neurobiol 33:237–252

Chen PL and Clandinin TR (2008) The cadherin Flamingo mediates level-dependent interactions that guide photoreceptor target choice in Drosophila. Neuron 58:26–33

Chen X and Gumbiner BM (2006) Paraxial protocadherin mediates cell sorting and tissue morphogenesis by regulating C-cadherin adhesion activity. J Cell Biol 174:301–313

Ciani L, Patel A, Allen ND et al. (2003) Mice lacking the giant protocadherin mFAT1 exhibit renal slit junction abnormalities and a partially penetrant cyclopia and anophthalmia phenotype. Mol Cell Biol 23:3575–3582

Curtin JA, Quint E, Tsipouri V et al. (2003) Mutation of Celsr1 disrupts planar polarity of inner ear hair cells and causes severe neural tube defects in the mouse. Curr Biol 13:1129–1133

D'Alterio C, Tran DD, Yeung MW et al. (2005) Drosophila melanogaster Cad99C, the orthologue of human Usher cadherin PCDH15, regulates the length of microvilli. J Cell Biol 171:549–558

Deckwerth TL, Elliott JL, Knudson CM et al. (1996) BAX is required for neuronal death after trophic factor deprivation and during development. Neuron 17:401–411

Dibbens LM, Tarpey PS, Hynes K et al. (2008) X-linked protocadherin 19 mutations cause female-limited epilepsy and cognitive impairment. Nat Genet 40:776–781

Elia LP, Yamamoto M, Zang K et al. (2006) p120 catenin regulates dendritic spine and synapse development through Rho-family GTPases and cadherins. Neuron 51:43–56

Erdmann B, Kirsch FP, Rathjen FG et al. (2003) N-cadherin is essential for retinal lamination in the zebrafish. Dev Dyn 226:570–577

Esumi S, Kakazu N, Taguchi Y et al. (2005) Monoallelic yet combinatorial expression of variable exons of the protocadherin-alpha gene cluster in single neurons. Nat Genet 37:171–176

Frank M, Ebert M, Shan W et al. (2005) Differential expression of individual gamma-protocadherins during mouse brain development. Mol Cell Neurosci 29:603–616

Frank M and Kemler R (2002) Protocadherins. Curr Opin Cell Biol 14:557–562

Fung S, Wang F, Chase M et al. (2008) Expression profile of the cadherin family in the developing Drosophila brain. J Comp Neurol 506:469–488

Gayet O, Labella V, Henderson CE et al. (2004) The b1 isoform of protocadherin-gamma (Pcdhgamma) interacts with the microtubule-destabilizing protein SCG10. FEBS Lett 578:175–179

Goldberg M, Wei M, Tycko B et al. (2002) Identification and expression analysis of the human mu-protocadherin gene in fetal and adult kidneys. Am J Physiol Renal Physiol 283:F454–F463

Haas IG, Frank M, Veron N et al. (2005) Presenilin-dependent processing and nuclear function of gamma-protocadherins. J Biol Chem 280:9313–9319

Halbleib JM and Nelson WJ (2006) Cadherins in development: cell adhesion, sorting, and tissue morphogenesis. Genes Dev 20:3199–3214

Hambsch B, Grinevich V, Seeburg PH et al. (2005) gamma -protocadherins: Presenilin-mediated release of C-terminal fragment promotes locus expression. J Biol Chem 280:15888–15897

Hasegawa S, Hamada S, Kumode Y et al. (2008) The protocadherin-alpha family is involved in axonal coalescence of olfactory sensory neurons into glomeruli of the olfactory bulb in mouse. Mol Cell Neurosci 38:66–79

Huntley GW and Benson DL (1999) Neural (N)-cadherin at developing thalamocortical synapses provides an adhesion mechanism for the formation of somatopically organized connections. J Comp Neurol 407:453–471

Inoue A and Sanes JR (1997) Lamina-specific connectivity in the brain: regulation by N-cadherin, neurotrophins, and glycoconjugates. Science 276:1428–1431

Iwai Y, Hirota Y, Ozaki K et al. (2002) DN-cadherin is required for spatial arrangement of nerve terminals and ultrastructural organization of synapses. Mol Cell Neurosci 19:375–388

Jontes JD, Emond MR, Smith SJ (2004) In vivo trafficking and targeting of N-cadherin to nascent presynaptic terminals. J Neurosci 24:9027–9034
Jontes JD and Phillips GR (2006) Selective stabilization and synaptic specificity: a new cell-biological model. Trends Neurosci 29:186–191
Junghans D, Haas IG and Kemler R (2005) Mammalian cadherins and protocadherins: about cell death, synapses and processing. Curr Opin Cell Biol 17:446–452
Junghans D, Heidenreich M, Hack I et al. (2008) Postsynaptic and differential localization to neuronal subtypes of protocadherin beta16 in the mammalian central nervous system. Eur J Neurosci 27:559–571
Jungling K, Eulenburg V, Moore R et al. (2006) N-cadherin transsynaptically regulates short-term plasticity at glutamatergic synapses in embryonic stem cell-derived neurons. J Neurosci 26:6968–6978
Kadowaki M, Nakamura S, Machon O et al. (2007) N-cadherin mediates cortical organization in the mouse brain. Dev Biol 304:22–33
Kaneko R, Kato H, Kawamura Y et al. (2006) Allelic gene regulation of Pcdh-alpha and Pcdh-gamma clusters involving both monoallelic and biallelic expression in single Purkinje cells. J Biol Chem 281:30551–30560
Kemler R (1993) From cadherins to catenins: cytoplasmic protein interactions and regulation of cell adhesion. Trends Genet 9:317–321
Kohmura N, Senzaki K, Hamada S et al. (1998) Diversity revealed by a novel family of cadherins expressed in neurons at a synaptic complex. Neuron 20:1137–1151
Kwiatkowski AV, Weis WI and Nelson WJ (2007) Catenins: playing both sides of the synapse. Curr Opin Cell Biol 19:551–556
Lee CH, Herman T, Clandinin TR et al. (2001) N-cadherin regulates target specificity in the Drosophila visual system. Neuron 30:437–450
Lee RC, Clandinin TR, Lee CH et al. (2003) The protocadherin Flamingo is required for axon target selection in the Drosophila visual system. Nat Neurosci 6:557–563
Lilien J and Balsamo J (2005) The regulation of cadherin-mediated adhesion by tyrosine phosphorylation/dephosphorylation of beta-catenin. Curr Opin Cell Biol 17:459–465
Lu B, Usui T, Uemura T et al. (1999) Flamingo controls the planar polarity of sensory bristles and asymmetric division of sensory organ precursors in Drosophila. Curr Biol 9:1247–1250
Manabe T, Togashi H, Uchida N et al. (2000) Loss of cadherin-11 adhesion receptor enhances plastic changes in hippocampal synapses and modifies behavioral responses. Mol Cell Neurosci 15:534–546
Marrs GS, Honda T, Fuller L et al. (2006) Dendritic arbors of developing retinal ganglion cells are stabilized by beta 1-integrins. Mol Cell Neurosci 32:230–241
Mast JD, Prakash S, Chen PL et al. (2006) The mechanisms and molecules that connect photoreceptor axons to their targets in Drosophila. Semin Cell Dev Biol 17:42–49
Moeller MJ, Soofi A, Braun GS et al. (2004) Protocadherin FAT1 binds Ena/VASP proteins and is necessary for actin dynamics and cell polarization. Embo J 23:3769–3779
Morishita H, Kawaguchi M, Murata Y et al. (2004) Myelination triggers local loss of axonal CNR/protocadherin alpha family protein expression. Eur J Neurosci 20:2843–2847
Morishita H and Yagi T (2007) Protocadherin family: diversity, structure, and function. Curr Opin Cell Biol 19:584–592
Murase S, Mosser E and Schuman EM (2002) Depolarization drives beta-Catenin into neuronal spines promoting changes in synaptic structure and function. Neuron 35:91–105
Murata Y, Hamada S, Morishita H et al. (2004) Interaction with protocadherin-gamma regulates the cell surface expression of protocadherin-alpha. J Biol Chem 279:49508–49516
Mutoh T, Hamada S, Senzaki K et al. (2004) Cadherin-related neuronal receptor 1 (CNR1) has cell adhesion activity with beta1 integrin mediated through the RGD site of CNR1. Exp Cell Res 294:494–508

Nagae S, Tanoue T and Takeichi M (2007) Temporal and spatial expression profiles of the Fat3 protein, a giant cadherin molecule, during mouse development. Dev Dyn 236:534–543
Nakayama M, Nakajima D, Yoshimura R et al. (2002) MEGF1/fat2 proteins containing extraordinarily large extracellular domains are localized to thin parallel fibers of cerebellar granule cells. Mol Cell Neurosci 20:563–578
Nern A, Zhu Y and Zipursky SL (2008) Local N-cadherin interactions mediate distinct steps in the targeting of lamina neurons. Neuron 58:34–41
Nollet F, Kools P and van Roy F (2000) Phylogenetic analysis of the cadherin superfamily allows identification of six major subfamilies besides several solitary members. J Mol Biol 299:551–572
Noonan JP, Grimwood J, Danke J et al. (2004) Coelacanth genome sequence reveals the evolutionary history of vertebrate genes. Genome Res 14:2397–2405
Obata S, Sago H, Mori N et al. (1995) Protocadherin Pcdh2 shows properties similar to, but distinct from, those of classical cadherins. J Cell Sci 108 (Pt 12):3765–3773
Okuda T, Yu LM, Cingolani LA et al. (2007) beta-Catenin regulates excitatory postsynaptic strength at hippocampal synapses. Proc Natl Acad Sci USA 104:13479–13484
Paradis S, Harrar DB, Lin Y et al. (2007) An RNAi-based approach identifies molecules required for glutamatergic and GABAergic synapse development. Neuron 53:217–232
Phillips GR, Tanaka H, Frank M et al. (2003) Gamma-protocadherins are targeted to subsets of synapses and intracellular organelles in neurons. J Neurosci 23:5096–5104
Poskanzer K, Needleman LA, Bozdagi O et al. (2003) N-cadherin regulates ingrowth and laminar targeting of thalamocortical axons. J Neurosci 23:2294–2305
Prakash S, Caldwell JC, Eberl DF et al. (2005) Drosophila N-cadherin mediates an attractive interaction between photoreceptor axons and their targets. Nat Neurosci 8:443–450
Prasad T, Wang X, Gray PA et al. (2008) A differential developmental pattern of spinal interneuron apoptosis during synaptogenesis: insights from genetic analyses of the protocadherin-© gene cluster. Development 135:4153–4164
Redies C, Treubert-Zimmermann U and Luo J (2003) Cadherins as regulators for the emergence of neural nets from embryonic divisions. J Physiol Paris 97:5–15
Redies C, Vanhalst K and Roy F (2005) delta-Protocadherins: unique structures and functions. Cell Mol. Life Sci 62:2840–2852
Reiss K, Maretzky T, Haas IG et al. (2006) Regulated ADAM10-dependent ectodomain shedding of gamma-protocadherin C3 modulates cell-cell adhesion. J Biol Chem 281:21735–21744
Rock R, Schrauth S and Gessler M (2005) Expression of mouse dchs1, fjx1, and fat-j suggests conservation of the planar cell polarity pathway identified in Drosophila. Dev Dyn 234:747–755
Rubio ME, Curcio C, Chauvet N et al. (2005) Assembly of the N-cadherin complex during synapse formation involves uncoupling of p120-catenin and association with presenilin 1. Mol Cell Neurosci 30:611–623
Saburi S and McNeill H (2005) Organising cells into tissues: new roles for cell adhesion molecules in planar cell polarity. Curr Opin Cell Biol 17:482–488
Saglietti L, Dequidt C, Kamieniarz K et al. (2007) Extracellular interactions between GluR2 and N-cadherin in spine regulation. Neuron 54:461–477
Sago H, Kitagawa M, Obata S et al. (1995) Cloning, expression, and chromosomal localization of a novel cadherin-related protein, protocadherin-3. Genomics 29:631–640
Salinas PC and Price SR (2005) Cadherins and catenins in synapse development. Curr Opin Neurobiol 15:73–80
Sano K, Tanihara H, Heimark RL et al. (1993) Protocadherins: a large family of cadherin-related molecules in central nervous system. Embo J 12:2249–2256
Scheiffele P, Fan J, Choih J et al. (2000) Neuroligin expressed in nonneuronal cells triggers presynaptic development in contacting axons. Cell 101:657–669

Schlichting K, Wilsch-Brauninger M, Demontis F et al. (2006) Cadherin Cad99C is required for normal microvilli morphology in Drosophila follicle cells. J Cell Sci 119:1184–1195

Schmitz C, Wacker I and Hutter H (2008) The Fat-like cadherin CDH-4 controls axon fasciculation, cell migration and hypodermis and pharynx development in *Caenorhabditis elegans*. Dev Biol 316:249–259

Schreiner D, Muller K and Hofer HW (2006) The intracellular domain of the human protocadherin hFat1 interacts with Homer signalling scaffolding proteins. FEBS Lett 580:5295–5300

Shen K and Bargmann CI (2003) The immunoglobulin superfamily protein SYG-1 determines the location of specific synapses in *C. elegans*. Cell 112:619–630

Shen K, Fetter RD and Bargmann CI (2004) Synaptic specificity is generated by the synaptic guidepost protein SYG-2 and its receptor, SYG-1. Cell 116:869–881

Shimoyama Y, Tsujimoto G, Kitajima M et al. (2000) Identification of three human type-II classic cadherins and frequent heterophilic interactions between different subclasses of type-II classic cadherins. Biochem J 349:159–167

Silva E, Tsatskis Y, Gardano L et al. (2006) The tumor-suppressor gene fat controls tissue growth upstream of expanded in the hippo signaling pathway. Curr Biol 16:2081–2089

Sperry RW (1963) Chemoaffinity in the orderly growth of nerve fiber patterns and connections. Proc Natl Acad Sci USA 50:703–710

Sugino H, Yanase H, Hamada S et al. (2004) Distinct genomic sequence of the CNR/Pcdhalpha genes in chicken. Biochem Biophys Res Commun 316:437–445

Suzuki SC, Furue H, Koga K et al. (2007) Cadherin-8 is required for the first relay synapses to receive functional inputs from primary sensory afferents for cold sensation. J Neurosci 27:3466–3476

Suzuki SC and Takeichi M (2008) Cadherins in neuronal morphogenesis and function. Dev Growth Differ 50 Suppl 1:S119–S130

Tada MN, Senzaki K, Tai Y et al. (2004) Genomic organization and transcripts of the zebrafish Protocadherin genes. Gene 340:197–211

Tai CY, Kim SA and Schuman EM (2008) Cadherins and synaptic plasticity. Curr Opin Cell Biol 20:567–575

Tai CY, Mysore SP, Chiu C et al. (2007) Activity-regulated N-cadherin endocytosis. Neuron 54:771–785

Takeichi M (2007) The cadherin superfamily in neuronal connections and interactions. Nat Rev Neurosci 8:11–20

Tanabe K, Takahashi Y, Sato Y et al. (2006) Cadherin is required for dendritic morphogenesis and synaptic terminal organization of retinal horizontal cells. Development 133:4085–4096

Tanaka H, Shan W, Phillips GR et al. (2000) Molecular modification of N-cadherin in response to synaptic activity. Neuron 25:93–107

Tang L, Hung CP and Schuman EM (1998) A role for the cadherin family of cell adhesion molecules in hippocampal long-term potentiation. Neuron 20:1165–75

Tanoue T and Takeichi M (2004) Mammalian Fat1 cadherin regulates actin dynamics and cell-cell contact. J Cell Biol 165:517–528

Tasic B, Nabholz CE, Baldwin KK et al. (2002) Promoter choice determines splice site selection in protocadherin alpha and gamma pre-mRNA splicing. Mol Cell 10:21–33

Tissir F, Bar I, Jossin Y et al. (2005) Protocadherin Celsr3 is crucial in axonal tract development. Nat Neurosci 8:451–457

Tissir F, De-Backer O, Goffinet AM et al. (2002) Developmental expression profiles of Celsr (Flamingo) genes in the mouse. Mech Dev 112:157–160

Togashi H, Abe K, Mizoguchi A et al. (2002) Cadherin regulates dendritic spine morphogenesis. Neuron 35:77–89

Uchida N, Honjo Y, Johnson KR et al. (1996) The catenin/cadherin adhesion system is localized in synaptic junctions bordering transmitter release zones. J Cell Biol 135:767–779

Usui T, Shima Y, Shimada Y et al. (1999) Flamingo, a seven-pass transmembrane cadherin, regulates planar cell polarity under the control of Frizzled. Cell 98:585–595

Vanhalst K, Kools P, Vanden Eynde E et al. (2001) The human and murine protocadherin-beta one-exon gene families show high evolutionary conservation, despite the difference in gene number. FEBS Lett 495:120–125
Wang X, Su H and Bradley A (2002) Molecular mechanisms governing Pcdh-gamma gene expression: evidence for a multiple promoter and cis-alternative splicing model. Genes Dev 16:1890–1905
Wang X, Weiner JA, Levi S et al. (2002) Gamma protocadherins are required for survival of spinal interneurons. Neuron 36:843–854
Weiner JA (2006) Protocadherins and Synapse Development. In: Dityatev A and El-Husseini A. (eds) Molecular mechanisms of synaptogenesis, Springer, New York
Weiner JA, Wang X, Tapia JC et al. (2005) Gamma protocadherins are required for synaptic development in the spinal cord. Proc Natl Acad Sci USA 102:8–14
Willecke M, Hamaratoglu F, Kango-Singh M et al. (2006) The fat cadherin acts through the hippo tumor-suppressor pathway to regulate tissue size. Curr Biol 16:2090–2100
Wu Q (2005) Comparative genomics and diversifying selection of the clustered vertebrate protocadherin genes. Genetics 169:2179–2188
Wu Q and Maniatis T (1999) A striking organization of a large family of human neural cadherin-like cell adhesion genes. Cell 97:779–790
Wu Q, Zhang T, Cheng JF et al. (2001) Comparative DNA sequence analysis of mouse and human protocadherin gene clusters. Genome Res 11:389–404
Wu S, Ying G, Wu Q et al. (2007) Toward simpler and faster genome-wide mutagenesis in mice. Nat Genet 39:922–930
Yagi T (2008) Clustered protocadherin family. Dev Growth Differ 50 Suppl 1:S131–S140
Yamagata K, Andreasson KI, Sugiura H et al. (1999) Arcadlin is a neural activity-regulated cadherin involved in long term potentiation. J Biol Chem 274:19473–19479
Yamagata M, Herman JP, Sanes JR (1995) Lamina-specific expression of adhesion molecules in developing chick optic tectum. J Neurosci 15:4556–4571
Yamagata M and Sanes JR (2008) Dscam and Sidekick proteins direct lamina-specific synaptic connections in vertebrate retina. Nature 451:465–469
Yamagata M, Weiner JA, Dulac C et al. (2006) Labeled lines in the retinotectal system: markers for retinorecipient sublaminae and the retinal ganglion cell subsets that innervate them. Mol Cell Neurosci 33:296–310
Yamagata M, Weiner JA and Sanes JR (2002) Sidekicks: synaptic adhesion molecules that promote lamina-specific connectivity in the retina. Cell 110:649–660
Yanase H, Sugino H and Yagi T (2004) Genomic sequence and organization of the family of CNR/Pcdhalpha genes in rat. Genomics 83:717–726
Yasuda S, Tanaka H, Sugiura H et al. (2007) Activity-induced protocadherin arcadlin regulates dendritic spine number by triggering N-cadherin endocytosis via TAO2beta and p38 MAP kinases. Neuron 56:456–471
Yu X and Malenka RC (2003) Beta-catenin is critical for dendritic morphogenesis. Nat Neurosci 6:1169–1177
Zhou L, Bar I, Achouri Y et al. (2008) Early forebrain wiring: genetic dissection using conditional Celsr3 mutant mice. Science 320:946–949
Zhu N, Lalla R, Eves P et al. (2004) Melanoma cell migration is upregulated by tumour necrosis factor-alpha and suppressed by alpha-melanocyte-stimulating hormone. Br J Cancer 90:1457–1463
Zhuo L, Theis M, Alvarez-Maya I et al. (2001) hGFAP-cre transgenic mice for manipulation of glial and neuronal function in vivo. Genesis 31:85–94
Zou C, Huang W, Ying G et al. (2007) Sequence analysis and expression mapping of the rat clustered protocadherin gene repertoires. Neuroscience 144:579–603

Chapter 8
Nectins and Nectin-Like Molecules in the Nervous System

Hideru Togashi, Hisakazu Ogita, and Yoshimi Takai

Abstract Nectins and nectin-like molecules (Necls) are immunoglobulin-like transmembrane cell-adhesion molecules that are expressed in various cell types. Nectins form homo- or hetero-*trans*-dimers in a Ca^{2+}-independent manner, causing cell–cell adhesion. Their heterotypic binding is much stronger than their homophilic binding. In epithelial cells in culture, cell–cell adhesion complexes are formed by nectins first and then cadherins are recruited to the nectin-based cell–cell adhesion sites to cooperatively form adherens junctions (AJs). Recent studies have revealed that nectins in cooperation with cadherins are also involved in the formation of synapses. In this chapter, we first describe the roles and modes of action of nectins and Necls in epithelial cells and fibroblasts and then in the formation and remodeling of synapses.

Keywords Nectin · Cadherin · Immunoglobulin · Adherens junctions · Puncta adherentia junctions · Synapse

8.1 Introduction

Synapses are a specialized form of intercellular junctions where the axon terminal of a neuron comes into functional contact with a target cell. The specificity and plasticity of synapses provide neurons with a structural and functional basis for the formation of the neuronal network system. Early ultrastructural studies showed that the synaptic junctional areas contain at least two types of adhesion structure (Spacek 1985, Peters et al. 1991) (Fig. 8.1A). The first is the transmitter release zone associated with synaptic vesicles, termed synaptic junctions (SJs). SJs are the actual sites of neurotransmission,

Y. Takai (✉)
Division of Molecular and Cellular Biology, Department of Biochemistry and Molecular Biology, Kobe University Graduate School of Medicine, 7-5-1 Kusunoki-cho, Chuo-ku, Kobe 650-0017 Japan
e-mail: ytakai@med.kobe-u.ac.jp

M. Hortsch, H. Umemori (eds.), *The Sticky Synapse*,
DOI 10.1007/978-0-387-92708-4_8,

including the presynaptic active zone, which is opposed to the postsynaptic density. The other is a symmetrical junctional structure termed puncta adherentia junctions (PAJs), defined by the two criteria that they have symmetric paramembranous dense materials and have no association with synaptic vesicles. PAJs are morphologically similar to adherens junctions (AJs) formed in epithelial cells and fibroblasts (Fig. 8.1) and are regarded as a neuronal form of AJs, also termed synaptic AJs (Uchida et al. 1996). These observations suggest that some basic properties of the assembly of junctional apparatuses may be shared between neurons and epithelial cells, but also that several functional and morphological features are unique to synapses. For example, AJs that are generally seen in epithelial cells are belt-like structures and continuously surround the cell (Fig. 8.1B). In contrast, synaptic AJs are punctuated structures and distributed irregularly at the peripheral zones of SJs (Fig. 8.1D). In epithelial cells, the junctional complex that consists of tight junctions (TJs) and AJs localizes at the apical region of the lateral plasma membrane. These junctional structures are typically aligned from the apical to the basal side. The molecular mechanisms of the formation of AJs play key roles in the establishment of the apical–basal polarity at cell–cell adhesion sites and the formation of TJs in

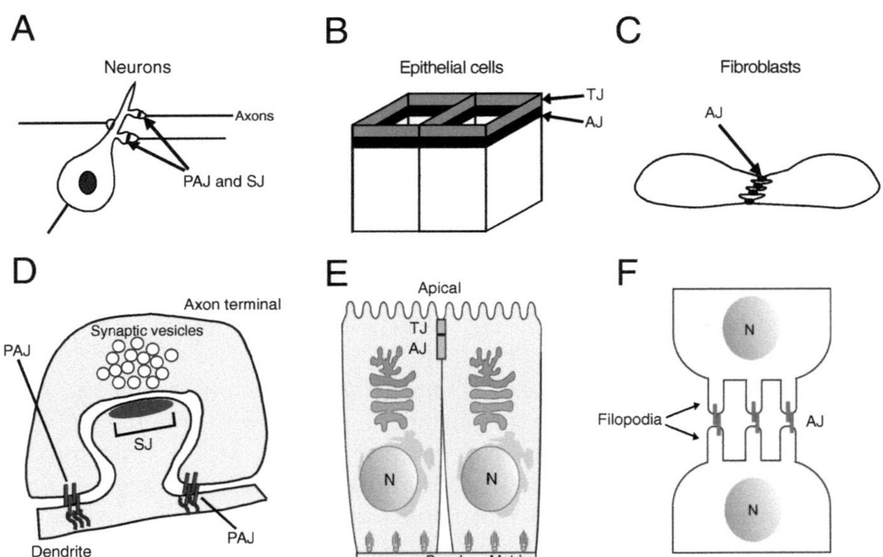

Fig. 8.1 Cell–cell junctions in neurons (**A**), epithelial cells (**B**), and fibroblasts (**C**). In neurons, at least two types of intercellular junctions, SJs and PAJs, are recognized. PAJs are punctuated structures similar to AJs in fibroblasts and distribute irregularly at the peripheral zones of SJs (**D**). In epithelial cells, TJs and AJs localize at the most apical region of the lateral membrane and appear as belt-like structures continuously surrounding the cell (**B**). These junctional structures are typically aligned from the apical to basal side of the cell (**E**). In fibroblasts, TJs are absent and AJs connect neighboring cells and show punctuated structures (**C**, **F**)

epithelial cells. However, structures corresponding to TJs do not exist in neural cells and the alignment of SJs and PAJs at synapses is not well defined.

The molecular mechanisms of intercellular junctions have been investigated most extensively in epithelial cells. AJs are symmetrically formed among identical cell types and are crucial for the stable association of neighboring epithelial cells. Without the activity of this adhesion system, epithelia fall apart and other cell–cell junctions, including TJs and desmosomes, do not form normally (Gumbiner and Simons 1986, Gumbiner et al. 1988). AJs are mediated by the homophilic interactions of the cell adhesion molecules (CAMs) cadherins (Takeichi 1991, Gumbiner 1996, Perez-Moreno et al. 2003). However, there is a lot of evidence that AJs are also formed by nectins. Nectins are Ca^{2+}-independent Ig-domain CAMs that play a key role in the organization of a variety of cell–cell junctions in cooperation with or independently of cadherins (Takai and Nakanishi 2003, Ogita and Takai 2008). Nectins, as well as cadherins are involved in the formation of PAJs (Honda et al. 2006) and in the proper association between axons and dendrites during synaptogenesis (Togashi et al. 2006). Furthermore, nectins regulate a variety of cellular activities including cell polarization, differentiation, movement, proliferation, and survival (Takai et al. 2008). Nectin-like molecules (Necls), which are CAMs structurally similar to nectins, may also have a variety of functions including cell adhesion, proliferation, and movement. For example, Necl-1 and -4 mediate axo-glial interactions, Schwann cell differentiation, and myelination (Maurel et al. 2007, Spiegel et al. 2007). Herein, we describe the roles and modes of action of nectins in the formation and remodeling of synapses. We also briefly describe the role of nectins in epithelial cells and fibroblasts, as it has been studied more extensively.

8.2 General Properties of Nectins and Necls

Nectins comprise a family with four members, nectin-1, -2, -3, and -4, each of which has two or three splice variants (Takahashi et al. 1999, Satoh-Horikawa et al. 2000, Reymond et al. 2001). All members have an extracellular region with three Ig-like loops, a single transmembrane region and a cytoplasmic region (Fig. 8.2A). Nectins are ubiquitously expressed in a variety of cells, including epithelial cells, fibroblasts, and neurons. Nectin-1 and -2 were originally isolated as the poliovirus receptor-related proteins, PRR-1/HveC and PRR-2/HveB, respectively (Geraghty et al. 1998, Warner et al. 1998), but were renamed nectins from the Latin word "necto," which means "to connect" (Takahashi et al. 1999).

Nectins have a conserved motif of four amino acid residues at their C-termini that binds to the PDZ domain of afadin (Fig. 8.2A). Afadin has multiple protein domains and is expressed in several alternatively spliced isoforms, which differ in their C-terminal regions. Herein, unless otherwise specified, afadin refers to the longest variant, l-afadin. In addition, nectin-1 and -3, but not nectin-2, bind

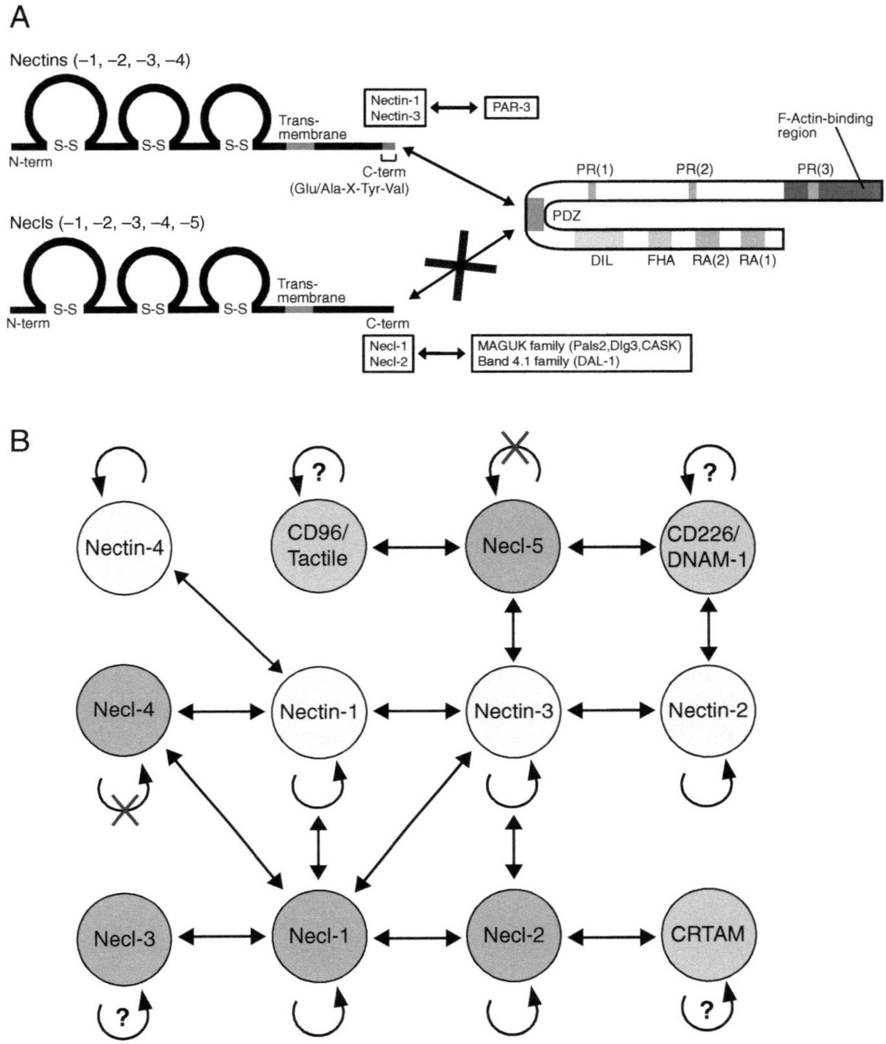

Fig. 8.2 Molecular structures and properties of nectins and Necls. (**A**) Both nectins and Necls contain three Ig-like loops in their extracellular region, a single transmembrane domain and a cytoplasmic tail. Nectin family members possess a consensus motif of four C-terminal amino acids that interact with the F-actin-binding protein afadin. This direct interaction involves the C-terminal motif of nectins and the PDZ domain of afadin, thereby linking nectins to the actin cytoskeleton. Necls are structurally similar to nectins, but do not directly bind afadin. Nectin-1 and -3 directly bind PAR-3, which is involved in cell polarization, while Necl-1 and -2 bind the members of the MAGUK and Band 4.1 families. (**B**) Nectins, Necls, and other Ig-like molecules homophilically (*looped arrows*) and heterophilically (*double arrows*) interact in *trans* with each other. MAGUK: membrane associated guanylate kinase, RA: Ras association domain, FHA: forkhead associated domain, DIL: dilute domain, PDZ: PDZ domain, PR: proline rich domain

PAR-3 (Takekuni et al. 2003) via their C-terminal four amino acids binding to the PAR-3 PDZ domain. The sequences of the C-terminal motif of nectin-1 and -3 are the same, whereas that of nectin-2 is different. This difference may determine the specific binding of PAR-3 to nectin-1 and -3. PAR-3 is one of the cell polarity proteins, which include PAR-6 and atypical protein kinase C (aPKC) and is required for apico-basal polarization of epithelial cells (Ohno 2001, Roh and Margolis 2003). Furthermore, nectins indirectly interact with other peripheral membrane proteins such as synaptic scaffolding molecule (S-SCAM), annexin II, and IQGAP1 (Katata et al. 2003, Yamada et al. 2003, Yamada et al. 2005, Yamada et al. 2006a). S-SCAM is a neural scaffolding protein, which interacts with many proteins including neuroligin, NMDA receptors, neural plakophilin-related armadillo-repeat protein/δ-catenin, a GDP/GTP exchange protein for Rap1 small G protein (PDZ-Rap-GEP), and β-catenin (Hirao et al. 1998, Ide et al. 1999, Nishimura et al. 2002). Annexin II, also termed calpactin I heavy chain, is a member of the annexin family of Ca^{2+}- and phospholipid-binding proteins and forms a heterotetrameric complex with S100A10, also termed calpactin I light chain (Gerke and Moss 2002). IQGAP1 is an F-actin-crosslinking protein known to be a downstream target of Rac and Cdc42 small G proteins (Noritake et al. 2005).

Necls are Ca^{2+}-independent Ig-like CAMs (Fig. 8.2A) and are genetically and structurally similar to nectins; Necls have an extracellular region with three Ig-like loops, a transmembrane region and a short cytoplasmic tail (Takai et al. 2003). In contrast to nectins, Necls do not bind afadin at their C-terminal regions. Necls comprise a family with five members, Necl-1 (TSLL1/Syn-CAM3), Necl-2 (IGSF4/RA175/SgIGSF/TSLC1/SynCAM1), Necl-3 (Syn-CAM2), Necl-4 (TSLL2/SynCAM4), and Necl-5 (Tage4/PVR/CD155). Although Necls have received many different names, here we will use the term Necls unless otherwise specified, as Necl-1 was first submitted to GenBank under this name. Although Necls do not bind afadin, many proteins that interact with Necls at their cytoplasmic regions have been identified. For instance, Necl-1 and Necl-2 bind Dlg3/MPP3, Pals2 and CASK, members of the membrane-associated guanylate kinase family, Necl-2 binds DAL-1, a member of the Band 4.1 family, and Necl-5 interacts with Tctex-1, a subunit of the motor-related protein dynein (Mueller et al. 2002, Yageta et al. 2002, Shingai et al. 2003).

8.3 Cell–Cell Adhesion Activity of Nectins and Necls

Each nectin first forms homo-*cis*-dimers and then forms homophilic or heterophilic *trans*-dimers (Fig. 8.2B). This property is different from that of cadherins, which *trans*-interact only homophilically in a Ca^{2+}-dependent manner. Nectin-1 heterophilically *trans*-interacts with nectin-3 and -4 and nectin-2 also *trans*-interacts with nectin-3. Of the various combinations of the homo- and hetero-*trans*-dimers, the *trans*-dimers formed between nectin-1 and -3 shows the strongest intercellular adhesion activity. For example, the Kd values of nectin-3 for nectin-1 and -2 are 2.3 and 360 nM, respectively, as estimated by surface plasmon resonance analysis (Ikeda et al. 2003). Nectins interact in *trans* not only with nectins but also with other Ig-domain molecules. Nectin-2 *trans*-interacts with CD226/DNAM1, a single membrane-spanning molecule possessing two Ig-like loops in its extracellular region. CD226/DNAM1 supports the differentiation and proliferation of T cells in which this molecule is mainly expressed (Chen et al. 2003, Shibuya et al. 2003).

Each member of the Necl family also interacts in *trans* homophilically with one another and heterophilically with other members of the Necl family and other Ig-domain molecules. Only Necl-4 and -5 do not exhibit a homophilical *trans* interaction (Fig. 8.2B). It is not known whether they form *cis*-dimers. Necl-1 homophilically *trans*-interacts with itself and heterophilically *trans*-interacts with nectin-1, -3 and Necl-2, but not with nectin-2 or Necl-5 (Kakunaga et al. 2005). Necl-2 also *trans*-interacts homophilically and heterophilically with nectin-3, Necl-1, and another Ig-like molecule CRTAM, which is reported to enhance the cytotoxicity of natural killer (NK) cells (Kennedy et al. 2000, Boles et al. 2005, Kakunaga et al. 2005). Necl-3 interacts with Necl-1. Necl-4 does not interact with itself, but heterophilically *trans*-interacts with Necl-1. Necl-5 does not homophilically *trans*-interact, but heterophilically *trans*-interacts with nectin-3 and other Ig-like molecules CD96/Tactile and CD226/DNAM-1. CD96/Tactile is expressed in T cells and triggers NK cell stimulation in association with integrin (Fuchs et al. 2004).

8.4 Nectins Form AJs Cooperatively with Cadherins

In epithelial cells, nectins and afadin are strictly concentrated at AJs, whereas cadherins are widely distributed from the apical to the basal side of the lateral plasma membrane, including AJs (Takai et al. 2003, Irie et al. 2004, Sakisaka et al. 2007). In the formation of AJs, cell–cell adhesions are first formed by nectins and subsequently cadherins are recruited to the nectin-based cell–cell adhesion sites. Nectins are associated with the actin cytoskeleton through afadin, as cadherins are associated with afadin through the catenin complex and its associated actin filament (F-actin)-binding proteins, α-actinin and vinculin (Fig. 8.3). Cadherins homophilically *trans*-interact with each other in a Ca^{2+}-dependent manner. With their cytoplasmic domain, classic cadherins bind β-catenin through their C-terminal regions and p120 catenin through their juxtamembrane segments. The cadherin-coupled β-catenin further associates with α-catenin, which then binds α-actinin and vinculin. The nectin–afadin system physically associates with the cadherin–catenin system through afadin and α-catenin. Both afadin and α-catenin directly interact with one another,

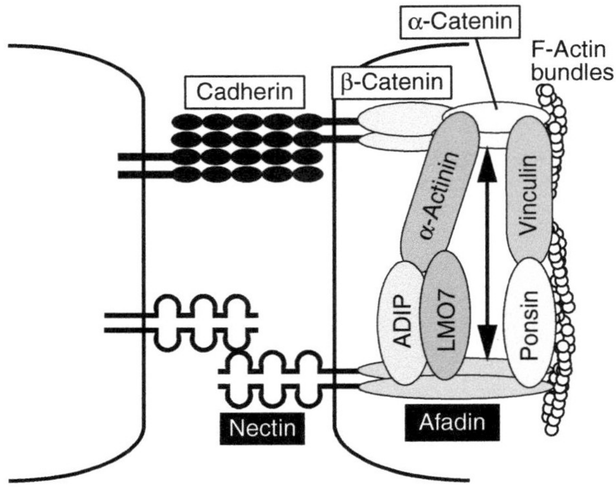

Fig. 8.3 Association of the nectin–afadin system with the cadherin–catenin system. In addition to the direct interaction between afadin and α-catenin, there are three connector units linking the nectin–afadin and cadherin–catenin systems: the ponsin–vinculin unit, the ADIP-α-actinin unit, and the LMO7-α-actinin unit. These connector units and the direct interaction between afadin and α-catenin, as well as the actin cytoskeleton, facilitate the recruitment of the cadherin–catenin system to the nectin-based cell–cell adhesion sites

although the affinity of these two molecules for each other is not high (Tachibana et al. 2000, Pokutta et al. 2002). Afadin and α-catenin are also indirectly associated through other F-actin binding proteins and adaptor proteins such as ponsin, vinculin, ADIP, LMO7, and α-actinin (Mandai et al. 1999, Asada et al. 2003, Ooshio et al. 2004) (Fig. 8.3). Of these proteins that are associating with nectins and cadherins, afadin, α-atenin, α-actinin, and vinculin are F-actin-binding proteins and participate in the association of these CAMs with each other and to the actin cytoskeleton.

In addition, *trans*-interactions between nectins induce the activation of intracellular signaling molecules at initial contact sites and the subsequent reorganization of the actin cytoskeleton. This also contributes to the association of the nectin–afadin system with the cadherin–catenin system. Upon *trans*-interaction, nectins first induce the activation of c-Src, a tyrosine kinase (Fukuhara et al. 2004). Activated c-Src then induces the activation of small G proteins Rap1, Cdc42, and Rac (Fig. 8.4). Cdc42 and Rac reorganize the IQGAP1-dependent actin cytoskeleton. This results in the recruitment of free cadherins and F-actin binding proteins that interact with cadherins to the nectin-based cell–cell adhesion sites. Rap1 binds to afadin, which then binds to p120 catenin associated with free cadherins (Hoshino et al. 2005). This Rap1-dependent binding of afadin to p120 catenin inhibits endocytosis of cadherins and enhances accumulation of free cadherins at the nectin-based cell–cell adhesion sites and the cell–cell adhesion activity of cadherins eventually results in the

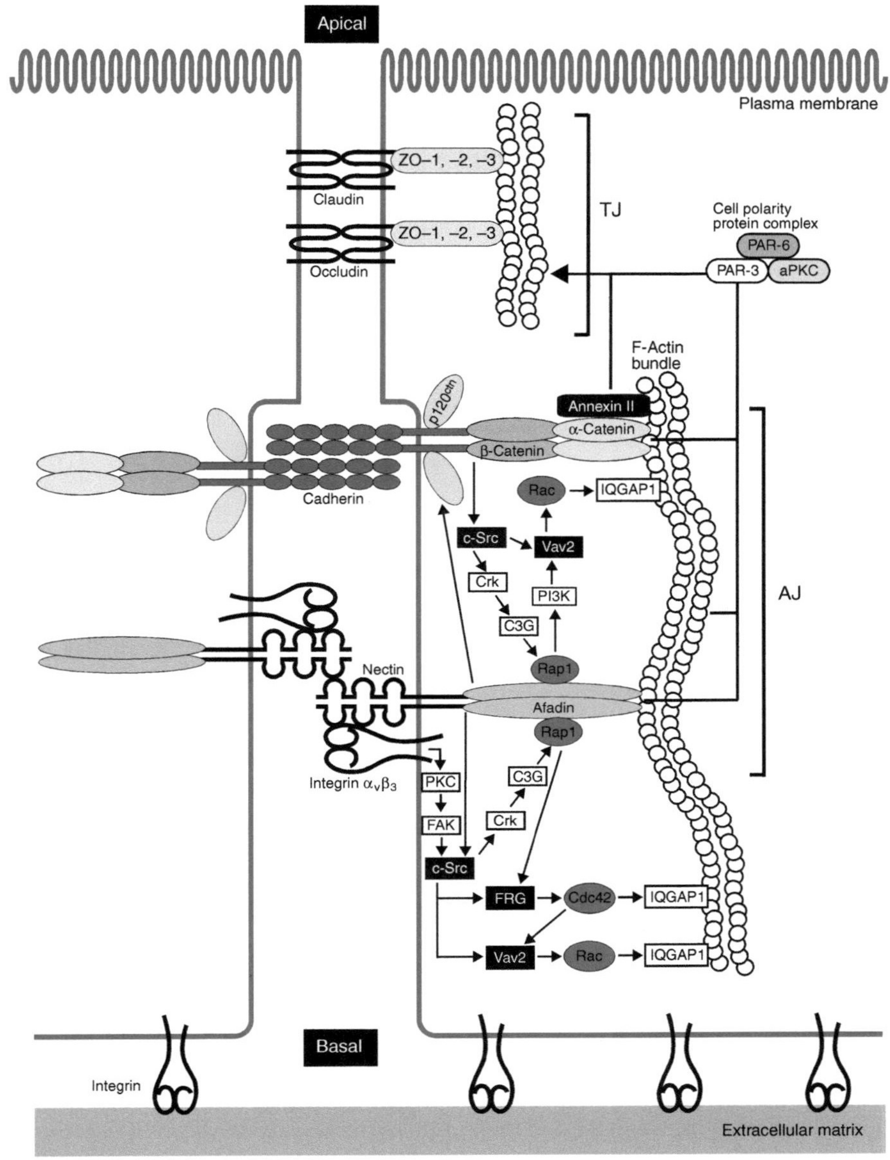

Fig. 8.4 Nectin- and cadherin-induced formation of AJs and TJs. *Trans*-interacting nectin at the initial cell–cell adhesion induces the activation of Rap1, Cdc42, and Rac through c-Src, Crk, C3G, FRG, and Vav2. This nectin-induced signaling is dependent on integrin $\alpha_v\beta_3$, which physically associates with nectin and its downstream signaling molecules PKC and FAK. Activated Cdc42 and Rac reorganize the actin cytoskeleton through IQGAP1 and recruit the cadherin–catenin system to the nectin-based cell–cell adhesion sites. During this phase, cadherin has only a weak adhesion activity; however, afadin interacting with activated Rap1 also associates with p120 catenin, leading to increased adhesion activity of cadherin and

formation of AJs (Hoshino et al. 2005, Sato et al. 2006). In addition, activated Cdc42 increases the number of filopodia and cell–cell adhesion sites, whereas activated Rac induces the formation of lamellipodia and efficiently seals cell–cell adhesion between filopodia like a zipper.

8.5 Interactions of Nectins with Other CAMs and a Growth Factor Receptor in Cell Adhesions

During the nectin-initiated formation of AJs, nectins interact in *cis* with integrin $\alpha_v\beta_3$, one of the CAMs that anchor cells to the extracellular matrix through their extracellular domain (Sakamoto et al. 2006). Although nectins are capable of interacting with both low- and high-affinity forms of integrin $\alpha_v\beta_3$, the interaction of nectins with the high-affinity form is crucial for the initial AJ formation step. The high-affinity form of integrin $\alpha_v\beta_3$ induces the activation of signaling molecules PKC and FAK (Ozaki et al. 2007), which participate in the nectin-induced activation of c-Src (Fig. 8.4). Thus, both nectin- and integrin $\alpha_v\beta_3$-initiated intracellular signaling are necessary for the formation of AJs and converge at the step of c-Src activation. However, after the establishment of AJs, the high-affinity form of integrin $\alpha_v\beta_3$ is gradually converted into the low-affinity form, which continues to interact with nectins at AJs. Nectins are involved in this conversion of integrin $\alpha_v\beta_3$. Nectins associate with and activate PTPμ, one of the receptor protein tyrosine phosphatases, at the cell–cell adhesion sites (Sakamoto et al. 2008). Activated PTPμ inhibits phosphatidylinositol phosphate kinase type Iγ90 (PIPKIγ90), the kinase which plays a key role in the activation of integrins by increasing the generation of phosphatidylinositol 4,5-bisphosphate and promoting the binding of talin to integrins (Martel et al. 2001, Di Paolo et al. 2002, Ling et al. 2002). Thus, when the nectin-based cell–cell junctions are formed, nectins inhibit integrin $\alpha_v\beta_3$ by negatively regulating PIPKIγ90 through PTPμ. Given that the high-affinity form of integrin $\alpha_v\beta_3$ upregulates cell movement and proliferation and thereby tend to cause the disruption of cell–cell junctions, inactivation of integrin $\alpha_v\beta_3$ after the formation of AJs seems to favor the maintenance of cell–cell junctions.

Cell survival is enhanced by growth factors such as PDGF. Nectin-3 and afadin were recently shown to positively regulate the PDGF-induced activation

Fig. 8.4 (continued) the establishment of cadherin-based AJs. The *trans*-interaction of cadherin induces the activation of Rap1 and Rac through c-Src, Crk, C3G, PI3K, and Vav2 to maintain cadherin-based AJs by reorganizing the actin cytoskeleton and inhibiting the endocytosis of cadherin. After the formation of AJs, both the nectin–afadin and cadherin–catenin systems cooperate to play an essential role in the formation of TJs, as well as AJs. The Par cell polarity protein complex, annexin II and the actin cytoskeleton are also involved in the formation of TJs

of the phosphatidylinositol 3-kinase (PI3K)/Akt pathway, a well-known pathway important for cell survival (Kanzaki et al. 2008). Formation of a complex with nectin-3 and afadin is important for activation of the PI3K/Akt pathway. These data suggest that the nectin–afadin complex plays a role in the cross talk between CAMs and growth factor receptors for cell survival.

Contact inhibition is a phenomenon where normal cells cease cell movement and proliferation after they grow to confluency (Abercrombie and Heaysman 1953). Nectins and Necl-5 function as important regulators of this process. In individually moving cells, Necl-5 accumulates at the leading edge of the cells together with integrin $\alpha_v\beta_3$ and PDGF receptor and enhances the integrin $\alpha_v\beta_3$- and PDGF receptor-mediated signaling of cell movement and proliferation (Minami et al. 2007a, Amano et al. 2008). However, when moving cells contact each other, Necl-5 *trans*-interacts with nectin-3 at the primordial cell–cell contact sites, which triggers the internalization of Necl-5 and results in the downregulation of Necl-5 from the cell surface (Fujito et al. 2005). In turn, this downregulation leads to a reduction in cell movement and proliferation by inhibiting the signaling pathways that are otherwise initiated by integrin $\alpha_v\beta_3$ and growth factor receptors. In contrast, contact inhibition is disrupted in transformed cells, in which the expression of Necl-5 is robustly upregulated (Minami et al. 2007b). Thus, the expression level of Necl-5 is critically correlated with contact inhibition. In addition, the *trans*-interaction of nectins, which occurs just after the Necl-5–nectin-3 interaction and the subsequent Necl-5 internalization, is also connected to contact inhibition. The nectin-induced inactivation of integrin $\alpha_v\beta_3$ contributes to the stabilization of cell–cell junctions and the prevention of cell movement (Sakamoto et al. 2008).

Recently, several lines of evidence have suggested that nectin-based cell–cell adhesion is essential for the formation of TJs (Kawakatsu et al. 2002, Honda et al. 2003, Takai et al. 2003, Takekuni et al. 2003, Irie et al. 2004, Sakisaka et al. 2007) (Fig. 8.4). In polarized epithelial cells, AJs are formed first, followed by TJs at the apical side of AJs, resulting in the formation of cell polarities. The formation of TJs is dependent on the formation of AJs. The molecular mechanism, by which TJ components are recruited to the apical side of the AJs, remains a mystery. However, it is known that afadin, ZO-1, and cell polarity proteins including PAR-3, atypical protein kinase C (aPKC), and PAR-6 are involved in this process. ZO-1 is associated with nectins through afadin and nectins are likely to recruit TJ components through afadin, ZO-1, and cell polarity proteins.

8.6 Involvement of Nectins and Cadherins in the Formation of Synapses

Nectins are also expressed in the neural systems. Recent studies have revealed that nectins also play a role in the interneuronal synapse formation. At the synapses between the mossy fiber terminals and the dendrites of pyramidal cells

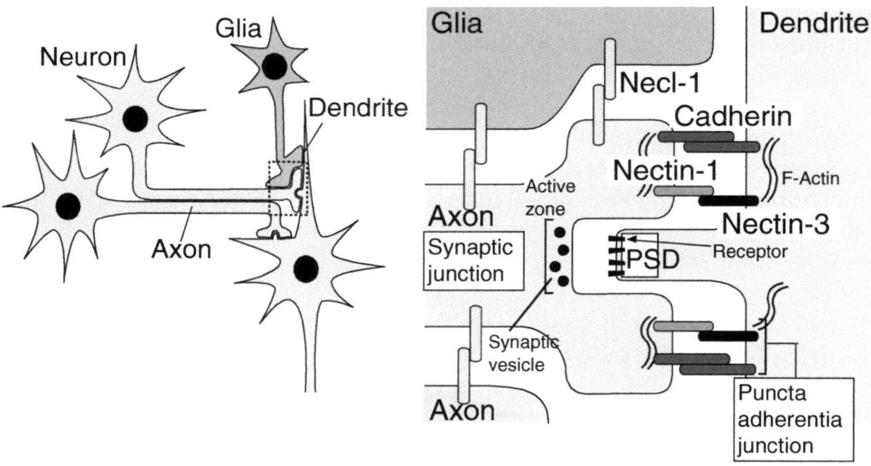

Fig. 8.5 Cell–cell junctions and contacts of neurons. Synapses are intercellular junctions that are typically formed between axons and dendrites of neurons. SJs are regarded as sites of neurotransmission and are associated with synaptic vesicles at presynaptic active zones where Ca^{2+} channels are localized and with PSDs where neurotransmitter receptors are localized. PAJs, which are not associated with synaptic vesicles or PSDs, appear to be ultrastructurally similar to AJs of epithelial cells. Nectin-1 and -3 localize asymmetrically to the presynaptic and postsynaptic sides, respectively, of the plasma membranes of PAJs. Cadherins are localized symmetrically at both sides. Necl-1 localizes to the contact sites between two axon terminals, between an axon terminal and an axonal shaft and between an axon terminal and glia cell processes

in the CA3 area of the hippocampus, both SJs and PAJs are highly developed (Fig. 8.5). Cadherins, β-catenin, αN-catenin, and afadin are localized symmetrically at PAJs, whereas nectin-1 and -3 are localized asymmetrically at the presynaptic and postsynaptic sides of PAJs, respectively. Accumulating evidence suggests that cadherins play roles in the formation of synapses. In cultured hippocampal neurons, the accumulation of the cadherin–catenin complex is also observed at early axo-dendritic filopodial contacts and is retained in many of the mature synapses during development. Initial contacts between synaptic partners are frequently established between axonal growth cones and dendritic filopodia extending from dendrites in vitro (Zhai et al. 2001, Ziv and Garner 2004). Cadherins and β-catenin are diffusely distributed along the length of free dendritic filopodia that are not in contact with axons, but rapidly accumulate at filopodia–axon contact sites (Togashi et al. 2002, Jontes et al. 2004). When compared with the control situation in the chick retinotectal tract, inhibiting the function of N-cadherin by the use of anti-N-cadherin antibodies results in the formation of widely separated synaptic clefts (Yamagata et al. 1995). Blockade of cadherin-6B activity using an anti-cadherin-6B antibody affects the distribution of PSD-95 proteins in cultured retinal neurons (Honjo et al. 2000). Blocking cadherin activity using a dominant-negative mutant results in the appearance of filopodia-like spines, an increase in the spine length and a decrease in the spine head width and also affects the organization of synapses (Togashi et al. 2002, Bozdagi et al. 2004). However, despite the evidence that cadherins are involved in the formation of synapses, cadherins are not sufficient to form synapses in vitro (Scheiffele et al. 2000, Sara et al. 2005), as the expression of N-cadherin in non-neuronal cells fails to induce presynaptic differentiation in pontine axons at the sites of contact. This is in contrast to other heterologous adhesion systems, such as the neurexin–neuroligin system (Scheiffele et al. 2000, Nam and Chen 2005, Sara et al. 2005), which were shown to be sufficient for the induction of synapse-like structures in vitro. In co-cultures with pontine explants, neuroligin-transfected non-neuronal HEK293 cells induce the clustering of synaptic vesicles in the pontine axons (Scheiffele et al. 2000; Sara et al. 2005). It is likely that many of these synaptic cell adhesion systems act in concert to coordinate the formation of synapses. Indeed, the cadherin–catenin complex interacts with S-SCAM, which plays a role in the localization of neuroligin to postsynaptic sites to facilitate synapse formation (Iida et al. 2004).

Accumulating evidence suggests that nectins are also involved in the formation of synapses. Inhibition of the nectin-based adhesion by an inhibitor of nectin-1 (gD or Nef-3) in cultured hippocampal neurons causes abnormal distribution patterns of nectin-1, -3, synaptophysin, and PSD-95, resulting in a decrease in the size of synapses and a concomitant increase in the number of synapses (Mizoguchi et al. 2002). gD is an extracellular fragment of an envelope glycoprotein D of herpes simplex virus type 1 fused to the Fc portion of IgG and similarly, Nef-3 is an extracellular fragment of nectin-3 fused to the Fc portion of IgG. Both gD and Nef-3 bind nectin-1 and inhibit the formation of the nectin-1-based intercellular adhesion. However, how inhibition of nectins decreases the size of synapses and increases the number of synapses remains to be elucidated. Genetic deletion analyses of nectins in mice suggest that nectins regulate the axo-dendritic spine contacts. In hippocampal neurons isolated from nectin-1 knockout mice, the dendritic spines are unusually elongated or deformed, resulting in a smaller spine head which exhibits a filopodia-like morphology (Fig. 8.6A) (Togashi et al. 2006). However, synaptic protein assembly more or less occurs normally in mature nectin-1-deleted neurons, although the puncta of synaptic proteins tends to be reduced in size (Togashi et al. 2006). In vivo analysis of the hippocampus in nectin-1 or -3 knockout mice also supports the idea that both nectin-1 and -3 are required for proper axo-dendritic contacts. The loss of expression of either nectin causes a reduction in the localization of the other nectin and in the sizes of the immunofluorescence signals for afadin and N-cadherin in the hippocampus (Honda et al. 2006). In addition, the number of PAJs at the synapses between the mossy fiber terminals and the dendrites of the CA3 pyramidal cells is reduced and abnormal mossy fiber trajectories are observed in both nectin-1 and -3 mutant mice (Honda et al. 2006). These data indicate that nectin-1 and -3 form hetero-*trans*-dimers at the PAJs of the mossy fiber synapses in hippocampus. It also suggests that the nectin-based PAJs cooperate with cadherins and are essential for the

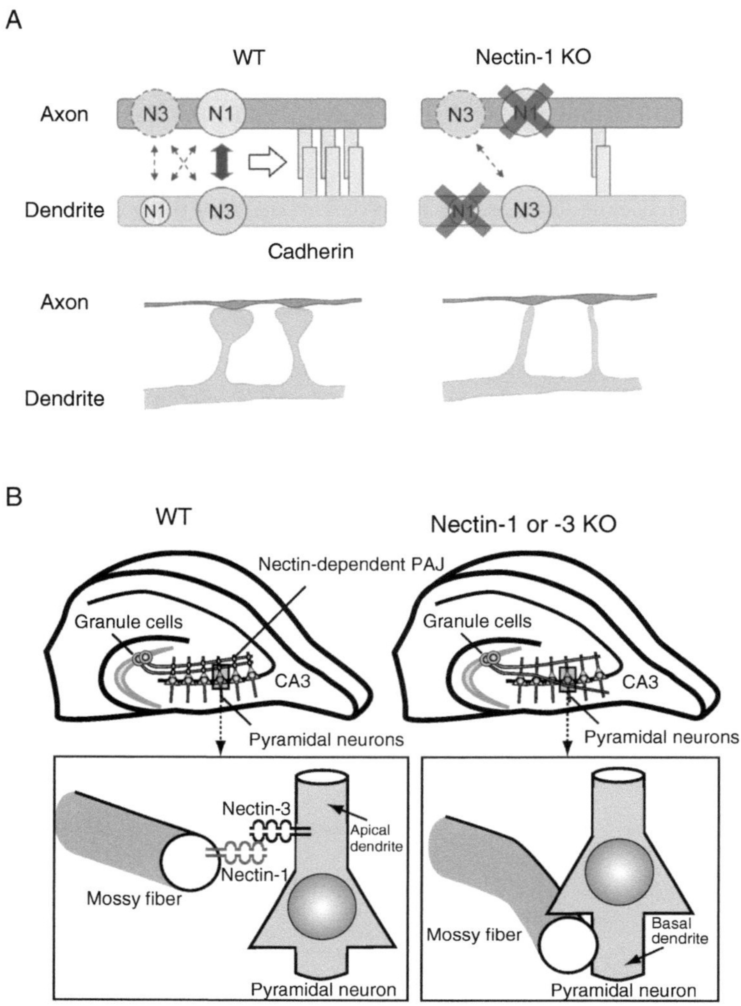

Fig. 8.6 The effects of genetic deletion of nectins in the brain. (**A**) In wild-type neurons, nectin-1 is abundant in the axon and interacts with nectin-3 in the dendrite. This heterophilic *trans*-interaction of nectin-1 and -3 promotes the homophilic cadherin–cadherin interactions to strengthen synaptic junctions. In the absence of nectin-1, only a basic level of cadherin interactions takes place. Homophilic interactions between nectin-3 and -3 may not be strong enough to sustain normal axo-dendritic contacts. As a result, their dendritic spines are unusually elongated or deformed, resulting in a filopodia-like morphology. Dotted arrows indicate possible weaker interactions between nectins. The axonal nectin-3 level appears to decrease with the maturation of neurons. (**B**) In the wild-type hippocampus, the mossy fibers are connected to the apical dendrites of the CA3 pyramidal cells by the nectin-based PAJs. In the hippocampi of nectin-$1^{-/-}$ or nectin-$3^{-/-}$ mice, the interactions of the mossy fibers with the apical dendrites of the CA3 pyramidal cells are lost because there are no nectin-based PAJs. As a result, the mossy fibers are misguided, resulting in abnormal mossy fiber trajectories

mechanical fixation of the mossy fibers to the apical dendrites of the CA3 pyramidal cells (Fig. 8.6B). These data concur with the observation that the contacts between the commissural axons and the floor plate cells are mediated by the hetero-*trans*-interaction between nectin-1 and -3 and are involved in the regulation of commissural axonal trajectories (Okabe et al. 2004). At the contact sites between the commissural axons and the floor plate cells, nectin-1 and -3 are asymmetrically localized. In vitro perturbation of the interaction between nectin-1 and -3 causes abnormal fasciculation of the commissural axons and an impairment of the contacts, resulting in a failure in longitudinal turns of the commissural axons at the contralateral sites of the hindbrain (Okabe et al. 2004). The genetic deletion analysis of afadin in mice also exhibited similar phenotypes as nectin mutant mice (unpublished data). Since perforated synapses in the hippocampus are only observed in afadin mutant mice, some phenotypic aspects are more severe than in nectin-1 or -3 mutant mice.

At least three developmental stages are recognized at the ultrastructural level during the maturation of synapses (Uchida et al. 1996, Mizoguchi et al. 2002) (Fig. 8.7). At the first developmental stage, nectins, afadin, catenins, and cadherins colocalize at cell–cell contact sites, which are probably the most primitive synapses where morphological differentiation between SJs and PAJs is not clearly observed. Appearance of the nectin–afadin and the cadherin–catenin systems precedes the membrane domain segregation of SJs and PAJs. At the second stage, when morphological differentiation between the two domains is complete, the nectin–afadin system localizes with αN-catenin to both SJs and PAJs. At the final stage, when SJs are formed exclusively on the spines that stem from dendrites, the

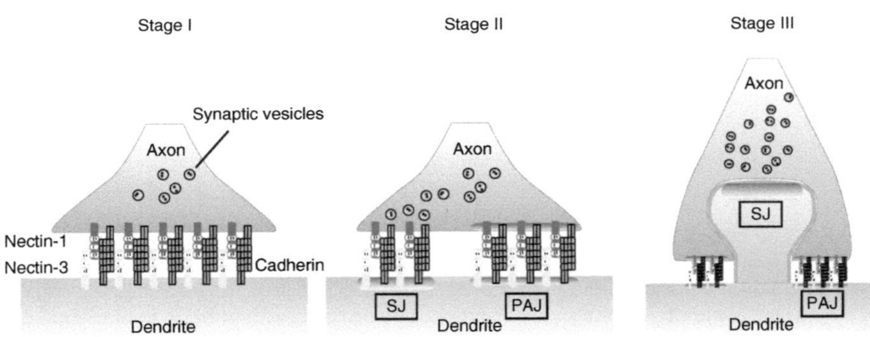

Fig. 8.7 Three developmental stages during the maturation of synapses. At the first developmental stage (*Stage I*), nectins and cadherins colocalize to cell–cell contact sites, which are probably the most primitive synapses where morphological differentiation between SJs and PAJs is not clearly observed. Appearance of the nectin–afadin and cadherin–catenin systems precedes the membrane domain segregation of SJs and PAJs. At the second stage (*Stage II*), when morphological differentiation between the two domains is complete, nectins and cadherins still localize both at SJs and PAJs. At the final stage (*Stage III*), when SJs are formed exclusively on the spines that stem from dendrites, the nectin–afadin system, together with the cadherin–catenin system, localizes to PAJs

nectin–afadin system, together with the cadherin–catenin system, localizes at PAJs. These data suggest that during the maturation of synapses, membrane domain specialization gradually proceeds and both the nectin–afadin and cadherin–catenin systems may participate in domain segregation and spine formation. The molecular mechanisms of the formation of PAJ and the neural membrane domain specialization may have some analogy with those found during the formation of the junctional complex in epithelial cells with respect to the dynamic localization patterns of the junctional proteins. The membrane domains, comprising SJs and PAJs, would then gradually become segregated, followed by a maturation of synapses as AJs and TJs are segregated. ZO-1 colocalizes with nectins and cadherins at PAJs (Inagaki et al. 2003), suggesting that ZO-1 plays a role in the segregation of the components of SJs and PAJs, as is described for the role of ZO-1 in the segregation of the components of AJs and TJs in epithelial cells.

8.7 Involvement of Nectins in the Selective Association between Axons and Dendrites

Axons attach to dendrites for synaptogenesis, but dendrites do not form stable contacts with each other, suggesting the presence of a mechanism that allows selective association. Firm contacts between axons and dendritic spines develop, but other types of contacts, such as dendro-dendritic contacts, are not stabilized. Nectin-1 in axons and nectin-3 in dendrites have recently been shown to play a critical role in the selective association of axons and dendrites. The heterophilic *trans*-interaction between nectin-1 and -3 preferentially takes place between axons and dendrites (Togashi et al. 2006), as this heterophilic *trans*-interaction is stronger than the homophilic *trans*-interaction of nectin-1 or that of nectin-3. This enhances accumulation of N-cadherin to axo-dendritic contacts and stabilizes synaptic contacts in cultured neurons. Nevertheless, overexpression of cadherins alone is not sufficient to induce the axo-dendritic interaction or aberrant neurite association (Togashi et al. 2006). These data demonstrate that cadherins alone can not initiate such specialized local adhesion contacts. Rather, local cadherin-based connections are formed by cooperation with the heterophilic adhesion between nectin-1 and -3 which distribute differentially between axons and dendrites. However, the exact mechanism of differential targeting of nectins has not been identified.

8.8 Possible Roles of Nectins and Cadherins in Synapse Remodeling

It is known that several types of mental retardation and cognitive disorders are associated with abnormalities in spine density and morphology (Purpura 1974, Fiala et al. 2002). A morphological change of synapses has also been implicated in learning and memory (Yuste and Bonhoeffer 2001). The shape of a spine is determined by the architecture of its actin cytoskeleton (Carlisle and Kennedy

2005). The formation of nectin- and cadherin-mediated cell–cell junctions is accompanied by a reorganization of the actin cytoskeleton through the activation of small G proteins such as Cdc42 and Rac. These observations suggest that the nectin- and cadherin-based interactions at synapses are involved in determining the spine shape by reorganizing the actin cytoskeleton in spines. Organization and dynamic remodeling of the postsynaptic density (PSD) are thought to be critical in postsynaptic signal transduction. Indeed, maintenance of the PSD structure involves the actin cytoskeleton (Allison et al. 1998). Recent studies indicate a variety of interactions between PSD scaffolding proteins and the actin cytoskeleton (Wyszynski et al. 1997, Bockers et al. 2001, Hering and Sheng 2003). These data suggest that the intact actin meshwork is important for the maintenance and remodeling of the PSD composition. In the hippocampus of afadin mutant mice, perforated synapses are frequently observed at an ultrastructural level (unpublished data). Perforated synapses have an interrupted PSD that is apposed to a single presynaptic terminal. The molecular details of how these perforated synapses are formed are still unknown. Interestingly, the induction of LTP also causes a rapid and transient increase in the proportion of perforated synapses (Geinisman et al. 1993, Luscher et al. 2000) (Fig. 8.8), which are thought to be transient components of synaptic activation (Sorra et al. 1998). Actin-dependent rapid shape changes of dendritic spines at excitatory synapses are thought to contribute to plasticity (Fischer et al. 2000). In this way, the nectin–afadin complex may play a role in the control of synaptic dynamics and contribute to the stability and plasticity of synaptic contacts.

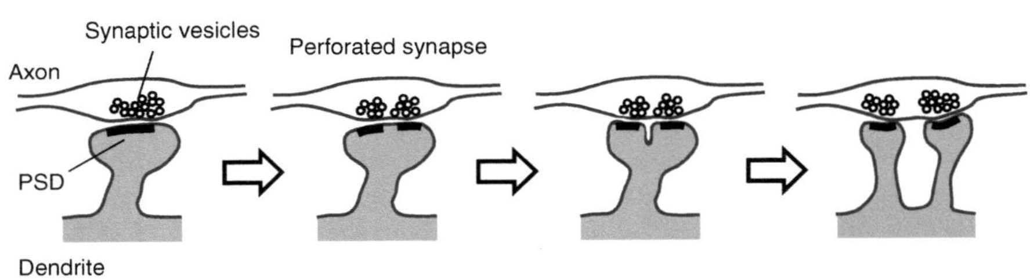

Fig. 8.8 Schematic diagram illustrating the morphologic changes that might be associated with synaptic plasticity. High-frequency stimulation is postulated to result in the transient occurrence of perforated synapses with larger PSDs. These might then be transformed into duplicated synapses

8.9 Involvement of Necls in the Formation of Various Types of Cell–Cell Junctions in the Central and Peripheral Nervous Systems

Necl-1 and -4 are specifically expressed in the central and peripheral nervous systems (CNS and PNS) and localize to contact sites along axons, nerve terminals and glial cell processes, axon bundles and myelinated axons. In the

PNS, neurons and Schwann cells express distinct sets of Necl proteins: axons highly express Necl-1 and -2, whereas Schwann cells express Necl-4 and -2. Two recent studies report that Necl-1 and -4 are internodal adhesion molecules that are critical for myelination in the PNS (Maurel et al. 2007, Spiegel et al. 2007). Binding of Necl-1 on axons specifically to Necl-4 on Schwann cells mediates axo-glial interactions, Schwann cell differentiation, and myelination. However, whether these Necls are involved in the myelination and formation of synapses in the CNS remains unknown.

Necl-2 is expressed in the granular layer of the cerebellum, as well as in the lungs and testes. Necl-2 is localized at the basolateral plasma membrane in epithelial cells, but not at specialized cell–cell junctions, such as TJs, AJs, and desmosomes. It plays important roles in epithelial cell adhesion and acts as a tumor suppressor in human non-small cell lung cancer (Kuramochi et al. 2001, Shingai et al. 2003). Deletion analysis of Necl-2 in mice shows that the adhesion of spermatocytes and spermatids to Sertoli cells is impaired and that Necl-2 is indispensable for their normal differentiation into mature spermatozoa (Yamada et al. 2006b). Recently, Biederer and co-workers reported SynCAM1 (synaptic cell adhesion molecule 1) as synaptic cell adhesion molecule in vertebrates and its localization at both pre- and postsynaptic specializations (Biederer et al. 2002). SynCAM1 is identical to Necl-2. An interaction between SynCAM1 from non-neuronal HEK293 cells and hippocampal neurons induces the development of functional excitatory presynaptic terminals at sites of contact (Biederer et al. 2002, Sara et al. 2005). The cytoplasmic tail of SynCAM1 includes a PDZ-domain protein-interaction sequence that binds CASK, Mint1, and syntenin. The expression of SynCAM1 protein is brain specific and temporally correlated with synaptogenesis. However, in contrast to our findings, Biederer et al. showed SynCAM1 to be specifically synthesized in the brain, whereas Necl-2 was found ubiquitously. We do not consider SynCAM1 as a synaptic CAM, because SynCAM1 is not concentrated at synapses. To date, the role of SynCAM1 in synapse formation in vivo remains unknown.

In addition to the CNS and the PNS, Necl-3 is expressed in several other tissues (Pellissier et al. 2007). It is expressed in ependymal cells and in myelinated axons and is present at the axon–oligodendrocyte interface indicating an involvement of Necl-3 in the adhesion between axons and oligodendrocytes. These data suggest that Necl-3 participates in the interactions between different cell types in the nervous system.

8.10 Conclusions and Perspectives

In this chapter, we have described recent findings on the molecular mechanisms of cell adhesion and neural circuit formation, focusing on nectins and Necls and their cross talk with other cell adhesion molecules and growth factor receptors.

A number of studies on the mode of action of nectins and cadherins are beginning to reveal the underlying mechanism of PAJ formation and the molecular details of the associations between axons and dendrites. However, our understanding of how synapses are formed and the molecules involved in this process remains incomplete. Cell–cell junctions between epithelial cells and at the leading edge of moving fibroblasts involve interactions of nectins or Necls with integrin $\alpha_v\beta_3$ and the PDGF receptor. These interactions induce the activation of various intracellular signals. However, it is unknown how these nectin- and Necl-mediated interactions and the associated intracellular signals might be involved in the assembly of synapses. Afferent axons continue to grow until they encounter and recognize their target cells. This phenomenon resembles contact inhibition of cell movement and proliferation observed in fibroblasts. However, it is also unclear whether the molecular mechanisms underlying contact inhibition also apply to this phenomenon. Further studies of nectins and Necls in the nervous system should address these issues in the future.

References

Abercrombie M and Heaysman JE (1953) Observations on the social behaviour of cells in tissue culture. I. Speed of movement of chick heart fibroblasts in relation to their mutual contacts. Exp Cell Res 5:111–131

Allison DW, Gelfand VI, Spector I et al. (1998) Role of actin in anchoring postsynaptic receptors in cultured hippocampal neurons: differential attachment of NMDA versus AMPA receptors. J Neurosci 18:2423–2436

Amano H, Ikeda W, Kawano S et al. (2008) Interaction and localization of Necl-5 and PDGF receptor beta at the leading edges of moving NIH3T3 cells: Implications for directional cell movement. Genes Cells 13:269–284

Asada M, Irie K, Morimoto K et al. (2003) ADIP, a novel Afadin- and alpha-actinin-binding protein localized at cell-cell adherens junctions. J Biol Chem 278:4103–4111

Biederer T, Sara Y, Mozhayeva M et al. (2002) SynCAM, a synaptic adhesion molecule that drives synapse assembly. Science 297:1525–1531

Bockers TM, Mameza MG, Kreutz MR et al. (2001) Synaptic scaffolding proteins in rat brain. Ankyrin repeats of the multidomain Shank protein family interact with the cytoskeletal protein alpha-fodrin. J Biol Chem 276:40104–40112

Boles KS, Barchet W, Diacovo T et al. (2005) The tumor suppressor TSLC1/NECL-2 triggers NK-cell and CD8+ T-cell responses through the cell-surface receptor CRTAM. Blood 106:779–786

Bozdagi O, Valcin M, Poskanzer K et al. (2004) Temporally distinct demands for classic cadherins in synapse formation and maturation. Mol Cell Neurosci 27:509–521

Carlisle HJ and Kennedy MB (2005) Spine architecture and synaptic plasticity. Trends Neurosci 28:182–187

Chen L, Xie X, Zhang X et al. (2003) The expression, regulation and adhesion function of a novel CD molecule, CD226, on human endothelial cells. Life Sci 73:2373–2382

Di Paolo G, Pellegrini L, Letinic K et al. (2002) Recruitment and regulation of phosphatidylinositol phosphate kinase type 1 gamma by the FERM domain of talin. Nature 420:85–89

Fiala JC, Spacek J and Harris KM (2002) Dendritic spine pathology: cause or consequence of neurological disorders? Brain Res Brain Res Rev 39:29–54

Fischer M, Kaech S, Wagner U et al. (2000) Glutamate receptors regulate actin-based plasticity in dendritic spines. Nat Neurosci 3:887–894
Fuchs A, Cella M, Giurisato E et al. (2004) Cutting edge: CD96 (tactile) promotes NK cell-target cell adhesion by interacting with the poliovirus receptor (CD155). J Immunol 172:3994–3998
Fujito T, Ikeda W, Kakunaga S et al. (2005) Inhibition of cell movement and proliferation by cell-cell contact-induced interaction of Necl-5 with nectin-3. J Cell Biol 171:165–173
Fukuhara T, Shimizu K, Kawakatsu T et al. (2004) Activation of Cdc42 by trans interactions of the cell adhesion molecules nectins through c-Src and Cdc42-GEF FRG. J Cell Biol 166:393–405
Geinisman Y, de Toledo-Morrell L, Morrell F et al. (1993) Structural synaptic correlate of long-term potentiation: formation of axospinous synapses with multiple, completely partitioned transmission zones. Hippocampus 3:435–445
Geraghty RJ, Krummenacher C, Cohen GH et al. (1998) Entry of alphaherpesviruses mediated by poliovirus receptor-related protein 1 and poliovirus receptor. Science 280:1618–1620
Gerke V and Moss SE (2002) Annexins: from structure to function. Physiol Rev 82:331–371
Gumbiner B and Simons K (1986) A functional assay for proteins involved in establishing an epithelial occluding barrier: identification of a uvomorulin-like polypeptide. J Cell Biol 102:457–468
Gumbiner B, Stevenson B and Grimaldi A (1988) The role of the cell adhesion molecule uvomorulin in the formation and maintenance of the epithelial junctional complex. J Cell Biol 107:1575–1587
Gumbiner BM (1996) Cell adhesion: the molecular basis of tissue architecture and morphogenesis. Cell 84:345–357
Hering H and Sheng M (2003) Activity-dependent redistribution and essential role of cortactin in dendritic spine morphogenesis. J Neurosci 23:11759–11769
Hirao K, Hata Y, Ide N et al. (1998) A novel multiple PDZ domain-containing molecule interacting with N-methyl-D-aspartate receptors and neuronal cell adhesion proteins. J Biol Chem 273:21105–21110
Honda T, Sakisaka T, Yamada T et al. (2006) Involvement of nectins in the formation of puncta adherentia junctions and the mossy fiber trajectory in the mouse hippocampus. Mol Cell Neurosci 31:315–325
Honda T, Shimizu K, Kawakatsu T et al. (2003) Cdc42 and Rac small G proteins activated by trans-interactions of nectins are involved in activation of c-Jun N-terminal kinase, but not in association of nectins and cadherin to form adherens junctions, in fibroblasts. Genes Cells 8:481–491
Honjo Y, Nakagawa S and Takeichi M (2000) Blockade of cadherin-6B activity perturbs the distribution of PSD-95 family proteins in retinal neurones. Genes Cells 5:309–318
Hoshino T, Sakisaka T, Baba T et al. (2005) Regulation of E-cadherin endocytosis by nectin through afadin, Rap1, and p120ctn. J Biol Chem 280:24095–24103
Ide N, Hata Y, Deguchi M et al. (1999) Interaction of S-SCAM with neural plakophilin-related Armadillo-repeat protein/delta-catenin. Biochem Biophys Res Commun 256:456–461
Iida J, Hirabayashi S, Sato Y et al. (2004) Synaptic scaffolding molecule is involved in the synaptic clustering of neuroligin. Mol Cell Neurosci 27:497–508
Ikeda W, Kakunaga S, Itoh S et al. (2003) Tage4/Nectin-like molecule-5 heterophilically trans-interacts with cell adhesion molecule nectin-3 and enhances cell migration. J Biol Chem 278:28167–28172
Inagaki M, Irie K, Deguchi-Tawarada M et al. (2003) Nectin-dependent localization of ZO-1 at puncta adhaerentia junctions between the mossy fiber terminals and the dendrites of the pyramidal cells in the CA3 area of adult mouse hippocampus. J Comp Neurol 460:514–524

Irie K, Shimizu K, Sakisaka T et al. (2004) Roles and modes of action of nectins in cell-cell adhesion. Semin Cell Dev Biol 15:643–656

Jontes JD, Emond MR and Smith SJ (2004) In vivo trafficking and targeting of N-cadherin to nascent presynaptic terminals. J Neurosci 24:9027–9034

Kakunaga S, Ikeda W, Itoh S et al. (2005) Nectin-like molecule-1/TSLL1/SynCAM3: a neural tissue-specific immunoglobulin-like cell-cell adhesion molecule localizing at non-junctional contact sites of presynaptic nerve terminals, axons and glia cell processes. J Cell Sci 118:1267–1277

Kanzaki N, Ogita H, Komura H et al. (2008) Involvement of the nectin-afadin complex in PDGF-induced cell survival. J Cell Sci 121:2008–2017

Katata T, Irie K, Fukuhara A et al. (2003) Involvement of nectin in the localization of IQGAP1 at the cell-cell adhesion sites through the actin cytoskeleton in Madin-Darby canine kidney cells. Oncogene 22:2097–2109

Kawakatsu T, Shimizu K, Honda T et al. (2002) Trans-interactions of nectins induce formation of filopodia and Lamellipodia through the respective activation of Cdc42 and Rac small G proteins. J Biol Chem 277:50749–50755

Kennedy J, Vicari AP, Saylor V et al. (2000) A molecular analysis of NKT cells: identification of a class-I restricted T cell-associated molecule (CRTAM). J Leukoc Biol 67:725–734

Kuramochi M, Fukuhara H, Nobukuni T et al. (2001) TSLC1 is a tumor-suppressor gene in human non-small-cell lung cancer. Nat Genet 27:427–430

Ling K, Doughman RL, Firestone AJ et al. (2002) Type I gamma phosphatidylinositol phosphate kinase targets and regulates focal adhesions. Nature 420:89–93

Luscher C, Nicoll RA, Malenka RC et al. (2000) Synaptic plasticity and dynamic modulation of the postsynaptic membrane. Nat Neurosci 3:545–550

Mandai K, Nakanishi H, Satoh A et al. (1999) Ponsin/SH3P12: an l-afadin- and vinculin-binding protein localized at cell-cell and cell-matrix adherens junctions. J Cell Biol 144:1001–1017

Martel V, Racaud-Sultan C, Dupe S et al. (2001) Conformation, localization, and integrin binding of talin depend on its interaction with phosphoinositides. J Biol Chem 276:21217–21227

Maurel P, Einheber S, Galinska J et al. (2007) Nectin-like proteins mediate axon Schwann cell interactions along the internode and are essential for myelination. J Cell Biol 178:861–874

Minami Y, Ikeda W, Kajita M et al. (2007a) Necl-5/poliovirus receptor interacts in cis with integrin $\alpha v\beta$ 3 and regulates its clustering and focal complex formation. J Biol Chem 282:18481–18496

Minami Y, Ikeda W, Kajita M et al. (2007b) Involvement of up-regulated Necl-5/Tage4/PVR/CD155 in the loss of contact inhibition in transformed NIH3T3 cells. Biochem Biophys Res Commun 352:856–860

Mizoguchi A, Nakanishi H, Kimura K et al. (2002) Nectin: an adhesion molecule involved in formation of synapses. J Cell Biol 156:555–565

Mueller S, Cao X, Welker R et al. (2002) Interaction of the poliovirus receptor CD155 with the dynein light chain Tctex-1 and its implication for poliovirus pathogenesis. J Biol Chem 277:7897–7904

Nam CI and Chen L (2005) Postsynaptic assembly induced by neurexin-neuroligin interaction and neurotransmitter. Proc Natl Acad Sci U S A 102:6137–6142

Nishimura W, Yao I, Iida J et al. (2002) Interaction of synaptic scaffolding molecule and Beta -catenin. J Neurosci 22:757–765

Noritake J, Watanabe T, Sato K et al. (2005) IQGAP1: a key regulator of adhesion and migration. J Cell Sci 118:2085–2092

Ogita H and Takai Y (2008) Cross-talk among integrin, cadherin, and growth factor receptor: roles of nectin and nectin-like molecule. Int Rev Cytol 265:1–54

Ohno S (2001) Intercellular junctions and cellular polarity: the PAR-aPKC complex, a conserved core cassette playing fundamental roles in cell polarity. Curr Opin Cell Biol 13:641–648

Okabe N, Shimizu K, Ozaki-Kuroda K et al. (2004) Contacts between the commissural axons and the floor plate cells are mediated by nectins. Dev Biol 273:244–256

Ooshio T, Irie K, Morimoto K et al. (2004) Involvement of LMO7 in the association of two cell-cell adhesion molecules, nectin and E-cadherin, through afadin and alpha-actinin in epithelial cells. J Biol Chem 279:31365–31373

Ozaki M, Ogita H and Takai Y (2007) Involvement of integrin-induced activation of protein kinase C in the formation of adherens junctions. Genes Cells 12:651–662

Pellissier F, Gerber A, Bauer C et al. (2007) The adhesion molecule Necl-3/SynCAM-2 localizes to myelinated axons, binds to oligodendrocytes and promotes cell adhesion. BMC Neurosci 8:90

Perez-Moreno M, Jamora C and Fuchs E (2003) Sticky business: orchestrating cellular signals at adherens junctions. Cell 112:535–548

Peters A, Palay SL and Webster Hd (1991) The fine structure of the nervous system : neurons and their supporting cells. Oxford University Press, New York

Pokutta S, Drees F, Takai Y et al. (2002) Biochemical and structural definition of the l-afadin- and actin-binding sites of alpha-catenin. J Biol Chem 277:18868–18874

Purpura DP (1974) Dendritic spine “dysgenesis” and mental retardation. Science 186:1126–1128

Reymond N, Fabre S, Lecocq E et al. (2001) Nectin4/PRR4, a new afadin-associated member of the nectin family that trans-interacts with nectin1/PRR1 through V domain interaction. J Biol Chem 276:43205–43215

Roh MH and Margolis B (2003) Composition and function of PDZ protein complexes during cell polarization. Am J Physiol Renal Physiol 285:F377–387

Sakamoto Y, Ogita H, Hirota T et al. (2006) Interaction of integrin α(v)β 3 with nectin. Implication in cross-talk between cell-matrix and cell-cell junctions. J Biol Chem 281:19631–19644

Sakamoto Y, Ogita H, Komura H et al. (2008) Involvement of nectin in inactivation of integrin αvβ 3 after the establishment of cell-cell adhesion. J Biol Chem 283:496–505

Sakisaka T, Ikeda W, Ogita H et al. (2007) The roles of nectins in cell adhesions: cooperation with other cell adhesion molecules and growth factor receptors. Curr Opin Cell Biol 19:593–602

Sara Y, Biederer T, Atasoy D et al. (2005) Selective capability of SynCAM and neuroligin for functional synapse assembly. J Neurosci 25:260–270

Sato T, Fujita N, Yamada A et al. (2006) Regulation of the assembly and adhesion activity of E-cadherin by nectin and afadin for the formation of adherens junctions in Madin-Darby canine kidney cells. J Biol Chem 281:5288–5299

Satoh-Horikawa K, Nakanishi H, Takahashi K et al. (2000) Nectin-3, a new member of immunoglobulin-like cell adhesion molecules that shows homophilic and heterophilic cell-cell adhesion activities. J Biol Chem 275:10291–10299

Scheiffele P, Fan J, Choih J et al. (2000) Neuroligin expressed in nonneuronal cells triggers presynaptic development in contacting axons. Cell 101:657–669

Shibuya K, Shirakawa J, Kameyama T et al. (2003) CD226 (DNAM-1) is involved in lymphocyte function-associated antigen 1 costimulatory signal for naive T cell differentiation and proliferation. J Exp Med 198:1829–1839

Shingai T, Ikeda W, Kakunaga S et al. (2003) Implications of nectin-like molecule-2/IGSF4/RA175/SgIGSF/TSLC1/SynCAM1 in cell-cell adhesion and transmembrane protein localization in epithelial cells. J Biol Chem 278:35421–35427

Sorra KE, Fiala JC and Harris KM (1998) Critical assessment of the involvement of perforations, spinules, and spine branching in hippocampal synapse formation. J Comp Neurol 398:225–240

Spacek J (1985) Relationships between synaptic junctions, puncta adhaerentia and the spine apparatus at neocortical axo-spinous synapses. A serial section study. Anat Embryol (Berl) 173:129–135

Spiegel I, Adamsky K, Eshed Y et al. (2007) A central role for Necl4 (SynCAM4) in Schwann cell-axon interaction and myelination. Nat Neurosci 10:861–869
Tachibana K, Nakanishi H, Mandai K et al. (2000) Two cell adhesion molecules, nectin and cadherin, interact through their cytoplasmic domain-associated proteins. J Cell Biol 150:1161–1176
Takahashi K, Nakanishi H, Miyahara M et al. (1999) Nectin/PRR: an immunoglobulin-like cell adhesion molecule recruited to cadherin-based adherens junctions through interaction with Afadin, a PDZ domain-containing protein. J Cell Biol 145:539–549
Takai Y, Irie K, Shimizu K et al. (2003) Nectins and nectin-like molecules: roles in cell adhesion, migration, and polarization. Cancer Sci 94:655–667
Takai Y and Nakanishi H (2003) Nectin and afadin: novel organizers of intercellular junctions. J Cell Sci 116:17–27
Takai Y, Miyoshi J, Ikeda W et al. (2008) Nectins and nectin-like molecules: roles in contact inhibition of cell movement and proliferation. Nat Rev Mol cell Biol 9:603–615
Takeichi M (1991) Cadherin cell adhesion receptors as a morphogenetic regulator. Science 251:1451–1455
Takekuni K, Ikeda W, Fujito T et al. (2003) Direct binding of cell polarity protein PAR-3 to cell-cell adhesion molecule nectin at neuroepithelial cells of developing mouse. J Biol Chem 278:5497–5500
Togashi H, Abe K, Mizoguchi A et al. (2002) Cadherin regulates dendritic spine morphogenesis. Neuron 35:77–89
Togashi H, Miyoshi J, Honda T et al. (2006) Interneurite affinity is regulated by heterophilic nectin interactions in concert with the cadherin machinery. J Cell Biol 174:141–151
Uchida N, Honjo Y, Johnson KR et al. (1996) The catenin/cadherin adhesion system is localized in synaptic junctions bordering transmitter release zones. J Cell Biol 135:767–779
Warner MS, Geraghty RJ, Martinez WM et al. (1998) A cell surface protein with herpesvirus entry activity (HveB) confers susceptibility to infection by mutants of herpes simplex virus type 1, herpes simplex virus type 2, and pseudorabies virus. Virology 246:179–189
Wyszynski M, Lin J, Rao A et al. (1997) Competitive binding of alpha-actinin and calmodulin to the NMDA receptor. Nature 385:439–442
Yageta M, Kuramochi M, Masuda M et al. (2002) Direct association of TSLC1 and DAL-1, two distinct tumor suppressor proteins in lung cancer. Cancer Res 62:5129–5133
Yamada A, Fujita N, Sato T et al. (2006a) Requirement of nectin, but not cadherin, for formation of claudin-based tight junctions in annexin II-knockdown MDCK cells. Oncogene 25:5085–5102
Yamada A, Irie K, Deguchi-Tawarada M et al. (2003) Nectin-dependent localization of synaptic scaffolding molecule (S-SCAM) at the puncta adherentia junctions formed between the mossy fibre terminals and the dendrites of pyramidal cells in the CA3 area of the mouse hippocampus. Genes Cells 8:985–994
Yamada A, Irie K, Hirota T et al. (2005) Involvement of the annexin II-S100A10 complex in the formation of E-cadherin-based adherens junctions in Madin-Darby canine kidney cells. J Biol Chem 280:6016–6027
Yamada D, Yoshida M, Williams YN et al. (2006b) Disruption of spermatogenic cell adhesion and male infertility in mice lacking TSLC1/IGSF4, an immunoglobulin superfamily cell adhesion molecule. Mol Cell Biol 26:3610–3624
Yamagata M, Herman JP and Sanes JR (1995) Lamina-specific expression of adhesion molecules in developing chick optic tectum. J Neurosci 15:4556–4571
Yuste R and Bonhoeffer T (2001) Morphological changes in dendritic spines associated with long-term synaptic plasticity. Annu Rev Neurosci 24:1071–1089
Zhai RG, Vardinon-Friedman H, Cases-Langhoff C et al. (2001) Assembling the presynaptic active zone: a characterization of an active one precursor vesicle. Neuron 29:131–143
Ziv NE and Garner CC (2004) Cellular and molecular mechanisms of presynaptic assembly. Nat Rev Neurosci 5:385–399

Chapter 9
The Down Syndrome Cell Adhesion Molecule

Hitesh Kathuria and James C. Clemens

Abstract The Down syndrome cell adhesion molecules (DSCAMs) are a structurally and functionally conserved family of cell surface receptors that play important roles in nervous system organization. These receptors are expressed on both axons and dendrites where they engage in isoform-specific binding interactions between DSCAM receptors on opposing cell surfaces. Massive alternative splicing of arthropod DSCAM transcripts greatly expands the complexity of the DSCAM family by endowing these organisms with the ability to produce tens of thousands of distinct receptor isoforms that undergo homophilic binding. In addition to homophilic binding, DSCAM extracellular domains serve as receptors for other proteins such as the attractant netrin-1. These diverse interaction properties allow DSCAMs to control a variety of nervous system patterning processes including axon path-finding and targeting, neurite branch segregation, self-recognition, and neurite tiling.

Keywords DSCAM · Ig domain · Alternative splicing · Axon guidance · Neuron · Synapse · *Drosophila*

9.1 Introduction

In 1998, Julie Korenberg's group at the Cedars-Sinai Research Institute identified a new member of the immunoglobulin superfamily of cell surface receptors (Yamakawa et al. 1998). Structurally, this protein was similar to previously identified proteins that function as cell adhesion molecules (CAMs), but it possesses a unique arrangement of domains within the extracellular region. The gene encoding this novel cell surface receptor is located on human chromosome 21 in band 21q22. This is significant because an increased expression of

---

J.C. Clemens (✉)
Department of Biochemistry, Purdue University, 175 S. University St., West Lafayette, IN 47907, USA
e-mail: jclemens@purdue.edu

M. Hortsch, H. Umemori (eds.), *The Sticky Synapse*,
DOI 10.1007/978-0-387-92708-4_9,

genes in this chromosomal region correlates with the manifestation of Down syndrome (Ds) phenotypes, including mental retardation (Korenberg et al. 1992, Delabar et al. 1993, Korenberg et al. 1994). Northern blot analysis and in situ hybridization studies revealed that this gene is broadly expressed within the nervous system (Yamakawa et al. 1998). Based on these observations, the protein was dubbed the Down syndrome cell adhesion molecule (DSCAM, pronounced Udē-es-kam) (Yamakawa et al. 1998).

Since this initial report, the DSCAM family has grown with the discovery of paralogs and homologs in mammals, fish, birds, insects, sea urchins, crustaceans, and mollusks. In chordates DSCAM is generally written in all capital letters, while non-chordate Dscams are written with only the first letter capitalized. DSCAM receptors are expressed throughout the nervous system during development and have been found to play widespread roles in patterning the nervous system, including axon targeting, neurite arborization, and branch segregation (Zipursky et al. 2006, Schmucker 2007). An additional role for hyper-variable arthropod Dscams has been described in innate immunity, but this role is beyond the scope of this chapter and will therefore not be described (for reviews on this subject, see Watson et al. 2005, Dong et al. 2006).

9.2 Identification of DSCAM Family Members

As mentioned previously, the founding member of the DSCAM family was identified as a novel member of the immunoglobulin superfamily and as a putative cell adhesion molecule (Yamakawa et al. 1998). Since the gene encoding this membrane receptor localizes to human chromosome band 21q22.2–22.3, a region that is critical for the neurological phenotypes of Ds, it was tempting to speculate that duplication of the DSCAM gene might play a causative role in the production of these phenotypes. The finding that DSCAM is expressed largely in the developing nervous system is consistent with a putative role in Ds. However, a subsequent study (Ronan et al. 2007) further demarcated the Ds critical region on chromosome 21 that is responsible for the generation of Ds-related phenotypes and determined that the DSCAM gene resided outside of this region. This suggests that DSCAM is unlikely to be a causative agent in the development of Down syndrome.

In an effort to further characterize DSCAM function, Yamakawa et al. (1998) isolated mouse homologs of DSCAM by using the human cDNA sequence as a probe to screen a mouse brain cDNA library. Tissue in situ hybridization of mouse DSCAM on mouse tissues revealed that DSCAM expression is largely localized to the central nervous system. The timing of its expression suggests that DSCAM may play a role during early nervous system development.

The third DSCAM family member that was identified came from the class Insecta. *Drosophila* Dscam was first characterized as an unknown tyrosine phosphorylated protein that physically associates with the Dock (Dreadlocks) SH2 domain (Schmucker et al. 2000). Dock is an adapter protein that comprises three SH3 domains and one SH2 domain and is required for axon guidance in flies (Clemens et al. 1996, Garrity et al. 1996). Since *dock* mutants exhibit defects in axon guidance during the development of the visual system in an adult fly (Garrity et al. 1996), as well as in the embryonic nervous system (Desai et al. 1999), it was proposed that Dock serves as a vital link that connects targeting receptors to downstream regulators of the actin cytoskeleton to control neural patterning. This hypothesis was further supported by Hing et al. (1999) with the discovery that a known regulator of the actin cytoskeleton, Pak (p21-activated kinase), physically and genetically interacts with dock to control axon guidance. These findings suggest that manipulating Pak signaling is one mechanism that insect Dscam family members employ to control connectivity within the nervous system. Additionally, a subsequent human DSCAM study found that DSCAM directly binds to Pak and stimulates Pak phosphorylation and activity (Li and Guan 2004). This suggests that DSCAM-mediated control of Pak activity appears to be a general property of all DSCAM receptors.

Following the identification of DSCAM genes in human, mouse, and fruit flies, DSCAM homologs have been identified primarily by sequence database comparisons in a number of other species. DSCAMs appear to be present in most animals that have a nervous system. The notable exception is the *Caenorhabditis elegans* genome, which lacks a DSCAM gene product that exhibits clear sequence conservation. Despite the growing number of DSCAM reports in other species, most of what we know about DSCAM function comes from *Drosophila*. Therefore, much of this chapter will center on *Drosophila* Dscam function with some additional examples of vertebrate DSCAM functions.

9.3 General Domain Structure

DSCAMs are type I cell surface transmembrane receptors that belong to the immunoglobulin (Ig) superfamily (Yamakawa et al. 1998). Members of the DSCAM protein family are made up of approximately 2,000 amino acids with an average molecular weight of 221 kDa. The amino acid sequence and domain structure of the extracellular region is conserved and encompasses ten Ig domains and six fibronectin type III modules (Fig. 9.1). Nine of the Ig domains are tandemly arrayed in the membrane distal (N-terminal) region of the extracellular domain. The six fibronectin type III modules are tandemly arrayed in the membrane proximal region with the tenth Ig domain located between fibronectin modules four and five. This unique arrangement of Ig and fibronectin domains is the hallmark that distinguishes DSCAM receptors from other Ig superfamily members.

The extracellular domain is connected to the cytoplasmic region of DSCAM by a single membrane-spanning domain. In the case of human DSCAM, it has

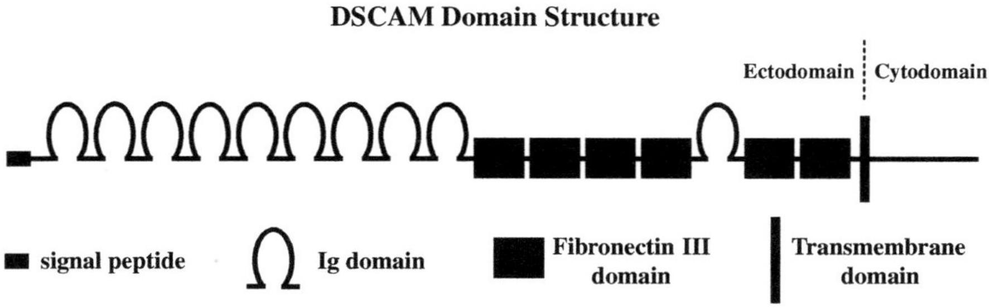

Fig. 9.1 DSCAM domain structure. The general domain structure of DSCAM receptor family members is shown. Symbols representing subdomains within DSCAMs are labeled below the receptor structure. The DSCAM ectodomain is separated from the cytodomain by a single transmembrane domain segment. The majority of DSCAM receptor sequences are located in the ectodomain, which is composed of an N-terminal signal peptide, ten Ig domains and six fibronectin type III domains in the order shown. The C-terminal cytodomain contains no catalytic domains or substantial sequence motifs

been shown that the transmembrane domain is necessary and sufficient for association of the DSCAM receptor with the netrin-binding receptor deleted in colorectal carcinomas (DCC) (Ly et al. 2008). It is unclear whether this is a general property of all DSCAM receptors or whether this is a property that is specific for human DSCAM.

The cytodomains of DSCAMs range in size from approximately 300 to 400 amino acids and contain no previously characterized catalytic domains or substantial sequence motifs. DSCAM family members within the same phylum tend to have cytoplasmic domain sequences that are fairly well conserved. However, comparison of cytoplasmic DSCAM sequences across the phyla (Arthropoda, Chordata, Platyhelminthes, Echinodermata, and Mollusca) reveals little sequence conservation. In general, DSCAM cytodomains contain multiple tyrosines, which are thought to serve as binding sites for SH2 domain-containing proteins such as Dock, as well as a C-terminal putative PDZ domain-binding site. PDZ (named after the first three letters of proteins containing this domain: PSD95, DlgA, and ZO-1) domain-containing proteins frequently bind to the C-terminal sequences of transmembrane receptors and serve as scaffolds that hold together signaling complexes (Ponting et al. 1997). No proteins have been reported to interact with this site in DSCAMs and it is currently unknown if this region functionally interacts with PDZ domain-containing proteins.

9.4 DSCAM Molecular Diversity

DSCAM transcripts usually undergo alternative splicing. Therefore (most or all), *DSCAM* genes typically express multiple protein isoforms. The extent and complexity of *DSCAM* gene transcript alternative splicing varies greatly among species. Non-arthropod *DSCAM* transcripts undergo little if any alternative splicing (Barlow et al. 2002). In contrast, unprecedented alternative splicing of some of the arthropod *Dscam* transcripts leads to the production of tens of thousands of different Dscam protein isoforms (Schmucker et al. 2000, Graveley et al. 2004, Watson et al. 2005, Brites et al. 2008). Thus far, in all arthropods examined (fruit fly, mosquito, honey bee, beetle and *Daphnia*) a Dscam gene has been detected that exhibits hyper-variable alternative splicing. However, Dscam paralogs that do not exhibit hyper-variable splicing have also been described in these organisms (Funada et al. 2007, Millard et al. 2007).

The *Drosophila Dscam* gene serves as an example to illustrate hyper-variable *Dscam* splicing and the impact this splicing has on Dscam receptor molecular diversity. The *Drosophila melanogaster* genome contains four *Dscam* genes. The first that was identified is called *Dscam* and the subsequently identified paralogs are known as *Dscam2*, *Dscam3*, and *Dscam4* (Millard et al. 2007). Of these four genes, only the (original) *Dscam* gene product is hyper-variable through the process of alternative splicing. *Dscam2* has been shown to encode two splice variants, while *Dscam3* and *Dscam4* are predicted to encode a single protein product (Millard et al. 2007).

The *Drosophila Dscam* gene contains 115 exons (Fig. 9.2). Twenty of these are considered as constant exons and are present in all *Dscam* transcripts, while 95 are variable and their inclusion in the Dscam mRNA is controlled by alternative splicing (Schmucker et al. 2000). Each mature Dscam mRNA comprises 24 exons: 20 constant exons and 4 variable exons (Fig. 9.2). The variable exons are exons 4, 6, 9, and 17, which are arrayed within the *Dscam* gene in linear clusters. Exon 4 has 12 alternatives, exon 6 has 48 alternatives, exon 9 has 33 alternatives, and exon 17 has 2 alternatives (Fig. 9.2). Alternative exons are included in the Dscam mRNA in a mutually exclusive fashion such that each mRNA will contain exactly one of each alternative exon 4, 6, 9, and 17. As a result, all Dscam mRNAs encode Dscam protein isoforms that share the same overall domain structure, but differ in the amino acid sequence at four distinct regions. The first half of Ig domain 2 is encoded by alternative exon 4, the first half of Ig domain 3 is encoded by alternative exon 6, Ig domain 7 is encoded by alternative exon 9, and the transmembrane domain is encoded by alternative exon 17.

This extraordinary example of alternative splicing lends *Drosophila Dscam* a unique advantage over other cell adhesion genes in that, by itself, the *Dscam* gene encodes a large family of molecularly diverse cell surface receptors. Mathematically, a total of 19,008 ($12 \times 48 \times 33$) different extracellular domains fused to one of two different transmembrane domains (resulting in 38,016 isoforms) can be produced by the fly *Dscam* gene (Schmucker et al. 2000). A standard nomenclature has been adopted by Dscam researchers to designate the usage of alternative exons. The particular variant within an alternative exon cluster is represented by a code consisting of two numbers separated by a decimal point. For example, the second variant within the alternative exon 4 array is represented as 4.2 in this scheme. To refer to a particular Dscam splice form that

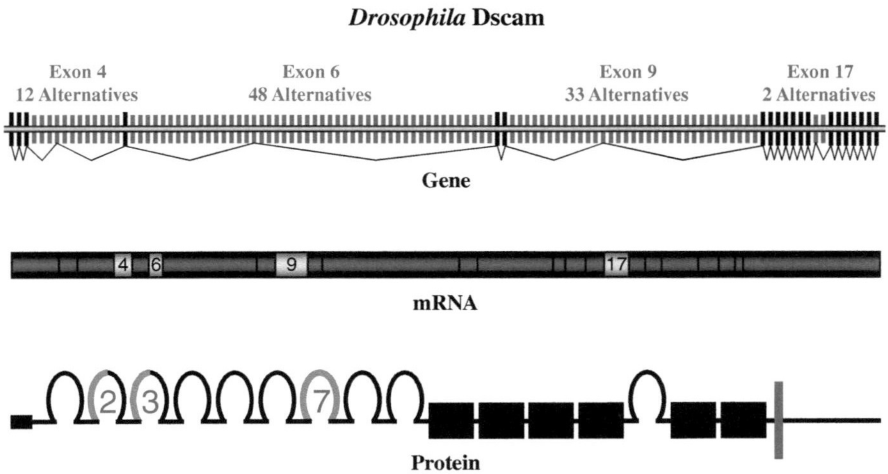

Fig. 9.2 ***Drosophila*** **Dscam gene, transcript, and protein structure**. The *Drosophila* Dscam gene, transcript, and protein are shown. Constant exons are represented as *black vertical bars* in the gene structure and *dark gray* segments in the mRNA structure. Variable exons are represented as shorter *gray bars* in the gene structure and as *light gray* segments in the mRNA structure. The constant exons are present in all Dscam mRNAs while only one of each variable domain exon is selected for inclusion in the Dscam mRNA. All Dscam protein isoforms share the same overall domain structure as shown, but differ in sequence in the *gray-colored* regions comprising the N-terminal halves of Ig domains 2 and 3, all of Ig domain 7, and the transmembrane domain

contains alternative exons 4.2, 6.20, 9.27, and 17.1 the first numbers of the code are dropped and the alternative exon variant numbers are listed sequentially and separated by decimal points: 2.20.27.1.

Dscam mRNA expression studies reveal that all alternative exons appear to be used with the exception of alternative exons 6.11 and 9.33, which have not been detected in any Dscam cDNA (Neves et al. 2004). Splicing of the alternative exons appears to be independent of each other such that random combinations of exons 4, 6, and 9 are produced. Individual neurons are predicted to contain 14–50 copies of the Dscam mRNA, and it is known that this population consists of a mixture of different splice forms (Neves et al. 2004).

In general, splicing of the alternative exons appears to be random; however, developmental and tissue-specific alternative exon trends have been observed. Within the alternative exon 4 cluster, exon 4.2 is rarely used in *Drosophila* embryos but its usage greatly increases in later developmental stages (Celotto and Graveley 2001, Neves et al. 2004). Splicing of alternative exon 6 appears to be random. The greatest splicing biases occur during alternative exon 9 selection. Alternative exons 9.6, 9.9, 9.13, 9.30, and 9.31 are highly favored in the embryonic stage, whereas during later stages of development a more even distribution of exon 9 variants are used (Neves et al. 2004). Exon 9 also appears to have a tissue-specific bias. For example, photoreceptor cells selectively use the same five variants that are prevalent in the embryo, while other neurons use a broader repertoire (Neves et al. 2004). At present the significance of the developmental and tissue-specific splicing preferences is not known.

9.5 Homophilic Interactions

The extracellular domain of a DSCAM receptor expressed on one cell surface can physically interact with the extracellular domain of a DSCAM receptor located on a neighboring cell membrane (Agarwala et al. 2000). These interactions, which span two cellular membranes, are referred to as *trans* interactions and only occur between two identical DSCAM receptor proteins (homophilic). For example, there are two DSCAM paralogs in chicken: DSCAM and DSCAML (Yamagata and Sanes 2008). Non-neuronal cultured cells expressing DSCAM will aggregate due to the *trans* homophilic interactions of DSCAM ectodomains. Similarly, cells expressing DSCAML will also form aggregates through DSCAML ectodomain *trans* homophilic interactions. However, cells expressing DSCAM do not form heterophilic interaction with cells expressing DSCAML (Yamagata and Sanes 2008).

This homophilic-specific binding property appears to be a conserved attribute shared by all DSCAM family members. Perhaps even more remarkable than the sheer number of *Drosophila* Dscam ectodomains (19,008) is the way these isoforms interact with each other. Dscam isoforms interact in *trans* (Fig. 9.3), but generally only do so when they contain identical sequences (alternative exons) at all three variable Ig domains (Wojtowicz et al. 2004, 2007). If they differ in one or more variable Ig domains (2, 3, and 7), no interaction is detectable. In some rare cases heterophilic interactions have been detected, but in each of these instances two of the variable domains are identical while the non-identical third domains are highly related to each other (Wojtowicz et al. 2007).

Because homophilic interaction relies on sequence identity at Ig domains 2, 3, and 7, it was proposed that these domains are major sites of interaction between Dscam monomers (Wojtowicz et al. 2004). Recent crystallographic and mutagenic studies support this hypothesis (Meijers et al. 2007, Wojtowicz et al. 2007, Sawaya et al. 2008). These structural studies have determined that specific sequences within the variable regions of Ig domains 2, 3, and 7 participate in binding interactions between identical Dscam splice forms. The individual variable domains of one monomer bind to the corresponding variable domain of the identical second monomer in an antiparallel configuration such that Ig 2 binds Ig domain 2, Ig 3 binds Ig 3, and Ig 7 binds Ig 7 (Fig. 9.3). Overall, the structure adopted by the first seven Ig domains of each interacting Dscam monomer resembles an S-shape in which Ig domains 2, 3, and 7 are roughly in a linear arrangement. The S-shape conformation is only observed

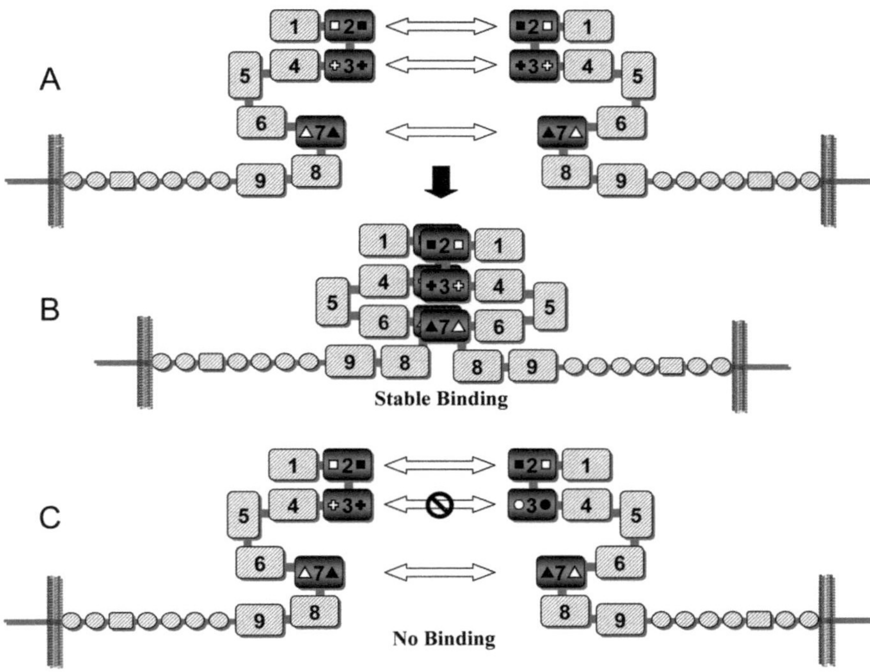

Fig. 9.3 Dscam isoform-specific homophilic interaction. A schematic representation of the Dscam receptor is shown. **(A–C)** Ig domains are represented as *rounded rectangles* and fibronectin domains as *ellipses*. Ig domains 1–9 are numbered. A representation of a lipid bilayer separates the cytodomain from the ectodomain. Ig domains 2, 3, and 7 contain antiparallel-binding determinants represented by *white* and *black geometric shapes*. These binding determinants interact such that *black* determinants bind to *white* determinants if and only if they are the same shape. **(A) Identical Dscam monomers**: Ig domains 1–4 pack into a horseshoe-shaped conformation while the other extracellular domains do not participate in higher order structures. **(B) Identical Dscam isoforms engaged in *trans* homophilic interaction**: The *black* determinants in Ig domains 2, 3, and 7 of each receptor bind to the respective *white* determinants present in the other receptor in an antiparallel manner. This results in the formation of an ordered "S"-shaped conformation involving Ig domains 1–7 of each monomer and stable receptor binding. **(C) Non-identical Dscam isoforms**: The two Dscam isoforms shown have identical Ig domains 2 and 7 but differ in Ig domain 3 as indicated by differently shaped binding determinants (*crosses* versus *circles*). This results in an inability to form a stable interaction between these different isoforms

when Dscam isoforms are participating in homophilic binding. When Dscam is monomeric, the distal Ig domains 1–4 exist in a horseshoe conformation (the top half of the S), while Ig domains 5–7 are in a flexible extended conformation (Meijers et al. 2007; Sawaya et al. 2008). This horseshoe conformation has been observed in other Ig superfamily members such as insect hemolin (Su et al. 1998) and the neural CAM axonin-1 (Freigang et al. 2000).

The most well-documented biological response of Dscam homophilic interaction is repulsion of neurites expressing identical (interacting) forms. The molecular mechanism of homophilic repulsion is unknown, but it is hypothesized to function in a stepwise manner. Initially, the identical ectodomains interact to create an adhesive complex. This in turn activates the Dscam cytodomain to initiate a signaling cascade that ultimately modifies the cytoskeleton to direct movement away from the sites of interaction (Hughes et al. 2007, Matthews et al. 2007, Soba et al. 2007). Indeed, mutation of the cytodomain leads to a stable adhesive interaction instead of repulsion (Hughes et al. 2007, Matthews et al. 2007, Soba et al. 2007). A second signal must also be generated to disrupt adhesion between the *trans* interacting ectodomains. It is thought that this occurs by one or more of the following mechanisms: a conformational shift, receptor internalization, or a proteolytic event.

In *Drosophila*, Dscam-mediated homophilic repulsion has been demonstrated to control neurite branch segregation (Zhan et al. 2004) and self-avoidance (Hughes et al. 2007, Matthews et al. 2007, Soba et al. 2007), while Dscam2-mediated homophilic repulsion contributes to neuronal tiling (Millard et al. 2007). In mice, DSCAM-mediated homophilic repulsion is involved in aspects of neuronal self-avoidance and tiling (Fuerst et al. 2008).

9.6 Branch Segregation and Self-Avoidance

Hyper-variable arthropod Dscam genes provide a molecular mechanism for neurons to distinguish self from non-self. In the case of the *Drosophila Dscam* gene, neurons select a nearly random population of 14–50 Dscam isoforms from a pool of roughly 38,000 possibilities. Because of this, neighboring neurons express a unique collection of Dscam isoforms and are consequently unlikely to contain any isoforms in common. Therefore, the Dscam isoforms expressed by an individual neuron serves as a molecular signature that can be used to differentiate itself from all other neurons (Hattori et al. 2007).

This signature is the molecular basis of self-avoidance in *Drosophila* and is used to help pattern the nervous system. Since only identical Dscam ectodomains interact, neighboring neurons do not repel each other because they do not express identical isoforms (Fig. 9.4A). Therefore, individual neurons can bundle into nerves (fasciculate) and grow along common pathways. If neurons that normally fasciculate are experimentally forced to express a Dscam isoform in common using a transgene, the neuron bundle will defasciculate and the individual neurons will move away from each other (Fig. 9.4B) (Schmucker et al. 2000, Zhan et al. 2004).

In the case of branched neurons, neurite branches display isoforms in common since each branch originates from the same cell body. Homophilic interactions between identical Dscam isoforms on each sister branch leads to the initiation of repulsion and the dispersion of sister branches. This mechanism

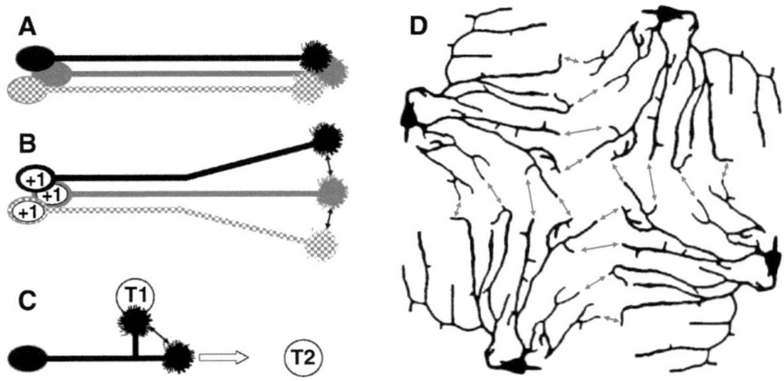

Fig. 9.4 Dscam-mediated repulsion: self-recognition versus tiling. (**A**–**C**) Neurons are represented schematically. Axons (*lines*) extending from neuron cell bodies (*ovals*) end in growth cones, which are complex motile structures that elaborate numerous filopodia. (**A**) Each neuron expresses and displays different Dscam populations on their plasma membranes as indicated by different shading patterns. Since these neurons do not express common Dscam isoforms, they do not recognize each other as "self". Their neurites do not repel each other and are able to fasciculate and grow along common pathways. (**B**) In addition to their normal unique combination of Dscam isoforms expressed from their endogenous *Dscam* genes (as in panel **A**), each neuron also expresses a Dscam isoform in common due to the presence of a transgene. Interaction between these common isoforms is interpreted as recognition of "self" and leads to neurite repulsion, which prevents them from growth along a common pathway. (**C**) In the case of axons that branch, each branch will contain the same Dscam isoforms allowing for self-recognition. This results in repulsion between the two growth cones and helps ensure that each branch seeks out different targets (T1 and T2) rather than both converging on the nearest target (T1). (**D**) The dendritic fields from four hypothetical neurons are shown. Self-recognition mediated by hyper-variable Dscam causes neurites originating from the same cell body to spread out and effectively cover a large spatial area. Tiling interactions (*gray double-headed arrows*) mediated by non-diverse Dscam receptors such as Dscam2 prevent overgrowth of the dendrites from one cell body into the area occupied by the dendrites coming from a different cell body

ensures that neurites emanating from the same cell body segregate properly to different targets instead of converging on a single target (Fig. 9.4C) (Wang et al. 2002, Zhan et al. 2004). It also allows dendritic branches of a neuron to fully disperse for the uniform exploration of a spatial field (Fig. 9.4D) (Zhu et al. 2006, Hughes et al. 2007, Matthews et al. 2007, Soba et al. 2007).

Dscam mutant neurons display defects in segregation of sister branches to their proper target fields. This can be illustrated within the mushroom bodies (MB), which are multi-lobed structures in the *Drosophila* brain that are involved in olfactory learning and memory. Axons extend from the MB cell bodies down a common pathway called the peduncle. At the distal end of the peduncle the MB axons branch. One branch targets to the dorsal lobe, while the second branch targets to the medial lobe. In Dscam mutant MB axons, the branches form but frequently target to the same lobe (Wang et al. 2002, Zhan et al. 2004).

Dscam mutant neurons display defects in patterning dendritic arborizations. Sensory neurons in the *Drosophila* larva body wall elaborate complex dendritic arborizations that spread out to evenly cover the sensory field. The highly branched dendrites of these neurons, known as dendritic arborization (da) neurons, never cross over each other in the plane of the body wall, elegantly illustrating the principle of self-avoidance (Fig. 9.4D). *Dscam* mutant da neurons still elaborate dendritic fields that display complex branch patterns; however, these branches cross over one another and fail to evenly disperse within the body wall (Hughes et al. 2007, Matthews et al. 2007, Soba et al. 2007).

Self-avoidance is a general property that likely functions in all species to properly disperse neurite branches. In arthropods hyper-variable Dscam genes have arisen to provide one mechanism to recognize and move away from self. Since *Drosophila* Dscam is expressed in most, if not all neurons during development, the single Dscam gene may be sufficient to confer self-avoidance properties to all neurites. Outside the arthropod lineage, a similar DSCAM diversity has not been detected; therefore, it is unlikely that DSCAMs are the only molecules responsible for this process. For example, DSCAM mutant mice display defects in self-avoidance of neurite arbors in two sub-populations of retina amacrine cells that normally express DSCAM protein (Fuerst et al. 2008). However, other retina amacrine cell populations exhibited normal neurite patterning in these mice. Therefore self-avoidance in non-arthropods may require the cell-type-specific expression of different repulsion receptor genes. Alternatively, a more general self-avoidance mechanism may have evolved in these organisms, which is yet to be elucidated.

9.7 Tiling

Tiling within the nervous system is the process of completely and evenly filling a spatial region with multiple synaptic domains of a particular neural class. Similar to tiles in a floor, the synaptic domains of each neuron are contained within a distinct region that approaches, but does not overlap with adjacent domains. The result of this is a uniform distribution of neural processes that are restricted to each individual domain. One possible mechanism to constrain neural processes to a particular domain (tile) is to utilize a repulsive receptor system. Receptors expressed on the processes of adjacent tiling neurons could prevent overlap of neurite fields by inducing repulsion upon receptor interaction (Fig. 9.4D).

Hyper-variable *Drosophila* Dscam is not a good candidate to perform this role. While Dscam diversity is well suited for self-avoidance, since individual neurons make different isoforms, Dscam is incapable of signaling repulsion between processes from different neurons. Non-variable Dscam receptors, on the other hand, which are expressed in a cell-type-specific manner, have been demonstrated to function in neuron tiling. In *Drosophila*, *Dscam2* gene function is required for tiling the L1 subclass of lamina neurons within the visual system (Millard et al. 2007). *Dscam2* mutant L1 neurons target the correct layer in the medulla, but extend processes laterally and invade neighboring columns (tiles). Presumably this is due to the lack of homophilic interaction-based repulsion between L1 neuron axons. A similar role for DSCAM-mediated tiling has been uncovered in mice (Fuerst et al. 2008). DSCAM is required in two sub-populations of retinal amacrine cells for proper dispersion of these neurons in a spatial field. In the absence of DSCAM, these neurons migrate together and inappropriately fasciculate.

In both the preceding examples, DSCAM family members serve as homotypic repulsion receptors, whose function is to preserve the spatial patterning (tiling) between members of the same neuron type. Since tiling is not self-avoidance, but rather avoidance of members within a particular class, it would follow that multiple tiling receptors must exist. This is because in some biological environments, different neuron classes will tile independently of each other within a shared spatial region (Hughes et al. 2007, Matthews et al. 2007, Soba et al. 2007). For example, da class III dendrites exhibit tiling in the *Drosophila* body wall. Da class IV dendrites also exhibit tiling in the same environment. Therefore, since class III dendrites do not repel those of class IV, the receptors that prevent class III dendrites from overlapping must be different from those that prevent class IV dendrites from overlapping. It is known that Dscam and Dscam2 are not responsible for da class IV tiling. An attractive possibility is that Dscam3 or 4 might play a role in tiling these neurons.

9.8 Non-repulsive DSCAM Functions

There is a growing body of evidence indicating that the DSCAM family of receptors can function in roles other than signaling homotypic repulsion. Loss-of-function studies in *Drosophila* suggest that Dscam is involved in aspects of axon target selection in both the olfactory system (Hummel et al. 2003) and mechanosensory neurons (Chen et al. 2006). The phenotypes observed in each of these studies are inconsistent with merely a loss of Dscam-mediated homophilic repulsion and hint at a possible instructive requirement of Dscam splice forms for target selection (Chen et al. 2006). For example, it might be the case that Dscam receptors signal attraction or even adhesion in response to either different ligands (non-Dscam) or changes in Dscam receptor molecular contexts.

Recent studies in chordates have uncovered new functions for DSCAMs. Yamagata and Sanes (2008) isolated chick Dscam and DscamL, orthologs of human DSCAM and DSCAML1, respectively, and have determined that these receptors function in a manner similar to sidekick receptors to pattern laminar arborizations in the chick retina (see Chapter 10 for a description of sidekick receptors). Like other Dscams, the ectodomains of these receptors were found to participate in homophilic, but not heterophilic, interactions. Instead of inducing repulsion between neurons, chick Dscam or DscamL homophilic interactions result in trans-synaptic adhesion between neurons at synaptic junctions within the inner plexiform layer (IPL) of the chick retina. Dscam protein is predominantly localized to the S5 sublamina of the IPL, while DscamL receptors localize to sublamina S1, S2, and S4. Depletion of Dscam expression disrupts the laminar patterning of S5 by causing the processes of affected neurons to extend beyond the S5 boundary. The processes of non-Dscam-expressing neurons in other sublamina were not affected by Dscam depletion. Moreover, ectopic expression of Dscam rerouted neuronal processes to the Dscam-positive layer S5. Similarly, ectopic expression of DscamL rerouted neurites to layers other than S3, a layer that is DscamL negative. Taken together, these data suggest that Dscam signaling is not limited to homotypic repulsion, but can also promote adhesion, and hint at putative roles for Dscams in synaptic specification or maintenance (Yamagata and Sanes 2008).

9.9 Non-DSCAM Interactions

During vertebrate spinal cord development commissural axons are attracted to the ventral midline due to expression of netrin-1 by the floor plate cells. Deleted in colorectal cancer (DCC) is a netrin-1 cell surface receptor expressed on commissural axons that function as a key mediator of the attractive and outgrowth promoting properties of netrin-1 (Keino-Masu et al. 1996). While impairment of DCC function blocks netrin-1-stimulated outgrowth of commissural axon explants, it does not completely block turning of these axons toward a netrin-1 source (Keino-Masu et al. 1996). These results suggest that an additional netrin receptor functions with DCC to induce axon turning toward the netrin-1 gradient.

In addition to homophilic interactions, the extracellular domain of rat DSCAM has been shown to engage in heterophilic interactions with netrin-1 (Ly et al. 2008). A protein truncation analysis revealed that the netrin-1-binding site is located within a region containing DSCAM Ig domains 7–9. Additionally, DSCAM and DCC form a receptor complex in the absence of netrin-1. This complex requires the transmembrane domain of DSCAM and it dissociates upon netrin-1 stimulation (Ly et al. 2008). Like knockdown of DCC, knockdown of DSCAM by siRNA does not completely block the turning response of rat commissural axons to netrin-1. However, simultaneous knockdown of both DSCAM and DCC results in a complete blockage of netrin-1-induced axon turning (Ly et al. 2008). Finally, ectopic expression of DSCAM in *Xenopus* spinal neurons is sufficient to confer a turning response to netrin-1 in the absence of DCC activity (Ly et al. 2008). These studies demonstrate that vertebrate DSCAM has additional heterophilic binding partners, which enable it to function as a netrin-1 receptor and mediate axon turning, much in the same way as DCC functions. A recent study in Drosophila has generated similar findings and suggests that netrin binding might be general property of all DSCAMs (Andrews et al.).

9.10 Concluding Remarks

It has been a decade since the discovery of the founding DSCAM family member. DSCAM genes have been identified in virtually all organisms that have a nervous system. During this time our understanding of DSCAM biochemical properties and biological functions has grown, but is far from complete. Thus far, all DSCAMs have been reported to engage in isoform-specific homophilic interactions. The biological consequences of these interactions appear to vary between homotypic repulsion and adhesion. This may simply reflect different species-specific roles that have evolved for DSCAMs since what we know about DSCAM function is being pieced together from experiments in a wide variety of organisms. An alternative and attractive hypothesis is that the signaling output from homotypic interactions is dependent on the signaling contexts in which these interactions occur. For example, it might be the case that axon–axon- or dendrite–dendrite-based DSCAM homotypic interactions lead to repulsion to segregate neurite branches, disperse arborizations, and tile receptive fields. On the other hand, axon–dendrite-based DSCAM homotypic interactions may be instructive for the selection and stabilization of synaptic partners as in the chicken IPL. It may even be the case that the hyper-variable arthropod Dscams play an instructive role in connection specificity as well, rather than simply serving as a means to distinguish self from non-self. Recent studies have added a new wrinkle to the DSCAM story with the discovery that DSCAMs also functions as netrin receptors. Finally the involvement of DSCAM in human Down syndrome and other nervous system structural disorders awaits critical assessment.

References

Agarwala KL, Nakamura S, Tsutsumi Y et al. (2000) Down syndrome cell adhesion molecule DSCAM mediates homophilic intercellular adhesion. Brain Res Mol Brain Res 79:118–126

Andrews GL, Tanglao S, Farmer WT, (2008) Dscam guides embryonic axons by Netrin-dependent and -independent functions. Development 135:3839–3848

Barlow GM, Micales B, Chen XN et al. (2002) Mammalian DSCAMs: roles in the development of the spinal cord, cortex, and cerebellum? Biochem Biophys Res Commun 293:881–891

Brites D, McTaggart S, Morris K et al. (2008) The Dscam homologue of the crustacean Daphnia is diversified by alternative splicing like in insects. Mol Biol Evol 25:1429–1439

brain, which is important for learning (Bliss et al. 2000, Law et al. 2003). Studies in chick embryos show that there are three critical periods for L1-CAM function in long-term memory formation (Tiunova et al. 1998). In the past decade, glia cells have emerged as a major component in the regulation of synaptic plasticity and further supporting this is the finding that overexpression of L1-CAM in astroctyes enhanced memory formation in chickens (Wolfer et al. 1998, Freeman 2005, Todd et al. 2006, Bains and Oliet 2007). Hence, in addition to their neuronal function, there may also be a glial function for L1-type family members contributing to memory formation.

Supplementing these behavioral studies there is also physiological evidence that L1-type proteins are important for synaptic transmission and plasticity. Long-term potentiation (LTP) is reduced in hippocampal slices that have been treated with antibodies to prevent the association of L1-CAM with NCAM (Luthl et al. 1994), as well as in mice that ectopically overexpress L1-CAM in astrocytes (Luthi et al. 1996). Surprisingly, LTP in the CA1 region and the dentate gyrus region of the hippocampus is normal in mice that lack L1-CAM constitutively or conditionally (Bliss et al. 2000, Law et al. 2003). However, basal excitatory synaptic transmission is increased in conditional knockout mice, but not in the constitutive L1-CAM knockout mice (Law et al. 2003). The cause for this is possibly due to a reduced input of GABAergic neurons onto the CA1 pyramidal neurons.

Though LTP induction in the CA1 region is normal in constitutively deficient L1-CAM mice, perisomatic inhibition of the CA1 neurons is reduced (Saghatelyan et al. 2004). The decreased inhibitory postsynaptic current in these neurons is likely to be caused by the loss of 30% of inhibitory active zones in these L1-CAM deficient mice (Saghatelyan et al. 2004). This suggests that L1-CAM has a function in the development or maintenance of active zones in GABAergic synapses. Strikingly, the loss of the L1-CAM paralog CHL1 had the opposite effect resulting in an increase in active zones and enhanced synaptic transmission at GABAergic synapses leading to a disruption of LTP in the CA1 region of the hippocampus (Nikonenko et al. 2006). This shows that different L1-type members can have oppositional roles in the same type of neurons and that the proper balance of their effects is crucial for maintaining the synaptic plasticity underlying learning and memory formation.

Some studies demonstrate that particular neuronal activities alter L1-type protein expression and processing. Continuous low-frequency stimulation (0.1 Hz) downregulates L1-CAM in mouse cell cultures of sensory neurons as well as dorsal root ganglion (DRG) neurons (Itoh et al. 1995, 1997). In addition, theta-burst stimulation which induces LTP in the hippocampus leads to dephosporylation of the FIGQY-motif resulting in the removal of L1-CAM from the cell surface via clathrin-mediated endocytosis in rats (Itoh et al. 2005). In contrast, application of K^+ and the *N*-methyl-d-aspartate (NMDA) to neuronal cell cultures increases the expression of L1-CAM in the cells (Scherer et al. 1992).

In the CA1–CA3 region of the hippocampus, L1-CAM is expressed in pyramidal neurons but only on the presynaptic side (Matsumoto-Miyai et al. 2003, Munakata et al. 2003, Nakamura et al. 2006). In the adult brain, L1-CAM is not found in large mature synaptic boutons, but only in small immature synaptic terminals, as well as so-called orphan boutons, which lack postsynaptic specializations (Nakamura et al. 2006). Activation of the NMDA receptor via theta-burst stimulation in the hippocampus of rats leads to the brief activation of the proteolytic enzyme neuropsin via protein kinases. The active form of neuropsin has been shown to cleave L1-CAM (Fig. 12.1), releasing its extracellular 180 kDa domain (Matsumoto-Miyai et al. 2003, Nakamura et al. 2006). In mice lacking neuropsin, an increased number of smaller immature synaptic boutons containing L1-CAM are present, but the total amount of mature synaptic boutons lacking L1-CAM is significantly reduced (Nakamura et al. 2006). These results suggest that the proteolytic processing of L1-CAM by neuropsin is part of the process required for the transformation of incipient synapses into mature synaptic terminals and is triggered by neuronal activity. These morphological changes are thought to be the cellular basis for memory formation (Collingridge and Bliss 1995, Krueger et al. 2003).

Despite a number of studies clearly providing evidence for a synaptic function of L1-type members, only a few studies have more recently started to shed light on the cellular mechanisms by which L1-type proteins affect the development and modulation of synapses. The following sections describe what is known so far about their functions in synapse targeting, formation, and maturation as well as synaptic transmission and signaling.

12.2.2 L1-Type Cell Adhesion Molecules in Synapse Targeting

In the central nervous system, a single neuron often receives multiple inputs from different neurons, and the input from a particular type of neuron is usually restricted to a particular subcellular compartment on the receiving neuron (Freund and Buzsaki 1996, Somogyi et al. 1998, Benson et al. 2001). An example of such compartmentalization is the Purkinje neuron which receives input from stellate neurons at the dendrites and input from basket neurons only on the axon initial segment (AIS) (Bayer and Altman 1987). The mechanism underlying the targeting of the synaptic terminals of GABAergic basket neurons, so-called pinceau synapses, to the AIS of glutamatergic Purkinje neurons has been shown to involve a particular splice variant of neurofascin, neurofascin 186 (NF186), but not L1-CAM or Nr-CAM (Hassel et al. 1997, Ango et al. 2004, Huang 2006). The targeting results from a subcellular gradient of NF186 which has its maximum concentration at the AIS of the Purkinje neuron and is dependent on NF186 binding to ankyrin-G on the intracellular side (Ango et al. 2004). A loss of the NF186 gradient results in guidance defects as well as in a reduction of basket neuron presynaptic terminals suggesting that NF186 is not only a cue for correct localization but also important for the stabilization of pinceau synapses (Ango et al. 2004). Neurofascin is expressed only postsynaptically and nothing is known about how it affects the presynaptic terminals. However, there have been recent studies investigating a function for neurofascin in the organization of the postsynaptic apparatus, which will be described in the next paragraph

12.2.3 L1-Type Cell Adhesion Molecules in Synaptogenesis

In the hippocampus, Gephyrin, a scaffolding protein, is important for clustering of the $GABA_A$ receptors at the postsynaptic side (Essrich et al. 1998, Fritschy et al. 2008). Studies from the Volkmer lab show that neurofascin is important for two early developmental steps of inhibitory synapse formation on the postsynaptic side (Burkarth et al. 2007). First, an alternative splice variant of neurofascin lacking the 5th fibronectin domain (NF166) has been shown to be important for Gephyrin clustering which in turn is necessary for the organization of the $GABA_A$ receptors (Burkarth et al. 2007). It is suggested that a heterophilic extracellular interaction of NF166 with an unknown interaction partner is translated into a postsynaptic intracellular signal that leads to Gephyrin clustering (Burkarth et al. 2007). In the second developmental step, these randomly distributed Gephyrin clusters on the pyramidal cell bodies need to be relocalized to the axon hillock where the presynaptic terminal will be formed. Knockdown experiments provide evidence that this relocalization event depends on NF166 as well (Burkarth et al. 2007). Though the mechanism underlying relocalization is unknown, it seems to be independent of ankyrin-G, which is important in the establishment of the NF186 gradient at the AIS of Purkinje neurons during synapse targeting in the cerebellum (Ango et al. 2004, Burkarth et al. 2007). In the hippocampus, it is seen that NF166 is expressed early during development and is involved in neurite outgrowth while NF186 is only expressed later during development primarily in mature neurons after Gephyrin clustering has occurred (Hassel et al. 1997). Interestingly, it has also been shown that overexpression of NF186 inhibits neurite outgrowth and Gephyrin clustering, but promotes synapse targeting and stabilization (Hassel et al. 1997, Burkarth et al. 2007). This further demonstrates that different neurofascin isoforms have different functions in development and synaptogenesis.

Contrary to a few earlier in vitro studies, recent experiments demonstrate a function for L1-CAM in the formation of cholinergic synapses (Mehrke et al. 1984, Godenschwege et al. 2006, Triana-Baltzer et al. 2006). At the chicken neuromuscular junction (NMJ) L1-CAM only has a presynaptic localization, while in nicotinic pathways of the CNS, namely the chick ciliary ganglion (CG), L1-CAM is localized to both sides of the synapse (Sanes et al. 1986, Triana-Baltzer et al. 2006). On the postsynaptic side of CG neurons, L1-CAM colocalizes with α7-nicotinic acetylcholine (ACh) receptors (Triana-Baltzer et al.

2006). Interestingly a separate study showed that at the adult rat NMJ, the FIGQY phosphorylated L1-CAM isoform also colocalizes with the ACh receptors (Jenkins et al. 2001). Expression of L1-CAM with the cytosolic domain replaced by GFP on the presynaptic side of the chick NMJ or the postsynaptic side of the nicotinic pathway in the CNS affects proper synapse formation in vivo and in vitro (Sanes et al. 1986, Godenschwege et al. 2006, Triana-Baltzer et al. 2006). In each case, there is a reduced clustering of synaptic vesicle proteins at the cell contact site suggesting that homophilic and/or heterophilic intercellular interaction of L1-CAM is involved in aligning components of the presynaptic side over the postsynaptic receptors (Sanes et al. 1986, Godenschwege et al. 2006, Triana-Baltzer et al. 2006). Similarly pre- and postsynaptic expression of neuroglian lacking the intracellular domain in the giant fiber (GF) pathway of *Drosophila* was found to disrupt the cholinergic giant synapse morphologically and functionally (Godenschwege et al. 2006, Allen and Murphey 2007). These dominant-negative experiments highlight the importance of outside-in signaling involving the intracellular domain of L1-type molecules in synapse formation. The intracellular output of L1-type CAMs that is important for synapse formation involves the FIGQY-motif. A missense mutation in the extracellular domain was found to reduce the phosphorylation at the tyrosine of the FIGQY-motif in *Drosophila* neuroglian, and these mutant specimens have disrupted GF synaptic terminals with a dramatically reduced number of synaptic vesicles (Godenschwege et al. 2006). In addition, a diminished number of microtubules and mitochondria are present at the synaptic terminal but their distribution is normal in the axon (Fig. 12.2). Similar to the overexpression of Nrg lacking the intracellular domain, the overexpression of a mutant Nrg construct with a FIGQF-motif in which the tyrosine is replaced with a phenylalanine results in a dominant-negative effect disrupting the morphology and function of the GF synapse (Godenschwege et al. 2006). The mutant FIGQF-motif binds ankyrin with reduced efficiency but is not able to associate with the microtubule-stabilizing protein doublecortin at all (Hortsch et al. 1998, Kizhatil et al. 2002). Hence a mechanism by which neuroglian aligns presynaptic structures over postsynaptic specializations may be by stabilizing microtubules locally in the synaptic terminal thereby targeting the delivery of presynaptic components important for synapse function, such as synaptic vesicles and mitochondria, to the appropriate place. This synaptic function is conserved in vertebrate L1-type CAMs, as the expression of human L1-CAM in *Drosophila* can fully substitute for the disrupted function in mutants that have a mutation in the extracellular domain resulting in reduced phosphorylation of the FIGQY-motif (Godenschwege et al. 2006). In contrast, vertebrate neurofascin and Nr-CAM are not able to compensate for the mutant Nrg protein demonstrating that this synaptic function is not common to all L1-type family members (Godenschwege et al. 2006). Phosphorylation of the FIGQY-motif is common to all L1-type proteins, but the trigger is likely to be different for various homologs. Hence, in vivo the signal transduction cascade is likely to involve not only a homophilic interaction but also a heterophilic interaction,

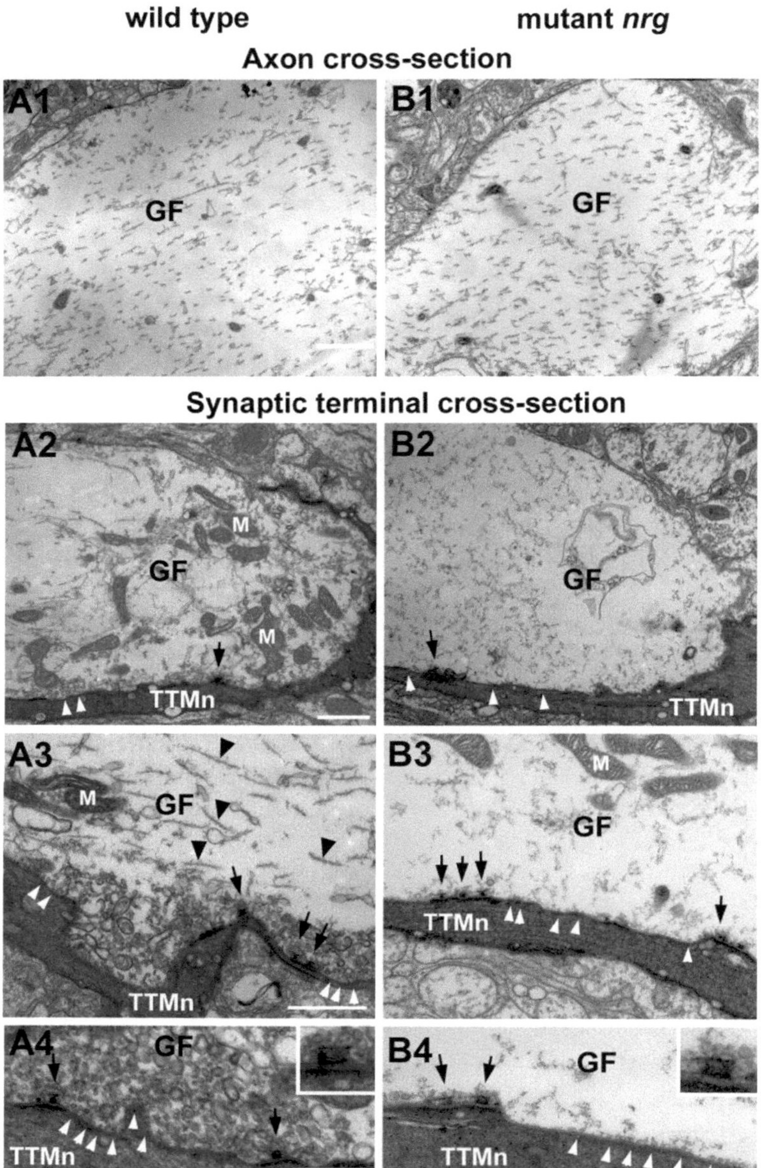

Fig. 12.2 Ultrastructural comparison between wild type and mutant *nrg* (adapted from Godenschwege et al. 2006). The cross-sections through axon and the synaptic terminal of the giant fiber in *Drosophila melanogaster* reveal dramatic differences at the synapse but not in the axon; *nrg* mutant presynaptic terminals display a lack of microtubules (*black arrowheads*), mitochondria (M), and synaptic vesicles. (A1) Thin section through a wild-type axon. (A2–A4) Ultrastructure of the presynaptic terminal in wild type. The electron micrograph of the contact region of the GF and the TTMn in wild-type specimens suggests the presence of

which with respect to cholinergic synapse formation is conserved in vertebrate L1-CAM but not in its paralogs.

As mentioned earlier, neurofascin plays a role in the development of GABAergic synapses, while L1-CAM has been shown to be involved in the organization of cholinergic synapses. Recent evidence suggests that Nr-CAM takes part in the structuring of glutamatergic synapses. The intracellular domain of Nr-CAM has a class I PDZ-binding motif, which is not present in its paralogs L1-CAM, CHL1, and neurofascin (Songyang et al. 1997). Two independent studies found that through this motif Nr-CAM binds to members of the membrane-associated guanylate kinase family (MAGUK), namely PSD95/SAP90, SAP97, and SAP102 (Davey et al. 2005, Dirks et al. 2006). These proteins are important for the organization of the postsynaptic side. While SAP90/PSD-95 and SAP102 are localized to the postsynaptic density (PSD) of type 1 glutamatergic synapses and associate with NMDA receptors, SAP97 interacts with α-amino-3-hydroxy-5-methyl-isoxazole-4-propionic acid (AMPA) receptors (Leonard et al. 1998, Seabold et al. 2003, Davey et al. 2005). In addition, it is noteworthy that SAP102 also has a presynaptic function in synaptic vesicle trafficking, targeting, and fusion (Sans et al. 2003) which suggests that the association of Nr-CAM with this postsynaptic protein could exert a presynaptic effect as well.

Studies indicate that other L1-type family members have a function at glutamatergic synapses. These studies provide further evidence for a temporal aspect for L1-type protein function. When its FIGQY-motif is phosphorylated, the invertebrate homolog neuroglian localizes to the early developing glutamatergic NMJ of *Drosophila* embryos (Jenkins et al. 2001). The protein is present before the synapses develop into the characteristic mature synaptic boutons of the larval NMJ at which stage it is no longer present (Jenkins et al. 2001). It has yet to be determined if this phosphorylated isoform is present pre- or postsynaptically at the developing *Drosophila* NMJ. However, it is interesting to note that in the Schaffer collateral CA1 hippocampal synapses L1-CAM appears to have a temporal presynaptic function in orphan boutons and is important for their transformation into mature synapses, though L1-CAM is not present at the synapse once it has matured (Nakamura et al. 2006).

Fig. 12.2 (continued) chemical and electrical synapses indicated by T-bars (*black arrows*, inset in A4) and the 1–5 nm intercellular space often associated with a single layer of vesicles (*white arrowheads*), respectively. (B1) Thin section through an axon in a mutant specimen. (B2–B4) Ultrastructure of nrg^{849} mutant synapses. There was a dramatic reduction of synaptic and pleomorphic vesicles, as well as of tubular structures in mutant presynaptic terminals. In addition, the number of mitochondria (M) and microtubules (*black arrowheads*) in nrg^{849} mutants appeared to be reduced, and they were not as closely associated with the synaptic contact region (B3) as in wild-type preparations. GF: giant fiber, TTMn: tergo-trochanteral motor neuron. Scale bar in (A1–A3) is 1 μm and in (A4) is 0.5 μm

12.2.4 L1-Type Cell Adhesion Molecules in Synaptic Transmission and Signaling

A number of experiments also demonstrate that L1-type family members have a developmental role not only in synaptogenesis but also in synaptic transmission and signaling. As mentioned in the previous section, postsynaptic L1-CAM colocalizes with ACh receptors (Jenkins et al. 2001, Triana-Baltzer et al. 2006). In the chick ciliary ganglion, L1-CAM regulates α7-nACh receptor signaling (Triana-Baltzer et al. 2006). Nicotine-induced activation of the α7-nACh receptor results in phosphorylation of transcription factor CREB, and co-activation of L1-CAM with antibodies or via a soluble extracellular L1-CAM domain further increases phospho-CREB (Triana-Baltzer et al. 2006). CREB is also phosphorylated through activation of MAP kinase and through calcium calmodulin-dependent protein kinase II, and the phosphorylation of CREB affects gene expression (Greenberg et al. 1986, Chang and Berg 2001). L1-CAM signaling has been shown to involve MAP kinases as well as other kinases, thereby affects the generation of phospho-CREB and alters downstream signaling through multiple pathways (Schaefer et al. 1999, Bliss et al. 2000, Forni et al. 2004, Islam et al. 2004, Nagaraj and Hortsch 2006, Whittard et al. 2006).

Finally, the intracellular domain of vertebrate CHL1 binds to ATPase Hsc70, which controls the uncoating of clathrin-coated synaptic vesicles (Bronk et al. 2001, Zinsmaier and Bronk 2001, Leshchyns'ka et al. 2006). Loss-of-function of CHL1 or interference of the association of CHL1 with the Hsc70 complex affects the ability of endocytosed synaptic vesicles to remove clathrin, and this results in an accumulation of clathrin-coated synaptic vesicles in the presynaptic terminal (Leshchyns'ka et al. 2006). Therefore, CHL1 has a function in regulating synaptic vesicle recycling and hence its disruption also affects the exocytosis of synaptic vesicles, which is essential for synaptic transmission (Bronk et al. 2001, Zinsmaier and Bronk 2001, Leshchyns'ka et al. 2006).

12.3 Conclusions and Outlook

Despite the limited number of studies that as been conducted to enumerate the roles played by L1-type CAMs at the synapse, the breadth of their synaptic functions described so far is astonishing. Various L1-type members have functions at many developmental stages ranging from synaptic targeting and synapse formation to synaptic transmission (Ango et al. 2004, Godenschwege et al. 2006, Leshchyns'ka et al. 2006, Triana-Baltzer et al. 2006). They have important functions at GABAergic, glutamatergic, and cholinergic synapses and are present either pre- or postsynaptically in both the CNS and the NMJ (Matsumoto-Miyai et al. 2003, Ango et al. 2004, Godenschwege et al. 2006, Nakamura et al. 2006, Triana-Baltzer et al. 2006, Burkarth et al. 2007). L1-type proteins have distinct functions in different types of neurons or sometimes even in the same neuron (Ango et al. 2004, Saghatelyan et al. 2004, Nikonenko et al. 2006). Hence it is not surprising that despite their similarities, different L1-type CAMs have different binding partners and that the same L1-type CAM has different functions depending on its phosphorylation status (Jenkins et al. 2001, Kizhatil et al. 2002, Davey et al. 2005, Dirks et al. 2006, Godenschwege et al. 2006). All these studies have been carried out in multiple model organisms, cell types, and tissues, and intriguingly, most have resulted in the identification of functions involving a single L1-type family member. Hence it is very likely that in the future more synaptic functions for L1-type proteins will be discovered, and the exact cellular mechanisms of many of the already identified functions will be further elucidated. It is most important to note that though there are well-studied conserved functions among all L1-type CAMs, it is imperative to focus on each L1-type CAM as a single entity and unearth its distinct array of functions, in order to get a complete understanding of the roles played by the L1-CAM family in synapse development.

Acknowledgments Tanja. A. Godenschwege is supported by RO1 HD050725-01A1, and Smitha B. Uthaman was supported by RO1 NS044609. The content is solely the responsibility of the authors and does not necessarily represent the official views of the National Institute of Child Health and Human Development or the National Institutes of Health.

References

Allen MJ and Murphey RK (2007) The chemical component of the mixed GF-TTMn synapse in Drosophila melanogaster uses acetylcholine as its neurotransmitter. Euro J Neurosci 26:439–445

Angeloni D, Wei MH and Lerman MI (1999) Two single nucleotide polymorphisms (SNPs) in the CALL gene for association studies with IQ. Psychiatric genetics 9(3):165–167

Ango F, di Cristo G, Higashiyama H et al. (2004) Ankyrin-based subcellular gradient of neurofascin, an immunoglobulin family protein, directs GABAergic innervation at purkinje axon initial segment. Cell 119:257–272

Arami S, Jucker M, Schachner M et al. (1996) The effect of continuous intraventricular infusion of L1 and NCAM antibodies on spatial learning in rats. Behav Brain Res 81:81–87

Bains JS and Oliet SH (2007) Glia: they make your memories stick! Trends Neurosci 30:417–424

Bayer SA and Altman J (1987) Directions in neurogenetic gradients and patterns of anatomical connections in the telencephalon. Prog Neurobiol 29:57–106

Benson DL, Colman DR and Huntley GW (2001) Molecules, maps and synapse specificity. Nature Rev 2:899–909

Bieber AJ, Snow PM, Hortsch M et al. (1989) *Drosophila* neuroglian: a member of the immunoglobulin superfamily with extensive homology to the vertebrate neural adhesion molecule L1. Cell 59:447–460

Blaess S, Kammerer RA and Hall H (1998) Structural analysis of the sixth immunoglobulin-like domain of mouse neural cell adhesion molecule L1 and its interactions with alpha(v)-beta3, alpha(IIb)beta3, and alpha5beta1 integrins. J Neurochem 71:2615–2625

Bliss T, Errington M, Fransen E et al. (2000) Long-term potentiation in mice lacking the neural cell adhesion molecule L1. Curr Biol 10:1607–1610

Bouley M, Tian MZ, Paisley K et al. (2000) The L1-type cell adhesion molecule neuroglian influences the stability of neural ankyrin in the Drosophila embryo but not its axonal localization. J Neurosci 20:4515–4523
Bronk P, Wenniger JJ, Dawson-Scully K et al. (2001) Drosophila Hsc70-4 is critical for neurotransmitter exocytosis in vivo. Neuron 30:475–488
Brümmendorf T, Kenwrick S and Rathjen FG (1998) Neural cell recognition molecule L1: from cell biology to human hereditary brain malformations. Curr Opin Neurobiol 8:87–97
Burkarth N, Kriebel M, Kranz EU et al. (2007) Neurofascin regulates the formation of gephyrin clusters and their subsequent translocation to the axon hillock of hippocampal neurons. Mol Cell Neurosci 36:59–70
Caspi M, Atlas R, Kantor A et al. (2000) Interaction between LIS1 and doublecortin, two lissencephaly gene products. Human Mol Genet 9:2205–2213
Castellani V, De Angelis E, Kenwrick S et al. (2002) Cis and trans interactions of L1 with neuropilin-1 control axonal responses to semaphorin 3A. EMBO J 21:6348–6357
Chang KT and Berg DK (2001) Voltage-gated channels block nicotinic regulation of CREB phosphorylation and gene expression in neurons. Neuron 32:855–865
Chen CL, Lampe DJ, Robertson HM et al. (1997) Neuroglian is expressed on cells destined to form the prothoracic glands of Manduca embryos as they segregate from surrounding cells and rearrange during morphogenesis. Dev Biol 181:1–13
Chen L, Ong B and Bennett V (2001) LAD-1, the Caenorhabditis elegans L1CAM homologue, participates in embryonic and gonadal morphogenesis and is a substrate for fibroblast growth factor receptor pathway-dependent phosphotyrosine-based signaling. J Cell Biol 154:841–855
Chen QY, Chen Q, Feng GY et al. (2005) Case-control association study of the close homologue of L1 (CHL1) gene and schizophrenia in the Chinese population. Schizophr Res 73:269–274
Collingridge GL and Bliss TV (1995) Memories of NMDA receptors and LTP. Trends Neurosci 18:54–56
Davey F, Hill M, Falk J et al. (2005) Synapse associated protein 102 is a novel binding partner to the cytoplasmic terminus of neurone-glial related cell adhesion molecule. J Neurochem 94:1243–1253
Davis JQ and Bennett V (1994) Ankyrin binding activity shared by the neurofascin/L1/NrCAM family of nervous system cell adhesion molecules. J Biol Chem 269:27163–27166
De Angelis E, MacFarlane J, Du JS et al. (1999) Pathological missense mutations of neural cell adhesion molecule L1 affect homophilic and heterophilic binding activities. EMBO J 18:4744–4753
De Angelis E, Watkins A, Schafer M et al. (2002) Disease-associated mutations in L1 CAM interfere with ligand interactions and cell-surface expression. Human Mol Genet 11:1–12
Dickson TC, Mintz CD, Benson DL et al. (2002) Functional binding interaction identified between the axonal CAM L1 and members of the ERM family. J Cell Biol 157:1105–1112
Dirks P, Thomas U and Montag D (2006) The cytoplasmic domain of NrCAM binds to PDZ domains of synapse-associated proteins SAP90/PSD95 and SAP97. Euro J Neurosci 24:25–31
Dubreuil RR, MacVicar G, Dissanayake S et al. (1996) Neuroglian-mediated cell adhesion induces assembly of the membrane skeleton at cell contact sites. J Cell Biol 133:647–655
Essrich C, Lorez M, Benson JA et al. (1998) Postsynaptic clustering of major GABAA receptor subtypes requires the gamma 2 subunit and gephyrin. Nature Neurosci 1:563–571
Falk J, Thoumine O, Dequidt C et al. (2004) NrCAM coupling to the cytoskeleton depends on multiple protein domains and partitioning into lipid rafts. Mol Biol Cell 15:4695–4709
Felding-Habermann B, Silletti S, Mei F et al. (1997) A single immunoglobulin-like domain of the human neural cell adhesion molecule L1 supports adhesion by multiple vascular and platelet integrins. J Cell Biol 139:1567–1581

Forni JJ, Romani S, Doherty P et al. (2004) Neuroglian and FasciclinII can promote neurite outgrowth via the FGF receptor Heartless. Mol Cell Neurosci 26:282–291

Fransen E, D'Hooge R, Van Camp G et al. (1998a) L1 knockout mice show dilated ventricles, vermis hypoplasia and impaired exploration patterns. Human Mol Genet 7:999–1009

Fransen E, Lemmon V, Van Camp G et al. (1995) CRASH syndrome: clinical spectrum of corpus callosum hypoplasia, retardation, adducted thumbs, spastic paraparesis and hydrocephalus due to mutations in one single gene, L1. Eur J Human Genet 3:273–284

Fransen E, Van Camp G, D'Hooge R et al. (1998b) Genotype-phenotype correlation in L1 associated diseases. J Med Genet 35:399–404

Fransen E, Van Camp G, Vits L et al. (1997) L1-associated diseases: clinical geneticists divide, molecular geneticists unite. Human Mol Genet 6:1625–1632

Freeman MR (2005) Glial control of synaptogenesis. Cell 120:292–293

Freund TF and Buzsaki G (1996) Interneurons of the hippocampus. Hippocampus 6:347–470

Frints SG, Marynen P, Hartmann D et al. (2003) CALL interrupted in a patient with non-specific mental retardation: gene dosage-dependent alteration of murine brain development and behavior. Human Mol Genet 12:1463–1474

Fritschy JM, Harvey RJ and Schwarz G (2008) Gephyrin: where do we stand, where do we go? Trends Neurosci 31(5):257–264

Garver TD, Ren Q, Tuvia S et al. (1997) Tyrosine phosphorylation at a site highly conserved in the L1 family of cell adhesion molecules abolishes ankyrin binding and increases lateral mobility of neurofascin. J Cell Biol 137:703–714

Gerrow K and EI-Husseini A (2206) Cell adhesion molecules at the synapse. Front Biosci 11:2400–2419

Godenschwege TA, Kristiansen LV, Uthaman SB et al. (2006) A conserved role for *Drosophila* Neuroglian and human L1-CAM in central-synapse formation. Curr Biol 16:12–23

Greenberg ME, Ziff EB and Greene LA (1986) Stimulation of neuronal acetylcholine receptors induces rapid gene transcription. Science (New York, NY) 234:80–83

Grumet M, Friedlander DR and Edelman GM (1993) Evidence for the binding of Ng-CAM to laminin. Cell Adhes Commun 1:177–190

Grumet M, Mauro V, Burgoon MP et al. (1991) Structure of a new nervous system glycoprotein, Nr-CAM, and its relationship to subgroups of neural cell adhesion molecules. J Cell Biol 113:1399–1412

Haspel J and Grumet M (2003) The L1CAM extracellular region: a multi-domain protein with modular and cooperative binding modes. Front Biosci 8:s1210–1225

Hassel B, Rathjen FG and Volkmer H (1997) Organization of the neurofascin gene and analysis of developmentally regulated alternative splicing. J Biol Chem 272:28742–28749

Holm J, Hillenbrand R, Steuber V et al. (1996) Structural features of a close homologue of L1 (CHL1) in the mouse: a new member of the L1 family of neural recognition molecules. Eur J Neurosci 8:1613–1629

Horstkorte R, Schachner M, Magyar JP et al. (1993) The fourth immunoglobulin-like domain of NCAM contains a carbohydrate recognition domain for oligomannosidic glycans implicated in association with L1 and neurite outgrowth. J Cell Biol 121:1409–1421

Hortsch M (1996) The L1 family of neural cell adhesion molecules: old proteins performing new tricks. Neuron 17:587–593

Hortsch M (2000) Structural and functional evolution of the L1 family: are four adhesion molecules better than one? Mol Cell Neurosci 15:1–10

Hortsch M, Bieber AJ, Patel NH et al. (1990) Differential splicing generates a nervous system-specific form of *Drosophila* neuroglian. Neuron 4:697–709

Hortsch M, Homer D, Malhotra JD et al. (1998) Structural requirements for outside-in and inside-out signaling by *Drosophila* neuroglian, a member of the L1 family of cell adhesion molecules. J Cell Biol 142:251–261

Huang ZJ (2006) Subcellular organization of GABAergic synapses: role of ankyrins and L1 cell adhesion molecules. Nature Neurosci 9:163–166

Hutcheson HB, Olson LM, Bradford Y et al. (2004) Examination of NRCAM, LRRN3, KIAA0716, and LAMB1 as autism candidate genes. BMC Med Genet 5:12

Islam R, Kristiansen LV, Romani S et al. (2004) Activation of EGF receptor kinase by L1-mediated homophilic cell interactions. Mol Biol Cell 15:2003–2012

Itoh K, Ozaki M, Stevens B et al. (1997) Activity-dependent regulation of N-cadherin in DRG neurons: differential regulation of N-cadherin, NCAM, and L1 by distinct patterns of action potentials. J Neurobiol 33:735–748

Itoh K, Shimono K and Lemmon V (2005) Dephosphorylation and internalization of cell adhesion molecule L1 induced by theta burst stimulation in rat hippocampus. Mol Cell Neurosci 29:245–249

Itoh K, Stevens B, Schachner M et al. (1995) Regulated expression of the neural cell adhesion molecule L1 by specific patterns of neural impulses. Science (New York, NY) 270:1369–1372

Jenkins SM and Bennett V (2001) Ankyrin-G coordinates assembly of the spectrin-based membrane skeleton, voltage-gated sodium channels, and L1 CAMs at Purkinje neuron initial segments. J Cell Biol 155:739–746

Jenkins SM, Kizhatil K, Kramarcy NR et al. (2001) FIGQY phosphorylation defines discrete populations of L1 cell adhesion molecules at sites of cell-cell contact and in migrating neurons. J Cell Sci 114:3823–3835

Kamiguchi H and Lemmon V (1997) Neural cell adhesion molecule L1: signaling pathways and growth cone motility. J Neurosci Res 49:1–8

Kamiguchi H and Lemmon V (2000) IgCAMs: bidirectional signals underlying neurite growth. Curr Opin Cell Biol 12:598–605

Kamiguchi H, Long KE, Pendergast M et al. (1998) The neural cell adhesion molecule L1 interacts with the AP-2 adaptor and is endocytosed via the clathrin-mediated pathway. J Neurosci 18:5311–5321

Kenwrick S, Watkins A and De Angelis E (2000) Neural cell recognition molecule L1: relating biological complexity to human disease mutations. Human Mol Genet 9:879–886

Kizhatil K, Wu YX, Sen A et al. (2002) A new activity of doublecortin in recognition of the phospho-FIGQY tyrosine in the cytoplasmic domain of neurofascin. J Neurosci 22:7948–7958

Krueger SR, Kolar A and Fitzsimonds RM (2003) The presynaptic release apparatus is functional in the absence of dendritic contact and highly mobile within isolated axons. Neuron 40:945–957

Kuhn TB, Stoeckli ET, Condrau MA et al. (1991) Neurite outgrowth on immobilized axonin-1 is mediated by a heterophilic interaction with L1(G4). J Cell Biol 115:1113–1126

Kulahin N, Li S, Hinsby A et al. (2008) Fibronectin type III (FN3) modules of the neuronal cell adhesion molecule L1 interact directly with the fibroblast growth factor (FGF) receptor. Mol Cell Neurosci 37:528–536

Kurumaji A, Nomoto H, Okano T et al. (2001) An association study between polymorphism of L1CAM gene and schizophrenia in a Japanese sample. Am J Med Genet 105:99–104

Law JW, Lee AY, Sun M et al. (2003) Decreased anxiety, altered place learning, and increased CA1 basal excitatory synaptic transmission in mice with conditional ablation of the neural cell adhesion molecule L1. J Neurosci 23:10419–10432

Leonard AS, Davare MA, Horne MC et al. (1998) SAP97 is associated with the alpha-amino-3-hydroxy-5-methylisoxazole-4-propionic acid receptor GluR1 subunit. J Biol Chem 273:19518–19524

Leshchyns'ka I, Sytnyk V, Richter M et al. (2006) The adhesion molecule CHL1 regulates uncoating of clathrin-coated synaptic vesicles. Neuron 52:1011–1025

Luthi A, Mohajeri H, Schachner M et al. (1996) Reduction of hippocampal long-term potentiation in transgenic mice ectopically expressing the neural cell adhesion molecule L1 in astrocytes. J Neurosci Res 46:1–6

Luthl A, Laurent JP, Figurov A et al. (1994) Hippocampal long-term potentiation and neural cell adhesion molecules L1 and NCAM. Nature 372:777–779

Matsumoto-Miyai K, Ninomiya A, Yamasaki H et al. (2003) NMDA-dependent proteolysis of presynaptic adhesion molecule L1 in the hippocampus by neuropsin. J Neurosci 23:7727–7736

Matzel LD, Babiarz J, Townsend DA, et al. (2008) Neuronal cell adhesion molecule deletion induces a cognitive and behavioral phenotype reflective of impulsivity. Genes, brain, and behavior 7(4):470–480

Mehrke G, Jockusch H, Faissner A et al. (1984) Synapse formation and synaptic activity in mammalian nerve-muscle co-culture are not inhibited by antibodies to neural cell adhesion molecule L1. Neurosci Lett 44:235–239

Mintz CD, Dickson TC, Gripp ML et al. (2003) ERMs colocalize transiently with L1 during neocortical axon outgrowth. J Comp Neurol 464:438–448

Munakata H, Nakamura Y, Matsumoto-Miyai K et al. (2003) Distribution and densitometry mapping of L1-CAM immunoreactivity in the adult mouse brain – light microscopic observation. BMC neuroscience 4:7

Nagaraj K and Hortsch M (2006) Phosphorylation of L1-type cell-adhesion molecules – ankyrins away! Trends in biochemical sciences 31:544–546

Nakamura Y, Tamura H, Horinouchi K et al. (2006) Role of neuropsin in formation and maturation of Schaffer-collateral L1cam-immunoreactive synaptic boutons. J Cell Sci 119:1341–1349

Nikonenko AG, Sun M, Lepsveridze E et al. (2006) Enhanced perisomatic inhibition and impaired long-term potentiation in the CA1 region of juvenile CHL1-deficient mice. Euro J Neurosci 23:1839–1852

Oleszewski M, Beer S, Katich S et al. (1999) Integrin and neurocan binding to L1 involves distinct Ig domains. J Biol Chem 274:24602–24610

Rathjen FG and Schachner M (1984) Immunocytological and biochemical characterization of a new neuronal cell surface component (L1 antigen) which is involved in cell adhesion. Embo J 3:1–10

Rathjen FG, Wolff JM, Chang S et al. (1987) Neurofascin: a novel chick cell-surface glycoprotein involved in neurite-neurite interactions. Cell 51:841–849

Ren Q and Bennett V (1998) Palmitoylation of neurofascin at a site in the membrane-spanning domain highly conserved among the L1 family of cell adhesion molecules. J Neurochem 70:1839–1849

Rose SP (1996) Cell adhesion molecules and the transition from short- to long-term memory. Journal of physiology, Paris 90(5-6):387–391

Saghatelyan AK, Nikonenko AG, Sun M et al. (2004) Reduced GABAergic transmission and number of hippocampal perisomatic inhibitory synapses in juvenile mice deficient in the neural cell adhesion molecule L1. Mol Cell Neurosci 26:191–203

Sakurai K, Migita O, Toru M et al. (2002) An association between a missense polymorphism in the close homologue of L1 (CHL1, CALL) gene and schizophrenia. Mol Psychiatry 7:412–415

Sakurai T, Ramoz N, Reichert JG et al. (2006) Association analysis of the NrCAM gene in autism and in subsets of families with severe obsessive-compulsive or self-stimulatory behaviors. Psychiatr Genet 16:251–257

Sanes JR, Schachner M and Covault J (1986) Expression of several adhesive macromolecules (N-CAM, L1, J1, NILE, uvomorulin, laminin, fibronectin, and a heparan sulfate proteoglycan) in embryonic, adult, and denervated adult skeletal muscle. J Cell Biol 102:420–431

Sans N, Prybylowski K, Petralia RS et al. (2003) NMDA receptor trafficking through an interaction between PDZ proteins and the exocyst complex. Nature cell biology 5(6):520–530

Schaefer AW, Kamei Y, Kamiguchi H et al. (2002) L1 endocytosis is controlled by a phosphorylation-dephosphorylation cycle stimulated by outside-in signaling by L1. J Cell Biol 157:1223–1232

Schaefer AW, Kamiguchi H, Wong EV et al. (1999) Activation of the MAPK signal cascade by the neural cell adhesion molecule L1 requires L1 internalization. J Biol Chem 274:37965–37973

Scherer M, Heller M and Schachner M (1992) Expression of the Neural Recognition Molecule L1 by Cultured Neural Cells is influenced by K+ and the Glutamate Receptor Agonist NMDA. Eur J Neurosci 4:554–562

Scholey AB, Mileusnic R, Schachner M et al. (1995) A role for a chicken homolog of the neural cell adhesion molecule L1 in consolidation of memory for a passive avoidance task in the chick. Learn Memory (Cold Spring Harbor, NY) 2:17–25

Seabold GK, Burette A, Lim IA et al. (2003) Interaction of the tyrosine kinase Pyk2 with the N-methyl-D-aspartate receptor complex via the Src homology 3 domains of PSD-95 and SAP102. J Biol Chem 278:15040–15048

Silletti S, Mei F, Sheppard D et al. (2000) Plasmin-sensitive dibasic sequences in the third fibronectin-like domain of L1-cell adhesion molecule (CAM) facilitate homomultimerization and concomitant integrin recruitment. J Cell Biol 149:1485–1502

Somogyi P, Tamas G, Lujan R et al. (1998) Salient features of synaptic organisation in the cerebral cortex. Brain Res 26:113–135

Songyang Z, Fanning AS, Fu C et al. (1997) Recognition of unique carboxyl-terminal motifs by distinct PDZ domains. Science (New York, NY) 275:73–77

Tiunova A, Anokhin KV, Schachner M et al. (1998) Three time windows for amnestic effect of antibodies to cell adhesion molecule L1 in chicks. Neuroreport 9:1645–1648

Todd KJ, Serrano A, Lacaille JC et al. (2006) Glial cells in synaptic plasticity. J Physiol, Paris 99:75–83

Triana-Baltzer GB, Liu Z and Berg DK (2006) Pre- and postsynaptic actions of L1-CAM in nicotinic pathways. Mol Cell Neurosci 33:214–226

Vallee RB, Tai C and Faulkner NE (2001) LIS1: cellular function of a disease-causing gene. Trends Cell Biol 11:155–160

Von Bohlen Und Halbach F, Taylor J and Schachner M (1992) Cell type-specific effects of the neural adhesion molecules L1 and N-CAM on diverse second messenger systems. Euro J Neurosci 4:896–909

Wang X, Zhang W, Cheever T et al. (2008) The C. elegans L1CAM homologue LAD-2 functions as a coreceptor in MAB-20/Sema2 mediated axon guidance. J Cell Biol 180:233–246

Welzl H and Stork (2003) Cell adhesion molecules: key players in memory consolidation? News Physiol Sci 18:147–150

Whittard JD, Sakurai T, Cassella MR et al. (2006) MAP kinase pathway-dependent phosphorylation of the L1-CAM ankyrin binding site regulates neuronal growth. Mol Biol Cell 17:2696–2706

Wolfer DP, Mohajeri HM, Lipp HP et al. (1998) Increased flexibility and selectivity in spatial learning of transgenic mice ectopically expressing the neural cell adhesion molecule L1 in astrocytes. Euro J Neurosci 10:708–717

Yamamoto M, Ueda R, Takahashi K et al. (2006) Control of axonal sprouting and dendrite branching by the Nrg-Ank complex at the neuron-glia interface. Curr Biol 16:1678–1683

Zinsmaier KE and Bronk P (2001) Molecular chaperones and the regulation of neurotransmitter exocytosis. Biochem Pharmacol 62:1–11

Chapter 13
Cell Adhesion Molecules of the NCAM Family and Their Roles at Synapses

Sylwia Owczarek, Lars V. Kristiansen, Michael Hortsch, and Peter S. Walmod

Abstract NCAM-type proteins modulate multiple neuronal functions, including the outgrowth and guidance of neurites, the formation, maturation, and plasticity of synapses, and the induction of both long-term potentiation and long-term depression. The ectodomains of NCAM proteins have a basic structure of five amino-terminal immunoglobulin (Ig), followed by two fibronectin type III (FnIII) modules. As a result of alternative splicing, many NCAM-type proteins exist in several isoforms, including both transmembrane and glycosylphosphatidylinositol (GPI)-anchored versions. Extracellularly, NCAM proteins mediate cell–cell adhesion through homophilic interactions and bind to growth factors, growth factor receptors, glutamate receptors, other CAMs, and components of the extracellular matrix. Intracellularly, NCAM-type proteins interact with various cytoskeletal proteins and regulators of intracellular signal transduction. A central feature of the synaptic function of NCAM proteins is the regulation of their extracellular interactions by adhesion-modulating glycoepitopes, their removal from the cell surface by endocytosis, and the elimination of their adhesion-mediating interactions by the proteolytic cleavage of their ectodomains. Although specific aspects of NCAM proteins have changed through evolution, core structural and functional features are conserved between NCAM-type proteins in vertebrates and invertebrates, demonstrating that the functions of this class of adhesive proteins are of general importance during nervous system formation.

Keywords apCAM · ATP · Fasciclin II · Immunoglobulin · NCAM · PSA

P.S. Walmod (✉)
Protein Laboratory, Department of Neuroscience and Pharmacology, Faculty of Health Sciences, University of Copenhagen, Panum Institute, Blegdamsvej 3, DK-2200, Copenhagen, Denmark
e-mail: psw@sund.ku.dk

M. Hortsch, H. Umemori (eds.), *The Sticky Synapse*,
DOI 10.1007/978-0-387-92708-4_13, © Springer Science+Business Media, LLC 2009

13.1 Members of the NCAM Family of Cell Adhesion Molecules

The mammalian neural cell adhesion molecule 1 (NCAM1) was the first cell adhesion molecule (CAM) to be identified (Jorgensen and Bock 1974, Rutishauser et al. 1976). Initially, mammalian NCAM1 was referred to as D2 or D2-CAM (Jacque et al. 1974, Jorgensen and Bock 1974, Noble et al. 1985) and is similar to cluster of differentiation (CD) antigen CD56 (HLDM 2008). However, since the discovery of an additional mammalian NCAM paralog, termed NCAM2, the original NCAM protein is correctly referred to as NCAM1 (Fu et al. 2006). Due to the simultaneous discovery of the mammalian *NCAM2* gene by three independent research groups in 1997, the derived protein is known by three names: NCAM2 (Paoloni-Giacobino et al. 1997), olfactory CAM (OCAM, Yoshihara et al. 1997), and Rb-8-neural cell adhesion molecule (RNCAM, Alenius and Bohm 1997). In the following sections, the mammalian members of the NCAM family will be referred to as NCAM1 and NCAM2, respectively.

Members of the NCAM family have been identified and studied in the nervous system of several invertebrate and non-mammalian vertebrates (Fig. 13.1). Like mammals, other vertebrates seem to express at least two NCAM family proteins. Thus, chicken (*Gallus gallus*) and clawed frog (*Xenopus sp.*) express homologs of both mammalian NCAM1 and NCAM2 (Murray et al. 1986a, Tonissen and Krieg 1993, Kulahin and Walmod 2008). Zebrafish (*Danio rerio*) expresses four NCAM family proteins, denoted zNCAM/NCAM1, zOCAM/NCAM2, zPCAM/NCAM3, and hypothetical protein LOC767777: zNCAM/NCAM1 and zpCAM/NCAM3 are homologs of mammalian NCAM1, and zOCAM/NCAM2 and LOC767777 are homologs of mammalian NCAM2, respectively (Mizuno et al. 2001, Strausberg et al. 2002, Bushell et al. 2008) (Fig. 13.1). In contrast to vertebrates, invertebrate species express only one NCAM-type protein (Fig. 13.1). In the roundworm *Caenorhabditis elegans*, the protein is termed NCAM-1 (Teichmann and Chothia 2000, Thierry-Mieg et al. 2008); in the snail *Aplysia californica*, it is denoted AplysiaCAM or ApCAM (Mayford et al. 1992); and in the leech *Hirudo medicinalis*, LeechCAM (Huang et al. 1997). Finally, in the *Drosophilidae* fruit fly family (specifically *Drosophila melanogaster*), and in other insects (e.g., the moth *Manduca sexta* and the mosquito *Aedes aegypti*), a single member of the NCAM family, denoted fasciclin II or FasII, is expressed (Grenningloh et al. 1991, Wright et al. 1999, Nene et al. 2007). The fact that vertebrate genomes encode more NCAM-type proteins than invertebrate genomes is consistent with reports for other conserved molecules, and it likely results from multiple genome duplication events during chordate evolution (Pebusque et al. 1998). Furthermore, zebrafish is believed, after the evolutionary separation from what later became tetrapods, to have undergone an additional genome duplication, but with only partial retention of the duplicated genes (Amores et al. 1998, Aparicio 2000, Zhou et al. 2001). Hence, the presence of four NCAM genes in zebrafish is not unexpected. For data of the phylogenetic

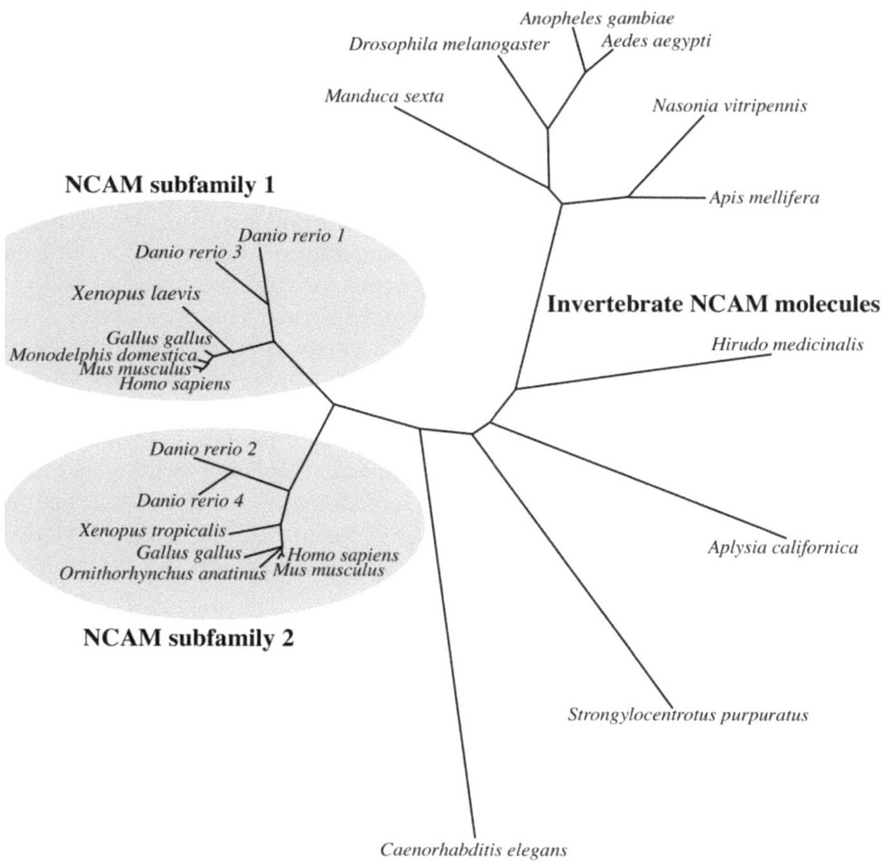

Fig. 13.1 Phylogenetic tree of NCAM-type genes in different species. With the exception of the *C. elegans* sax-3 protein, all other 23 polypeptides, which were selected for this phylogenetic analysis, conform to the five Ig plus two FnIII module structures of NCAM-type proteins. Polypeptide sequences, which cover four and a half of the Ig and both FnIII modules, were used for a protein sequence comparison. The multiple sequence alignment was performed using the online version of the MAFFT program (http://align.bmr.kyushu-u.ac.jp/mafft/online/server/). An unrooted phylogenetic tree was constructed using the Promlk and the Drawtree subroutines of the Phylip v3.65 program package (Felsenstein 1981). The genomes of the included invertebrate species [*Caenorhabditis elegans* (nematode; NP_741748), *Aplysia californica* (California sea slug; AAK14901), *Hirudo medicinalis* (medicinal leech; AAC47655), *Nasonia vitripennis* (jewel wasp; XP_001606611), *Anopheles gambiae* (malaria mosquito; XP_312112), *Aedes aegypti* (yellow fever mosquito; XP_001659793), *Drosophila melanogaster* (fruit fly; P34082), *Apis mellifera* (honey bee; XP_393339), *Manduca sexta* (tobacco hornworm; AAD17919), and *Strongylocentrotus purpuratus* (purple sea urchin; XP_001177682)] contain only one NCAM-type gene. In contrast, most vertebrate species, such as *Gallus gallus* (chicken; XP_001234122 and XP_425540), *Mus musculus* (mouse; NP_001074914 and NP_035084), and *Homo sapiens* (human; NP_000606 and NP_004531), have two distinct NCAM-type genes, with the *Danio rerio* (zebrafish; NP_571277, NP_571905, NP571906, and NP_001070212) genome encoding a total of four

relationships between different members of the NCAM family, see Jie et al. (1999) and Kristiansen and Hortsch (2008).

13.2 Structure of NCAM Family Proteins

All genes of the NCAM family encode transmembrane proteins, which consist of an ectodomain that is followed by a transmembrane region and a cytosolic domain (Fig. 13.2). However, the general protein domain structure of NCAM proteins is often varied as a result of alternative splicing. This produces multiple protein isoforms that are generated from a single NCAM gene. For many species, the potential presence of multiple splice variants has not been studied in detail. Therefore, in the following section, only a selection of the potential types of NCAM proteins across various species is presented. One of the most universally observed results of NCAM alternative splicing is the generation of molecules consisting of the NCAM ectodomain, which is attached to the membrane by a glycosylphosphatidylinositol (GPI) anchor. Hence, GPI-anchored NCAM isoforms have been described for the mammalian

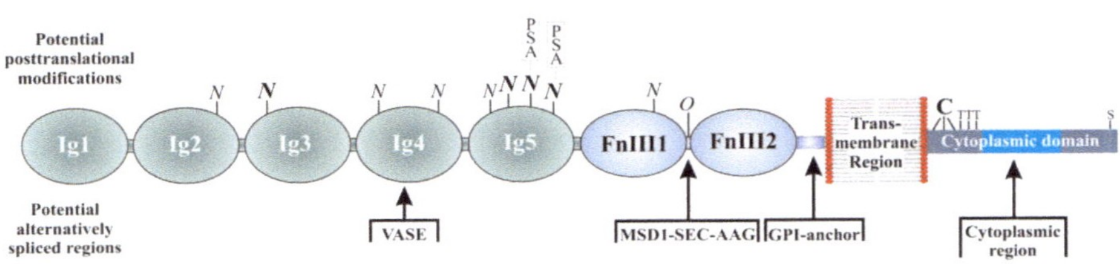

Fig. 13.2 Schematic drawing of the organization of NCAM-type proteins. All NCAM proteins consist of an ectodomain composed of five Ig and two FnIII protein folds. All the indicated alternatively spliced regions are found in mammalian NCAM1, whereas NCAM2 proteins only contain few of these alternatively spliced inserts (see text). The posttranslational modifications indicated at the top of the figure are representatives for vertebrate NCAM1 and NCAM2 proteins. ***N***: N-glycosylation site conserved in all NCAM1 and NCAM2 proteins. *N*: N-glycosylation site conserved in most NCAM1 and/or NCAM2 proteins. O: O-glycosylation site conserved in NCAM1. PSA: polysialylation sites in NCAM1. These N-glycosylation sites can also be glycosylated with non-PSA glycoepitopes. **C**: conserved palmitoylation sites in NCAM1 and NCAM2 (Little et al. 1998). T: conserved threonine phosphorylation sites (Little et al. 2001). S: conserved serine phosphorylation site (Polo-Parada et al. 2005). For a more detailed discussion of NCAM2 protein modification see Kulahin and Walmod (2008)

Fig. 13.1 (continued) separate NCAM polypeptides. As only partial genomic sequences are currently available for *Xenopus tropicalis* (Western clawed frog; Xentr4|393939), *Xenopus laevis* (African clawed frog; NP_001081296), *Ornithorhynchus anatinus* (duck-billed platypus; XP_001511913), and *Monodelphis domestica* (gray short-tailed opossum; XP_001381275), it appears likely that these genomes also contain a second NCAM-type gene

NCAM1 (Cunningham et al. 1987) and NCAM2 proteins (Alenius and Bohm 1997, Paoloni-Giacobino et al. 1997, Yoshihara et al. 1997), *Xenopus sp.* NCAMs (Krieg et al. 1989), ApCAM (Mayford et al. 1992), and fasciclin II (reviewed by Kristiansen and Hortsch 2008). Interestingly, in *Manduca sexta*, the transmembrane fasciclin II form is exclusively expressed by neuronal cells in the CNS, whereas the GPI-linked fasciclin II isoform is expressed in different glial cell types where it likely functions as an adhesion substrate (Wright and Copenhaver 2001, Higgins et al. 2002).

Alternative splicing of several exons in the region that encodes the cytoplasmic domain of NCAM gives rise to protein isoforms with short and long cytoplasmic domains, respectively. In mammalian NCAM1, the cytoplasmic domains have a length of ~119 and ~386 amino, respectively (Barthels et al. 1988, Tacke and Goridis 1991, The UniProt Consortium 2008, UniProtKB/Swiss-Prot accession number P13595). The same pattern of alternative splicing is observed in chicken (Murray et al. 1986b, Owens et al. 1987) and *Xenopus* (Krieg et al. 1989), whereas the alternative splicing of the *Drosophila* transmembrane fasciclin II proteins is entirely different. It generates two alternatively spliced cytoplasmic domains that differ in the inclusion/exclusion of an amino acid motif that is rich in proline, glutamic acid, serine, and threonine residues (PEST, Lin et al. 1994) (reviewed by Kristiansen and Hortsch 2008). The above-described alternative splicing generates up to three main NCAM isoforms: one GPI-anchored and two transmembrane forms. These isoforms are referred to as small surface-domain polypeptide (ssd), small intracellular-domain polypeptide (sd), and large intracellular-domain polypeptide (ld). This nomenclature was originally proposed by Hemperly et al. (1986). However, for mammalian NCAM1, these terms are often substituted by NCAM1-120, -140, and -180, respectively, the numbers referring to the apparent molecular weights of the glycosylated NCAM1 isoforms. This classification was first suggested by He et al. (1986). The different NCAM1 spliceforms exhibit different cellular and subcellular distributions. Thus, studies of the cerebellum and the hippocampus of adult rodents have shown that NCAM-120 is not detectable in synaptosomal membranes, whereas NCAM-140 is expressed pre- and postsynaptically, and NCAM-180 is found predominantly postsynaptically (Persohn et al. 1989, Schuster et al. 2001). However, NCAM-180 can also be expressed presynaptically, and studies in chicken have shown that the expression of NCAM-180 is important for the organization of proteins at the presynaptic terminal (Polo-Parada et al. 2004). Furthermore, the pre- and postsynaptic expression of NCAM1-180 is increased, and its synaptic distribution is modulated in response to induction of long-term potentiation (LTP) (Schuster et al. 1998, Fux et al. 2003). Studies of motor neurons have shown that the presynaptic targeting of NCAM1-140, but not of NCAM1-180 requires the presence of NCAM1 at the postsynaptic membrane (Hata, Polo-Parada and Landmesser 2007).

The ectodomains of NCAM proteins invariably consist of five N-terminal, membrane-distal immunoglobulin (Ig) homology modules (denoted Ig1-5), which are followed by two membrane-proximal fibronectin type III (FnIII) homology modules (denoted FnIII1 and FnIII2). The mammalian NCAM1 ectodomain also exists in multiple isoforms, which are generated by the alternative inclusion of six small exons that are present in the original transcript. In the region encoding the Ig4-module, a small exon termed VASE or π, which encodes a 10 amino acids long sequence, is located between exons 7 and 8 (Small and Akeson 1990). Five additional small exons are located in the so-called muscle-specific domain (MSD) between exons 12 and 13, a region encoding a linker between FnIII1 and FnIII2. The first three of these exons, MSD1a, -b, and -c, encode regions of 5, 15, and 14 amino acid residues, respectively (Dickson et al. 1987, Santoni et al. 1989, Thompson et al. 1989, Hamshere et al. 1991, Reyes et al. 1991, Barthels et al. 1992). MSD1c is followed by the SEC exon, which includes a stop codon and therefore, when expressed, leads to the generation of a truncated, secreted NCAM1 isoform. This NCAM isoform encompasses only the six most N-terminal modules of the ectodomain (Gower et al. 1988). The SEC exon is followed by the AAG exon, which forms only a single nucleotide triplet (Santoni et al. 1989, Reyes et al. 1991). Chicken NCAM1 can contain the VASE insert, as well as four distinct amino acid segments, which are encoded by exons in the MSD region (Prediger et al. 1988, Barthels et al. 1992). In zebrafish, zNCAM/NCAM1 contains an optional 10 amino acids long VASE region, whereas zPCAM/NCAM3 can contain two VASE sequences; a short, identical to that found in zNCAM, and a longer, with 8 additional amino acids in the N-terminal part (Mizuno et al. 2001). *Xenopus* NCAM1 also exists in multiple isoforms with a VASE insert and/or two inserts encoded by exons in the MSD region (Zorn and Krieg 1992).

Aggregation assays demonstrated that VASE-containing NCAM1 proteins preferentially form *trans* interactions with other VASE-containing NCAM1 proteins, whereas NCAM1 lacking the VASE segment preferentially binds other NCAM1 molecules that are also missing the VASE sequence (Chen et al. 1994). The inclusion of the VASE region reduces NCAM1-mediated neurite outgrowth (Doherty et al. 1992, Liu et al. 1993, Saffell et al. 1994) and during the development of the chick nervous system, the expression of VASE-containing NCAM1 correlates inversely with the expression of PSA-NCAM1 in some brain regions. Thus, the appearance of VASE-containing NCAM1 in chick cerebellum and tectum coincides with a reduced expression of PSA-NCAM1. However, the presence of the VASE segment does not prevent the attachment of PSA to the NCAM1 protein (Oka et al. 1995). Furthermore, a study of senescence-accelerated mice has shown that an increase in the expression of VASE-containing NCAM1 in the hippocampus correlates with an impairment of spatial learning and memory (Qin et al. 2005). This suggests that the presence of the VASE regions in NCAM molecules specifically affects the formation or maturation of synapses.

In contrast to the numerous potential splice variants of the vertebrate NCAM1 ectodomain, the ectodomains of NCAM2 or non-vertebrate NCAM1 homologs are generally not affected by differential splicing processes (Paoloni-Giacobino et al. 1997, Yoshihara et al. 1997, Kristiansen and Hortsch

2008). However, a GPI-anchored isoform of *Xenopus* NCAM2 has been reported to contain an MSD, but not the VASE exon-encoded sequence (Kudo et al. 1998).

13.3 Posttranslational Modifications of NCAM Family Proteins

Several glycosylation sites are present in the ectodomain of NCAM-type proteins. Hence, zebrafish NCAM1 and NCAM2-homology proteins contain 4 and 6 glycosylation sites, respectively, whereas mammalian NCAM1 and NCAM2 contain 6–8 sites for *N*-linked glycosylation, and one site for *O*-linked glycosylation in the MSD region of mammalian NCAM1 (Walsh et al. 1989) (see Kulahin and Walmod 2008). Similarly, several potential sites for *N*-linked glycosylation are conserved between the *Manduca* and *Drosophila* fasciclin II protein sequences (Wright et al. 1999). In addition to N- and *O*-linked glycosylation, vertebrate NCAM1 proteins are heavily glycosylated with the unusual carbohydrate polysialic acid (PSA or PolySia). Modification by PSA, which consists of a linear chain of alpha 2,8 linked *N*-acetylneuraminic acid is spatially and temporally regulated (Hildebrandt et al. 2007, Gascon et al. 2008, Hildebrandt et al. 2008). As described in detail in later sections, the presence or absence of PSA is of major functional importance. Invertebrate NCAM-type proteins are not modified by PSA, indicating that this modification occurred as a separate evolutionary event after the split from the arthropod lineage.

NCAM1 expresses the glycoepitope HNK-1 (Walsh et al. 1989, Liedtke et al. 2001, Ong et al. 2002, Wuhrer et al. 2003), as well as a number of non-PSA/non-HNK-1 glycosylation moieties (Key and Akeson 1990, Key and Akeson 1991, Pestean et al. 1995, Dowsing et al. 1997). Glycosylation has been studied extensively in leech (Huang et al. 1997, Jie et al. 1999, reviewed by Tai and Zipser 2002). LeechCAM is glycosylated with several glycoepitopes, some of which are attached to specific glycosylation sites on the polypeptide. Furthermore, some glycoepitopes are specifically expressed by LeechCAM in the PNS, but not in the CNS, and others only by a subset of PNS neurons (Huang et al. 1997, Jie et al. 1999). Similar region-specific glycosylation modifications have also been observed for NCAM1, which in addition to the regulated expression of the PSA moiety is also modified in the olfactory system by specific carbohydrate groups (Key and Akeson 1990, Key and Akeson 1991, Pestean et al. 1995, Dowsing et al. 1997). The spatially and temporally regulated expression of specific glycoepitopes on NCAM family proteins is believed to be important for the formation and maintenance of synapses. For instance, studies of LeechCAM indicate that the transformation of cell–ECM interactions during sprouting of afferent neurons to cell–cell interactions at later stages, when the afferent neurons make contact with central neurons, is accompanied by a mannose-to-galactose change in its glycosylation pattern. Thus, mannosidic glycoepitopes seems to promote growth and dynamics of neurites and synapses, whereas galactosidic glycoepitopes serves to stabilize the formed connections (reviewed by Tai and Zipser 2002).

Intracellular posttranslational modifications of NCAM family proteins have mainly been studied in mammalian NCAM1 proteins and include additional membrane attachment through palmitoylations of 2–4 highly conserved cysteine residues (Little et al. 1998). This posttranslational modification regulates the subcellular distribution of NCAM1 molecules and is necessary for NCAM1-140-mediated neurite outgrowth via non-receptor protein tyrosine kinases (Niethammer et al. 2002). Recently, it was demonstrated that fibroblast growth factor 2 (FGF2), through an FGF receptor (FGFR)-mediated signaling cascade, activates the palmitoyl transferase DHHC-7, which palmitoylates NCAM1-140 and -180 (Ponimaskin et al. 2008). Consequently, FGF2 stimulation of FGFR can induce the subcellular relocation of transmembrane NCAM1 isoforms to lipid rafts and thereby promotes NCAM1-mediated neurite outgrowth (Ponimaskin et al. 2008). Three palmitoylations sites are conserved between human NCAM1 and NCAM2 (Kulahin and Walmod 2008). However, whether NCAM2 is also palmitoylated is currently unknown.

Transmembrane isoforms of NCAM1 can be phosphorylated at serine and threonine residues (Gennarini et al. 1984, Lyles et al. 1984, Sorkin et al. 1984). In NCAM1, phosphorylation of one or more threonine residues is important for the NCAM-mediated activation of the NF-κB transcription factor (Little et al. 2001). Furthermore, all transmembrane vertebrate NCAM1 proteins, and some, but not all transmembrane vertebrate NCAM2 proteins have a highly conserved serine residue located in their C-terminus (Polo-Parada et al. 2005, Kulahin and Walmod 2008). In NCAM1, phosphorylation of this serine residue has been shown to regulate mobilization or exocytosis of synaptic vesicles through a process that is believed to involve myosin light-chain kinase and myosin II (Polo-Parada et al. 2005). Although the sequence is present in both NCAM1-140 and -180, only NCAM1-180 seems to be involved in this synapse-specific function, possibly due to different subcellular localizations of the two NCAM1-isoforms, and mice lacking only NCAM1-180 demonstrate dysfunctional neurotransmission (Polo-Parada et al. 2004, 2005). In apCAM, phosphorylation of a threonine residue that is located in a mitogen-activated protein kinase (MAPK) phosphorylation consensus sequence is required for the internalization of transmembrane apCAM (Bailey et al. 1997), a process that will be described in more detail later.

13.4 Extracellular Interaction Partners of NCAM Family Proteins

All NCAM proteins engage in homophilic protein–protein interactions, including *cis* (a molecule binding to another molecule located in the same plasma membrane) and/or *trans* interactions (a molecule binding to another molecule located in an opposing plasma membrane or cell). The capacity to bind homophilically appears to be an evolutionary well-conserved feature of NCAM-type proteins (Hoffman et al. 1984). A large number of models have been proposed for *trans*-homophilic NCAM1 interactions. However, the best-described model of homophilic NCAM1 interactions are derived from protein structure studies of recombinant NCAM1 Ig1–Ig2 and Ig1–Ig2–Ig3 that have been obtained by X-ray crystallography (Kasper et al. 2000, Soroka et al. 2003). These structures reveal a reciprocal *cis* interaction between the Ig1–Ig2 modules of two NCAM1 molecules (Kasper et al. 2000, Soroka et al. 2003) and further suggest the formation of zipper-like multimolecular clusters through reciprocal Ig1–Ig3, Ig2–Ig3, and Ig2–Ig2 *trans* interactions (reviewed by Walmod et al. 2004, Kiselyov et al. 2005, Soroka et al. 2008). NCAM2 is known to form homophilic *trans* interactions (Yoshihara et al. 1997), but the protein modules involved in these interactions have not been identified. Recently, a crystal structure of the Ig1-module of human NCAM2 was published that reveals a dimerization of the Ig1-modules by a so-called domain swapping mechanism, in which the two N-terminal β-strands are interchanged (Rasmussen et al. 2008). This type of dimerization is known from *trans* homophilic interactions of Cadherins (Boggon et al. 2002, Chen et al. 2005) but has not previously been observed for CAMs of the immunoglobulin superfamily.

NCAM family proteins interact extracellularly with a large number of different molecules. Whereas NCAM1-mediated interactions have been studied in much detail (Table 13.1), no heterophilic interactions involving NCAM2 have currently been identified. NCAM1 interacts with other CAMs, including the cellular Prion protein (PrP^c), TAG-1, and L1-CAM. Interaction with PrP^c occurs within the three most membrane-proximal modules of NCAM1 (Schmitt-Ulms et al. 2001). As PrP^c is a GPI-anchored membrane protein,

Table 13.1 Extracellular binding partners of NCAM-type proteins

Binding partner	NCAM type (protein domains)	References
Proteoglycans and other ECM molecules		
Agrin	Vertebrate NCAM1	Bixby et al. (2002)
Collagens I–VI and IX	Vertebrate NCAM1	Probstmeier et al. (1989) and Probstmeier et al. (1992)
Collagen XVIII	Vertebrate NCAM1	Storms et al. (1996)
Heparin	Vertebrate NCAM1 (Ig2)	Cole and Akeson (1989)
Neurocan	Vertebrate NCAM1	Friedlander et al. (1994)
Phosphacan (RPTPζ/β)	Vertebrate NCAM1	Milev et al. (1995)
Cell adhesion molecules		
Fasciclin II	Fasciclin II	Grenningloh et al. (1991)
L1-CAM	Vertebrate NCAM1 (Ig4)	Horstkorte et al. (1993)
NCAM1	Vertebrate NCAM1 (Ig1, -2, and -3)	Soroka et al. (2003)

Table 13.1 (continued)

Binding partner	NCAM type (protein domains)	References
NCAM2	Vertebrate NCAM2	Yoshihara et al. (1997)
Prion protein	Vertebrate NCAM1 (Ig5, FnIII1 and -2)	Schmitt-Ulms et al. (2001)
TAG-1	Vertebrate NCAM1	Milev et al. (1996)
Growth factors, growth factor receptors, and receptor tyrosine kinases		
FGFR	Vertebrate NCAM1	Kiselyov et al. (2003)
GDNF	Vertebrate NCAM1	Paratcha et al. (2003)
GFRα1	Vertebrate NCAM1	Paratcha et al. (2003)
Heartless	*Drosophila* Fasciclin II	Forni et al. (2004)
MuSK	*Danio rerio* NCAM1	Bushell et al. (2008)
Neurturin	Vertebrate NCAM1	Paratcha et al.(2003)
Persephin	Vertebrate NCAM1	Paratcha et al. (2003)
Other		
ATP	Vertebrate NCAM1 (FnIII2)	Dzhandzhugazyan and Bock (1997) and Kiselyov et al. (2003)
AMPA receptors	Vertebrate NCAM1 (Ig5 – PSA)	Vaithianathan et al. (2004)
NMDA receptors	Vertebrate NCAM1 (Ig5 – PSA)	Hammond et al. (2006)
T- and L-type VDCCs	Vertebrate NCAM1	Bodrikov et al. (2008)
Rabies virus	Vertebrate NCAM1	Thoulouze et al. (1998)

Extracellular binding partners for NCAM-type proteins are shown. Some of the interactions might be indirect, since they only have been demonstrated by immunoprecipitation experiments. The interactions between NCAM1 and VDCC have been demonstrated by immunoprecipitation and might as well represent intracellular interactions.

this interaction increases the fraction of NCAM1 in lipid rafts and promotes, like palmitoylation, NCAM1-mediated signaling and neurite outgrowth (Santuccione et al. 2005). The biological significance of NCAM1's interaction with the Ig-CAMs TAG-1 (axonin-1/contactin-2) and L1-CAM is not known (Milev et al. 1996). The interaction between L1-CAM and NCAM1 is mediated by a lectin-like sequence in NCAM1 Ig4 and carbohydrates on L1-CAM and has been reported to facilitate homophilic L1-CAM interactions and axon outgrowth (Horstkorte et al. 1993, Kristiansen et al. 1999).

NCAM1 binds a number of extracellular matrix (ECM) components, including heparan sulphate proteoglycans (HSPGs; heparin, agrin; collagen XVIII), chondroitin sulfate proteoglycans (CSPGs; neurocan, phosphacan), and collagens (type I–VI and IX) (reviewed by Nielsen et al. 2008). Heparin binds to the NCAM1 Ig2 module (Cole and Akeson 1989) at a heparin-binding domain that overlaps with one of the homophilic NCAM1-binding sites. This indicates that heparin can modulate NCAM1's homophilic interactions. Abrogation of HSPG–NCAM1 interactions by mutation of the heparin-binding domain in NCAM1 or by enzymatic removal of HSPGs reduces the degree of synaptophysin immunoreactivity in cultures of hippocampal neurons. This suggests that this interaction is of pivotal importance for the synaptogenic activity of NCAM1 (Dityatev et al. 2004). Agrin is a key regulator of synaptogenesis at the neuromuscular junction (reviewed by Ngo et al. 2007), where it is involved in postsynaptic aggregation of acetylcholine receptors (reviewed by Martin 2002). The effect of agrin is mediated by muscle-specific kinase (MuSK), a receptor tyrosine kinase. Interestingly, it was recently found that zebrafish NCAM1 binds to MuSK extracellularly (Bushell et al. 2008). However, whether NCAM1 can activate MuSK signaling and how the NCAM1–MuSK interaction contributes to the formation of neuromuscular junctions has not been investigated. Agrin can also regulate axonal guidance by inhibiting neurite outgrowth. Although NCAM1 interacts with agrin, this binding does not inhibit NCAM1-mediated neurite outgrowth (Bixby et al. 2002). Therefore, the function of the NCAM1–agrin interaction remains unclear. The CSPGs neurocan and phosphacan, which is a secreted splice variant of receptor protein tyrosine phosphatase (RPTP) ζ/β (reviewed by Beltran and Bixby 2003), also bind to NCAM1 (see Nielsen et al. 2008 for review). Both supposedly bind to NCAM1's Ig2 module at a binding site for sulphated chondroitin polymers. This binding site overlaps the heparin-binding pocket and one of the homophilic NCAM1 interaction sites. These interactions thereby potentially influence both homophilic and other heterophilic NCAM1 interactions (Kulahin et al. 2005). Supporting this assertion, both neurocan and phosphacan have been shown to inhibit homophilic *trans*-NCAM1 binding (Grumet et al. 1993). Nevertheless, the biological significance of these NCAM1–CSPGs interactions is not clear. In the hippocampus, the expression of neurocan is regulated by synaptic activity (Schwarzacher et al. 2006), and molecules like neurocan and phosphacan have been proposed to generate protective barriers that facilitate neurite extension and fasciculation. The degradation of these barriers is a prerequisite for subsequent synaptic maturation or plasticity (Okamoto et al. 2001, Murakami and Ohtsuka 2003). It is therefore likely that NCAM1, through its interactions with neurocan and phosphacan, contributes to the formation and maintenance of such barriers.

Together with the muscle-specific α1 subunit of nicotinic acetylcholine receptors (nAChRs), and the p75 neurotrophin receptor (p75NTR), NCAM1 is known to bind rabies virus (RABV) (Thoulouze et al. 1998, Hotta et al. 2007). RABV enters the nervous system via neuromuscular junctions, and in the CNS RABV infection of neurons occurs primarily at synaptic junctions. Since the RABV-binding nAChR is specifically expressed in muscles, it has been proposed that this receptor facilitates neuronal RABV infection by concentrating RABV particles at neuromuscular junctions. As p75NRT is not expressed at neuromuscular junctions, the subsequent infection of neurons is believed to be facilitated by NCAM1 (reviewed by Lafon 2005).

NCAM1 interacts with several growth factors and growth factor receptors, including the FGFR (Williams et al. 1994), the glial cell line-derived neurotrophic factor (GDNF) family ligands, GDNF, neurturin and persephin, the GDNF family receptor $\alpha 1$ (GFRα1), and possibly also GFRα2 and GFRα4 (Paratcha et al. 2003). The ectodomain of FGFR is composed of three Ig-modules, and the NCAM1–FGFR interaction involves the NCAM1 FnIII2 and the two most membrane-proximal modules of FGFR (Kochoyan et al. 2008). The interaction between NCAM1 and FGFR initiates FGFR-dependent signaling and is modulated by extracellular adenosine 5′-triphosphate (ATP) and by PSA glycosylation of NCAM1 (see below) (reviewed by Kiselyov et al. 2005). FGFR-mediated NCAM1 signaling is also observed in *Drosophila*, where fasciclin II-mediated neurite outgrowth requires the presence of the FGFR heartless (Forni et al. 2004). GDNF binds to NCAM1's Ig3 module (Sjostrand et al. 2007), whereas GFRα1 binds NCAM1's Ig4 module. GDNF, GFRα, and NCAM1 form complexes, which initiate downstream signaling. However, GFRα1 and NCAM1 can also interact independently of GDNF. GDNF–NCAM1 interactions do not interfere with *trans*-homophilic NCAM1 cell adhesion, whereas GFRα1–NCAM1 interactions do (Sjostrand and Ibanez 2008).

13.5 Intracellular Interaction Partners of NCAM Family Proteins

The cytoplasmic domains of transmembrane NCAM1 isoforms interact with a number of intracellular proteins (Table 13.2). The first identified interaction partner was spectrin (Pollerberg et al. 1987). This protein binds F-actin and

Table 13.2 Cytosolic NCAM-binding partners

Binding partner	NCAM protein isoform	References
α-Actinin 1	Vertebrate NCAM1-140/180	Buttner et al. (2003)
α-Adaptin	Vertebrate NCAM1	Minana et al. (2001)
α- and β-Tubulin	Vertebrate NCAM1-140/180	Buttner et al. (2003)
β-Actin	Vertebrate NCAM1-180	Buttner et al. (2003)
CAMAP	*Aplysia* TM-apCAM	Lee et al. (2007)
CaMKIIα	Vertebrate NCAM1	Bodrikov et al. (2008)
Clathrin	Vertebrate NCAM1	Minana et al. (2001)
Disc-large (Dlg)	*Drosophila* Fasciclin II	Zito et al. (1997)
LANP/PHAP-1	Vertebrate NCAM1-140/180	Buttner et al. (2005)
MAP 1A	Vertebrate NCAM1-180	Buttner et al. (2003)
PP1	Vertebrate NCAM1-140/180	Buttner et al. (2005)
PP2A	Vertebrate NCAM1-140/180	Buttner et al. (2005)
Phospholipase Cγ	Vertebrate NCAM1-140/180	Buttner et al. (2005)
PKCδ	Vertebrate NCAM1	Bodrikov et al. (2008)
RPTPα	Vertebrate NCAM1-140/180	Bodrikov et al. (2005)
ROKα	Vertebrate NCAM1-180	Buttner et al. (2003)
Spectrin	Vertebrate NCAM1-140/180	Pollerberg et al. (1987)
Syndapin	Vertebrate NCAM1-140/180	Buttner et al. (2005)
TOAD-64/CRMP-62	Vertebrate NCAM1-180	Buttner et al. (2005)
Tropomyosin	Vertebrate NCAM1-180	Buttner et al. (2003)

Intracellular binding partners for NCAM-type proteins are shown. Some of the interactions might be indirect, since they only have been demonstrated by immunoprecipitation experiments.

thereby links NCAM1 to the membrane cytoskeleton. Furthermore, spectrin binds protein kinase Cβ (PKCβ) (Rodriguez et al. 1999), one of the major downstream effectors of NCAM mediate neurite outgrowth (Kolkova et al. 2005). Activation of NCAM1 by antibodies or recombinant NCAM1 protein fragments induces the aggregation of NCAM1–spectrin–PKCβ complexes in lipid rafts (Leshchyns'ka et al. 2003). NCAM–spectrin complex formation is also important for synapse development and organization. Thus, removal of spectrin in *Drosophila* has been shown to result in the absence or disorganized expression of fasciclin II at presynaptic membranes (Pielage et al. 2005). Furthermore, postsynaptic NCAM–spectrin complexes bind the *N*-methyl-D-aspartate (NMDA) receptor subunits NR1 and NR2B and Ca^{2+}/calmodulin-dependent protein kinase II α (CaMKIIα), and the NCAM-mediated accumulation of these complexes is required for proper synapse formation (Sytnyk et al. 2006).

Affinity chromatography has revealed a large number of additional cytosolic NCAM1-binding partners. These include additional components of the cytoskeleton (α- and β-tubulin, microtubule-associated protein MAP 1A, β-actin, α-actinin 1, tropomyosin, and rhoA-binding kinase α (ROKα)), of which only α- and β-tubulin and α-actinin 1 bind NCAM1-140. Furthermore, turned on after division-64/collapsin response mediator protein of relative molecular mass of 62 kDa (TOAD-64/CRMP-62), a protein involved in axonal outgrowth and pathfinding (Minturn et al. 1995, Wang and Strittmatter 1996), specifically binds NCAM-180, and the serine/threonine phosphatases PP1 and PP2A, leucine-rich acidic nuclear protein (LANP/PHAP-1; a potent inhibitor of PP2A) (Li et al. 1996), syndapin (a dynamic-bind protein involved in vesicular trafficking) (reviewed by Kessels and Qualmann 2004), and phospholipase Cγ (PLCγ), bind both NCAM-140 and -180 (Buttner et al. 2003, 2005, reviewed by Buttner and Horstkorte 2008). However, the importance of the respective protein interactions for the function of NCAM1 has not been investigated. Other studies have demonstrated that mammalian NCAM1 can be immunoprecipitated with T- and L-type voltage-dependent Ca^{2+} channels (VDCC), and these channels co-localize with NCAM1 at growth cones. The association of NCAM1 with VDCCs leads to an influx of Ca^{2+} and thereby to the activation of calmodulin-dependent protein kinase IIα (CaMKIIα), which also can be co-immunoprecipitated with NCAM1 (Bodrikov et al. 2008). The activation of CaMKIIα is important for the NCAM1-regulated activation of RPTPα. RPTPα binds NCAM1 intracellularly, and this interaction results in a spectrin-dependent translocation of the two proteins to lipid rafts, where the interaction between CaMKIIα and NCAM1 promotes a CaMKIIα-dependent activation of RPTPα. The expression of RPTPα is of pivotal importance for the NCAM1-mediated activation of the Src-related non-receptor kinase p59Fyn in relation to NCAM1-mediated neurite outgrowth (Bodrikov et al. 2005). Whether these signaling pathways play a role in synaptic plasticity remains to be established.

A two-hybrid screen indicated that the C-terminus of *Drosophila* fasciclin II binds the PDZ domains of the discs-large (Dlg) protein. This interaction is both necessary and sufficient for the subcellular targeting of fasciclin II to synapses (Zito et al. 1997). Vertebrate NCAM-type proteins do not seem to form similar interactions.

13.6 NCAM-Mediated Intracellular Signaling Pathways

The above-described extracellular and intracellular interactions that involve NCAM proteins direct NCAM proteins to specific subcellular regions in the plasma membrane. Furthermore, several of these interactions serve as transmitters of NCAM-mediated intracellular signaling. NCAM-mediated signaling has predominantly been investigated for vertebrate NCAM1 and is described in detail elsewhere (Ditlevsen and Kolkova 2008, Ditlevsen et al. 2008). Briefly, NCAM1 stimulates intracellular signaling through VDCCs, heterotrimeric G proteins, non-receptor tyrosine kinases, and receptor tyrosine kinases. The NCAM1-mediated activation of VDCCs in combination with Fyn-activation is described above (Bodrikov et al. 2008). VDCCs can also be activated in combination with signaling through FGFR (Kiryushko et al. 2006). Several studies demonstrate that NCAM1-mediated signaling also includes activation of heterotrimeric G proteins (Schuch et al. 1989, Doherty et al. 1991, Sandig et al. 1994, Williams et al. 1994, Hansen et al. 2007). However, how NCAM1 modulates the activity of heterotrimeric G proteins is currently unknown. NCAM1-mediated signaling through non-receptor tyrosine kinases involves the Src-related kinase Fyn and the focal adhesion kinase (FAK) and can be facilitated through the interaction between NCAM1 and RPTPα or through interactions between NCAM1, GDNF, and GFRα1 (Paratcha et al. 2003). As described above, NCAM1-mediated signaling through receptor tyrosine kinases is activated by the direct interaction between NCAM1 and FGFR (Kiselyov et al. 2003). Downstream signaling pathways that are induced by NCAM1-mediated signaling events includes the Ras-MAPK pathway, the PI3K-Akt pathway, PLCγ-mediated activation of Ca^{2+}-signaling and PKC proteins, and cAMP-mediated activation of PKA. Transcription factors that are activated by NCAM1 include CREB, Fos, and NKκB (Ditlevsen et al. 2008).

13.7 Effects of Extracellular ATP on NCAM Function

In addition to its role in intracellular signal transduction, ATP also acts as a neurotransmitter in several regions of the nervous system, including the dorsal horn, hippocampus, locus coeruleus, medial habenula, and somatosensory cortex, where it is released with or without other neurotransmitters, such as γ-aminobutyric acid (GABA), noradrenaline, and glutamate (Pankratov et al. 2006, Abbracchio et al. 2009). Furthermore, ATP release occurs from astrocytes and from damaged and dying cells and thereby increases the general concentrations of extracellular ATP and its metabolites above normal levels (Rathbone et al. 1999, Volonte et al. 2003, Abbracchio et al. 2009). ATP acts on both pre- and postsynaptic receptors, including the P2X and P2Y receptors and nicotinic (nAChRs) and muscarinic acetylcholine receptors (mAChRs). Furthermore, ADP, which is generated by ectonucleotidases during the metabolization of ATP, activates additional P2Y receptors (Volonte et al. 2003, Abbracchio et al. 2009, Zimmermann 2008).

Extracellular ATP has been shown to bind directly to NCAM1 (Dzhandzhugazyan and Bock 1997, Kiselyov et al. 2003). In NCAM1, the ATP-binding site is located in the F-G loop of the FnIII2 module (Kiselyov et al. 2003), which includes a nucleotide-binding Walker A motif/P-loop (G/AxxxxGKT/S, where x denotes any amino acid) (Saraste et al. 1990). Interestingly, NCAM1 also possesses a weak ATPase activity (Dzhandzhugazyan and Bock 1993). In vitro, extracellular ATP inhibits *trans*-homophilic NCAM1-mediated cell aggregation in a dose-dependent manner. ATP also inhibits NCAM1-mediated neurite outgrowth that is induced by *trans*-homophilic NCAM1 interactions in neurons, which are grown on top of NCAM1-expressing fibroblasts. These effects are independent of ATP hydrolysis, as the same response is observed with non-hydrolysable ATP analogs. This indicates that binding of ATP, rather than ATP hydrolysis elicits these effects (Skladchikova et al. 1999). The ATP-binding region in NCAM1 overlaps with one of the FGFR-binding regions, and surface plasmon resonance (SPR) experiments demonstrated that ATP inhibits the interaction between NCAM1 and FGFR (Kiselyov et al. 2003). Thus, ATP is a negative regulator of NCAM1-mediated neurite outgrowth by inhibiting the NCAM1-mediated activation of FGFR. How ATP hinders the *trans*-homophilic NCAM1 interactions is not clear.

Soluble NCAM1 can be generated by cleaving the GPI anchor of NCAM1-120 by phosphatidylinositol-specific phospholipase C (PI-PLC) (Sadoul et al. 1986) and by cleavage of the NCAM1 from the membrane by the serine protease tissue-type plasminogen activator (tPA) (Endo et al. 1998, Endo et al. 1999b) or A disintegrin and metalloprotease (ADAM)17/tumour necrosis factor alpha converting enzyme (TACE) (Kalus et al. 2006) (Diestel et al. 2005, Hubschmann et al. 2005, reviewed by Secher 2008). The cleavage of NCAM1 is mediated by extracellular ATP in a manner that does not involve ATP binding by NCAM1, but do require the hydrolysis of ATP molecules by the extracellular proteases (Hubschmann et al. 2005). In single-cell cultures of neurons grown at low density on plastic, extracellular ATP induces a weak dose-dependent increase in neurite outgrowth in a manner that involves ATP hydrolysis (Skladchikova et al. 1999). The stimulation of neurite outgrowth observed in this assay may be explained by alterations in cell–substrate interactions induced by the ATP-mediated cleavage of NCAM1. Consistent with this interpretation the inhibition of metalloprotease activity reduces neurite outgrowth from cerebellar microexplant cultures (Kalus et al. 2006) and increases NCAM1-mediated neuronal aggregation (Hubschmann et al. 2005) In conclusion, extracellular ATP can inhibit NCAM1-mediated cell aggregation both in an ATP hydrolysis-independent manner (by binding to NCAM1 and thereby abrogating homophilic NCAM1 interactions and/or NCAM1–FGFR interactions) and in an ATP hydrolysis-dependent manner (by stimulating extracellular proteases that cleave NCAM1). Consequently, ATP can inhibit NCAM1-mediated neurite outgrowth induced by *trans*-homophilic NCAM1 interactions in co-cultures, but stimulate neurite outgrowth in single-cell cultures. Increased neurite outgrowth in response to extracellular ATP is most likely the result of changes in the adhesive properties of neurites and growth cones in response to NCAM1 cleavage. Since extracellular ATP antagonizes *trans*-homophilic NCAM1 interactions, an increase in extracellular ATP at synapses might result in a transient reduction of synaptic cell–cell interactions and hence promote synaptic plasticity. However, extracellular ATP concentrations are increased not only by general nerve activity but also by neuronal stress, e.g., hypoxia, ischemia, and seizures (Rathbone et al. 1999, Abbracchio et al. 2009). Interestingly, an increase of soluble NCAM1 has been found in several pathological conditions, including cancer, schizophrenia, and various mood disorders (reviewed by Secher 2008). However, the significance these observations and the association to altered synaptic ATP in pathological conditions is not known.

A corresponding ATP-binding Walker A motif does not exist in the FnIII2 module of NCAM2, but NCAM2 contains a similar motif in a different region of its FnIII2 module (see Kulahin and Walmod 2008). However, whether any NCAM2 function is regulated by ATP remains to be determined. The Walker A motifs in NCAM1 and NCAM2 are present in NCAM proteins from mammals, chicken, and clawed frog, and is present in zebrafish NCAM3. Non-vertebrate NCAM proteins do not have any Walker A motif in their FnIII module.

13.8 Regulatory Roles for Polysialic Acid in NCAM1 Function

As mentioned above, vertebrate NCAM1 can be glycosylated with the unusual carbohydrate (PSA). PSA can be attached to two N-glycosylation sites in Ig5. However, the FnIII1 module is necessary for a correct polysialylation of the Ig5 module, probably because it contains a docking motif recognized by the enzymes responsible for the attachment of PSA to NCAM1 (Mendiratta et al. 2005, Mendiratta et al. 2006, Colley 2008).

The synthesis of PSA is catalyzed by the two enzymes 2,8-polysialyltransferases type II (ST8SiaII/STX) and type IV (ST8SiaIV/PST). The expressions of these enzymes overlap spatially and temporally, but the genes encoding these proteins are independently regulated at the transcriptional level. Thus, overexpression of the transcription factor Pax3 induces NCAM1 polysialylation through an increase in the expression of ST8SiaII but not of ST8SiaIV (Mayanil et al. 2000, 2001). Both enzymes attach PSA to NCAM1 independently of each other. Whereas ST8SiaII is the major NCAM1 polysialyltransferase during embryonic development, ST8SiaIV mainly is expressed in the adult brain (Phillips et al. 1997, Hildebrandt et al. 1998, Ong et al. 1998). However, the two enzymes polysialylate NCAM1 differently and synergistically (Kitazume-Kawaguchi et al. 2001, Angata et al. 2002). An ancestor of all ST8Sia genes has been identified in *Cephalochordata* (lancelets) (Harduin-Lepers et al. 2005). In zebrafish, which expresses both ST8SiaII and ST8SiaIV, only ST8SiaII seems to be important for NCAM polysialylation, suggesting that the specificity of both enzymes have diverged during evolution (Marx et al. 2007).

In mouse, the process of polysialylation starts shortly after the first appearance of NCAM1 at embryonic day 8–8.5, and the level of NCAM1-associtated PSA expression peaks during the perinatal phase (Probstmeier et al. 1994). Thereafter the expression decreases gradually, and consequently PSA-NCAM1 is sometimes referred to as embryonic (or E)-NCAM1 as opposed to non-PSA-NCAM1, which is referred to as adult (or A)-NCAM1 (Edelman and Chuong 1982). However, in brain regions that retain a neurogenic capacity or are associated with synaptic plasticity, high levels of PSA-NCAM1 are maintained throughout adult life (Seki and Arai 1993).

PSA expression is in part regulated by the level of polysialyltransferase enzymatic activity. For example, in the developing chick ciliary ganglion, the activities of the polysialyltransferases are regulated by the concentration of Ca^{2+} in intracellular Ca^{2+} stores (Bruses and Rutishauser 1998). Moreover, synaptic activation modulates PSA expression through the activation of NMDA receptors and of nitric oxide (NO)-cGMP-mediated signaling. Thus, electrical stimulation of brainstem slices decreases PSA-NCAM1 expression, and inhibition of either NMDA receptor activity or neuronal NO synthase activity prevents this decrease (Bouzioukh et al. 2001). The degree of PSA-NCAM1 is also negatively regulated by PKCδ, which is downregulated in the hippocampus during memory consolidation. This decrease of PKCδ coincides with a transient increase of PSA-NCAM1 expression (Gallagher et al. 2000, 2001). PKCδ can be co-immunoprecipitated with NCAM1, but NCAM1 apparently does not affect the activity of PKCδ (Bodrikov et al. 2008). The expression of PSA is of pivotal importance for a range of NCAM1 functions, including the migration and differentiation of neuronal progenitor cells, the targeting and outgrowth of axons, synaptogenesis, synaptic plasticity, and CNS repair (reviewed by Gascon et al. 2007, El Maarouf and Rutishauser 2008, Gascon et al. 2008, Rutishauser 2008). Knockdown mice lacking ST8SiaII or ST8SiaIV, respectively, are viable and have no apparent phenotypes (Eckhardt et al. 2000, Angata et al. 2004). However, the double knockout of both ST8SiaII and ST8SiaIV is lethal. These mice appear normal at birth, but their postnatal development is severely retarded, and more than 80% die before the age of 4 weeks (Weinhold et al. 2005, Angata et al. 2007). Several of the abnormalities observed in ST8SiaII/ST8SiaIV double knockout animals are absent in ST8SiaII/ST8SiaIV/NCAM1 triple knockout animals, and the triple knockout is not lethal. This suggests that the lethality observed for ST8SiaII/ST8SiaIV double knockout mice is the result of NCAM1 gain-of-function effects, such as increased adhesion since PSA decreases the adhesive function of NCAM (see below) (Weinhold et al. 2005).

The presence of PSA regulates the distance between neighboring cells and their attachment between each other and to the ECM. The anti-adhesive property of PSA is a consequence of its physical properties: multiple negative charges and a hydrated volume that is much larger than that of its carrier protein NCAM1. Furthermore, the PSA backbone is highly flexible, thus reducing its binding to proteins. As a consequence, PSA occupies a large space between cells without forming contacts with other cell surface or ECM components (Rutishauser 2008). Since PSA masks NCAM1, the expression of NCAM1-associated PSA modulates extracellular NCAM1 interactions in a temporally and spatially regulated manner. The balance between PSA-NCAM1 and non-PSA-NCAM1 is critical for the timing of neuronal differentiation. When PSA is enzymatically removed from progenitor cells migrating along the subventricular zone (SVZ) to the olfactory bulb, these cells not only lose their ability to move but also undergo premature neuronal differentiation and form inappropriate synaptic contacts (Petridis 2004). Consistently, the size of the olfactory bulb is markedly reduced in NCAM1-deficient mice. This is a result of migration deficits of SVZ-derived interneuron precursors (Tomasiewicz et al. 1993). During axonal growth, the presence of PSA along axons seems to prevent inappropriate synapse formation (Seki and Rutishauser 1998, El Maarouf and Rutishauser 2003). PSA also prevents myelination, and PSA expression is downregulated on axons and oligodendrocyte precursors prior to the onset of myelination (Fewou et al. 2007, Jakovcevski et al. 2007).

PSA is an important modulator of learning and memory formation. In NCAM1-deficient mice, contextual memory is impaired. It is partially restored by the application of PSA-NCAM1, but not of NCAM1 without PSA. Likewise, PSA-NCAM1 application to hippocampal slice cultures that are derived from NCAM1-deficient mice restores normal level of LTP, whereas the application of recombinant PSA-NCAM1 or free PSA to wild-type mice before initiation of a fear conditioning paradigm disrupts formation of hippocampus-dependent contextual memory. These observations demonstrate that PSA-NCAM1 is involved in both the formation and the late consolidation of emotional memory (Senkov et al. 2006). Removal of PSA with the enzyme endoneuraminidase-N (endo-N) abrogates NCAM-mediated synapse formation (Dityatev et al. 2004). In the CA1, but not the CA3 region, of the hippocampus, suppression of ST8SiaIV-mediated polysialylation results in a reduction or an elimination of LTP and LTD (Muller et al. 1996, Eckhardt et al. 2000). In contrast, mice lacking ST8SiaII, do not show changes in synaptic plasticity, but do demonstrate anatomical changes including the formation of ectopic synapses in the hippocampus (Angata et al. 2004). These and other studies suggest that synaptic plasticity depend on a balance between the adhesive properties of pre- and postsynaptic membranes, which are modulated by the ratio between PSA-NCAM1 and non-PSA-NCAM1. Furthermore, the requirements for PSA-NCAM1 relative to non-PSA-NCAM1 seem to be synapse-type dependent (Rutishauser et al. 1988, Gascon et al. 2007). For instance, ST8SiaIV knockout mice demonstrate deficits in spatial and reversal learning, when tested in a water maze, whereas their auditory fear conditioning is unaffected (Markram et al. 2007). Moreover, the expression of PSA-NCAM1 in relation to learning and memory differs between brain regions and the memory/learning task. Thus, spatial learning tested in water maze is accompanied by increased PSA-NCAM1 expression in the ventromedial and insular prefrontal cortex cortex, whereas passive avoidance conditioning results in a downregulation of PSA-NCAM1 in the dorsomedial prefrontal cortex (Ter Horst et al. 2008). In addition to the anti-adhesive effects of PSA, the glycosylation can affect synaptic function through direct interaction with glutamate receptors. Thus PSA-NCAM1 acts as a competitive antagonist of NMDA receptors, specifically inhibiting NMDA receptor-mediated currents, but prolong the open channel time of AMPA receptors, thereby stimulating AMPA receptor-mediated currents (Vaithianathan et al. 2004, Hammond et al. 2006). Furthermore, recent data suggest that PSA-NCAM1 regulate neurotrophin signaling through a modulation of the low-affinity neurotrophin receptor $p75^{NTR}$ (reviewed by Gascon et al. 2007).

The level of PSA-NCAM1 is altered in some neuropsychiatric and neurodegenerative diseases. For instance, PSA-NCAM1 levels are decreased in the hippocampus of individuals with schizophrenia (Barbeau et al. 1995), and one of the single nucleotide polymorphisms (SNPs) that is associated with cognitive impairment is in the NCAM1 FnIII1 module (Lewis et al. 2003, Sullivan et al. 2007), which contains the binding site that is recognized by the enzymes responsible for the attachment of PSA (Close et al. 2003, Mendiratta et al. 2005, 2006). Moreover, SNPs in ST8SiaII and ST8SiaIV are reported as susceptibility loci for schizophrenia (Barbeau et al. 1995). Current treatment of patients with Alzheimer's disease includes the use of cholinoesterase inhibitors, which enhance cholinergic memory consolidation. Alzheimer's patients and rats that were chronically exposed to these drugs demonstrate an increased PSA-NCAM1 expression in the hippocampus (Jin et al. 2004, Murphy et al. 2006).

13.9 NCAM Protein in Long-Term Potentiation and Long-Term Depression

The mechanisms underlying learning-related synaptic plasticity are remarkably conserved. Thus, the involvement of NCAM-type proteins in mediating these processes have been successfully studied using simple model systems, especially in the marine snail *Aplysia* and in the fruit fly (*Drosophila*). The sensitization of the gill and siphon withdrawal reflex in *Aplysia* is considered an elementary form of non-associative learning. Following a single training session, the reflex produces short-term sensitization that lasts from minutes to hours, and five training sessions induce long-term sensitization that persists several weeks (Pinsker et al. 1970). Short- and long-term sensitizations result in enhanced neurotransmission at monosynaptic connections between sensory neurons (SNs) and target cells, including a specific excitatory motor neuron, denoted L7 (Montarolo et al. 1986, Rayport and Schacher 1986). In dissociated *Aplysia* neuronal cell cultures, application of serotonin (5-HT) induces short- and long-term changes in synaptic transmission that are similar to those observed during in vivo sensitization. These changes, which are referred to as short- and long-term facilitation, STF and LTF, respectively, therefore appear to involve enhanced 5-HT transmission at specific synaptic sites (Montarolo et al. 1986). In dissociated *Aplysia* neurons, 5-HT-induced LTF is accompanied by endocytosis of apCAM, the *Aplysia* NCAM ortholog, from the surface of the presynaptic SNs (Bailey et al. 1992, Hu et al. 1993), and an increased number of sensory branches and varicosities contacting the major axons of the L7 target motor neuron (Zhu et al. 1994). The 5-HT induced decrease in apCAM expression on SNs subsequently facilities defasciculation of SN neurites (Peter et al. 1994) and formation of new synaptic connections (Zhu et al. 1994). Interestingly, a treatment with Phe-Met-Arg-Phe-amid (FMRFamid), which induces long-term depression (LTD), has opposite effects, inducing a reduction of apCAM expression on motor neurons and a loss of SN neurites and varicosities (Montarolo et al. 1988, Schacher and Montarolo 1991).

Transmembrane (TM), but not GPI-anchored apCAM, is endocytosed following synaptic activation. This internalization requires the phosphorylation of mitogen-activated protein kinase (MAPK), which is a prerequisite for the generation of LTF (Bailey et al. 1997, Martin et al. 1997). This suggests a selective involvement of TM-apCAM in this process. Accordingly, the overexpression of TM-apCAM inhibits both the functional and structural changes in relation to LTF (the increase in EPSP amplitude and the change in spine shape, respectively), whereas the overexpression of GPI-anchored apCAM has no effect on LTF. However, the overexpression of various apCAM deletion mutants, including proteins lacking the ectodomain or the cytoplasmic domain, only inhibited the structural changes related to the formation of LTF (Han et al. 2004). These observations suggest that apCAM affect the functional and structural changes associated with LTF through different mechanisms (Han et al. 2004). I nterestingly, SNs predominantly express GPI-anchored apCAM, whereas TM-apCAM is the main isoform in L7 motor neurons. Consequently, 5-HT-induced endocytosis of TM-apCAM in SNs during LTF amplifies cell-specific differences in isoform expression, whereas FMRF-amid-induced endocytosis of TM-apCAM in the L7 motor neuron reduces these differences. The endocytosis of TM-apCAM also leads to alterations in gene transcription. This mechanism is initiated by an interaction between the cytoplasmic domain of TM-apCAM and CAM-associated protein (CAMAP). Subsequently, CAMAP becomes phosphorylated by protein kinase A (PKA), which leads to the dissociation of CAMAP from apCAM, and the translocation of CAMAP to the nucleus. In the nucleus, CAMAP binds the transcription factor CREB1 and induces the activation of ApC/EBP, which in *Aplysia* is required for the consolidation of LTF (Alberini et al. 1994, Lee et al. 2007).

As it can be easily manipulated at the genetic level, *Drosophila* is a uniquely suited model system for studying the molecular mechanisms of synapse formation and of synaptic plasticity (Hebbar et al. 2006). Using this system, the roles of regulated FasII expression and FasII-mediated cell adhesion/signaling during synaptogenesis and experience-dependent synaptic changes have been studied in detail (Davis et al. 1996, Koh et al. 2002, Kristiansen and Hortsch 2008, Schwenkert et al. 2008). In particular, the formation and the continued changes of the *Drosophila* neuromuscular junction (NMJ) have been shown to be highly regulated and to be dependent on synaptic activity (Sweatt 2001, Hoeffer et al. 2003, Sigrist et al. 2003). Consistent with results from *Aplysia*, alterations in the balance between the pre- and postsynaptic expression of the *Drosophila* NCAM homolog, fasciclin II, regulates the formation of synaptic contacts. Hence, specific overexpression of recombinant fasciclin II disrupts synaptic connections (Baines et al. 2002), a process that involves synaptic recycling of fasciclin II (Mathew et al. 2003). The mechanisms involved in synapse development and the role of FasII during this process has been studied thoroughly in an in vivo system using the *Drosophila* NMJ as a model for synapse formation (Brunner and O'Kane 1997, Zito et al. 1997). Synaptogenesis of the NMJ starts with the establishment of initial synaptic contacts, which develop between axons and receptive areas of the target muscle (Suzuki et al. 2000, see Chapter 2). Following the establishment of immature synapses, a process that does not depend on fasciclin II expression, the phase of synaptic maturation and growth begins. During this phase, the continued synaptic development depends on the balanced pre- and postsynaptic expression of fasciclin II (Schuster et al. 1996). Interestingly, in addition to the regulation of cell–cell adhesion, the synaptic maturation process involves a reorganization of cytoskeletal components through a fasciclin II-mediated interaction with the PSD95-like *Drosophila* protein DLG (Budnik 1996, Zito et al. 1997, Rivlin et al. 2004, Kohsaka et al. 2007). Furthermore, the synapse maturation requires the activation of intracellular signaling cascades, including the CREB- and CaMKII-dependent pathways (Kazama et al. 2007, Kristiansen and Hortsch 2008).

Studies on mammalian NCAM1 have demonstrated that this protein can be internalized by both clathrin-dependent and clathrin-independent endocytosis (Minana et al. 2001). Furthermore, endocytosis of NCAM1-140 (Diestel et al. 2005) and NCAM1-180 (Foley et al. 2000) is regulated by ubiquitylation, a process that appears to involve a similar phosphorylation of the cytoplasmic NCAM1 domain as of apCAM. The majority of endocytosed NCAM1 molecules are eventually recycled to the plasma membrane (Diestel et al. 2005). In neurons, downregulation of NCAM1-180 has been shown to occur 3–4 h after passive avoidance response training, and a blockage of NCAM1 internalization by a treatment with an NCAM-binding peptide is accompanied by amnesia. This suggests that downregulation of synaptic NCAM1 is of central importance for memory consolidation (Foley et al. 2000). In contrast to NCAM1, NCAM2 has a conserved internalization signal, suggesting that NCAM2 is also endocytosed, but probably by a different mechanism than NCAM1 (Kulahin and Walmod 2008). As described above, the ectodomain of NCAM1 can be removed by extracellular proteolysis. This phenomenon has been observed in vivo in response to the activation of NMDA receptors during the induction of long-term potentiation and in response to the induction of seizures by stimulation of kainate receptors in the hippocampus (Fazeli et al. 1994, Hoffman et al. 1998, Endo et al. 1999a). Therefore, proteolysis might play an important role in synaptic alterations, including those occurring during experience-induced plasticity.

13.10 Conclusions

Despite the complex and diverse functions NCAM proteins carry out in the nervous system, evolutionary conserved themes for their regulation appear to exist. As described above, the activities of NCAM proteins are modulated by a number of mechanisms. The cytoplasmic domains of the proteins are posttranslationally modified (Little et al. 1998, Gennarini et al. 1984, Lyles et al. 1984, Sorkin et al. 1984, Polo-Parada et al. 2005). These modifications not only regulate the intracellular interactions mediated by NCAM proteins, but also affect the subcellular distribution of these proteins (Niethammer et al. 2002). Furthermore, extracellular interactions that are mediated by the NCAM ectodomains are regulated in several ways. NCAM1 adhesion is also regulated by the posttranslational attachment of PSA, by the proteolytic cleavage of the ectodomain, by endocytosis, and by the regulated expression of the VASE sequence motif. The importance of these regulatory mechanisms is emphasized by the observation that NCAM-mediated adhesion in invertebrates is also modified by both endocytosis and extracellular cleavage of NCAM. Moreover, although invertebrates do not express PSA-NCAM, a modulation of NCAM function through the regulated expression of glycoepitopes has also been observed in leeches (Tai and Zipser 2002). Whereas the attachment of PSA to NCAM1 is a complex and highly regulated process, the endocytosis or the cleavage of NCAM proteins is induced in response to more immediate changes. Endocytosis removes NCAM proteins completely from the cell surface, but only for a limited time period, as the majority of NCAM molecules are immediately recycled (Diestel et al. 2005). Furthermore, the endocytosis of NCAM seems to be connected to an induction of gene transcription (Lee et al. 2007). In contrast, the extracellular cleavage of NCAM is an irreversible process, although it does not immediately remove the NCAM ectodomains from the extracellular space. The exact function of soluble NCAM molecules is currently unknown. However, these NCAM fragments can serve as competitive inhibitors of homophilic NCAM-mediated adhesion (Olsen et al. 1993). Hence, transgenic mice that overexpress soluble NCAM1 exhibit a decrease in the number of presynaptic terminals in the prefrontal cortex and the amygdala and display an impaired memory in contextual and cued fear conditioning (Pillai-Nair et al. 2005). This mechanism may be connected to some of the pathophysiological changes in schizophrenia, Alzheimer's disease, and multiple sclerosis (reviewed by Secher 2008). How soluble and membrane-attached NCAM proteins might regulate adhesion and signaling in these diseases will need to be addressed by future studies.

Acknowledgments The work was supported by "Augustinus Fonden," "Lundbeckfonden" (project R19-A2087), The Danish Medical Research Council (project 271-07-0558), and the European Commission 7th Framework Programme (International Reintegration Grant 231108).

References

Abbracchio MP, Burnstock G, Verkhrastky A et al. (2009) Purinergic signalling in the nervous system: an overview. Trends Neurosci. 32:19–29

Alberini CM, Ghirardi M, Metz R et al. (1994) C/EBP is an immediate-early gene required for the consolidation of long-term facilitation in Aplysia. Cell 76:1099–1114

Alenius M and Bohm S (1997) Identification of a novel neural cell adhesion molecule-related gene with a potential role in selective axonal projection. J Biol Chem 272:26083–26086

Amores A, Force A, Yan YL et al. (1998) Zebrafish hox clusters and vertebrate genome evolution. Science 282:1711–1714

Angata K, Huckaby V, Ranscht B et al. (2007) Polysialic acid-directed migration and differentiation of neural precursors are essential for mouse brain development. Mol Cell Biol 27:6659–6668

Angata K, Long JM, Bukalo O et al. (2004) Sialyltransferase ST8Sia-II assembles a subset of polysialic acid that directs hippocampal axonal targeting and promotes fear behavior. J Biol Chem 279:32603–32613

Angata K, Suzuki M and Fukuda M (2002) ST8Sia II and ST8Sia IV polysialyltransferases exhibit marked differences in utilizing various acceptors containing oligosialic acid and short polysialic acid. The basis for cooperative polysialylation by two enzymes. J Biol Chem 277:36808–36817

Aparicio S (2000) Vertebrate evolution: recent perspectives from fish. Trends Genet 16:54–56

Bailey CH, Chen M, Keller F et al. (1992) Serotonin-mediated endocytosis of apCAM: an early step of learning-related synaptic growth in Aplysia. Science 256:645–649

Bailey CH, Kaang BK, Chen M et al. (1997) Mutation in the phosphorylation sites of MAP kinase blocks learning-related internalization of apCAM in Aplysia sensory neurons. Neuron 18:913–924

Baines RA, Seugnet L, Thompson A et al. (2002) Regulation of synaptic connectivity: levels of Fasciclin II influence synaptic growth in the Drosophila CNS. J Neurosci 22:6587–6595

Barbeau D, Liang JJ, Robitalille Y et al. (1995) Decreased expression of the embryonic form of the neural cell adhesion molecule in schizophrenic brains. Proc Natl Acad Sci USA 92:2785–2789

Barthels D, Vopper G, Boned A et al. (1992) High degree of NCAM diversity generated by alternative RNA splicing in brain and muscle. Eur J Neurosci 4:327–337

Barthels D, Vopper G and Wille W (1988) NCAM-180, the large isoform of the neural cell adhesion molecule of the mouse, is encoded by an alternatively spliced transcript. Nucleic Acids Res 16:4217–4225

Beltran PJ and Bixby JL (2003) Receptor protein tyrosine phosphatases as mediators of cellular adhesion. Front Biosci 8:d87–99

Bixby JL, Baerwald-De la Torre K, Wang C et al. (2002) A neuronal inhibitory domain in the N-terminal half of agrin. J Neurobiol 50:164–179

Bodrikov V, Leshchyns'ka I, Sytnyk V et al. (2005) RPTPalpha is essential for NCAM-mediated p59fyn activation and neurite elongation. J Cell Biol 168:127–139

Bodrikov V, Sytnyk V, Leshchyns'ka I et al. (2008) NCAM induces CaMKIIalpha-mediated RPTPalpha phosphorylation to enhance its catalytic activity and neurite outgrowth. J Cell Biol 182:1185–1200

Boggon TJ, Murray J, Chappuis-Flament S et al. (2002) C-cadherin ectodomain structure and implications for cell adhesion mechanisms. Science 296:1308–1313

Bouzioukh F, Tell F, Jean A et al. (2001) NMDA receptor and nitric oxide synthase activation regulate polysialylated neural cell adhesion molecule expression in adult brainstem synapses. J Neurosci 21:4721–4730

Brunner A and O'Kane CJ (1997) The fascination of the Drosophila NMJ. Trends Genet 13:85–87

Bruses JL and Rutishauser U (1998) Regulation of neural cell adhesion molecule polysialylation: evidence for nontranscriptional control and sensitivity to an intracellular pool of calcium. J Cell Biol 140:1177–1186

Budnik V (1996) Synapse maturation and structural plasticity at Drosophila neuromuscular junctions. Curr Opin Neurobiol 6:858–867

Bushell KM, Sollner C, Schuster-Boeckler B et al. (2008) Large-scale screening for novel low-affinity extracellular protein interactions. Genome Res 18:622–630

Buttner B and Horstkorte R (2008) Intracelluar ligands of NCAM. Neurochem Res [Epub ahead of print]. DOI: 10.1007/s11064-008-9592-1

Buttner B, Kannicht C, Reutter W et al. (2003) The neural cell adhesion molecule is associated with major components of the cytoskeleton. Biochem Biophys Res Commun 310:967–971

Buttner B, Kannicht C, Reutter W et al. (2005) Novel cytosolic binding partners of the neural cell adhesion molecule: mapping the binding domains of PLC gamma, LANP, TOAD-64, syndapin, PP1, and PP2A. Biochemistry 44:6938–6947

Chen A, Haines S, Maxson K et al. (1994) VASE exon expression alters NCAM-mediated cell-cell interactions. J Neurosci Res 38:483–492

Chen CP, Posy S, Ben-Shaul A et al. (2005) Specificity of cell-cell adhesion by classical cadherins: critical role for low-affinity dimerization through beta-strand swapping. Proc Natl Acad Sci USA 102:8531–8536

Close BE, Mendiratta SS, Geiger KM et al. (2003) The minimal structural domains required for neural cell adhesion molecule polysialylation by PST/ST8Sia IV and STX/ST8Sia II. J Biol Chem 278:30796–30805

Cole GJ and Akeson R (1989) Identification of a heparin binding domain of the neural cell adhesion molecule N-CAM using synthetic peptides. Neuron 2:1157–1165

Colley KJ (2008) Structural basis for the polysialylation of the neural cell adhesion molecule. Neurochem Res [Epub ahead of print]. DOI: 10.1007/s11064-008-9652-6

Consortium TU (2008) The universal protein resource (UniProt). Nucleic Acids Res 36:D190–195

Cunningham BA, Hemperly JJ, Murray BA et al. (1987) Neural cell adhesion molecule: structure, immunoglobulin-like domains, cell surface modulation, and alternative RNA splicing. Science 236:799–806

Davis GW, Schuster CM and Goodman CS (1996) Genetic dissection of structural and functional components of synaptic plasticity. III. CREB is necessary for presynaptic functional plasticity. Neuron 17:669–679

Dickson G, Gower HJ, Barton CH et al. (1987) Human muscle neural cell adhesion molecule (N-CAM): identification of a muscle-specific sequence in the extracellular domain. Cell 50:1119–1130

Diestel S, Hinkle CL, Schmitz B et al. (2005) NCAM140 stimulates integrin-dependent cell migration by ectodomain shedding. J Neurochem 95:1777–1784

Ditlevsen DK and Kolkova K (2008) Signaling pathways involved in NCAM-induced neurite outgrowth. Neurochem Res [Epub ahead of print]. DOI: 10.1007/s11064-008-9768-8

Ditlevsen DK, Povlsen GK, Berezin V et al. (2008) NCAM-induced intracellular signaling revisited. J Neurosci Res 86:727–743

Dityatev A, Dityateva G, Sytnyk V et al. (2004) Polysialylated neural cell adhesion molecule promotes remodeling and formation of hippocampal synapses. J Neurosci 24:9372–9382

Doherty P, Ashton SV, Moore SE et al. (1991) Morphoregulatory activities of NCAM and N-cadherin can be accounted for by G protein-dependent activation of L- and N-type neuronal Ca2+ channels. Cell 67:21–33

Doherty P, Moolenaar CE, Ashton SV et al. (1992) The VASE exon downregulates the neurite growth-promoting activity of NCAM 140. Nature 356:791–793

Dowsing B, Puche A, Hearn C et al. (1997) Presence of novel N-CAM glycoforms in the rat olfactory system. J Neurobiol 32:659–670

Dzhandzhugazyan K and Bock E (1993) Demonstration of (Ca(2+)-Mg2+)-ATPase activity of the neural cell adhesion molecule. FEBS Lett 336:279–283

Dzhandzhugazyan K and Bock E (1997) Demonstration of an extracellular ATP-binding site in NCAM: functional implications of nucleotide binding. Biochemistry 36:15381–15395

Eckhardt M, Bukalo O, Chazal G et al. (2000) Mice deficient in the polysialyltransferase ST8SiaIV/PST-1 allow discrimination of the roles of neural cell adhesion molecule protein and polysialic acid in neural development and synaptic plasticity. J Neurosci 20:5234–5244

Edelman GM and Chuong CM (1982) Embryonic to adult conversion of neural cell adhesion molecules in normal and staggerer mice. Proc Natl Acad Sci USA 79:7036–7040

El Maarouf A and Rutishauser U (2003) Removal of polysialic acid induces aberrant pathways, synaptic vesicle distribution, and terminal arborization of retinotectal axons. J Comp Neurol 460:203–211

El Maarouf A and Rutishauser U (2008) Use of PSA-NCAM in repair of the central nervous system. Neurochem Res [Epub ahead of print]. DOI: 10.1007/s11064-008-9635-7

Endo A, Hashimoto K, Takada Y et al. (1999a) The activation of the tissue plasminogen activator-plasmin system induced in the mouse hippocampus after injection of trimethyltin: possible proteolysis of highly polysialylated NCAM. Jpn J Physiol 49:463–466

Endo A, Nagai N, Urano T et al. (1998) Proteolysis of highly polysialylated NCAM by the tissue plasminogen activator-plasmin system in rats. Neurosci Lett 246:37–40

Endo A, Nagai N, Urano T et al. (1999b) Proteolysis of neuronal cell adhesion molecule by the tissue plasminogen activator-plasmin system after kainate injection in the mouse hippocampus. Neurosci Res 33:1–8

Fazeli MS, Breen K, Errington ML et al. (1994) Increase in extracellular NCAM and amyloid precursor protein following induction of long-term potentiation in the dentate gyrus of anaesthetized rats. Neurosci Lett 169:77–80

Felsenstein J (1981) Evolutionary trees from DNA sequences: a maximum likelihood approach. J Mol Evol 17:368–376

Fewou SN, Ramakrishnan H, Bussow H et al. (2007) Down-regulation of polysialic acid is required for efficient myelin formation. J Biol Chem 282:16700–16711

Foley AG, Hartz BP, Gallagher HC et al. (2000) A synthetic peptide ligand of neural cell adhesion molecule (NCAM) IgI domain prevents NCAM internalization and disrupts passive avoidance learning. J Neurochem 74:2607–2613

Forni JJ, Romani S, Doherty P et al. (2004) Neuroglian and Fasciclin II can promote neurite outgrowth via the FGF receptor Heartless. Mol Cell Neurosci 26:282–291

Friedlander DR, Milev P, Karthikeyan L et al. (1994) The neuronal chondroitin sulfate proteoglycan neurocan binds to the neural cell adhesion molecules Ng-CAM/L1/NILE and N-CAM, and inhibits neuronal adhesion and neurite outgrowth. J Cell Biol 125:669–680

Fu M, Vohra BP, Wind D et al. (2006) BMP signaling regulates murine enteric nervous system precursor migration, neurite fasciculation, and patterning via altered Ncam1 polysialic acid addition. Dev Biol 299:137–150

Fux CM, Krug M, Dityatev A et al. (2003) NCAM180 and glutamate receptor subtypes in potentiated spine synapses: an immunogold electron microscopic study. Mol Cell Neurosci 24:939–950

Gallagher HC, Murphy KJ, Foley AG et al. (2001) Protein kinase C delta regulates neural cell adhesion molecule polysialylation state in the rat brain. J Neurochem 77:425–434

Gallagher HC, Odumeru OA and Regan CM (2000) Regulation of neural cell adhesion molecule polysialylation state by cell-cell contact and protein kinase C delta. J Neurosci Res 61:636–645

Gascon E, Vutskits L and Kiss JZ (2007) Polysialic acid-neural cell adhesion molecule in brain plasticity: from synapses to integration of new neurons. Brain Res Rev 56:101–118

Gascon E, Vutskits L and Kiss JZ (2008) The Role of PSA-NCAM in Adult Neurogenesis. Neurochem Res

Gennarini G, Rougon G, Deagostini-Bazin H et al. (1984) Studies on the transmembrane disposition of the neural cell adhesion molecule N-CAM. A monoclonal antibody recognizing a cytoplasmic domain and evidence for the presence of phosphoserine residues. Eur J Biochem 142:57–64

Gower HJ, Barton CH, Elsom VL et al. (1988) Alternative splicing generates a secreted form of N-CAM in muscle and brain. Cell 55:955–964

Grenningloh G, Rehm EJ and Goodman CS (1991) Genetic analysis of growth cone guidance in Drosophila: fasciclin II functions as a neuronal recognition molecule. Cell 67:45–57

Grumet M, Flaccus A and Margolis RU (1993) Functional characterization of chondroitin sulfate proteoglycans of brain: interactions with neurons and neural cell adhesion molecules. J Cell Biol 120:815–824

Hammond MS, Sims C, Parameshwaran K et al. (2006) Neural cell adhesion molecule-associated polysialic acid inhibits NR2B-containing *N*-methyl-D-aspartate receptors and prevents glutamate-induced cell death. J Biol Chem 281:34859–34869

Hamshere M, Dickson G and Eperon I (1991) The muscle specific domain of mouse N-CAM: structure and alternative splicing patterns. Nucleic Acids Res 19:4709–4716

Han JH, Lim CS, Lee YS et al. (2004) Role of Aplysia cell adhesion molecules during 5-HT-induced long-term functional and structural changes. Learn Mem 11:421–435

Hansen RK, Christensen C, Korshunova I et al. (2007) Identification of NCAM-binding peptides promoting neurite outgrowth via a heterotrimeric G-protein-coupled pathway. J Neurochem 103:1396–1407

Harduin-Lepers A, Mollicone R, Delannoy P et al. (2005) The animal sialyltransferases and sialyltransferase-related genes: a phylogenetic approach. Glycobiology 15:805–817

Hata K, Polo-Parada L and Landmesser LT (2007) Selective targeting of different neural cell adhesion molecule isoforms during motoneuron myotube synapse formation in culture and the switch from an immature to mature form of synaptic vesicle cycling. J Neurosci 27:14481–14493

He HT, Barbet J, Chaix JC et al. (1986) Phosphatidylinositol is involved in the membrane attachment of NCAM-120, the smallest component of the neural cell adhesion molecule. Embo J 5:2489–2494

Hebbar S, Hall RE, Demski SA et al. (2006) The adult abdominal neuromuscular junction of Drosophila: a model for synaptic plasticity. J Neurobiol 66:1140–1155

Hemperly JJ, Murray BA, Edelman GM et al. (1986) Sequence of a cDNA clone encoding the polysialic acid-rich and cytoplasmic domains of the neural cell adhesion molecule N-CAM. Proc Natl Acad Sci USA 83:3037–3041

Higgins MR, Gibson NJ, Eckholdt PA et al. (2002) Different isoforms of fasciclin II are expressed by a subset of developing olfactory receptor neurons and by olfactory-nerve glial cells during formation of glomeruli in the moth Manduca sexta. Dev Biol 244:134–154

Hildebrandt H, Becker C, Murau M et al. (1998) Heterogeneous expression of the polysialyltransferases ST8Sia II and ST8Sia IV during postnatal rat brain development. J Neurochem 71:2339–2348

Hildebrandt H, Muhlenhoff M and Gerardy-Schahn R (2008) Polysialylation of NCAM. Neurochem Res [Epub ahead of print]. DOI: 10.1007/s11064-008-9724-7

Hildebrandt H, Muhlenhoff M, Weinhold B et al. (2007) Dissecting polysialic acid and NCAM functions in brain development. J Neurochem 103 Suppl 1:56–64

HLDM (2008) http://www.hcdm.org/MoleculeInformation/tabid/54/Default.aspx.

Hoeffer CA, Sanyal S and Ramaswami M (2003) Acute induction of conserved synaptic signaling pathways in Drosophila melanogaster. J Neurosci 23:6362–6372

Hoffman KB, Larson J, Bahr BA et al. (1998) Activation of NMDA receptors stimulates extracellular proteolysis of cell adhesion molecules in hippocampus. Brain Res 811:152–155

Hoffman S, Chuong CM and Edelman GM (1984) Evolutionary conservation of key structures and binding functions of neural cell adhesion molecules. Proc Natl Acad Sci USA 81:6881–6885

Horstkorte R, Schachner M, Magyar JP et al. (1993) The fourth immunoglobulin-like domain of NCAM contains a carbohydrate recognition domain for oligomannosidic glycans implicated in association with L1 and neurite outgrowth. J Cell Biol 121:1409–1421

Hotta K, Motoi Y, Okutani A et al. (2007) Role of GPI-anchored NCAM-120 in rabies virus infection. Microbes Infect 9:167–174

Hu Y, Barzilai A, Chen M et al. (1993) 5-HT and cAMP induce the formation of coated pits and vesicles and increase the expression of clathrin light chain in sensory neurons of aplysia. Neuron 10:921–929

Huang Y, Jellies J, Johansen KM et al. (1997) Differential glycosylation of tractin and LeechCAM, two novel Ig superfamily members, regulates neurite extension and fascicle formation. J Cell Biol 138:143–157

Hubschmann MV, Skladchikova G, Bock E et al. (2005) Neural cell adhesion molecule function is regulated by metalloproteinase-mediated ectodomain release. J Neurosci Res 80:826–837

Jacque CM, Jorgensen OS and Bock E (1974) Quantitative studies on the brain specific antigens S-100, GFA, 14-3-2, D1, D2, D3 and C1 in Quaking mouse. FEBS Lett 49:264–266

Jakovcevski I, Mo Z and Zecevic N (2007) Down-regulation of the axonal polysialic acid-neural cell adhesion molecule expression coincides with the onset of myelination in the human fetal forebrain. Neuroscience 149:328–337

Jie C, Zipser B, Jellies J et al. (1999) Differential glycosylation and proteolytical processing of LeechCAM in central and peripheral leech neurons. Biochim Biophys Acta 1452:161–171

Jin K, Peel AL, Mao XO et al. (2004) Increased hippocampal neurogenesis in Alzheimer's disease. Proc Natl Acad Sci USA 101:343–347

Jorgensen OS and Bock E (1974) Brain specific synaptosomal membrane proteins demonstrated by crossed immunoelectrophoresis. J Neurochem 23:879–880

Kalus I, Bormann U, Mzoughi M et al. (2006) Proteolytic cleavage of the neural cell adhesion molecule by ADAM17/TACE is involved in neurite outgrowth. J Neurochem 98:78–88

Kasper C, Rasmussen H, Kastrup JS et al. (2000) Structural basis of cell-cell adhesion by NCAM. Nat Struct Biol 7:389–393

Kazama H, Nose A and Morimoto-Tanifuji T (2007) Synaptic components necessary for retrograde signaling triggered by calcium/calmodulin-dependent protein kinase II during synaptogenesis. Neuroscience 145:1007–1015

Kessels MM and Qualmann B (2004) The syndapin protein family: linking membrane trafficking with the cytoskeleton. J Cell Sci 117:3077–3086

Key B and Akeson RA (1990) Olfactory neurons express a unique glycosylated form of the neural cell adhesion molecule (N-CAM). J Cell Biol 110:1729–1743

Key B and Akeson RA (1991) Delineation of olfactory pathways in the frog nervous system by unique glycoconjugates and N-CAM glycoforms. Neuron 6:381–396

Kiryushko D, Korshunova I, Berezin V et al. (2006) Neural cell adhesion molecule induces intracellular signaling via multiple mechanisms of Ca2+ homeostasis. Mol Biol Cell 17:2278–2286

Kiselyov VV, Skladchikova G, Hinsby AM et al. (2003) Structural basis for a direct interaction between FGFR1 and NCAM and evidence for a regulatory role of ATP. Structure 11:691–701

Kiselyov VV, Soroka V, Berezin V et al. (2005) Structural biology of NCAM homophilic binding and activation of FGFR. J Neurochem 94:1169–1179

Kitazume-Kawaguchi S, Kabata S and Arita M (2001) Differential biosynthesis of polysialic or disialic acid Structure by ST8Sia II and ST8Sia IV. J Biol Chem 276:15696–15703

Kochoyan A, Poulsen FM, Berezin V et al. (2008) Structural basis for the activation of FGFR by NCAM. Protein Sci 17:1698–1705

Koh YH, Ruiz-Canada C, Gorczyca M et al. (2002) The Ras1-mitogen-activated protein kinase signal transduction pathway regulates synaptic plasticity through fasciclin II-mediated cell adhesion. J Neurosci 22:2496–2504

Kohsaka H, Takasu E and Nose A (2007) In vivo induction of postsynaptic molecular assembly by the cell adhesion molecule Fasciclin2. J Cell Biol 179:1289–1300

Kolkova K, Stensman H, Berezin V et al. (2005) Distinct roles of PKC isoforms in NCAM-mediated neurite outgrowth. J Neurochem 92:886–894

Krieg PA, Sakaguchi DS and Kintner CR (1989) Primary structure and developmental expression of a large cytoplasmic domain form of Xenopus laevis neural cell adhesion molecule (NCAM). Nucleic Acids Res 17:10321–10335

Kristiansen LV and Hortsch M (2008) Fasciclin II: the NCAM Ortholog in Drosophila melanogaster. Neurochem Res [Epub ahead of print]. DOI: 10.1007/s11064-007-9566-8

Kristiansen LV, Marques FA, Soroka V et al. (1999) Homophilic NCAM interactions interfere with L1 stimulated neurite outgrowth. FEBS Lett 464:30–34

Kudo M, Takayama E, Tadakuma T et al. (1998) Molecular cloning of ssd-form neural cell adhesion molecules (N-CAMs) as the major form in Xenopus heart. Biochem Biophys Res Commun 245:127–132

Kulahin N, Rudenko O, Kiselyov V et al. (2005) Modulation of the homophilic interaction between the first and second Ig modules of neural cell adhesion molecule by heparin. J Neurochem 95:46–55

Kulahin N and Walmod PS (2008) The neural cell adhesion molecule NCAM2/OCAM/RNCAM, a close relative to NCAM. Neurochem Res [Epub ahead of print]. DOI: 10.1007/s11064-008-9614-z

Lafon M (2005) Rabies virus receptors. J Neurovirol 11:82–87

Lee SH, Lim CS, Park H et al. (2007) Nuclear translocation of CAM-associated protein activates transcription for long-term facilitation in Aplysia. Cell 129:801–812

Leshchyns'ka I, Sytnyk V, Morrow JS et al. (2003) Neural cell adhesion molecule (NCAM) association with PKCbeta2 via betaI spectrin is implicated in NCAM-mediated neurite outgrowth. J Cell Biol 161:625–639

Lewis CM, Levinson DF, Wise LH et al. (2003) Genome scan meta-analysis of schizophrenia and bipolar disorder, part II: schizophrenia. Am J Hum Genet 73:34–48

Li M, Makkinje A and Damuni Z (1996) Molecular identification of I1PP2A, a novel potent heat-stable inhibitor protein of protein phosphatase 2A. Biochemistry 35:6998–7002

Liedtke S, Geyer H, Wuhrer M et al. (2001) Characterization of N-glycans from mouse brain neural cell adhesion molecule. Glycobiology 11:373–384

Lin DM, Fetter RD, Kopczynski C et al. (1994) Genetic analysis of Fasciclin II in Drosophila: defasciculation, refasciculation, and altered fasciculation. Neuron 13:1055–1069

Little EB, Crossin KL, Krushel LA et al. (2001) A short segment within the cytoplasmic domain of the neural cell adhesion molecule (N-CAM) is essential for N-CAM-induced NF-kappa B activity in astrocytes. Proc Natl Acad Sci USA 98:2238–2243

Little EB, Edelman GM and Cunningham BA (1998) Palmitoylation of the cytoplasmic domain of the neural cell adhesion molecule N-CAM serves as an anchor to cellular membranes. Cell Adhes Commun 6:415–430

Liu L, Haines S, Shew R et al. (1993) Axon growth is enhanced by NCAM lacking the VASE exon when expressed in either the growth substrate or the growing axon. J Neurosci Res 35:327–345

Lyles JM, Linnemann D and Bock E (1984) Biosynthesis of the D2-cell adhesion molecule: post-translational modifications, intracellular transport, and developmental changes. J Cell Biol 99:2082–2091

Markram K, Gerardy-Schahn R and Sandi C (2007) Selective learning and memory impairments in mice deficient for polysialylated NCAM in adulthood. Neuroscience 144:788–796

Martin KC, Michael D, Rose JC et al. (1997) MAP kinase translocates into the nucleus of the presynaptic cell and is required for long-term facilitation in Aplysia. Neuron 18:899–912

Martin PT (2002) Glycobiology of the synapse. Glycobiology 12:1R–7R

Marx M, Rivera-Milla E, Stummeyer K et al. (2007) Divergent evolution of the vertebrate polysialyltransferase Stx and Pst genes revealed by fish-to-mammal comparison. Dev Biol 306:560–571

Mathew D, Popescu A and Budnik V (2003) Drosophila amphiphysin functions during synaptic Fasciclin II membrane cycling. J Neurosci 23:10710–10716

Mayanil CS, George D, Freilich L et al. (2001) Microarray analysis detects novel Pax3 downstream target genes. J Biol Chem 276:49299–49309

Mayanil CS, George D, Mania-Farnell B et al. (2000) Overexpression of murine Pax3 increases NCAM polysialylation in a human medulloblastoma cell line. J Biol Chem 275:23259–23266

Mayford M, Barzilai A, Keller F et al. (1992) Modulation of an NCAM-related adhesion molecule with long-term synaptic plasticity in Aplysia. Science 256:638–644

Mendiratta SS, Sekulic N, Hernandez-Guzman FG et al. (2006) A novel alpha-helix in the first fibronectin type III repeat of the neural cell adhesion molecule is critical for N-glycan polysialylation. J Biol Chem 281:36052–36059

Mendiratta SS, Sekulic N, Lavie A et al. (2005) Specific amino acids in the first fibronectin type III repeat of the neural cell adhesion molecule play a role in its recognition and polysialylation by the polysialyltransferase ST8Sia IV/PST. J Biol Chem 280:32340–32348

Milev P, Maurel P, Haring M et al. (1996) TAG-1/axonin-1 is a high-affinity ligand of neurocan, phosphacan/protein-tyrosine phosphatase-zeta/beta, and N-CAM. J Biol Chem 271:15716–15723

Milev P, Meyer-Puttlitz B, Margolis RK et al. (1995) Complex-type asparagine-linked oligosaccharides on phosphacan and protein-tyrosine phosphatase-zeta/beta mediate their binding to neural cell adhesion molecules and tenascin. J Biol Chem 270:24650–24653

Minana R, Duran JM, Tomas M et al. (2001) Neural cell adhesion molecule is endocytosed via a clathrin-dependent pathway. Eur J Neurosci 13:749–756

Minturn JE, Fryer HJ, Geschwind DH et al. (1995) TOAD-64, a gene expressed early in neuronal differentiation in the rat, is related to unc-33, a C. elegans gene involved in axon outgrowth. J Neurosci 15:6757–6766
Mizuno T, Kawasaki M, Nakahira M et al. (2001) Molecular diversity in zebrafish NCAM family: three members with different VASE usage and distinct localization. Mol Cell Neurosci 18:119–130
Montarolo PG, Goelet P, Castellucci VF et al. (1986) A critical period for macromolecular synthesis in long-term heterosynaptic facilitation in Aplysia. Science 234:1249–1254
Montarolo PG, Kandel ER and Schacher S (1988) Long-term heterosynaptic inhibition in Aplysia. Nature 333:171–174
Muller D, Wang C, Skibo G et al. (1996) PSA-NCAM is required for activity-induced synaptic plasticity. Neuron 17:413–422
Murakami T and Ohtsuka A (2003) Perisynaptic barrier of proteoglycans in the mature brain and spinal cord. Arch Histol Cytol 66:195–207
Murphy KJ, Foley AG, O'Connell A W et al. (2006) Chronic exposure of rats to cognition enhancing drugs produces a neuroplastic response identical to that obtained by complex environment rearing. Neuropsychopharmacology 31:90–100
Murray BA, Hemperly JJ, Prediger EA et al. (1986a) Alternatively spliced mRNAs code for different polypeptide chains of the chicken neural cell adhesion molecule (N-CAM). J Cell Biol 102:189–193
Murray BA, Owens GC, Prediger EA et al. (1986b) Cell surface modulation of the neural cell adhesion molecule resulting from alternative mRNA splicing in a tissue-specific developmental sequence. J Cell Biol 103:1431–1439
Nene V, Wortman JR, Lawson D et al. (2007) Genome sequence of Aedes aegypti, a major arbovirus vector. Science 316:1718–1723
Ngo ST, Noakes PG and Phillips WD (2007) Neural agrin: a synaptic stabiliser. Int J Biochem Cell Biol 39:863–867
Nielsen J, Kulahin N and Walmod PS (2008) Extracellular protein interactions mediated by the neural cell adhesion molecule, NCAM: heterophilic interactions between NCAM and cell adhesion molecules, extracellular matrix proteins, and viruses. Neurochem Res [Epub ahead of print]. DOI: 10.1007/s11064-008-9761-2
Niethammer P, Delling M, Sytnyk V et al. (2002) Cosignaling of NCAM via lipid rafts and the FGF receptor is required for neuritogenesis. J Cell Biol 157:521–532
Noble M, Albrechtsen M, Moller C et al. (1985) Glial cells express N-CAM/D2-CAM-like polypeptides in vitro. Nature 316:725–728
Oka S, Bruses JL, Nelson RW et al. (1995) Properties and developmental regulation of polysialyltransferase activity in the chicken embryo brain. J Biol Chem 270:19357–19363
Okamoto M, Sakiyama J, Kurazono S et al. (2001) Developmentally regulated expression of brain-specific chondroitin sulfate proteoglycans, neurocan and phosphacan, in the postnatal rat hippocampus. Cell Tissue Res 306:217–229
Olsen M, Krog L, Edvardsen K et al. (1993) Intact transmembrane isoforms of the neural cell adhesion molecule are released from the plasma membrane. Biochem J 295 (Pt 3):833–840
Ong E, Nakayama J, Angata K et al. (1998) Developmental regulation of polysialic acid synthesis in mouse directed by two polysialyltransferases, PST and STX. Glycobiology 8:415–424
Ong E, Suzuki M, Belot F et al. (2002) Biosynthesis of HNK-1 glycans on O-linked oligosaccharides attached to the neural cell adhesion molecule (NCAM): the requirement for core 2 beta 1,6-N-acetylglucosaminyltransferase and the muscle-specific domain in NCAM. J Biol Chem 277:18182–18190

Owens GC, Edelman GM and Cunningham BA (1987) Organization of the neural cell adhesion molecule (N-CAM) gene: alternative exon usage as the basis for different membrane-associated domains. Proc Natl Acad Sci USA 84:294–298

Pankratov Y, Lalo U, Verkhratsky A et al. (2006) Vesicular release of ATP at central synapses. Pflugers Arch 452:589–597

Paoloni-Giacobino A, Chen H and Antonarakis SE (1997) Cloning of a novel human neural cell adhesion molecule gene (NCAM2) that maps to chromosome region 21q21 and is potentially involved in Down syndrome. Genomics 43:43–51

Paratcha G, Ledda F and Ibanez CF (2003) The neural cell adhesion molecule NCAM is an alternative signaling receptor for GDNF family ligands. Cell 113:867–879

Pebusque MJ, Coulier F, Birnbaum D et al. (1998) Ancient large-scale genome duplications: phylogenetic and linkage analyses shed light on chordate genome evolution. Mol Biol Evol 15:1145–1159

Persohn E, Pollerberg GE and Schachner M (1989) Immunoelectron-microscopic localization of the 180 kD component of the neural cell adhesion molecule N-CAM in postsynaptic membranes. J Comp Neurol 288:92–100

Pestean A, Krizbai I, Bottcher H et al. (1995) Identification of the Ulex europaeus agglutinin-I-binding protein as a unique glycoform of the neural cell adhesion molecule in the olfactory sensory axons of adults rats. Neurosci Lett 195:117–120

Peter N, Aronoff B, Wu F et al. (1994) Decrease in growth cone-neurite fasciculation by sensory or motor cells in vitro accompanies downregulation of Aplysia cell adhesion molecules by neurotransmitters. J Neurosci 14:1413–1421

Petridis AK, El-Maarouf A, and Rutishauser U (2004) Polysialic acid regulates cell contact-dependent neuronal differentiation of progenitor cells from the subventricular zone. Dev Dyn. 230:675–684

Phillips GR, Krushel LA and Crossin KL (1997) Developmental expression of two rat sialyltransferases that modify the neural cell adhesion molecule, N-CAM. Brain Res Dev Brain Res 102:143–155

Pielage J, Fetter RD and Davis GW (2005) Presynaptic spectrin is essential for synapse stabilization. Curr Biol 15:918–928

Pillai-Nair N, Panicker AK, Rodriguiz RM et al. (2005) Neural cell adhesion molecule-secreting transgenic mice display abnormalities in GABAergic interneurons and alterations in behavior. J Neurosci 25:4659–4671

Pinsker H, Kupfermann I, Castellucci V et al. (1970) Habituation and dishabituation of the gill-withdrawal reflex in Aplysia. Science 167:1740–1742

Pollerberg GE, Burridge K, Krebs KE et al. (1987) The 180-kD component of the neural cell adhesion molecule N-CAM is involved in cell-cell contacts and cytoskeleton-membrane interactions. Cell Tissue Res 250:227–236

Polo-Parada L, Bose CM, Plattner F et al. (2004) Distinct roles of different neural cell adhesion molecule (NCAM) isoforms in synaptic maturation revealed by analysis of NCAM 180 kDa isoform-deficient mice. J Neurosci 24:1852–1864

Polo-Parada L, Plattner F, Bose C et al. (2005) NCAM 180 acting via a conserved C-terminal domain and MLCK is essential for effective transmission with repetitive stimulation. Neuron 46:917–931

Ponimaskin E, Dityateva G, Ruonala MO et al. (2008) Fibroblast growth factor-regulated palmitoylation of the neural cell adhesion molecule determines neuronal morphogenesis. J Neurosci 28:8897–8907

Prediger EA, Hoffman S, Edelman GM et al. (1988) Four exons encode a 93-base-pair insert in three neural cell adhesion molecule mRNAs specific for chicken heart and skeletal muscle. Proc Natl Acad Sci USA 85:9616–9620

Probstmeier R, Bilz A and Schneider-Schaulies J (1994) Expression of the neural cell adhesion molecule and polysialic acid during early mouse embryogenesis. J Neurosci Res 37:324–335

Probstmeier R, Fahrig T, Spiess E et al. (1992) Interactions of the neural cell adhesion molecule and the myelin-associated glycoprotein with collagen type I: involvement in fibrillogenesis. J Cell Biol 116:1063–1070
Probstmeier R, Kuhn K and Schachner M (1989) Binding properties of the neural cell adhesion molecule to different components of the extracellular matrix. J Neurochem 53:1794–1801
Qin S, Zheng F, Chen GH et al. (2005) Variable alternative spliced exon (VASE)-containing and VASE-lacking neural cell adhesion molecule in the dorsal and ventral hippocampus of SAMP8 mice. J Neurosci Res 80:838–844
Rasmussen KK, Kulahin N, Kristensen O et al. (2008) Crystal structure of the Ig1 domain of the neural cell adhesion molecule NCAM2 displays domain swapping. J Mol Biol 382:1113–1120
Rathbone MP, Middlemiss PJ, Gysbers JW et al. (1999) Trophic effects of purines in neurons and glial cells. Prog Neurobiol 59:663–690
Rayport SG and Schacher S (1986) Synaptic plasticity in vitro: cell culture of identified Aplysia neurons mediating short-term habituation and sensitization. J Neurosci 6:759–763
Reyes AA, Small SJ and Akeson R (1991) At least 27 alternatively spliced forms of the neural cell adhesion molecule mRNA are expressed during rat heart development. Mol Cell Biol 11:1654–1661
Rivlin PK, St Clair RM, Vilinsky I et al. (2004) Morphology and molecular organization of the adult neuromuscular junction of Drosophila. J Comp Neurol 468:596–613
Rodriguez MM, Ron D, Touhara K et al. (1999) RACK1, a protein kinase C anchoring protein, coordinates the binding of activated protein kinase C and select pleckstrin homology domains in vitro. Biochemistry 38:13787–13794
Rutishauser U (2008) Polysialic acid in the plasticity of the developing and adult vertebrate nervous system. Nat Rev Neurosci 9:26–35
Rutishauser U, Acheson A, Hall AK et al. (1988) The neural cell adhesion molecule (NCAM) as a regulator of cell-cell interactions. Science 240:53–57
Rutishauser U, Thiery JP, Brackenbury R et al. (1976) Mechanisms of adhesion among cells from neural tissues of the chick embryo. Proc Natl Acad Sci USA 73:577–581
Sadoul K, Meyer A, Low MG et al. (1986) Release of the 120 kDa component of the mouse neural cell adhesion molecule N-CAM from cell surfaces by phosphatidylinositol-specific phospholipase C. Neurosci Lett 72:341–346
Saffell JL, Walsh FS and Doherty P (1994) Expression of NCAM containing VASE in neurons can account for a developmental loss in their neurite outgrowth response to NCAM in a cellular substratum. J Cell Biol 125:427–436
Sandig M, Rao Y and Siu CH (1994) The homophilic binding site of the neural cell adhesion molecule NCAM is directly involved in promoting neurite outgrowth from cultured neural retinal cells. J Biol Chem 269:14841–14848
Santoni MJ, Barthels D, Vopper G et al. (1989) Differential exon usage involving an unusual splicing mechanism generates at least eight types of NCAM cDNA in mouse brain. Embo J 8:385–392
Santuccione A, Sytnyk V, Leshchyns'ka I et al. (2005) Prion protein recruits its neuronal receptor NCAM to lipid rafts to activate p59fyn and to enhance neurite outgrowth. J Cell Biol 169:341–354
Saraste M, Sibbald PR and Wittinghofer A (1990) The P-loop–a common motif in ATP- and GTP-binding proteins. Trends Biochem Sci 15:430–434
Schacher S and Montarolo PG (1991) Target-dependent structural changes in sensory neurons of Aplysia accompany long-term heterosynaptic inhibition. Neuron 6:679–690
Schmitt-Ulms G, Legname G, Baldwin MA et al. (2001) Binding of neural cell adhesion molecules (N-CAMs) to the cellular prion protein. J Mol Biol 314:1209–1225
Schuch U, Lohse MJ and Schachner M (1989) Neural cell adhesion molecules influence second messenger systems. Neuron 3:13–20

Schuster CM, Davis GW, Fetter RD et al. (1996) Genetic dissection of structural and functional components of synaptic plasticity. I. Fasciclin II controls synaptic stabilization and growth. Neuron 17:641–654

Schuster T, Krug M, Hassan H et al. (1998) Increase in proportion of hippocampal spine synapses expressing neural cell adhesion molecule NCAM180 following long-term potentiation. J Neurobiol 37:359–372

Schuster T, Krug M, Stalder M et al. (2001) Immunoelectron microscopic localization of the neural recognition molecules L1, NCAM, and its isoform NCAM180, the NCAM-associated polysialic acid, beta1 integrin and the extracellular matrix molecule tenascin-R in synapses of the adult rat hippocampus. J Neurobiol 49:142–158

Schwarzacher SW, Vuksic M, Haas CA et al. (2006) Neuronal hyperactivity induces astrocytic expression of neurocan in the adult rat hippocampus. Glia 53:704–714

Schwenkert I, Eltrop R, Funk N et al. (2008) The hangover gene negatively regulates bouton addition at the Drosophila neuromuscular junction. Mech Dev 125:700–711

Secher T (2008) Soluble NCAM. Neurochem Res [Epub ahead of print]. DOI: 10.1007/s11064-008-9743-4

Seki T and Arai Y (1993) Distribution and possible roles of the highly polysialylated neural cell adhesion molecule (NCAM-H) in the developing and adult central nervous system. Neurosci Res 17:265–290

Seki T and Rutishauser U (1998) Removal of polysialic acid-neural cell adhesion molecule induces aberrant mossy fiber innervation and ectopic synaptogenesis in the hippocampus. J Neurosci 18:3757–3766

Senkov O, Sun M, Weinhold B et al. (2006) Polysialylated neural cell adhesion molecule is involved in induction of long-term potentiation and memory acquisition and consolidation in a fear-conditioning paradigm. J Neurosci 26:10888–109898

Sigrist SJ, Reiff DF, Thiel PR et al. (2003) Experience-dependent strengthening of Drosophila neuromuscular junctions. J Neurosci 23:6546–6556

Sjostrand D, Carlsson J, Paratcha G et al. (2007) Disruption of the GDNF binding site in NCAM dissociates ligand binding and homophilic cell adhesion. J Biol Chem 282: 12734–12740

Sjostrand D and Ibanez CF (2008) Insights into GFRalpha1 regulation of neural cell adhesion molecule (NCAM) function from structure-function analysis of the NCAM/GFRalpha1 receptor complex. J Biol Chem 283:13792–13798

Skladchikova G, Ronn LC, Berezin V et al. (1999) Extracellular adenosine triphosphate affects neural cell adhesion molecule (NCAM)-mediated cell adhesion and neurite outgrowth. J Neurosci Res 57:207–218

Small SJ and Akeson R (1990) Expression of the unique NCAM VASE exon is independently regulated in distinct tissues during development. J Cell Biol 111:2089–2096

Sorkin BC, Hoffman S, Edelman GM et al. (1984) Sulfation and phosphorylation of the neural cell adhesion molecule, N-CAM. Science 225:1476–1478

Soroka V, Kasper C and Poulsen FM (2008) Structural biology of NCAM. Neurochem Res [Epub ahead of print]. DOI: 10.1007/s11064-008-9837-z

Soroka V, Kolkova K, Kastrup JS et al. (2003) Structure and interactions of NCAM Ig1-2-3 suggest a novel zipper mechanism for homophilic adhesion. Structure 11:1291–1301

Storms SD, Kim AC, Tran BH et al. (1996) NCAM-mediated adhesion of transfected cells to agrin. Cell Adhes Commun 3:497–509

Strausberg RL, Feingold EA, Grouse LH et al. (2002) Generation and initial analysis of more than 15,000 full-length human and mouse cDNA sequences. Proc Natl Acad Sci USA 99:16899–16903

Sullivan PF, Keefe RS, Lange LA et al. (2007) NCAM1 and neurocognition in schizophrenia. Biol Psychiatry 61:902–910

Suzuki E, Rose D and Chiba A (2000) The ultrastructural interactions of identified pre- and postsynaptic cells during synaptic target recognition in Drosophila embryos. J Neurobiol 42:448–459

Sweatt JD (2001) The neuronal MAP kinase cascade: a biochemical signal integration system subserving synaptic plasticity and memory. J Neurochem 76:1–10

Sytnyk V, Leshchyns'ka I, Nikonenko AG et al. (2006) NCAM promotes assembly and activity-dependent remodeling of the postsynaptic signaling complex. J Cell Biol 174:1071–1085

Tacke R and Goridis C (1991) Alternative splicing in the neural cell adhesion molecule pre-mRNA: regulation of exon 18 skipping depends on the 5′-splice site. Genes Dev 5: 1416–1429

Tai MH and Zipser B (2002) Sequential steps of carbohydrate signaling mediate sensory afferent differentiation. J Neurocytol 31:743–754

Teichmann SA and Chothia C (2000) Immunoglobulin superfamily proteins in Caenorhabditis elegans. J Mol Biol 296:1367–1383

Ter Horst JP, Loscher JS, Pickering M et al. (2008) Learning-associated regulation of polysialylated neural cell adhesion molecule expression in the rat prefrontal cortex is region-, cell type- and paradigm-specific. Eur J Neurosci 28:419–427

Thierry-Mieg D, Thierry-Mieg J, Potdevin M et al. (2008) The AceView genes. www.ncbi.nlm.nih.gov/IEB/Research/Acembly november 7, 2008

Thompson J, Dickson G, Moore SE et al. (1989) Alternative splicing of the neural cell adhesion molecule gene generates variant extracellular domain structure in skeletal muscle and brain. Genes Dev 3:348–357

Thoulouze MI, Lafage M, Schachner M et al. (1998) The neural cell adhesion molecule is a receptor for rabies virus. J Virol 72:7181–7190

Tomasiewicz H, Ono K, Yee D et al. (1993) Genetic deletion of a neural cell adhesion molecule variant (N-CAM-180) produces distinct defects in the central nervous system. Neuron 11:1163–1174

Tonissen KF and Krieg PA (1993) Two neural-cell adhesion molecule (NCAM)-encoding genes in Xenopus laevis are expressed during development and in adult tissues. Gene 127:243–247

Vaithianathan T, Matthias K, Bahr B et al. (2004) Neural cell adhesion molecule-associated polysialic acid potentiates alpha-amino-3-hydroxy-5-methylisoxazole-4-propionic acid receptor currents. J Biol Chem 279:47975–47984

Volonte C, Amadio S, Cavaliere F et al. (2003) Extracellular ATP and neurodegeneration. Curr Drug Targets CNS Neurol Disord 2:403–412

Walmod PS, Kolkova K, Berezin V et al. (2004) Zippers make signals: NCAM-mediated molecular interactions and signal transduction. Neurochem Res 29:2015–2035

Walsh FS, Parekh RB, Moore SE et al. (1989) Tissue specific O-linked glycosylation of the neural cell adhesion molecule (N-CAM). Development 105:803–811

Wang LH and Strittmatter SM (1996) A family of rat CRMP genes is differentially expressed in the nervous system. J Neurosci 16:6197–6207

Weinhold B, Seidenfaden R, Rockle I et al. (2005) Genetic ablation of polysialic acid causes severe neurodevelopmental defects rescued by deletion of the neural cell adhesion molecule. J Biol Chem 280:42971–42977

Williams EJ, Furness J, Walsh FS et al. (1994) Activation of the FGF receptor underlies neurite outgrowth stimulated by L1, N-CAM, and N-cadherin. Neuron 13:583–594

Williams EJ, Walsh FS and Doherty P (1994) Tyrosine kinase inhibitors can differentially inhibit integrin-dependent and CAM-stimulated neurite outgrowth. J Cell Biol 124:1029–1037

Wright JW and Copenhaver PF (2001) Cell type-specific expression of fasciclin II isoforms reveals neuronal-glial interactions during peripheral nerve growth. Dev Biol 234:24–41
Wright JW, Snyder MA, Schwinof KM et al. (1999) A role for fasciclin II in the guidance of neuronal migration. Development 126:3217–3228
Wuhrer M, Geyer H, von der Ohe M et al. (2003) Localization of defined carbohydrate epitopes in bovine polysialylated NCAM. Biochimie 85:207–218
Yoshihara Y, Kawasaki M, Tamada A et al. (1997) OCAM: a new member of the neural cell adhesion molecule family related to zone-to-zone projection of olfactory and vomeronasal axons. J Neurosci 17:5830–5842
Zhou R, Cheng H and Tiersch TR (2001) Differential genome duplication and fish diversity. Reviews in Fish Biology and Fisheries 11:331–337
Zhu H, Wu F and Schacher S (1994) Aplysia cell adhesion molecules and serotonin regulate sensory cell-motor cell interactions during early stages of synapse formation in vitro. J Neurosci 14:6886–6900
Zimmermann H (2008) ATP and acetylcholine, equal brethren. Neurochem Int 52:634–648
Zito K, Fetter RD, Goodman CS et al. (1997) Synaptic clustering of Fascilin II and Shaker: essential targeting sequences and role of Dlg. Neuron 19:1007–1016
Zorn AM and Krieg PA (1992) Developmental regulation of alternative splicing in the mRNA encoding Xenopus laevis neural cell adhesion molecule (NCAM). Dev Biol 149:197–205

Chapter 14
MHC Class I Function at the Neuronal Synapse

Sebastian Thams and Staffan Cullheim

Abstract The major histocompatibility complex class I (MHC class I) encodes a family of immune recognition molecules acting as ligands at immune synapses, thereby conveying adaptive and innate immunity. MHC class I proteins are expressed by all nucleated vertebrate cells, including neurons, and can mediate immune clearance of neurotropic viruses in the CNS. Intriguingly, there are now indications for a non-immune role of MHC class I signalling at neuronal synapses. Thus, neuronal MHC class I expression has been linked to synaptic plasticity and the modulation of neuronal functions. Up to this day, two MHC class I receptors have been identified at the neuronal synapse, which provide a molecular basis for MHC class I-dependent signalling across the synapse. Possible clinical implications of MHC class I expression in the CNS are currently being investigated in relation to neurodevelopmental and neurodegenerative diseases, in which synaptic dysfunction is a cardinal feature.

Keywords Major histocompatibility complex · β_2-microglobulin · Synapse

14.1 Background

The major histocompatibility complex (MHC) is a conserved genomic region in jawed vertebrates (Flajnik and Kasahara 2001), which contains structurally related genes and encode widely expressed cell surface molecules, such as the MHC class I proteins. Classical MHC class I α-polypeptides (class Ia), which are transmembrane proteins encoded by few highly polymorphic genes, associate with a β_2-microglobulin (β_2m) polypeptide and contain unique peptide-binding clefts that can bind 8–10 amino acid peptides. Non-classical MHC class I

S. Thams (✉)
Department of Neuroscience, Karolinska Institutet, Retzius väg 8,
171 77 Stockholm, Sweden
e-mail: sebastian.thams@ki.se

M. Hortsch, H. Umemori (eds.), *The Sticky Synapse*,
DOI 10.1007/978-0-387-92708-4_14,

α-polypeptides (class Ib) are encoded by a number of oligomorphic genes, some of which are expressed independently of β_2m or a peptide fragment or both. The number of non-classical MHC class I genes varies substantially between species (Niedermann et al. 1995, Janeway and Travers 1997, Niedermann et al. 1997).

MHC class I molecules are assembled in the endoplasmic reticulum (ER). Correct folding and surface expression of the class Ia molecules is dependent on the association with β_2m and loading of peptide. Generally, peptides that bind MHC class I molecules are derived from the cytoplasm and transferred into the ER molecules by the transporter associated with antigen processing (TAP, consisting of the TAP1 and TAP2 subunits) (Janeway and Travers 1997). In the absence of β_2m, MHC class Ia molecules are trapped in the ER. Moreover, in the absence of TAP, MHC class I molecules are unstable and only a fraction is transported to the cell surface (Ljunggren et al. 1990). Peptides which are presented by MHC class I molecules are generated from cytoplasmic proteins through the degradation by the proteasome complex. Thus, the mature MHC class I molecules on the cell surface present a peptide repertoire, which reflects the protein metabolism of the cell. Importantly, MHC class I molecules are expressed by all nucleated cells in jawed vertebrates (Janeway and Travers 1997, Flajnik and Kasahara 2001).

The function of MHC class I molecules has been extensively studied in the adaptive immune system, where they facilitate thymus-derived cytotoxic T-lymphocyte (CTL) surveillance of tissues for intracellular infections and malignant transformations. CTLs carry clone specific T-cell receptors (TCRs), generated through the somatic recombination of genes, which are specific for unique MHC class I-peptide combinations. In the thymus, developing thymus-derived lymphocytes (T lymphocytes) that react with MHC class I molecules presenting endogenous peptides are eliminated, whereas T lymphocytes with a weak affinity for MHC class I molecules are selected. In this way, a repertoire of T lymphocytes with the ability to recognize MHC class I molecules presenting foreign peptides, such as those derived from intracellular pathogens, is generated (Janeway and Travers 1997). Complementing CTL-mediated immunity, natural killer (NK) cells can eliminate cells with downregulated surface expression of the MHC class I (Janeway and Travers 1997). This subset of lymphocytes conveys innate immune functions relying on rapidly evolving germline encoded receptors that bind MHC class I molecules more or less independently of peptide presentation (Zinkernagel and Doherty 1974, Karre et al. 1986, Ljunggren and Karre 1990, Bryceson et al. 2006). The NK-cell receptors for MHC class I molecules can be of an inhibitory or activating type. Comparisons between the immune system and the nervous system can be made with regard to the expression of common molecules and the interactions through synaptic complexes (Dustin and Colman 2002). This chapter discusses the putative role of MHC class I molecules in neuronal synaptic function.

14.2 MHC Class I Expression in Neurons

For many years the central nervous system (CNS) was perceived as an 'immune privileged' site, due to the low or absent expression of MHC class I molecules in adult neurons, the lack of rejection of foreign tissue transplanted into the CNS, the restricted migration of most immune cells across the blood–brain barrier and the high local expression of immunosuppressive soluble factors (e.g. interleukin 10, transforming growth factor β, etc.) (Becher et al. 1998, Galea et al. 2007). An argument that further strengthened the notion of CNS immune privilege due to MHC class I paucity was the finding that neuronal MHC class I immunoreactivity could not be detected in mice overexpressing interferon (IFN)-γ under an astrocytic promoter (Horwitz et al. 1999). IFN- γ is a powerful inducer of MHC class I expression and therefore high astrocyte-driven secretion of this cytokine is expected to induce neuronal MHC class I expression. Even though several publications demonstrated the presence of MHC class I mRNAs in neurons under certain conditions (Corriveau et al. 1998, Linda et al. 1998, Lidman et al. 1999, Linda et al. 1999, Huh et al. 2000, Kimura and Griffin 2000), the surface expression pattern of MHC class I proteins remained unconfirmed until recently due to difficulties in staining MHC class I molecules in vivo.

However, evidence for the cell surface expression of MHC class I in neuronal subpopulations is accumulating (Corriveau et al. 1998, Redwine et al. 2001, Hoftberger et al. 2004, Peng et al. 2006, Rolleke et al. 2006, Ishii and Mombaerts 2008, Zohar et al. 2008). Neuronal upregulation of MHC class I proteins and associated molecules is seen as a part of the tissue response to neurotropic infections (Bilzer and Stitz 1994, Kimura and Griffin 2000, Foster et al. 2002) and in Rasmussen encephalitis, a human epileptic disorder (Bien et al. 2002). Accordingly, neurons infected with pathogens or affected by epileptic changes can be targeted by MHC class I-dependent immune-mediated cytotoxicity (Bilzer and Stitz 1994, Rivera-Quinones et al. 1998, Bien et al. 2002, Mendez-Fernandez et al. 2003), thereby providing evidence for the existence of functional MHC class I molecules at the neuronal cell surface. Remarkably, neuronal MHC class I molecules appear to also participate in processes other than immunity. The constitutive expression of different MHC class Ia and Ib mRNAs is restricted to certain brain regions or neuronal subpopulations (Corriveau et al. 1998, Lidman et al. 1999, Linda et al. 1999, Loconto et al. 2003). Furthermore, neuronal MHC class I expression is regulated by neuronal activity and has been linked to activity-dependent neuronal plasticity (Neumann et al. 1997, Corriveau et al. 1998). Its specific regional expression (Boulanger and Shatz 2004) implies that the role for MHC class I proteins in neurons is not purely immunological as this would require a more general expression.

Neuronal MHC class I expression and its regulation has been studied both in vitro and in vivo in different systems. In vitro, hippocampal neurons have been shown to express both MHC class I mRNA and protein in response to IFN-γ or tetradotoxin (TTX)-treatment (Neumann et al. 1995, 1997). Co-localization was seen between MHC class I protein clusters and postsynaptic densities (PSD-95) (Goddard et al. 2007) (Fig. 14.1), indicating an association with excitatory synapses. Similarly, IFN-γ treatment of embryonic rat motoneuron cultures induced MHC class Ia expression in clusters along neurites (Linda et al. 1998). In vivo, classical and non-classical MHC class I mRNAs are constitutively expressed in the olfactory bulb, cerebral cortex, hippocampus, substantia nigra and motor nuclei in the brainstem and spinal cord (Corriveau et al. 1998, Linda et al. 1998, Lidman et al. 1999, Linda et al. 1999, Dulac and Torello 2003, Loconto et al. 2003, Zohar et al. 2008). In addition to the regulation by neuronal activity, neurotropic infections and cytokines, MHC class I mRNA is also upregulated by neuronal injury, e.g. by axotomy (Linda et al. 1998, Lidman et al. 1999, Linda et al. 1999). At the protein level, neuronal MHC class I immunoreactivity has been described in vivo in the rodent, the simian and the human CNS (Corriveau et al. 1998, Linda et al. 1998, Redwine et al. 2001, Bien et al. 2002, Loconto et al. 2003, Peng et al. 2006, Rolleke et al. 2006). However, a precise subcellular localization of the MHC class I protein in postsynaptic elements is yet to be described in vivo.

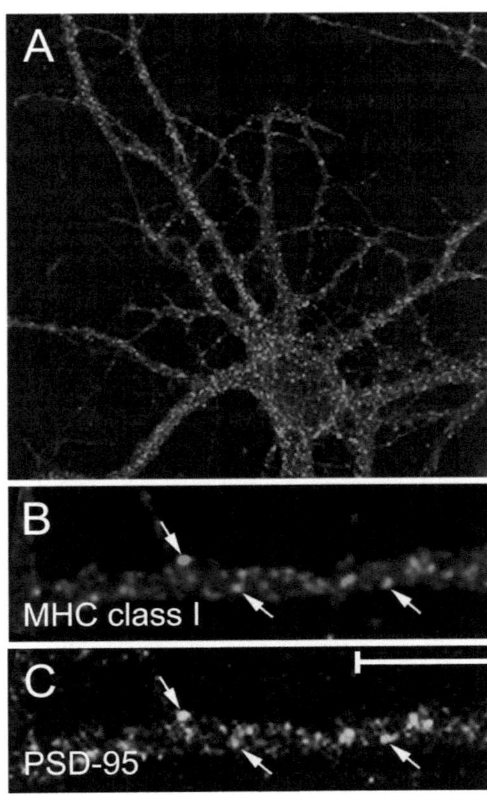

Fig. 14.1 A confocal micrograph showing a cultured hippocampal neuron stained with a monoclonal anti-MHC class I antibody (**A**). At higher magnification (**B**) one can observe a cluster-like staining pattern of MHC class I staining that co-localizes with the postsynaptic marker PSD-95 (**C**, *arrows*). Modified from Goddard et al. (2007) with permission, copyright (2007) National Academy of Sciences, USA

Although it is difficult to assess whether MHC class I molecules expressed on neurons present peptides in vivo, a few in vitro and in vivo studies suggest that certain neuronal populations possess such capacity (Medana et al. 2000, 2001, Mendez-Fernandez et al. 2003, Zohar et al. 2008). However, in certain systems, such as the vomeronasal organ (VNO) neuronal MHC class I function appears to be independent of peptide presentation (Loconto et al. 2003, Olson et al. 2005).

14.2.1 Surface Expression of Neuronal MHC Class I Molecules

A selection of adult neurons constitutively express MHC class I mRNA. Although several studies demonstrated an MHC class I immunoreactivity, many of the available antibodies fail to label in neuronal somata. When immunoreactivity is detected it is generally sparse and does not provide a precise subcellular localization of the molecules. Furthermore, under conditions when MHC class I mRNA is strongly upregulated the immunoreactivity in neuronal cell bodies remains low or moderate. Here, we discuss possible reasons for the difficulties in staining neuronal MHC class I proteins in vivo and why there sometimes appears to be a discrepancy between MHC class I mRNA levels and immunoreactivity.

One explanation for the low MHC class I immunoreactivity in neuronal cell bodies might be a low MHC class I protein level, which could be caused by the constitutive transport of MHC class I molecules into axons or dendrites, thereby depleting the cell body of protein. It has been shown that MHC class I protein is in fact transported and possibly confined to dendrites and axons in the spinal cord (Rivera-Quinones et al. 1998), in the vomeronasal system (Ishii and Mombaerts 2008) and in hippocampal cultures (Medana et al. 2001). Moreover, MHC class I mRNA can be transported into the dendritic tree of pyramidal hippocampal neurons in vivo, where it might be targeted for localized protein synthesis (Zhong et al. 2006).

Secretion of MHC class I proteins into the neuropil presents an alternative mechanism. This has been described in the immune system, where the extracellular domains of β_2m-MHC class I complexes are cleaved at the level of the plasma membrane by the proteolytic presenilin/γ-secretase complex, which is also present in the CNS, and thus released into the extracellular space (Carey et al. 2007). Moreover, MHC class I complexes have been shown to exist in soluble form in the blood (Demaria and Bushkin 2000, MacKay et al. 2006). It is yet to be examined whether this is also true for cerebrospinal fluid, which may contain secreted soluble MHC class I molecule from neuronal cells.

One might also consider other mechanisms of regulating MHC class I surface expression, which have been originally described in the immune system. For instance, it is known that MHC class I molecules can interact with receptors on the same cell through *cis*-interactions, thereby preventing their binding to receptors on other cells (Held and Mariuzza 2008). If a majority of all MHC class I molecules are bound to receptors on the same cell, this might block the available epitopes, thus increasing the detection threshold for certain anti-MHC class I antibodies, which are used for tissue-staining protocols. *Cis*-interactions involving MHC class I molecules have been shown in the mouse immune system, e.g. for the paired immunoglobulin receptor B (PIR-B), the leucocyte immunoglobulin-like receptor (LILR) and Ly49 (Held and Mariuzza 2008) (see Section 14.4). It was recently demonstrated that *cis*-interactions regulate the formation of synapses mediated by neuroligin-1/neurexin-1β interactions in hippocampal cultures (Taniguchi et al. 2007).

14.3 Link to Synaptic Function

14.3.1 Synaptic Plasticity in the Developing and Adult Brain

The refinement of retinogeniculate projections into segregated eye-specific layers in the lateral geniculate nucleus (LGN) is an activity-dependent process, during which neuronal inputs compete for stabilization by Hebbian plasticity. Strikingly, the expression of MHC class I genes is regulated in an activity-dependent manner in the LGN at this developmental stage (Corriveau et al. 1998). However, the cellular and subcellular expression of MHC class I molecules in the LGN remains to be further characterized.

The precise function of MHC class I proteins in the LGN was initially unclear. However, studies with mice deficient in β_2m or β_2m and TAP1 ($\beta_2m^{-/-}$, TAP1/$\beta_2m^{-/-}$; lacking subsets or a majority of all MHC class I; in this chapter we call them as MHC class I-deficient mice) supported the idea of their involvement in activity-dependent connectivity. In MHC class I-deficient mice a larger region of the LGN was occupied by projections from the ipsilateral eye, which may indicate the existence of redundant synaptic terminals (Huh et al. 2000). Another explanation could be an aberrant or shifted retinogeniculate projection pattern. Moreover, the MHC class I-deficient mice display an enhanced LTP and an absent LTD in the hippocampus after stimulation of Schaffer collaterals (SC) in slice preparations (Huh et al. 2000).

At the morphological level, the lack of MHC class I protein appears to influence the basic properties of synapses in the hippocampus, e.g. their number of synaptic vesicles and perforated PSDs (Goddard et al. 2007). In addition, a moderate change in the average presynaptic terminal size was seen in hippocampal cultures from MHC class I-deficient mice and this correlated to an enlargement of glutamatergic terminals (Goddard et al. 2007). The electrophysiological properties of neurons lacking MHC class I molecules were studied in hippocampal cultures and slice preparations from the visual cortex. These recordings showed an increase in miniature excitatory postsynaptic currents in the absence of MHC class I molecules, indicating an altered basal synaptic function (Goddard et al. 2007).

When considering these results, one should keep in mind that the mice studied above lack MHC class I proteins in all cells and thus have a constitutive immune deficiency. As reported in the literature, a compromised adaptive immunity can affect brain development and plasticity (Cohen et al. 2006, Ziv et al. 2006). It should also be pointed out that even though TAP1/β_2m$^{-/-}$ mice lack functional CTLs and NK cells, their immunological phenotype is not as severe as expected (Muller et al. 1992). For solving this conundrum, mice deficient in the recombination-activating gene 1 (RAG1$^{-/-}$) (Janeway and Travers 1997, Flajnik and Kasahara 2001) were studied with regard to their brain plasticity (Huh et al. 2000). RAG1$^{-/-}$ mice express MHC class I proteins, but have a severe immunodeficiency due to the lack of both thymic-derived and bone marrow-derived lymphocyte function (Janeway and Travers 1997). No synaptic phenotype similar to that of MHC class I-deficient mice was found in the RAG1$^{-/-}$ mice. Nevertheless, the construction of a conditional knockout mouse would be required to fully resolve this issue.

Human MHC class I (termed human leucocyte antigen, HLA) deficiency is a rare condition that is caused by either a transcription defect in the HLA and the β_2m gene or mutations in the TAP transporter (de la Salle et al. 2002). The first group has a moderate reduction in HLA levels and is virtually asymptomatic. The second group almost completely lacks HLA, but its immune defect is surprisingly mild (de la Salle et al. 2002). Whether the TAP-deficient patients have a neurological phenotype is unknown and the examination of such patients could provide important insights into the phenotype that accompanies MHC class I deficiencies.

14.3.2 Synaptic Elimination in the Axotomized Spinal Cord

Already over a decade ago adult spinal motoneurons were reported to constitutively express mRNAs encoding MHC class I molecules, as well as the β_2m polypeptide (Linda et al. 1998, 1999). However, the function of these proteins in motoneurons was unknown. The mRNA levels for both genes are upregulated after peripheral nerve transection (Linda et al. 1998, 1999), indicating an involvement of MHC class I molecules in regeneration. Since MHC class I was reported to be involved in synaptic refinement in the visual system (Huh et al. 2000), it was of interest to study the role of the molecule in synaptic plasticity of axotomized spinal motoneurons. As a part of the retrograde response to axotomy, the motoneurons lose a majority of their synaptic inputs in a process commonly referred to as 'synaptic stripping'. This is a highly selective process, which may protect the neurons from glutamate-induced excitotoxicity, as indicated by preferential removal of excitatory terminals (Linda et al. 2000). Thus, the severed neurons mostly preserve clusters of inhibitory terminals on the cell bodies (Linda et al. 2000), thereby favouring inhibition and subsequently inactivation during the recovery period.

Interestingly, in mice lacking a majority of all MHC class I molecules (i.e. $\beta_2m^{-/-}$ mice) more presynaptic terminals were eliminated from motoneurons than in wild-type mice after axotomy (Fig. 14.2). The surplus of eliminated synapses predominantly consisted of presynaptic terminals containing flat

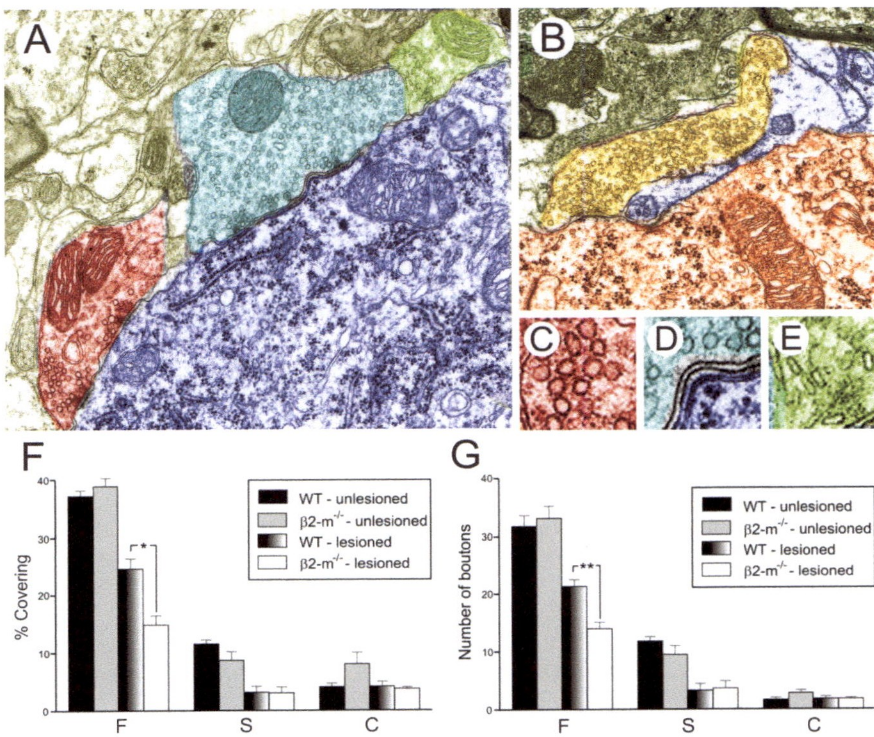

Fig. 14.2 (**A**) An electron micrograph in pseudocolour showing the three morphological types of presynaptic terminals that are found on the motoneuron somata (*dark blue*), based on *vesicle shape* in glutaraldehyde-fixed tissue. The synaptic terminal in *red* with only *spherical vesicles*, termed S-type, is glutamatergic (see **C**). The *light blue* middle terminal with a characteristic subsynaptic cistern (see **D**) is a cholinergic C-bouton, which serves as a motoneuron marker. The *upper green* terminal with flat vesicles (see **E**), termed F-type, contains glycine and GABA (Ornung et al. 1998). (**B**) An electron micrograph in pseudocolour showing how a F-type terminal marked in *yellow* is removed from the surface of an axotomized motoneuron (*red–brown*). Note the glial process (*blue*) that invades the synaptic cleft. (**F**) and (**G**) Histograms showing synaptic covering and number of synapses/100 μm of surface length in unlesioned and axotomized motoneurons at one week post-operatively in Wt compared to mice lacking functional MHC class I ($\beta_2m^{-/-}$). F-type (F), S-type (S) and C-boutons (**C**) were analysed. Note the preferential removal of F-type terminals in the MHC class I-deficient animals. Modified from Oliveira et al. (2004) with permission, copyright (2007) National Academy of Sciences, USA

synaptic vesicles (Fig. 14.2). Ultrastructural studies of glutaraldehyde-fixed spinal cords have shown that such terminals contain inhibitory amino acid neurotransmitters (Ornung et al. 1994, 1996, 1998). Fewer differences were seen when analyzing terminals containing only spherical vesicles (Fig. 14.2), which are likely to be glutamatergic (Ornung et al. 1998). This would imply that the balance between inhibition and excitation is shifted in axotomized $\beta_2m^{-/-}$ motoneurons when they are compared to wild-type mice (Oliveira et al. 2004), which may affect their recovery. Supporting this hypothesis, axotomized $\beta_2m^{-/-}$ motoneurons displayed a less successful axonal regeneration (Oliveira et al. 2004), possibly due to a less favourable environment in the spinal cord. Another explanation could be an impaired interaction between the distal parts of axons and non-neuronal cells in the nerve stump, e.g. Schwann cells, fibroblasts or immune cells. No clear difference in the number, covering or composition of synapses on motoneuron somata was detected between uninjured wild-type and MHC class I-deficient animals. However, this only indicates that the total number of synapses on the soma is similar, but it still has to be determined whether these synapses originate from the correct presynaptic neurons and if they display functional abnormalities.

In conclusion, the effects of MHC class I signalling appear to vary between different neuronal systems. In some cases, MHC class I molecules are important for the selective retraction of redundant synaptic terminals during pruning, whereas in other cases they appear to be involved in maintaining essential inputs during 'synaptic stripping'. In the immune system, depending on the target cell MHC class I molecules interact with a repertoire of both activating and inhibitory receptors and induce completely different cellular responses, as summarized in Fig. 14.3. This could also explain the diverse functions of MHC class I molecules that are seen in the CNS.

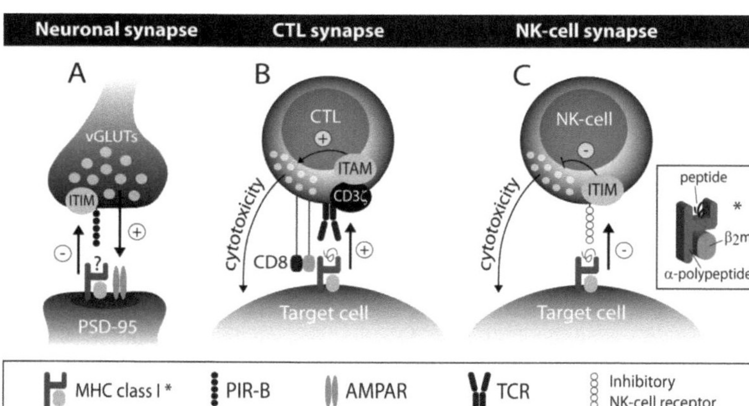

Fig. 14.3 A schematic figure showing MHC class I-dependent synapses in the nervous system and the immune system. (**A**) An excitatory presynaptic terminal (*top*) is depicted in connection to a dendritic spine (*bottom*). Normal synaptic transmission is conveyed through vesicular

14.4 Putative Neuronal MHC Class I Receptors

After the first exciting reports of neuronal functions for MHC class I molecules, the search for putative receptors for MHC class I molecules in the CNS started. Naturally, MHC class I-restricted immune cells can enter the CNS as discussed, but there are also indications that resident non-immunological CNS cells express MHC class I receptors. When drawing parallels to the immune system, the T-cell (i.e. T lymphocyte) receptor (TCR) is an obvious candidate. Studies in CD3ζ deficient mice (CD3$\zeta^{-/-}$), which lack the intracellular-signalling component of the TCR, showed a similar phenotype as MHC class I-deficient mice, when comparing the developing retinogeniculate projections and synaptic plasticity in the hippocampus (Huh et al. 2000). CD3ζ is the signalling subunit of the TCR-complex and CD3ζ mRNA is expressed in the developing brain in a pattern similar to MHC class I mRNA (Corriveau et al. 1998). When the TCR is activated through interaction with a proper MHC class I ligand, transmembrane signalling is conveyed by the CD3ζ-associated immune tyrosine-activating motif (ITAM). Because of negatively charged amino acid residues, it is unlikely that CD3 protein is expressed at the cell surface without a receptor with matching positively charged residues (Janeway and Travers 1997). However, despite reports that TCR subunit β mRNA is expressed in the developing and adult brain (Syken and Shatz 2003), it is unclear whether the TCR-complex is present since neither TCR nor CD3 immunoreactivity has been demonstrated in neurons. A possible explanation for the synaptic phenotype seen in CD3$\zeta^{-/-}$ mice could either be that it is paired with a so far

Fig. 14.3 (continued) glutamate release, which activates, e.g. postsynaptic AMPA-receptors (AMPAR), leading to excitation (+) in the postsynaptic neuron. Neuronal MHC class I molecules are co-localized in vitro with postsynaptic density 95 (PSD-95) in association with glutamatergic presynaptic terminals (vGLUTs = vesicular glutamate transporters). It is not known whether synaptic MHC class I molecules present processed peptides in vivo or not. The first MHC class I receptor reported in neuronal cells was PIR-B, which is localized to presynaptic terminals. PIR-B has immune tyrosine inhibitory motifs (ITIMs) and thus conveys an inhibitory signal (–) in the opposite direction across the synaptic cleft. It may possibly lead to an inhibition or the retraction of the presynaptic terminal. At the cytotoxic T-lymphocyte (CTL) synapse (**B**), MHC class I molecules on target cells present a peptide fragment to a CTL, leading to an activating signal (+) if the TCR recognized the peptide fragment as foreign. The CD8 molecule serves as co-receptor, which binds to the MHC class I α-polypeptide, enabling an MHC class I-TCR interaction. The TCR is associated with the CD3 complex, which contains an immune tyrosine-activating motif (ITAM) on the CD3ζ subunit. Upon activation, the CTL kills the target cell, e.g. by releasing cytotoxic granulae, thus mediating an adaptive immune response. At the natural killer (NK) cell synapse (**C**), MHC class I molecules on the target cell interacts with a display of NK-cell receptors, in this case of inhibitory type. If MHC class I molecules are expressed on the cell surface to a normal extent, an inhibitory signal is conveyed across the synapse through ITIM-associated NK-cell receptors, thereby preventing degranulation. If, however, the levels of MHC class I molecules are decreased on the cell surface, the NK cell will release its cytotoxic granules and kill the target cell, thus mediating innate immunity

uncharacterized MHC class I receptor or that it results from general immune deficiency. So far TCR-deficient animals have not been studied with regard to their ability to remodel synapses.

Other known binding partners for MHC class I molecules in the vertebrate immune system include PIR-B, LILR, killer cell immunoglobulin-like receptors (KIR) and Ly49 for class Ia and CD94/NKG2 and $\gamma\delta$-TCR for class Ib MHC class I (Janeway and Travers 1997, Held and Mariuzza 2008). PIR-B is the first documented presynaptic MHC class I receptor candidate that is expressed in the CNS and exhibits a mutant phenotype matching the suggested function of neuronal MHC class I molecules. PIR-B is expressed in the cerebral cortex, hippocampus, olfactory bulb and cerebellum and has been shown in vitro to associate with presynaptic markers, such as synaptophysin and synapsin (Syken et al. 2006). In the visual system PIR-B appears to regulate and limit ocular dominance plasticity, as indicated by mononuclear enucleation experiments in PIR-B deficient mice (Syken et al. 2006). The PIR-B molecule contains cytoplasmic immune tyrosine inhibitory motifs (ITIMs). This means that there could be a two-way signalling across the synapse, with an excitatory signal that is mediated postsynaptically by AMPA-receptors and an inhibitory signal that is mediated presynaptically by PIR-B (Fig. 14.3). In this way, retraction of inappropriate synapses in the developing brain would be mediated by an inhibitory signal that is triggered by an MHC class I-PIR-B interaction across the synapse.

Another MHC class I receptor candidate was reported in a recent publication showing the expression of a class of Ly-49 receptors in cortical neurons in vitro by immunoreactivity and in vivo by in situ hybridization (Zohar et al. 2008). Ly-49 is a receptor family that mediates both inhibitory and activating effects in the immune system. Ly-49 is expressed on the pre-, as well as the postsynaptic side on both axons and dendrites. Experimental blocking of Ly-49 with antibodies in neuronal cultures results in an increased number of presynaptic puncta and suppressed neurite growth. This suggests that Ly-49 receptor somehow restricts synapse formation and promotes neurite branching through its interaction with MHC class I molecules at the synaptic and neuritic level.

mRNAs encoding other MHC class I receptors that are found in the CNS include the KIR-like receptor (KIRL), which is expressed in plastic brain regions such as the olfactory bulb, hippocampus and the rostroventral migratory stream (Bryceson et al. 2005). However, the function of this receptor in the CNS is currently unknown.

Finally, another interesting MHC class I receptor-associated molecule with known expression in the CNS is DAP12, also known as killer cell activating receptor-associated protein (KARAP) (Roumier et al. 2004). DAP12/KARAP is structurally similar to CD3ζ and serves as an adaptor molecule for several NK-cell-associated MHC class I receptors, such as NKG2D and non-MHC class I receptors like the triggering receptor expressed on myeloid cells (TREM2) (Takahashi et al. 2005). DAP12/KARAP is expressed by microglia during development (Roumier et al. 2004) and one paper reports positive

Chapter 16
Ephrins and Eph Receptor Tyrosine Kinases in Synapse Formation

Catherine E. Krull and Daniel J. Liebl

Abstract Here, we discuss what is known about the function of ephrins and Eph receptors in synapse formation in the peripheral nervous system and central nervous system. Ephrins have been shown to be present and functional at synapses at the neuromuscular junction and in the brain; evidence at the neuromuscular junction implicates ephrins in the topographic mapping of synapses on certain muscles. Also, certain Eph receptors also function in synapse formation but less information is known about them including their distribution and function. In addition, future directions are defined. Together, these data implicate ephrins, as well as Ephs, strongly in synaptic development.

Keywords Synaptogenesis · Eph family · PNS · CNS

16.1 Introduction

Eph receptor tyrosine kinases (RTKs) and their ligands, the ephrins, were first recognized for their roles in axon pathfinding, cell migration, and angiogenesis (Krull et al. 1997, Wang and Anderson 1997, Smith et al. 1997, Adams et al. 1999, Helmbacher et al. 2000, Eberhart et al. 2000, Eberhart et al. 2002, Kullander et al. 2001, Eberhart et al. 2004, Sahin et al. 2005). However, these molecules are now thought to play central roles in the formation of synapses as well (Donoghue et al. 1996, Feng et al. 2000, Murai et al. 2003, Grunwald et al. 2004, Rodenas-Ruano et al. 2006, Bourgin et al. 2007). Here, in addition to providing pertinent background, recent data will be reviewed that define a key function for these molecules in the formation of synapses in the peripheral nervous system (PNS) and the central nervous system (CNS). Information on other aspects of Eph–ephrin function can be found in several reviews (Pasquale

C.E. Krull (✉)
Biologic and Materials Sciences, University of Michigan, 5211 Dental, 1011 N. University Ave., Ann Arbor, MI 48109-1078, USA
e-mail: krullc@umich.edu

M. Hortsch, H. Umemori (eds.), *The Sticky Synapse*,
DOI 10.1007/978-0-387-92708-4_16,

2008, Arvanitis and Davy 2008, Himanen et al. 2007, Kullander and Klein 2002, Murai and Pasquale 2004, Wilkinson 2001).

16.2 The Eph Family: Description

Several members of the Eph family of receptor tyrosine kinases (RTKs) and their membrane-associated ligands, the ephrins, have been implicated in patterning in the developing nervous system and angiogenesis (Wilkinson 2001, Table 16.1). Based on their ligand-binding specificities, the Eph RTKs are subdivided into two subclasses, EphAs, which interact primarily with GPI-linked ephrins (i.e., ephrin-As), and EphBs, which bind to and activate transmembrane ephrins (i.e., ephrin-Bs) (Gale et al. 1996, but see Himanen et al. 2004). The extracellular region of the Eph receptors includes an N-terminal globular domain that binds ephrin, a cysteine-rich domain, and two fibronectin type III repeats (Yamaguchi and Pasquale 2004). The tyrosine kinase domain is

Table 16.1 Known vertebrate Eph family members

	Members of the Eph family
Eph receptor tyrosine kinases	Ephrin ligands
EphA1	Ephrin-A1
EphA2	Ephrin-A2
EphA3	*Ephrin-A3*
EphA4	Ephrin-A4
EphA5	*Ephrin-A5*
EphA6	Ephrin-A6
EphA7	
EphA8	
EphA9	
EphA10	
EphB1	*Ephrin-B1*
EphB2	*Ephrin-B2*
EphB3	*Ephrin-B3*
EphB4	
EphB5	
EphB6	

Originally, the Eph family was divided into two subgroups: EphA receptor tyrosine kinases (RTKs) bound primarily ephrin-As; EphB RTKs bound ephrin-Bs (Eph Nomenclature Committee). However, there is a significant amount of cross talk between the two subgroups in biological events, so that EphA RTKs interact with ephrin-Bs, and EphB RTKs interact with ephrin-As. Investigators are urged to examine whether an Eph or ephrin from a different subgroup is involved in the biological event being examined. Molecule names emphasized by an outline font format are implicated in synapse formation.

located below a juxtamembrane region, followed by a sterile-α-motif (SAM) and a type II PSD-95/Disc large/ZO-1 (PDZ)-binding motif at the carboxyl terminus (for review, Kullander and Klein 2002, Fig. 16.1). Importantly, Eph receptors can undergo homo- or heterodimerization (Freywald et al. 2002), which is regulated directly by the extracellular cysteine-rich domain (Lackmann et al. 1998) and the SAM motif (Stapleton et al. 1999, Thanos et al. 1999) or indirectly through PDZ protein interactions (Fanning and Anderson 1999).

Localization of ephrins to the cell surface is critical for their ability to activate their receptors (Davis et al. 1994). Because of their membrane association, Eph RTK–ephrin interactions require cell–cell contact that initiates signals from both the ligand- and the receptor-expressing cells. The requirement for bidirectional signaling has been demonstrated best for ephrin-Bs and EphB2 (Henkemeyer et al. 1996, Cowan and Henkemeyer. 2001, Lu et al. 2001). Studying ephrin-A signaling has proved more difficult, as they do not have direct contact with the cytoplasm because of their GPI anchor. However, previous studies suggest that signaling is mediated by the localization of

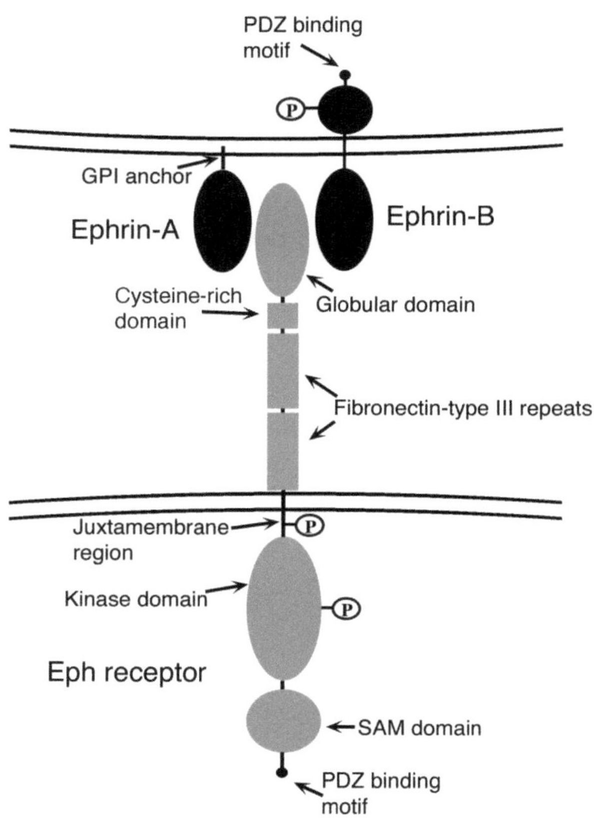

Fig. 16.1 General protein domain organization of Ephrins and Eph receptor tyrosine kinases. A schematic diagram that shows ephrins (*top cell*) interacting with a canonical Eph receptor tyrosine kinase (*bottom cell*), including important domains. Note that in most systems, Eph receptors and ephrins are expressed in the same cell; the function of this co-localization is not well understood

GPI-anchored ephrins in cholesterol-rich microdomains in the cell membrane (Davy et al. 1999, Sargiacomo et al. 1993, Shenoy-Scaria et al. 1994). How signaling works exactly to alter an axon's response from adhesion to repulsion is not well understood, but is thought to involve trans-endocytosis and enzymatic cleavage (Hattori et al. 2000, Zimmer et al. 2003, Marston et al. 2003, Irie et al. 2005, Lin et al. 2008).

Several mouse mutants have been generated that lack particular Ephs or ephrins (Kullander et al. 2001, Egea et al. 2005, Henkemeyer et al. 1996, Feng et al. 2000, Grunwald et al. 2004). Mutant analyses have shown that certain Eph receptors and ephrins are required for the formation of axonal commissures, patterning of the retinocollicular system, synapse formation, and synaptic plasticity (Henkemeyer et al. 1996, Cheng et al. 1995, Feng et al. 2000, Kullander et al. 2001, McLaughlin et al. 2003, Grunwald et al. 2004, Mendes et al. 2006, Rodenas-Ruano et al. 2006).

There are additional factors, including Ret and GDNF, that cooperate with Ephs and ephrins for axon pathfinding. However, exactly how they function together remains elusive (Kramer et al. 2006). Furthermore, it is not known whether Ephs/ephrins work with other axon guidance molecules, such as semaphorins/neuropilins and plexins (Chauvet et al. 2007, Huber et al. 2005, Giger et al. 2000); it is becoming more and more pressing to know how the growth cone of a neuron responds to multiple guidance cues as it navigates to its targets and forms synapses.

16.3 Ephrins Control Neuromuscular Topography and Synapse Formation in the PNS

Before any molecule is implicated in synapse formation, it must be localized to either presumptive synapses or their targets (i.e., muscles or other neurons) (Murai and Pasquale 2004). Localization to synapses or to targets is determined by many methods including antibody staining (Garcia et al. 2007, Hu et al. 2007, Kania et al. 2000). Needless to say, in situ hybridization usually provides limited information, as the transcript of interest localizes to the nucleus and this approach cannot provide information regarding the localization of the protein to the growth cone or developing synapses.

In the PNS, using subtractive hybridization, Donoghue and colleagues found that ephrin-A5 (AL-1) is expressed by rostral skeletal muscles and inhibits the outgrowth from caudal (lumbar) neurons in a mouse model system (Donoghue et al. 1996). Wang, Chadaram and colleagues explored this finding further, revealing that spinal cord neurites maintain their preference for muscle membranes derived from their corresponding axial level using *in vitro* stripe assays (Wang et al. 2001; Chadaram et al. 2007). For the first time, these authors showed that topographic outgrowth of motor neurons occurs within a single motor pool onto a single target muscle *in vitro*. Overexpression of ephrin-A5 degrades the topographic specificity of the gluteus maximus muscle (Lampa et al. 2004). Feng and his colleagues established that ephrin-A5 was required in mouse for selective synaptogenesis in the PNS *in vivo* (Feng et al. 2000). Their results showed that motor axons and their target muscles possess complementary molecular labels that vary with rostrocaudal position and that, therefore, bias synaptic connections in favor of matched partners (i.e., motor axons, muscle). Feng and colleagues first demonstrated this pattern by comparing ephrin-A RNAs expression between rostral versus caudal muscles, using RT-PCR and Northern blot analysis. They also assayed the cellular source of ephrin-As using fusion protein staining. While this approach does not identify which ephrin protein is expressed, it did demonstrate whether ephrin-A proteins were expressed collectively by myotubes. Importantly, no ephrin-Bs were expressed by adult PNS synapses (Yamaguchi and Pasquale 2004), indicating that ephrin-As are present selectively at the right time and place to influence synapse formation. Feng and colleagues then evaluated the differential sensitivity of motor neurons from rostral versus caudal levels to ephrin-A5 expressed by fibroblasts. Their results showed that ephrin-A5 inhibited growth from both rostral and caudal neurons, but that caudal neurons were on average more sensitive to ephrin-A5. Also, shallow gradients of ephrin-A5 were formed and the outgrowth of rostral versus caudal neurons was measured. Here too, the growth of rostral versus caudal neurons was inhibited, but caudal neurons were better able to detect differences in ephrin concentration. The authors then generated mice where ephrin-A5 was overexpressed in certain muscle fibers. The rostrocaudal axis of the gluteus muscle pool was dramatically degraded in mice with excess ephrin-A5 compared to wild-type mice. Abolishing the gradient of ephrin-A5 on muscle led to disruptions in the neuromuscular map, demonstrating that ephrin-A5 can affect specific synaptic connections. To determine whether endogenous ephrins were required for this process, the authors examined diaphragm muscles from mice that lacked ephrin-A2 and ephrin-A5. Here, topographic mapping of synapses onto muscle was also defective. Collectively, these data indicate that muscle-derived ephrins act in selective synaptic connections between nerve and muscle partners (Feng et al. 2000). How ephrins act specifically to regulate nerve–muscle synapses is obviously the next question.

As for Eph receptors, EphA4 has been knocked out by two research groups (Helmbacher et al. 2000, Dottori et al. 1998). One set of mice that lack EphA4 has defects in the dorsoventral choice made by motor axons (Helmbacher et al. 2000). However, it is unclear whether synapse formation is also affected in these mice. Moreover, the general organization of muscle appears normal in these mice. Mice that lack other EphAs have also been reported, but it is not known whether there are defects in neuromuscular synaptogenesis (Irie et al. 2008, Shintani et al. 2006).

16.4 Ephs and Ephrins in Synapse Formation in the CNS

Although there is sufficient evidence that ephrins act during synapse formation in the PNS, there is some controversy regarding their activities in the CNS. The controversy stems from whether Ephs and ephrins are involved in synapse formation or synaptic function and which receptor or ligand acts in specific cell types.

Grunwald and colleagues recently examined the localization of ephrin-B2 and its function in synaptic plasticity in the hippocampus (Grunwald et al. 2001, 2004). Using in situ hybridization and antibody staining, Grunwald and colleagues found that ephrin-B2 and ephrin-B3 are localized postsynaptically to CA3–CA1 synapses in the hippocampus. To directly visualize ephrin-B2 at CA1–CA3 synapses, these authors then performed electron microscopy using a specific antibody against ephrin-B2. Collectively, these results showed that ephrin-B2 protein localizes postsynaptically. Rodenas-Ruano and colleagues also demonstrated that ephrin-B3 is localized to both axons and dendrites and is located to the postsynaptic membrane in CA3–CA1 synapses (Rodenas-Ruano et al. 2006). One recent report identifies ephrin-B2 protein in the cell bodies of neurons in the adult mouse brain, suggesting it may have roles in the formation of axonal–somatic inhibitory synapses (Migani et al. 2007).

Subsequently, Grunwald and colleagues conditionally removed ephrin-B2 from the postnatal forebrain and examined synapse formation using electron microscopy (Grunwald et al. 2004). Interestingly, they found no difference in synapse number or morphology compared to control mice. In addition, the authors examined hippocampal neurons from mice that lacked EphA4, a high-affinity receptor for ephrin-B2, and control mice and found no difference in synaptogenesis. Together, these findings indicate that ephrin-B2 is not required for synapse formation in the hippocampus.

Grunwald and colleagues then analyzed basal synaptic transmission in both mutant and control embryos (Grunwald et al. 2004). Again, no difference in the extracellular recordings of the CA3–CA1 pathway in the hippocampus of mutant and control mice was apparent. However, a strong reduction in LTP was observed after the application of theta-burst stimulation, an experimental paradigm that is used to induce long-lasting LTP. Together, these data show that ephrin-B2 is required for hippocampal LTP. In addition, using similar approaches, these authors and others examined the contribution of ephrin-B3 to LTP and found that it too was required for early stages of LTP (Grunwald et al. 2004, Rodenas-Ruano et al. 2006). Moreover, LTD was also affected in these mice that lacked ephrin-B2 and ephrin-B3, compared to controls. Finally, due to its known role as a high-affinity receptor for ephrin-B2 and ephrin-B3, EphA4 mutant mice were examined for their role in LTP (Gale et al. 1996). Based on localization and functional analyses, EphA4 appears to be a critical interaction partner of ephrin-B2 and ephrin-B3 in the early phase of LTP (Grunwald et al. 2001, Grunwald et al. 2004). Interestingly, mice that lack the signaling portion of ephrin-B3, but retain their ability to activate Eph receptors, do not exhibit a similar reduction in LTP, suggesting that these mechanisms may involve Eph receptor signaling and/or multiple ephrins (Rodenas-Ruano et al. 2006).

Using three different *in vivo* approaches, Lim and colleagues examined the effect of the presynaptic ephrin-B reverse intracellular signaling on synapse formation in the retinotectal system (Lim et al. 2008). First, soluble fusion proteins (EphB2-Fc and ephrin-B1-Fc) were infused into the brain, followed by assaying synaptic morphology. Second, plasmid constructs that encoded ephrin-B1 or a mutated form of ephrin-B1 were electroporated into the retina, followed by an analysis of presynaptic development. Third, plasmids that encoded a mutated form of EphB2 were electroporated into the tectum, to interfere with endogenous EphB–ephrin-B signaling, and the effects on retinotectal synapses were examined. Their results show that activating ephrin-B1 signaling in retinal ganglion cells promotes the formation of presynaptic specializations without affecting overall axon growth. In contrast, reducing reverse signaling slows this process. In addition, these authors demonstrated that EphB–ephrin-B interactions affect synaptic function, including synaptic plasticity.

There is other tantalizing data that suggest that EphB receptors participate in synaptogenesis between neurons. Henkemeyer and colleagues have reported that EphB2 acts postsynaptically by modulating glutamate receptors (Henkemeyer et al. 2003). There is additional data that support the postsynaptic function of EphB2. In cultured hippocampal neurons, EphB2 interacts directly with NMDA receptors and thus modulates the influx of calcium into postsynaptic neurons. In addition, postsynaptic EphB2 receptors interact with the PDZ-domain protein GRIP at the mossy fiber synapses and mediate long-term potentiation (LTP). EphB receptors are known to co-distribute with a nicotinic acetylcholine receptor subtype (a7-nAChR) and regulate nicotinic downstream signaling in isolated parasympathetic ganglia neurons *in vitro* (Liu et al. 2008). Signaling is enhanced by the activation of EphB receptors interacting with a7-nAChRs on the neurons. However, interactions between EphB receptors/ephrin-B1 and a7-nAChRs are not direct and it is not known whether EphB/ephrin-B1 may also have a role in cell death in this system.

Henkemeyer, Hoogenraad and their colleagues have shown that EphB (EphB1, EphB2, EphB3) receptors are necessary for the formation of dendritic spines, small protrusions on the dendrite's surface that receive most of excitatory synaptic input (Henkemeyer et al. 2003, Hoogenraad et al. 2005). Their data strongly indicate that EphB receptors, to varying degrees, are involved in the morphogenesis of dendritic spines and synapse formation in the mouse hippocampus. Previously, EphB receptors were shown to be likely molecules that transduced extracellular signals into the dendrite to induce spine morphogenesis (Ethell et al. 2001). Indeed, EphBs are found in the postsynaptic structures of dendrites (Torres et al. 1998, Buchert et al. 1999) and ephrin-Bs have been localized to axons (Henkemeyer et al. 1996). At least four members of the

Eph receptors (EphB1, EphB2, EphB3, EphB4) are expressed in the hippocampus (Liebl et al. 2003). Furthermore, presynaptic specializations are induced in HEK-293 cells that overexpress EphB2 in co-cultured cortical neurons (Kayser et al. 2006). In addition, there is a link between ephrin-B/EphB signaling and Rho-GEF kalirin-7 in spine morphogenesis in cultured hippocampal neurons (Penzes et al. 2003, see Table 16.1).

EphB receptors also promote the formation of spines by activating Rho family GEFs, including kalirin-7 (see above) and Tiam1 (Penzes et al. 2003, Tolias et al. 2007). Rho GTPases, including Rac1, RhoA, and Cdc42, are key regulators of the actin cytoskeleton and are known to play critical roles in spine morphogenesis. Rho family GTPases act as molecular switches, cycling between an active GTP-bound state and an inactive GDP-bound state. The activity of Rho family GTPases is controlled tightly by guanine nucleotide exchange factors (GEFs). Tolias and colleagues determined that the Rac1 GEF Tiam1 mediates EphB receptor-dependent dendritic spine development (Tolias et al. 2007). Activation of EphB receptors by ephrin-B ligands induces the tyrosine phosphorylation of Tiam1 and recruitment of Tiam1 to EphB complexes that contain NMDA-type glutamate receptors. Knockdown of Tiam1 or expression of a dominant-negative Tiam1 blocks signaling, and most importantly, dendritic spine formation induced by ephrin-B1 stimulation. Indeed, Dalva and colleagues demonstrated previously that ephrin-B activation of EphBs leads to the recruitment of NMDA receptors to an EphB-containing complex and influences the number of pre- and postsynaptic specializations (Dalva et al. 2000). More recently, Kayser and colleagues showed that loss of EphBs results in reduced filopodial motility and abnormal synapse formation in cultured neurons and brain slices (Kayser et al. 2008). Rescue experiments during different developmental times show that EphBs are required for synaptogenesis, but only when filopodia are plenty and motile (Kayser et al. 2008).

In contrast, EphA receptors appear to promote spine retraction (Murai et al. 2003). How do EphA receptors perform this task mechanistically? Fu and colleagues recently showed that Cdk5 regulates EphA4-mediated spine retraction through an ephexin1-dependent mechanism (Fu et al. 2007). Blocking Cdk5 activity inhibits ephrin-A1-triggered spine retraction and a reduction in the frequency of mESPCs at hippocampal synapses. Activating EphA4 results in the recruitment of Cdk5 to EphA4, leading to the activation of Cdk5. Then, together, these factors activate ephexin1, a Rho-GEF, leading to spine retraction. In another example, EphA4 seems to regulate dendritic spine remodeling by affecting β1-integrin signaling pathways (Bourgin et al. 2007). The model proposed by Bourgin and colleagues is that EphA4 interferes with integrin signaling pathways that stabilize dendritic spines, thus modulating synaptic interactions with the extracellular milieu.

The situation may be more complex than first appreciated. Murai and colleagues have shown that astrocytes express ephrin-A3, a membrane-bound ligand for EphA4, and that it regulates the morphology of dendritic spines in the hippocampus (Murai et al. 2003). Their data suggest that local activation of

EphA receptors on dendritic spines occurs by astrocytic ephrin-A3. Time-lapse imaging studies showed that dendritic protrusions are highly active (Dailey and Smith, 1996) and the idea has emerged that a subset of motile protrusions are stabilized and transformed into stable spines with cell–cell contacts. Murai and colleagues suggested that local astrocytic contacts may be important for maintaining stable spines and regulate the maturation of these dendritic protrusions. Nishida and Okabe further examined the formation of spines, using pyramidal neurons in hippocampal slices, and tested the role of ephrins and Ephs in this process (Nishida and Okabe 2007). The authors observed frequent contact between astrocytes and dendritic protrusions. Furthermore, the authors found that perturbation of Rac1 signaling in astrocytes to suppress astrocytic motility results in the development of abnormal and thin dendritic protrusions and that blocking Eph and ephrin signaling reduces the lifetime of newly generated dendritic protrusions with astrocytic contacts. Lauterbach and Klein showed that EphB2 receptor-containing vesicles are released by hippocampal neurons to co-cultured glial cells, suggesting that glial cells can trans-endocytose full-length EphB2 from neighboring neurons (Lauterbach and Klein 2006). Together, these results suggest that glial cells express ephrin-A3 as well as EphB2 receptors and thus should be considered as important “players” at the synapse.

16.5 Summary and Future Directions

Ephrins and Eph receptors play a crucial role in synapse formation in the PNS and CNS. Interestingly, the analyses of mouse mutants indicate that many members of this family are required for synapse formation. However, a substantial body of data is still missing:

1. How exactly do ephrins and their Eph receptors work at a mechanistic level to regulate synapse formation?
2. How do ephrins and Eph receptors, expressed by the same cell, function together? Do all receptors have the capacity to work in *cis* and in *trans*?
3. What is the role of trans-endocytosis in the process of synapse formation?
4. Are ephrins involved in dendrite versus axon targeting (i.e., neuronal polarity)?

Answers to these questions await future investigations in this very open and fertile area.

References

Adams RH, Wilkinson GA, Weiss C et al. (1999) Roles of ephrinB ligands and EphB receptors in cardiovascular development: demarcation of arterial/venous domains, vascular morphogenesis, and sprouting angiogenesis. Genes Dev 13:295–306

Arvanitis D and Davy A (2008) Eph/ephrin signaling: networks. Genes Dev 22:416–429
Buchert M, Schneider S, Meskenaite V et al. (1999) The junction-associated protein AF-6 interacts and clusters with specific Eph receptor tyrosine kinases at specialized sites of cell-cell contact in the brain. J Cell Biol 144:361–371
Bourgin C, Murai KK, Richter M et al. (2007) The EphA4 receptor regulates dendritic spine remodeling by affecting beta1-integrin signaling pathways. J Cell Biol 178:1295–1307
Chadaram SR, Laskowski MB and Madison RD (2007) Topographic specificity within membranes of a single muscle detected in vitro. J Neurosci 27:13938–13948
Chauvet S, Cohen S, Yoshida Y et al. (2007) Gating of Sema3E/PlexinD1 signaling by neuropilin-1 switches axonal repulsion to attraction during brain development. Neuron 56:807–822
Cheng HJ, Nakamoto M, Bergemann AD et al. (1995) Complementary gradients in expression and binding of ELF-1 and Mek4 in development of the topographic retinotectal projection map. Cell 82:371–381
Cowan CA and Henkemeyer M (2001) The SH2/SH3 adaptor Grb4 transduces B-ephrin reverse signals. Nature 413:174–179
Dailey ME and Smith SJ (1996) The dynamics of dendritic structure in developing hippocampal slices. J Neurosci 16:2983–2994
Dalva MB, Takasu MA, Lin MZ et al. (2000) EphB receptors interact with NMDA receptors and regulate excitatory synapse formation. Cell 103:945–956
Davis S, Gale NW, Aldrich TH et al. (1994) Ligands for the Eph-related receptor tyrosine kinases that require membrane attachment or clustering for activity. Science 266:816–819
Davy A, Gale NW, Murray EW et al. (1999) Compartmentalized signaling by GPI-anchored ephrin-A5 requires the Fyn tyrosine kinase to regulate cellular adhesion. Genes Dev 13:3125–3135
Donoghue MJ, Merlie J and Sanes JR (1996) The Eph kinase ligand AL-1 is expressed by rostral muscles and inhibits outgrowth from caudal neurons. Mol Cell Neurosci 8:185–198
Dottori M, Hartley L, Galea M et al. (1998) EphA4 (Sek1) receptor tyrosine kinase is required for the development of the corticospinal tract. PNAS 95:13248–13253
Eberhart J, Swartz M, Koblar SA et al. (2000) Expression of EphA4, ephrin-A2 and ephrin-A5 during axon outgrowth to the hindlimb indicates potential roles in pathfinding. Dev Neurosci 22:237–250
Eberhart J, Barr J, O'Connell S et al. (2004) Ephrin-A5 exerts positive or inhibitory effects on distinct subsets of EphA4-positive motor neurons. J Neurosci 24:1070–1078
Eberhart J, Swartz ME, Koblar SA et al. (2002) EphA4 constitutes a population-specific guidance cue for motor neurons. Dev Biol 247:89–101
Egea J, Nissen UV, Dufour A et al. (2005) Regulation of EphA4 kinase activity is required for a subset of axon guidance decisions suggesting a key role for receptor clustering in Eph function. Neuron 47:515–528
Ethell IM, Irie F, Kalo MS et al. (2001) EphB/syndecan-2 signaling in dendritic spine morphogenesis. Neuron 31:1001–1013
Fanning AS and Anderson JM (1999) PDZ domains: fundamental building blocks in the organization of protein complexes at the plasma membrane. J Clin Invest 103:767–772
Feng G, Laskowski MB, Feldheim DA et al. (2000) Roles for ephrins in positionally selective synaptogenesis between motor neurons and muscle fibers. Neuron 25:295–306
Freywald A, Sharfe N and Roifman CM (2002) The kinase-null EphB6 receptor undergoes transphorylation in a complex with EphB1. J Biol Chem 277:3823–3828
Fu W-Y, Chen Y, Sahin M et al. (2007) Cdk5 regulates EphA4-mediated dendritic spine retraction through an ephexin1-dependent mechanism. Nat Neurosci 10:67–76
Gale NW, Flenniken A, Compton DC et al. (1996) Elk-3, a novel transmembrane ligand for the Eph family of receptor tyrosine kinases, expressed in embryonic floor plate, roof plate, and hindbrain segments. Oncogene 13:1343–1352

Garcia SM, Casanueva MO, Silva MC et al. (2007) Neuronal signaling modulates protein homeostasis in Caenorhabditis elegans post-synaptic muscle cells. Genes Dev 21:3006–3016
Giger RJ, Cloutier JF, Sahay A et al. (2000) Neuropilin-2 is required in vivo for selective axon guidance responses to secreted semaphorins. Neuron 25:29–41
Grunwald IC, Korte M, Wolfer D et al. (2001) Kinase-independent requirement of EphB2 receptors in hippocampal synaptic plasticity. Neuron 32:1027–1040
Grunwald IC, Korte M, Adelmann G et al. (2004) Hippocampal plasticity requires postsynaptic ephrin-Bs. Nat Neurosci 7:33–40
Hattori M, Osterfield M and Flanagan JG (2000) Regulated cleavage of a contact-mediated axon repellent. Science 289:1360–1365
Helmbacher F, Schneider-Maunoury S, Topilko P et al. (2000) Targeting of the EphA4 tyrosine kinase receptor affects dorsal/ventral pathfinding of limb motor axons. Development 127:3313–3324
Henkemeyer M, Orioli D, Henderson JT et al. (1996) Nuk controls pathfinding of commissural axons in the mammalian central nervous system. Cell 86:35–46
Henkemeyer M, Itkis OS, Ngo M et al. (2003) Multiple EphB receptor tyrosine kinases shape dendritic spines in the hippocampus. J Cell Biol 163:1313–1326
Himanen JP, Saha N and Nikolov DB (2007) Cell-cell signaling via Eph receptors and ephrins. Curr Opin Cell Biol 19:534–542
Himanen JP, Chumley MJ, Lackmann M et al. (2004) Repelling class discrimination: ephrin-A5 binds to and activates EphB2 receptor signaling. Nat Neurosci 7:501–509
Hoogenraad CC, Milstein AD, Ethell IM et al. (2005) GRIP1 controls dendrite morphogenesis by regulating EphB receptor trafficking. Nat Neurosci 8:906–915
Hu JY, Chen Y and Schacher S (2007) Multifunctional role of protein kinase C in regulating the formation and maturation of specific synapses. J Neurosci 27:11712–11724
Huber AB, Kania A, Tran TS et al. (2005) Distinct roles for secreted semaphoring signaling in spinal motor axon guidance. Neuron 48:949–964
Irie F, Okuno M, Pasquale EB et al. (2005) EphrinB-EphB signaling regulates clathrin-mediated endocytosis through tyrosine phosphorylation of synaptojanin 1. Nat Cell Biol 7:454–456
Irie F, Okumo M, Matsumoto K et al. (2008) Heparan sulfate regulates ephrin-A3/EphA receptor signaling. PNAS 34:12307–12312
Kayser MS, McClelland AC, Hughes EG et al. (2006) Intracellular and trans-synaptic regulation of glutamatergic synaptogenesis by EphB receptors. J Neurosci 26:12152–12164.
Kayser MS, Nolt MJ and Dalva MB (2008) EphB receptors couple dendritic filopodia motility to synapse formation. Neuron 59:56–69
Kania A, Johnson RL and Jessell TM (2000) Coordinate roles for LIM homeobox genes in directing dorsoventral trajectory of motor axons in the vertebrate limb. Cell 102:161–173
Kramer ER, Knott L, Su F et al. (2006) Cooperation between GDNF/Ret and ephrinA/EphA4 signals for motor axon pathway selection in the limb. Neuron 50:35–47
Krull CE, Lansford R, Gale NW et al. (1997) Interaction of Eph-related receptors and ligands confer rostrocaudal pattern to trunk neural crest migration. Curr Biol 7:571–580
Kullander K and Klein R (2002) Mechanisms and functions of Eph and ephrin signaling. Nat Rev Mol Cell Biol 3:475–486
Kullander K, Mather NK, Diella F et al. (2001) Kinase-dependent and kinase-independent functions of EphA4 receptors in major axon tract formation in vivo. Neuron 29:73–84
Lackmann M, Oates AC, Dottori M et al. (1998) Distinct subdomains of the EphA3 receptor mediate ligand binding and receptor dimerization. J Biol Chem 273:20228–20237
Lampa SJ, Potluri S, Norton AS et al. (2004) Ephrin-A5 overexpression degrades topographic specificity in the mouse gluteus maximus muscle. Dev Brain Res 153:271–274
Lauterbach J and Klein R (2006) Release of full-length EphB2 receptors from hippocampal neurons to cocultured glial cells. J Neurosci 26:11575–11581

Liebl DJ, Morris CJ, Henkemeyer M et al. (2003) mRNA expression of ephrins and Eph receptor tyrosine kinases in the neonatal and adult mouse central nervous system. J Neurosci Res 71:7–22
Lim BK, Matsuda N and Poo MM (2008) Ephrin-B reverse signaling promotes structural and functional synaptic maturation in vivo. Nat Neurosci 11:160–169
Lin KT, Sloniowski S, Ethell DW et al. (2008) Ephrin-B2 induced cleavage of EphB2 receptor is mediated by matrix metalloproteinases to trigger cell repulsion. J Biol Chem. (Available online)
Liu Z, Conroy WG, Stawicki TM et al. (2008) EphB receptors co-distribute with a nicotinic receptor subtype and regulate nicotinic downstream signaling in neurons. Mol Cell Neurosci 38:236–244
Lu Q, Sun EE, Klein RS et al. (2001) Ephrin-B reverse signaling is mediated by a novel PDZ-RGS protein and selectively inhibits G protein-coupled chemoattraction. Cell 105:69–79
Marston DJ, Dickinson S and Nobes CD (2003) Rac-dependent trans-endocytosis of ephrinBs regulates Eph-ephrin contact repulsion. Nat Cell Biol 5:879–888
McLaughlin T, Hindges R, Yates PA et al. (2003) Bifunctional action of ephrin-B1 as a repellent and attractant to control bidirectional branch extension in dorsal-ventral retinotopic mapping. Development 130:2407–2418
Mendes SW, Henkemeyer M and Liebl DJ (2006) Multiple Eph receptors and B-class ephrins regulate midline crossing of corpus callosum fibers in the developing mouse forebrain. J Neurosci 26:882–892
Migani P, Bartlett C, Dunlop S et al. (2007) Ephrin-B2 immunoreactivity distribution in the adult mouse brain. Brain Res 1182:60–72
Murai KK, Nguyen LN, Irie F et al. (2003) Control of hippocampal dendritic spine morphology through ephrin-A3/EphA4 signaling. Nat Neurosci 6:153–160
Murai KK and Pasquale EB (2004) Eph receptors, ephrins and synaptic function. The Neuroscientist 10:304–314
Nishida H and Okabe S (2007) Direct astrocytic contacts regulate local maturation of dendritic spines. J Neurosci 27:331–340
Pasquale EB (2008) Eph-ephrin bidirectional signaling in physiology and disease. Cell 133:38–52
Penzes P, Beeser A, Chernoff J et al. (2003) Rapid induction of dendritic spine morphogenesis by trans-synaptic ephrinB-EphB receptor activation of the Rho-GEF kalirin. Neuron 37:263–274
Rodenas-Ruano A, Perez-Pinzon MA, Green EJ et al. (2006) Distinct roles for ephrin-B3 in the formation and function of hippocampal synapses. Dev Biol 292:34–45
Sahin M, Greer PL, Lin MZ et al. (2005) Eph-dependent tyrosine phosphorylation of ephexin1 modulates growth cone collapse. Neuron 46:191–204
Sargiacomo M, Sudol M, Tang Z et al. (1993) Signal transducing molecules and glycosyl-phosphatidylinositol-linked proteins form a caveolin-rich insoluble complex in MDCK cells. J Cell Biol 122:789–807
Shenoy-Scaria AM, Dietzen DJ, Kwong J et al. (1994) Cysteine3 of Src family protein tyrosine kinase determines palmitoylation and localization in caveolae. J Cell Biol 126:353–363
Shintani T, Ihara M, Sakuta H et al. (2006) Eph receptors are negatively controlled by protein tyrosine phosphatase receptor type O. Nat Neurosci 9:761–769
Smith A, Robinson V, Patel K et al. (1997) The EphA4 and EphB1 receptor tyrosine kinases and ephrin-B2 ligand regulate targeted migration of brachial neural crest cells. Curr Biol 7:561–570
Stapleton D, Balan I, Pawson T et al. (1999) The crystal structure of an Eph receptor SAM domain reveals a mechanism for modular dimerization. Nat Struct Biol 6:44–49
Tolias KF, Bikoff JB, Kane CG et al. (2007) The Rac1 guanine nucleotide exchange factor Tiam1 mediates EphB receptor-dependent dendritic spine development. PNAS 104:7265–7270

Torres R, Firestein BL, Dong H et al. (1998) PDZ proteins bind, cluster and synaptically co-localize with Ephs and their ephrin ligands. Neuron 21:1227–1229
Thanos CD, Faham S, Goodwill KE et al. (1999) Monomeric structure of the human EphB2 sterile alpha motif domain. J Biol Chem 274:37301–37306
Wang HU and Anderson DJ (1997) Eph family transmembrane ligands can mediate repulsive guidance of trunk neural crest migration and motor axon outgrowth. Neuron 18:383–396
Wang H, Chadaram SR, Norton AS et al. (2001) Development of inhibition by ephrin-A5 on outgrowth of embryonic spinal motor neurites. J Neurobiol 47:233–243
Wilkinson DG (2001) Multiple roles of EPH receptors and ephrins in neural development. Nat Rev Neurosci 2:155–164
Yamaguchi Y and Pasquale EB (2004) Eph receptors in the adult brain. Curr Opin Neurobiol 14:288–296
Zimmer M, Palmer A, Kohler J et al. (2003) EphB-ephrinB bi-directional endocytosis terminates adhesion allowing contact mediated repulsion. Nat Cell Biol 5:869–878

Chapter 17
Neurexins and Neuroligins: A Synaptic Code for Neuronal Wiring That Is Implicated in Autism

Alexander A. Chubykin

Abstract Neurexins and neuroligins are neuronal cell adhesion molecules (CAMs), which are specifically expressed in the nervous system and particularly in the brain. Presynaptic neurexins interact with postsynaptic neuroligins in a Ca^{2+}-dependent manner and form trans-synaptic adhesion complexes, which are implicated in synaptogenesis. Differential splicing of neurexin and neuroligin transcripts results in a large variety of isoforms. These isoforms differ in their binding affinities and ligand specificities and are differentially distributed in the central nervous system. This may reflect a synaptic code, which along with other neuronal adhesion molecules determines the wiring diagram of neuronal connections in the brain. Mouse genetic studies suggest that neurexins and neuroligins are involved in the stabilization of transient, rather than the formation of de novo synapses. Mutations in neurexin and neuroligin genes, as well as their downstream signaling molecules, have been identified in patients with mental retardation and autism spectrum disorders. Mice with corresponding mouse mutations represent new promising experimental models, which display the typical phenotypes associated with these diseases. The first results obtained from these models suggest that a disruption of the balance between excitatory and inhibitory neurotransmission (E/I) is one of the potential pathophysiological mechanisms for autism and mental retardation.

Keywords Neurexin · Neuroligin · Synapse formation · Synapse stabilization · Autism

17.1 Neurexins: Genes and Proteins Structure

Neurexins were originally discovered during the search for the vertebrate receptor of α-latrotoxin (α-LTX), a neurotoxin from the venom of black widow spider (Ushkaryov et al. 1992). When α-LTX binds to neurexin-1α, it

A.A. Chubykin (✉)
The Picower Institute for Learning and Memory, Massachusetts Institute of Technology, 77 Massachusetts Avenue, 46-3301, Cambridge, MA 02139, USA
e-mail: chubykin@mit.edu

M. Hortsch, H. Umemori (eds.), *The Sticky Synapse*,

DOI 10.1007/978-0-387-92708-4_17,

induces exhaustive release of neurotransmitters from presynaptic nerve terminals. This suggested a presynaptic localization for neurexins, which was later confirmed by subsequent studies (Dean et al. 2003, Chubykin et al. 2005).

Three neurexin genes are expressed in mammals. Each gene produces two principal forms, the longer α- and the shorter β-neurexin isoform (Rowen et al. 2002, Tabuchi and Südhof 2002). These genes are highly conserved among vertebrate species and exhibit identical exon/intron structures. With the sizes of ~1.1 and of ~1.6 Mb, the neurexin-1 (NRXN1) and the neurexin-3 (NRXN3) genes are unusually large. In contrast, the neurexin-2 (NRXN2) gene spans only ~110 kb (Tabuchi and Südhof 2002). The human neurexin genes are located at 2p16 (NRXN1), 11q12 (NRXN2), and 14q32 (NRXN3), respectively. Structurally, all neurexins are composed of alternatively spliced extracellular domains, a single transmembrane region (TMR), and a short cytoplasmic tail. Thus they resemble typical cell-surface receptors. α- and β-neurexins differ only in their N-terminal extracellular sequences and share identical C-termini, including the O-glycosylation site close to the TMR, the TMR itself, and the cytoplasmic region. α-Neurexin contains six laminin/neurexin/sex hormone-binding globulin (LNS) domains, which are separated by three epidermal growth factor-like (EGF) repeats, and five splice sites SS#1–SS#5, while β-neurexin contains a small β-neurexin-specific sequence, which is followed by only one LNS domain and two splice sites SS#4 and SS#5 (Fig. 17.1a).

17.2 Neurexin Genes Are Differentially Expressed

So far only a general characterization of the expression pattern of six major neurexins (1α, 1β, 2α, 2β, 3α, and 3β) has been published (Ullrich et al. 1995). The existing evidence suggests that these neurexin genes have differential pattern of expression. In situ hybridizations with oligonucleotides specific for the six major neurexins have shown that neurexin 1α is expressed in all brain areas with the highest level in the claustrum, anterior thalamic nuclei and deep cerebellar nuclei. Neurexin 1β is expressed mostly in the cortical layers 2 and 3, the thalamus, and specific parts of the hippocampus. Neurexin 2α is expressed only in specific subpopulations of the cortical layers 2, 4, and 6, the thalamus and the cerebellum. The distribution of neurexin 2β is more uniform with higher levels in the superficial layers of the cortex and the cerebellum. Finally, neurexin 3α has low expression in most of the brain except for the very superficial and the infragranular layers of the cortex, the striatum, septal nuclei, and the reticular thalamic nucleus (RN). Neurexin 3β is expressed uniformly in most brain regions (see Table 17.1) (Ullrich et al. 1995).

The distinctive anatomical features of the hippocampus and the olfactory bulb provide a way to characterize expression of different neurexin genes in cells forming highly specific synapses. The hippocampus is organized into two sheets

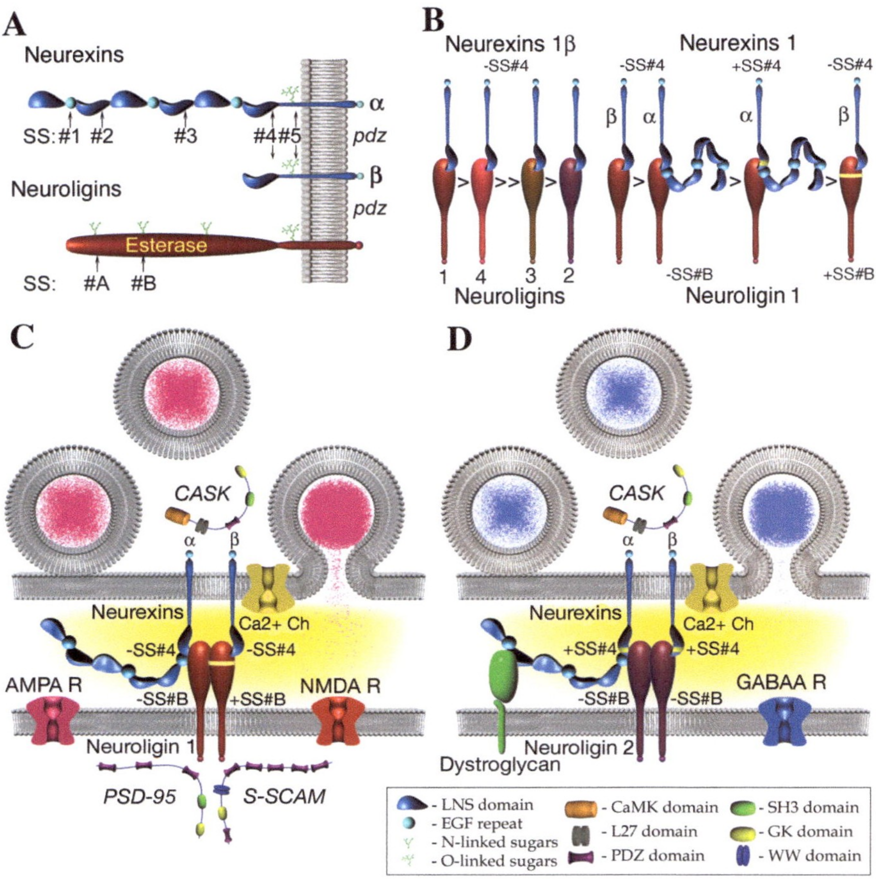

Fig. 17.1 Neurexins and neuroligins. (**a**) The structure of neurexins and neuroligins. SS–splice sites, EGF – epidermal growth factor-like repeats, LNS – laminin/neurexin/sex hormone-binding globulin domains, pdz (*small letters*) – PSD-95-Dlg-ZO homology-binding motif. (**b**) Binding specificities of neurexins to neuroligins. The binding affinity of neurexin 1β-SS#4 to neuroligin 1 is greater than to neuroligin 4 and much greater than to neuroligin 3. The lowest binding affinity is to neuroligin 2. Binding of neurexins 1α-SS#4 or +SS#4 to neuroligin 1-SS#B is weaker than neurexin 1β-SS#4, but stronger than neurexin 1β + SS#4. (**c**) Scheme of a glutamatergic excitatory synapse. α- and β-neurexins lacking splice site SS#4 insert interact with neuroligin 1 with or without splice site #B insert. Presynaptically, neurexins interact with CASK via their PDZ-binding motif and the CASK PDZ domain. Postsynaptically, neuroligins interact with PSD-95 and S-SCAM via their PDZ-binding motif and the corresponding PDZ domains. CaMK – Ca^{2+}/calmodulin-dependent kinase domain, L27–L27 domain, PDZ (capitalized)–PSD-95-Dlg-ZO homology domain, SH3–Src homology 3 domain, GK–guanylate kinase domain, WW–WW domain, Ca2+ Ch–calcium channel, AMPA R–AMPA receptor, NMDA R – NMDA receptor. (**d**) Scheme of a GABAergic inhibitory synapse. α-neurexins with splice site SS#4 insert bind neuroligin 2 lacking splice site SS#B insert via their sixth LNS domain, and dystroglycan via their second LNS domain. β-neurexins interact only with neuroligin 2, and not dystroglycan. GABAA R – $GABA_A$ receptor

Table 17. 1 Distribution of neurexins in brain

		Hippocampus					Olfactory bulb			
		Pyramidal cells			Interneurons					
Genes	Brain	CA1	CA3	DG	CA1–CA3	DG	Glom	EPL	MCL	GCL
Neurexin 1α	All brain; enriched in the claustrum, anterior thalamic nuclei, deep cerebellar nuclei	+	+	+ +	+	+ +	+	–	–	+
Neurexin 1β	Cortical layers 2/3, the thalamus, parts of the hippocampus	–	+ +	+	–	+ +	+ +	+ +	+ + +	+
Neurexin 2α	Subpopulations of cortical layers 2, 4, 6, the thalamus, the cerebellum	+	+	+	+ +	+ +	–	–	–	+
Neurexin 2β	All brain; enriched in the superficial layers of the cortex, the cerebellum	+ +	+ +	+ + +	+	+ +	+ + +	+	+ + +	+ + +
Neurexin 3α	Superficial and infragranular layers of the cortex, the striatum, septal nuclei, the reticular thalamic nucleus	–	+ + +	–	+ + +	+ + +	+	–	+ + +	+ + +
Neurexin 3β	All brain	+ + +	+ + +	+ + +	+ +	+ + +	+ + +	+	+ + +	+ + +

DG, dentate gyrus; Glom, glomerular layer; EPL, external plexiform layer; MCL, mitral cell layer; GCL, granular cell layer.
–, no expression; +, low expression; + +, medium expression; + + +, high espression.

of neurons folded into each other. They are called the dentate gyrus and Ammon's horn. Ammon's horn consists of four regions called CA1–CA4, where CA stands for *cornu Ammonis*, which means 'Ammon's horn' in Latin. The hippocampus is characterized by a unidirectional three-synaptic organization. Input from the entorhinal cortex (EC) is called perforant path (PP) and is received in the Dentate Gyrus (DG) and the CA3. In addition to this input, the CA3 neurons also receive input from the DG via the mossy fibers (MF). The CA3 neurons send axons to CA1 via the Schaffer collateral pathway (SC). The CA1 cells also receive direct input from the perforant path, although these synapses are located on the distal apical dendrites.

In the hippocampus, there is a clear difference between CA1 and CA3 cells in neurexin gene expression. CA1 pyramidal cells and interneurons have no detectable neurexin 1β mRNA and CA1 pyramidal cells and dentate gyrus granule cells lack neurexin 3α. In contrast, pyramidal cells of CA3 co-express all six neurexin isoforms. Interestingly, interneurons express variable levels of different neurexins in different areas, but neurexin 3α exhibits the highest expression level (Table 17.1).

In the olfactory bulb, different neuronal cell types are organized in layers making identification of these cells unambiguous. Neurexin 1α is expressed mostly in periglomerular cells, but is almost absent from mitral and tufted cells. In contrast, neurexin 1β is highly enriched in mitral and tufted cells. Neurexin 2α has a very low level of expression in the olfactory bulb, whereas neurexin 2β is highly expressed, particularly in inhibitory periglomerular and granule cells. Neurexins 3α and 3β are both expressed in granule cells, but only neurexin 3β is present in periglomerular cells (Table 17.1) (Ullrich et al. 1995).

The presence or absence of the splice site SS#4 determines the specificity of the neurexin–neuroligin interaction. Thus, changes in the distribution of these isoforms may support the role of neurexin as a specific synaptic code. Preliminary data using in situ hybridizations with the oligonucleotides specific for neurexins 1 and 2 with or without SS#4 indicate that expression levels of neurexins 1 and 2 with SS#4 are much higher in the striatum, substantia nigra, and cerebellar nuclei, while in the CA3 region of the hippocampus the expression levels are reversed with higher expression levels of neurexins 1 and 2 without SS#4. In the hippocampus, the most prominent isoform is neurexin 3 without SS#4, which is expressed mostly in the pyramidal cells of CA1–CA4 (Ichtchenko et al. 1995).

17.3 Dystroglycan and Neurexophilin – Neurexin-Interacting Proteins with Unknown Functions

Neurexins have three different protein-binding partners: dystroglycans, neurexophilins, and neuroligins (Petrenko et al. 1996, Sugita et al. 2001). The alternative splicing of neurexin transcripts determines the specificity of these interactions.

Dystroglycans represent a family of CAMs that associate with dystrophins and are present at inhibitory central nervous synapses and neuromuscular junctions (NMJs). Dystroglycans bind α-neurexins, but not β-neurexins, in a Ca^{2+}-dependent manner. The binding site for dystroglycans is localized to the second LNS domain of α-neurexin (Fig. 17.1a). In addition to neurexin, dystroglycans are also receptors for agrin, laminin, and perlecan. Agrin also contains an LNS domain and is also alternatively spliced at amino acid residues highly conserved between agrin and α-neurexin. Dystroglycans are responsible for organizing the extracellular matrix (ECM) and facilitate the clustering of acetylcholine receptors (AChR) at the NMJ. Impairments of dystroglycan glycosylation abolish the interaction with its ligands and are causally implicated in several muscular dystrophies, such as muscle–eye–brain disease (MEB), Walker–Warburg syndrome (WWS), Fukuyama congenital muscular dystrophy (FCMD), and congenital muscular dystrophy 1C and 1D (MDC1C and MDC1D), which are also often accompanied by mental retardation (Barresi and Campbell 2006). Although much is already known about dystroglycan, the importance of its interaction with neurexin for synaptic integrity or function remains a mystery.

Neurexophilin was originally purified in a tight complex with neurexin 1α (Petrenko et al. 1996). Based on in situ hybridization results, neurexophilin is a secreted glycoprotein, which is processed from a pre-propeptide and expressed at high levels primarily in interneurons containing gamma-aminobutyric acid (GABA) neurotransmitter (Petrenko et al. 1996). The function of neurexophilin is currently still unknown, but it might be related to its expression in inhibitory interneurons.

17.4 Neuroligins: Genes and Proteins Structure

Neuroligins are postsynaptic CAMs that bind to both α- and β-neurexins (Ichtchenko et al. 1995, 1996, Boucard et al. 2005). Neuroligins are composed of a large extracellular N-terminal sequence that is homologous to the α/β-hydrolase fold domain of acetylcholinesterase (Ichtchenko et al. 1995), an O-linked sugar-rich domain, a single transmembrane region, and a short cytoplasmic tail. There are two differentially spliced sites, SS#A and SS#B, in the esterase homology domain of neuroligins (Fig. 17.1a). Neuroligin 1 and neuroligin 3 have two alternatively used inserts for SS#A (A1 and A2), whereas all other neuroligins (2, 4, 5, and 4*) have only one alternatively used insert corresponding to A2 (Bolliger et al. 2008). Insert at SS#B has been identified only in neuroligin 1, where it regulates the binding specificity of α- and β-neurexins.

Rodents have four neuroligin genes (Ichtchenko et al. 1996) and five neuroligin genes have been identified in humans (Bolliger et al. 2001). Interestingly, the three rodent neuroligin genes have more than 98% amino acid sequence identity to their human orthologs, while the fourth rodent neuroligin (NLGN4*) gene is significantly different and has diverged during evolution from its corresponding human neuroligin 4 gene (Bolliger et al. 2008). Human neuroligin genes are localized to 3q26 (NLGN1), 17p13 (NLGN2), Xq13 (NLGN3), Xp22.3 (NLGN4) and Yq11.2 (NLGN5 or NLGN4Y). All neuroligins are expressed in the central nervous system, where neuroligin 1 is localized to the excitatory glutamatergic synapses, while neuroligin 2 is localized to the inhibitory GABAergic synapses (Song et al. 1999, Varoqueaux et al. 2004). Although recent findings suggest that neuroligin 3 might also be present in some GABAergic synapses, it is mostly localized to glutamatergic synapses (Budreck and Scheiffele 2007).

The extracellular structure of neuroligin is similar to acetylcholinesterase. However, the amino acids, corresponding to the catalytic triad of acetylcholinesterase, are not conserved in neuroligins resulting in a loss of its enzymatic activity. In addition, the substrate entrance site is completely inaccessible for the substrate in neuroligins (Arac et al. 2007, Koehnke et al. 2008). Like acetylcholinesterase, the extracellular domains of neuroligins interact and form constitutive dimers (Comoletti et al. 2003). Two helices from each monomer form a four-helix bundle at the binding interface and the dimerization process is primarily driven by hydrophobic interactions. Dimerization of postsynaptic neuroligins is probably important for the stable association with presynaptic neurexins and/or for the activation of presynaptic signaling via neurexin and consequential neurexin clustering. The significance of neuroligin dimerization is underscored by the observation, that mutations of the putative dimerization sequences in the four-helix bundle at the binding interface of the neuroligin 1 monomer abolishes its ability to form synapses (Dean et al. 2003). Although the EF hands within the neuroligin structure were hypothesized to be Ca^{2+}-binding sites, so far the experimental data have not been able to support this hypothesis (Arac et al. 2007).

17.5 Splicing of Both Neurexins and Neuroligins Determines Affinity and Specificity of Their Interaction

Neurexins and neuroligins bind to each other and form heterotypic intercellular junctions, which are regulated by the alternative splicing of the primary transcripts encoding both molecules (Ichtchenko et al. 1995, 1996, Comoletti et al. 2003). This interaction is Ca^{2+} dependent, highly hydrophilic, and requires water for stabilization of the neurexin–neuroligin complex. Therefore, it is highly dependent on the Ca^{2+} and the ionic concentrations of the environment (Chen et al. 2008). Binding to neuroligin is mediated by the last LNS domain of α-neurexins or by the sole LNS domain of β-neurexins. Structurally, an LNS domain is composed of two seven-stranded β-sheets forming a jelly roll with structural similarity to lectins (Rudenko et al. 1999). Recent studies suggest that the structural basis of the neurexin–neuroligin interaction specificity is determined by the selection of differentially spliced exons in the neurexin and neuroligin transcript, respectively. In a neurexin molecule, SS#2, SS#3, and SS#4 splice sites in different LNS domains all map to the loops that surround the Ca^{2+}-binding site and together form a so-called ‘hypervariable surface’ of the LNS domains (Rudenko et al. 2001, Arac et al. 2007, Fabrichny et al. 2007, Chen et al. 2008, Shen et al. 2008). This ‘hypervariable surface’ directly interacts with the corresponding binding sites of neuroligins. Mutations of the amino acids composing these hypervariable surfaces disrupt the interaction of neurexin with neuroligin and completely abolish synaptogenic activity (Graf et al. 2006). The binding of Ca^{2+} in the center of the hypervariable surface is essential for establishing the complex with neuroligin and plays a key structural role in stabilizing the complex. The presence of SS#4 insert results in major structural rearrangements of the LNS domain. Specifically, part of the extended loop close to the Ca^{2+}-binding site attains a helical conformation and provides additional protein contact points to the Ca^{2+} ion increasing the affinity of Ca^{2+} ion binding. In addition, SS#4 is located in close proximity to one of the salt bridges formed by the interaction between neurexin and neuroligin (Chen et al. 2008). Thus, the insertion at splice site SS#4 probably changes the Ca^{2+}-binding affinity, as well as rearranges the topology of the ‘hypervariable surface’. In contrast, the presence of insert at SS#B in the neuroligin protein could impede these rearrangements and disrupt the neighboring salt bridge, which is necessary for the formation of the neurexin 1 β-SS#4:neuroligin 1 + SS#B complex (Shen et al. 2008).

Recent surface plasmon resonance (SPR) experiments have revealed that neurexin 1β-SS#4 binds all neuroligin 1 isoforms, but has the highest affinity for neuroligin 1-SS#B (Comoletti et al. 2006). Its interaction with neuroligin 2 (regardless whether the SS#A insert is included) is two orders of magnitude weaker than that with neuroligin 1 (so far, the SS#B insert has only been identified in neuroligin 1). Neurexin 1β-SS#4 also binds neuroligin 3 and neuroligin 4. The preference for its interaction with these molecules is NL1>NL4>>NL3>NL2 (Fig. 17.1b). On the other hand, the presence of the SS#4 insert favors binding of neurexin 1β to any neuroligins without the SS#B insert, and particularly to neuroligin 2.

Similar to neurexin splice site SS#4, neuroligin splice site SS#B determines specificity of neuroligin binding to presynaptic neurexins. It is a master switch between a more ‘promiscuous’ form that binds both α- and β-neurexins when the insert at the splice site is absent and a more selective form that only binds to β-neurexins when this insert is present (Boucard et al. 2005) (Fig. 17.1b,c).

Revisiting the expression profile data for neuroligins and neurexins, neuroligin 1 and neuroligin 3 are localized mostly to excitatory synapses, and neurexin 1β-SS#4 preferentially interacts with these ‘excitatory’ neuroligins. Neuroligin 2 is localized to inhibitory synapses and neurexin 1β + SS#4 preferentially interacts with this ‘inhibitory’ neuroligin (Song et al. 1999, Budreck and Scheiffele 2007). The expression pattern of neurexins + SS#4 and –SS#4 is consistent with these findings. In situ hybridizations have shown that neurexins including the SS#4 sequence have higher levels of expression in the striatum, substantia nigra, and cerebellar nuclei, which predominantly have inhibitory neurons (Oertel and Mugnaini 1984), while neurexins without the SS#4 sequence are enriched in the pyramidal cell layers of the hippocampus which contains mostly excitatory neurons (neurexins 1 and 2-SS#4 in CA3, neurexin 3-SS#4 in CA1–CA4) (Ichtchenko et al. 1995).

17.6 The Role of Neurexins and Neuroligins in Synapse Formation and Stabilization

17.6.1 In Vitro Synapse Formation Assays

The function of neurexins and neuroligins has long been unclear until it was demonstrated in an artificial synapse formation assay. These results demonstrated that neuroligin 1, when expressed in non-neuronal HEK293 cells, can induce presynaptic differentiation in co-cultured pontine explants (Scheiffele et al. 2000). Neuroligin expression alone was sufficient to induce presynaptic differentiation of neurons. The presynaptic terminals formed by the neurons in these co-cultures were morphologically indistinguishable from regular neuronal synapses. They contain synaptic vesicles, which are filled with neurotransmitters, and can be released by hypertonic solutions or by high potassium buffers. In co-cultures with neuronal cells, the co-expression of neuroligins and glutamate receptors in non-neuronal cells formed 'minimal synapse'. Lipophilic FM dye stainings and direct whole-cell patch clamp recordings of glutamate currents in these non-neuronal cells have shown that these 'minimal synapses' are fully functional. Such preparations facilitated the characterization of different glutamate receptors subunit kinetics, as well as some of the presynaptic properties exhibited by these 'minimal synapses' (Fu et al. 2003, Sara et al. 2005).

The results that were obtained from these artificial synapse formation assays suggest that the exposure of the neurons to large enough amounts of neuroligin may be sufficient to induce formation of the presynaptic terminals. This conclusion was supported by the finding that substrate-immobilized recombinant neuroligin 1 promotes the clustering of synaptic vesicle antigens in cultured neurons (Dean et al. 2003). Similarly, antibody cross-linking of epitope-tagged neurexins, which were transfected into neurons, brings the neurexin molecules together independently of neuroligins and causes the same effect (Dean et al., 2003). Together with the requirement of neuroligin dimerization for synaptogenesis, these findings raised a hypothesis, that presynaptic clustering of neurexins is involved in either the formation or the stabilization of presynaptic terminals. This raised the question whether the neurexin–neuroligin signaling is bidirectional. As neurexins could similarly induce postsynaptic differentiation in the dendrites of the neurons, which contact the non-neuronal neurexin-expressing cells (Graf et al. 2004), trans-synaptic neurexin–neuroligin complexes appear to function bidirectionally.

Artificial synapse formation assays revealed how the neurexin–neuroligin splice code regulates the specificity of synaptogenesis. Culturing dissociated hippocampal neurons with non-neuronal cells, which were transfected with neurexin 1β-SS#4 constructs, results in the clustering of both the PSD-95 excitatory and the gephyrin inhibitory synaptic marker. This suggests an induction of both excitatory and inhibitory postsynaptic differentiations by neurexin 1β-SS#4 (Graf et al. 2006). On the other hand, neurexin 1β + SS#4 preferentially promotes the clustering of the gephyrin and does not induce PSD-95 clustering (Boucard et al. 2005, Chih et al. 2006, Comoletti et al. 2006, Graf et al. 2006). It is important to note though that the lack of preference for excitatory postsynaptic clustering by the neurexin 1β-SS#4 can be explained by the low sensitivity of the artificial synapse formation assays. The quantity of the recombinant protein that is expressed on the surface of a non-neuronal cell is many times higher than in normal neurons. Therefore, even a low-affinity interaction between particular neurexin and neuroligin isoforms is potentially able to induce the clustering of corresponding synaptic markers. The extent of synaptogenesis is correlated with the amounts of neuroligin exposed on the surface of a dendrite to the axons. This suggests that the surface levels of neuroligin in vivo must be tightly controlled (Chubykin et al. 2005).

17.6.2 Intracellular Signaling of Neurexins and Neuroligins

The bidirectional neurexin–neuroligin signaling involves intracellular interactions between the C-terminal regions of neurexins and neuroligins with synaptic scaffolding molecules. On the presynaptic side, the PDZ-binding motif of the C-terminal neurexin region binds Ca^{2+}/calmodulin-activated Ser-Thr kinase (CASK). CASK is an unusual protein kinase and a member of the membrane-associated guanylate kinases (MAGUK) family of adaptor proteins (Hata et al. 1996, Mukherjee et al. 2008). Several different interacting partners of CASK have been identified. First, it interacts stoichiometrically with Velis and Mint-1 (Butz et al. 1998). In addition, it also binds protein 4.1. Together with neurexins this complex was shown to be a potent nucleator center for actin polymerization (Biederer and Südhof 2001). CASK can also bind Ca^{2+} and K^+ channels and potentially clusters these channels at the presynaptic terminal (Maximov, Südhof and Bezprozvanny 1999, Leonoudakis et al. 2004). K^+ channels can be phosphorylated by CASK and thereby induce an enhancement of K^+ currents (Marble et al. 2005). CASK can also phosphorylate the neurexin C-terminal region, a process that is dependent on neuronal activity. After an inhibition of *N*-methyl-D-aspartic acid (NMDA) receptors and Na^+ channels the CASK-dependent phosphorylation of β-neurexin is increased more than twofold (Mukherjee et al. 2008).

The importance of CASK for clustering Ca^{2+} channels at the presynaptic terminal is suggested from the findings that α-neurexin knockout mice have impaired presynaptic Ca^{2+} channel functions. While the total number of the surface Ca^{2+} channels is unchanged in these animals, Ca^{2+} currents are severely suppressed. In addition, the deletion of α-neurexin genes severely impaired the evoked synaptic inhibitory transmission in the neocortex and the excitatory transmission in the brainstem. Furthermore, the frequencies of both inhibitory and excitatory miniature postsynaptic currents (mPSCs or minis) are drastically decreased in α-neurexin knockout animals. These changes in synaptic physiology are associated with a twofold decrease in the density of inhibitory GABAergic terminals in both the brainstem and the neocortex (Missler et al. 2003). The splice site code theory may provide an explanation why only the density of GABAergic synapses is decreased. α-neurexins only interact with neuroligins-SS#B and dystroglycans (specific for GABAergic synapses), while β-neurexins bind both neuroligins-SS#B and neuroligins + SS#B. Insert at the splice site SS#B has been identified only in neuroligin 1, therefore, the neuroligin 1 isoform with the SS#B sequence is unable to bind α-neurexins. Consequently, a subpopulation of inhibitory synapses that expresses neuroligin 2-SS#B and interacts with α-neurexins would be eliminated, while the subpopulation of excitatory synapses that expresses neuroligin 1 + SS#B and interacts only with β-neurexins would be spared.

Considering the involvement of CASK, it is tempting to hypothesize that α-neurexin may provide a nucleation core for the formation of the presynaptic machinery and that CASK binding is necessary for the successive clustering of the Ca^{2+} channels close to the proper synaptic vesicle fusion sites. Once this recruitment is accomplished, CASK might modulate Ca^{2+} channels function by phosphorylation of the channel protein (Atlas 2001). The deletion of α-neurexins may then result in a mislocalization of Ca^{2+} channels at the presynaptic terminals (Missler et al. 2003).

The C-terminal region of all neuroligins contains a PSD-95-Dlg-ZO homology (PDZ)-binding motif, which can interact with the third PDZ domain of PSD-95 and the second PDZ domain of synaptic scaffolding molecule (S-SCAM). S-SCAM is also known as membrane-associated guanylate kinase with inverted domain organization (MAGI) (Fig. 17.1c). Interestingly, S-SCAM also binds the proline-rich region of neuroligins, which is located upstream of the PDZ-binding motif and this binding might be involved in the proper localization and trafficking of neuroligins to synapses (Iida et al. 2004). In addition to PSD-95 and S-SCAM, neuroligins also interact with the Shank protein in yeast-two-hybrid assays. Shanks represent another family of scaffolding proteins that is involved in spine formation and metabotropic glutamate receptor recruitment (Sheng and Kim 2000, Meyer et al. 2004). PDZ domains of PSD-95 and S-SCAM also bind the C-terminal sequences of K^+ channels and NMDA receptors, and thus may potentially mediate clustering of K^+ channels and NMDA receptors in the postsynaptic density (Irie et al. 1997, Hirao et al. 1998).

17.6.3 The Link Between Cell Adhesion and Synaptic Plasticity

Specific localization of neuroligin 1, 3 and neuroligin 2 to excitatory and inhibitory synapses, respectively, suggests that their selective expression may dictate the type of synapse formed. Indeed, overexpression of neuroligin 1 and 2 in primary dissociated neurons selectively increases either the density of glutamatergic or of GABAergic synapses, respectively. This in turn raises the frequency of corresponding miniature postsynaptic currents (mPSCs) and the evoked response, either excitatory for neuroligin 1 or inhibitory for neuroligin 2 expression. Interestingly, neuroligin 1 overexpression has stronger effect on the NMDA receptor-mediated excitatory postsynaptic currents (EPSCs) than the alpha-amino-3-hydroxy-5-methyl-4-isoxazolepropionic acid (AMPA) receptor-mediated EPSCs and results in an increase of the NMDA/AMPA ratio (Prange et al. 2004, Chubykin et al. 2007). This suggests that neuroligin 1 may induce the formation of so-called 'silent synapses' that lack AMPA receptors, which may become stabilized by synaptic activity and 'activated' by AMPA receptor recruitment. The observation that neurexin 1β, which is expressed in PC12 cells, induces the clustering of PSD-95 and of NMDA, but not of AMPA receptors in contacted dendrites of the co-cultured hippocampal neurons, supports this hypothesis. AMPA receptors may be later recruited into these contacts by glutamate application or by the neuronal overexpression of a constitutively active form of calmodulin kinase II (Nam and Chen 2005). Alternatively, NMDA currents could be regulated by neuroligin 1 through its interaction with PSD-95 or S-SCAM and the subsequent recruitment and stabilization of NMDA receptors at the postsynaptic density.

Consistent with the overexpression studies in cultured neurons, transgenic mice, which overexpress neuroligin 2, have an increased frequency of mIPSCs and a decreased balance of excitatory to inhibitory neurotransmission (E/I) (Hines et al. 2008). Experiments with acute cortical and hippocampal slice preparations from neuroligin knockout mice have also shown results consistent with culture experiments. In neuroligin 1 knockout mice the NMDA receptor-mediated neurotransmission is decreased, while inhibitory neurotransmission remains unaffected. At the same time, in neuroligin 2 knockout mice the amplitude of IPSCs is decreased, while there is no change in excitatory neurotransmission (Chubykin et al. 2007). In addition, acute suppression of neuroligin 1 expression in lateral amygdala using lentiviral-mediated delivery of small hairpin RNA (shRNA) against neuroligin 1 resulted in the specific decrease of NMDA receptor-mediated EPSCs and the subsequent decrease of NMDA/AMPA ratio in thalamo-amygdala synapses of principal neurons. There is no change in the intrinsic electrophysiological properties or the input–output curves. Furthermore, acute suppression of neuroligin 1 expression abolishes the induction of NMDA receptor-dependent form of long-term potentiation (LTP) in the amygdala. This in turn resulted in deficits of both contextual- and cued-fear memory. These results suggest that neuroligin 1 expression is necessary for NMDA receptor-mediated neurotransmission and for synaptic plasticity, which is involved in the storage of long-term memory in the amygdala (Kim et al. 2008b).

Interestingly, recent reports further support the role of the neurexin–neuroligin trans-synaptic complex in bidirectional signaling. The neurexin–neuroligin complex alters presynaptic short-term plasticity by increasing the sensitivity of the presynaptic release machinery to the extracellular Ca^{2+} concentration (Futai et al. 2007). These findings are consistent with the previous observations that in α-neurexins knockout mice the presynaptic Ca^{2+} channel function is impaired (Missler et al. 2003). This signaling process could be important for synchronizing synapse stabilization and elimination between pre- and postsynaptic sites. Furthermore, it might also provide a potential feedback mechanism for homeostatic changes.

Such homeostatic changes might account for the observation that neuroligin and α-neurexin knockout mice still form synapses. Thus, the neurexin–neuroligin trans-synaptic complexes do not appear to play a role in the initial formation of synapses, but rather in their activity-dependent stabilization (Varoqueaux et al. 2006). This activity-dependent stabilization involves an intricate interplay between the Ca^{2+} channels, CASK, and neurexins on the presynaptic side. On the postsynaptic side, NMDA receptors, PSD-95, or S-SCAM cooperate with excitatory neuroligins (Fig. 17.1c) and GABA receptors and dystroglycans with inhibitory neuroligins (Fig. 17.1d).

17.7 Neurexin and Neuroligin Gene Polymorphisms in Autism Spectrum Disorders and Mental Retardation

Autism is a severe disorder of the nervous system, which is characterized by difficulties in social interactions and is often accompanied by learning disabilities. Autistic patients also have perceptual processing abnormalities, which are expressed in a hypersensitivity to auditory and tactile stimuli (Kootz et al. 1981). They also exhibit impairments in executive functions and motor control, procedural, emotional and social memory (Squire and Zola 1996). The mechanisms underlying the development and progression of autism are still unknown. There is a significant evidence of a genetic etiology for autism. The rate of occurrence of autism is about 1/500 and twin studies show that the concordance rate is 70–90% in monozygotic twins and 0–10% for dizygotic twins (Steffenburg et al. 1989, Blasi et al. 2006, Szatmari et al. 2007). The incidence of disease for male versus female affected is 4:1 for autism and 8:1 for Asperger syndrome. Male predisposition to these disorders may be partially explained by the growing number of identified X-linked mutations. This includes mutations in two X-linked genes encoding neuroligins NLGN3 and NLGN4 in siblings with autism spectrum disorders (Jamain et al. 2003). A number of mutations in the NLGN3 and the NLGN4 gene cause pathological phenotypes. A missense mutation in human neuroligin 3 (R451C) and two frameshift mutations in neuroligin 4 have been identified in patients with familial forms of autism, mental retardation, and/or Asperger syndrome (Jamain et al. 2003, Zoghbi 2003, Laumonnier et al. 2004). As the protein is truncated in the middle of the acetylcholinesterase domain, the neuroligin 4 mutation probably acts as a null allele.

The R471C mutation in rat neuroligin 3 has been well characterized. This mutation (which corresponds to the human R451C substitution) results in an unpaired cysteine residue and the partial retention of neuroligin 3 protein in the endoplasmic reticulum (Chih et al. 2004, Comoletti et al. 2004). Since neuroligin 3 can form heterodimers with neuroligins 1 and 2 (Comoletti et al. 2006, Budreck and Scheiffele 2007), this might explain a dominant-negative gain-of-function effect, which is induced by this mutation. Neuroligin 3 is localized mostly to glutamatergic synapses. Therefore, the R451C substitution mutation may result in formation of dimers between the mutant neuroligin 3 form with wild-type neuroligin 1, thereby disrupting the balance of excitatory versus inhibitory synapse formation and stabilization. This hypothesis is supported by the findings that overexpression of the mutant protein in dissociated neurons results in a massive decrease of synapse density and, consequently, of synaptic transmission. This mutation specifically impairs the balance of excitatory to inhibitory neurotransmission (E/I balance) (Chubykin et al. 2007). This disruption of the E/I balance has recently been proposed as a potential mechanism of autism pathophysiology. The characterization of the R451C mutation in mice has given further support to this hypothesis. However in contrast to the human mutation, in the mouse model the shift in the E/I balance was caused by an increase in inhibitory, rather than a decrease in excitatory neurotransmission. In addition, these R451C mutant mice also have an increased density of GABAergic synapses. This is consistent with the observed changes in synaptic physiology. One of the potential explanations for these discrepancies between humans and mice could be compensatory homeostatic changes.

The recent identification of various new mutations in the neurexin/neuroligin synapse formation/stabilization pathway suggests that this pathway may hold significant insights into the pathophysiology of autism spectrum disorders and mental retardation. In addition, it may provide new candidates for the mutation screening (also see Chapter 6). Neurexin autism mutations have recently been discovered (Szatmari et al. 2007, Kim et al. 2008a). In addition, a gene for contactin-associated protein-like 2 (CNTNAP2), a member of the superfamily of neurexin-like proteins, has also been identified as an autism-susceptibility gene (Baumgartner et al. 1996, Alarcon et al. 2008, Bakkaloglu et al. 2008). Mutations in the CASK gene have been reported in patients with the FG syndrome, which is characterized by developmental abnormalities and mental retardation (Piluso et al. 2003). Also mutations in the Shank 3 gene have recently been found in patients with autism spectrum disorders (Durand et al. 2007). CASK binds to the C-terminal region of neurexins, while Shanks are putative interacting proteins for neuroligins. Interestingly, Shanks have been shown to be involved in spine formation and metabotropic glutamate receptor recruitment (Sheng and Kim 2000, Meyer et al. 2004). Metabotropic glutamate receptors are important for various forms of synaptic plasticity, including long-term depression (LTD). Loss of fragile X mental retardation protein (FMRP) causes fragile X syndrome in humans and increases protein synthesis-dependent synaptic plasticity, which is mediated by metabotropic glutamate receptors (Bear et al. 2004). Thus, there are currently two major theories of autism grouped by the types of genes mutated in the disease: the first group includes mutations of the proteins in the neurexin/neuroligin pathway, the second group includes mutations of the proteins involved in various forms of pre- and post-synaptic plasticity, for example, FMRP (Huber et al. 2002), methyl CpG-binding protein 2 (MECP2) (Moretti et al. 2006), and metabotropic glutamate receptors (Serajee et al. 2003). Shank proteins may represent a bridge between the two theories of autism, which involve synapse formation/maintenance and synaptic plasticity.

One way to unify these two hypotheses to explain the mechanism of autism is to postulate that mechanisms of synaptic plasticity involve neurexin/neuroligin signaling in selecting which synapses are stabilized or will be eliminated. Consequently, an impairment of signaling at any step of this pathway may potentially result in mental retardation or autism. In line with this hypothesis, autism mutants of the proteins involved in different steps of the neurexin/neuroligin pathway manifest similar impairments in synaptic morphology and physiology. They also often exhibit similar behavior characteristics of the disorder, such as impaired social interactions and reduced ultrasound vocalizations (Jamain et al. 2008). Surprisingly, in addition to these impairments, some of the autism models show improvements in certain types of memory, i.e., an enhanced spatial learning and memory, which results in an improved performance in Morris water maze tests (Tabuchi et al. 2007, Hung et al. 2008).

Many of the newly discovered mutations in neurexin and neuroligin genes, as well as their signaling partners represent de novo mutations. Germ-line mutations increase with age, thus increasing the risk for older parents to have children with autism. This might contribute to the recent reported increase of autism cases (Sebat et al. 2007, Zhao et al. 2007).

17.8 Conclusions, the Concept of a Synaptic Code and Future Directions

The alternative splicing of the various neurexin transcripts may yield more than 1000 protein isoforms. Although alternative splicing has been reported for other genes, most of those cases result in less than 100 variants. Such a high degree of receptor diversity is one of the well-established mechanisms to achieve a large number of unique specificities for molecular interactions. In nervous system it may allow for an almost unlimited potential for combinatorial variations of different neurexin isoforms in different cell types and possibly even individual cells.

Consequently, more detailed characterization of the various neurexin and neuroligin isoforms and their expression patterns will be essential for our understanding of the postulated neurexin–neuroligin-dependent synaptic code. Understanding this neurexin–neuroligin-dependent code might help us to learn how the brain is wired by specific connections between different cell types and by individual neurexin–neuroligin pairs, which specify different types of synapses. More modern, high-throughput, sensitive assays will shed more light on synaptic complexity in vertebrate brains. One of the predictions would be that the expression of neurexins–SS#4 is mostly in excitatory neurons, and neurexins + SS#4 in inhibitory neurons. An individual neuron may express either neurexin-SS#4 or neurexin + SS#4 isoform, but not both. The SS#4 splice site could represent one of the potential markers of the neuronal type. Consequently, investigation of how the splicing is regulated becomes very important, particularly, identification of the *cis* elements in the neurexin and neuroligin genes, as well as the trans-acting splicing factors interacting with these elements.

A growing body of evidence suggests that the activity-dependent regulation of neurexins and neuroligins at both excitatory and inhibitory synapses may be involved in synaptic plasticity. However, we will need a considerably better and more detailed knowledge of how synaptic activity is regulated by neurexins and neuroligins at the molecular level, before we will understand the role of these molecules in higher cognitive functions, such as memory formation and behavior.

Acknowledgments The author thanks Thomas C. Südhof and Dilja Krueger for their helpful comments on the manuscript.

References

Alarcon M, Abrahams BS, Stone JL et al. (2008) Linkage, association, and gene-expression analyses identify CNTNAP2 as an autism-susceptibility gene, Am J Hum Genet 82:150–159

Arac D, Boucard AA, Ozkan E et al. (2007) Structures of neuroligin-1 and the neuroligin-1/neurexin-1 beta complex reveal specific protein-protein and protein-Ca2+ interactions, Neuron 56:992–1003

Atlas D (2001) Functional and physical coupling of voltage-sensitive calcium channels with exocytotic proteins: ramifications for the secretion mechanism, J Neurochem 77:972–985

Bakkaloglu B, O'Roak BJ, Louvi A et al. (2008) Molecular cytogenetic analysis and resequencing of contactin associated protein-like 2 in autism spectrum disorders, Am J Hum Genet 82:165–173

Barresi R and Campbell KP (2006) Dystroglycan: from biosynthesis to pathogenesis of human disease, J Cell Sci 119:199–207

Baumgartner S, Littleton JT, Broadie K et al. (1996) A Drosophila neurexin is required for septate junction and blood-nerve barrier formation and function, Cell 87:1059–1068

Bear MF, Huber KM and Warren ST (2004) The mGluR theory of fragile X mental retardation, Trends Neurosci 27:370–377
Biederer T and Südhof TC (2001) CASK and protein 4.1 support F-actin nucleation on neurexins, J Biol Chem 276:47869–47876
Blasi F, Bacchelli E, Pesaresi G et al. (2006) Absence of coding mutations in the X-linked genes neuroligin 3 and neuroligin 4 in individuals with autism from the IMGSAC collection, Am J Med Genet B Neuropsychiatr Genet 141:220–221
Bolliger MF, Frei K, Winterhalter KH et al. (2001) Identification of a novel neuroligin in humans which binds to PSD-95 and has a widespread expression, Biochem J 356:581–588
Bolliger MF, Pei J, Maxeiner S et al. (2008) Unusually rapid evolution of Neuroligin-4 in mice, Proc Natl Acad Sci USA 105:6421–6426
Boucard AA, Chubykin AA, Comoletti D et al. (2005) A splice code for trans-synaptic cell adhesion mediated by binding of neuroligin 1 to alpha- and beta-neurexins, Neuron 48:229–236
Budreck EC and Scheiffele P (2007) Neuroligin-3 is a neuronal adhesion protein at GABAergic and glutamatergic synapses, Eur J Neurosci 26:1738–1748
Butz S, Okamoto M and Südhof TC (1998) A tripartite protein complex with the potential to couple synaptic vesicle exocytosis to cell adhesion in brain, Cell 94:773–782
Chen X, Liu H, Shim AH et al. (2008) Structural basis for synaptic adhesion mediated by neuroligin–neurexin interactions, Nat Struct Mol Biol 15:50–56
Chih B, Afridi SK, Clark L et al. (2004) Disorder-associated mutations lead to functional inactivation of neuroligins, Hum Mol Genet 13:1471–1477
Chih B, Gollan L and Scheiffele P (2006) Alternative splicing controls selective trans-synaptic interactions of the neuroligin–neurexin complex, Neuron 51:171–178
Chubykin AA, Atasoy D, Etherton MR et al. (2007) Activity-dependent validation of excitatory versus inhibitory synapses by neuroligin-1 versus neuroligin-2. Neuron 54:919–931
Chubykin AA, Liu X, Comoletti D et al. (2005) Dissection of synapse induction by neuroligins: effect of a neuroligin mutation associated with autism, J Biol Chem 280:22365–22374
Comoletti D, De Jaco A, Jennings LL et al. (2004) The Arg451Cys-neuroligin-3 mutation associated with autism reveals a defect in protein processing, J Neurosci 24:4889–4893
Comoletti D, Flynn R, Jennings LL et al. (2003) Characterization of the interaction of a recombinant soluble neuroligin-1 with neurexin-1beta, J Biol Chem 278:50497–50505
Comoletti D, Flynn RE, Boucard AA et al. (2006) Gene selection, alternative splicing, and post-translational processing regulate neuroligin selectivity for beta-neurexins, Biochemistry 45:12816–12827
Dean C, Scholl FG, Choih J et al. (2003) Neurexin mediates the assembly of presynaptic terminals, Nat Neurosci 6:708–716
Durand CM, Betancur C, Boeckers TM et al. (2007) Mutations in the gene encoding the synaptic scaffolding protein SHANK3 are associated with autism spectrum disorders, Nat Genet 39:25–27
Fabrichny IP, Leone P, Sulzenbacher G et al. (2007) Structural analysis of the synaptic protein neuroligin and its beta-neurexin complex: determinants for folding and cell adhesion, Neuron 56:979–991
Fu Z, Washbourne P, Ortinski P et al. (2003) Functional excitatory synapses in HEK293 cells expressing neuroligin and glutamate receptors, J Neurophysiol 90:3950–3957
Futai K, Kim MJ, Hashikawa T et al. (2007) Retrograde modulation of presynaptic release probability through signaling mediated by PSD-95-neuroligin, Nat Neurosci 10:186–195
Graf ER, Kang Y, Hauner AM et al. (2006) Structure function and splice site analysis of the synaptogenic activity of the neurexin-1 beta LNS domain, J Neurosci 26:4256–4265
Graf ER, Zhang X, Jin SX et al. (2004) Neurexins induce differentiation of GABA and glutamate postsynaptic specializations via neuroligins, Cell 119:1013–1026

Hata Y, Butz S and Südhof TC (1996) CASK: a novel dlg/PSD95 homolog with an N-terminal calmodulin-dependent protein kinase domain identified by interaction with neurexins, J Neurosci 16:2488–2494

Hines RM, Wu L, Hines DJ et al. (2008) Synaptic imbalance, stereotypies, and impaired social interactions in mice with altered neuroligin 2 expression, J Neurosci 28:6055–6067

Hirao K, Hata Y, Ide N et al. (1998) A novel multiple PDZ domain-containing molecule interacting with N-methyl-D-aspartate receptors and neuronal cell adhesion proteins, J Biol Chem 273:21105–21110

Huber KM, Gallagher SM, Warren ST et al. (2002) Altered synaptic plasticity in a mouse model of fragile X mental retardation, Proc Natl Acad Sci USA 99:7746–7750

Hung AY, Futai K, Sala C et al. (2008) Smaller dendritic spines, weaker synaptic transmission, but enhanced spatial learning in mice lacking Shank1. J Neurosci 28:1697–1708

Ichtchenko K, Hata Y, Nguyen T et al. (1995) Neuroligin 1: a splice site-specific ligand for beta-neurexins, Cell 81:435–443

Ichtchenko K, Nguyen T and Südhof TC (1996) Structures, alternative splicing, and neurexin binding of multiple neuroligins, J Biol Chem 271:2676–2682

Iida J, Hirabayashi S, Sato Y et al. (2004) Synaptic scaffolding molecule is involved in the synaptic clustering of neuroligin, Mol Cell Neurosci 27:497–508

Irie M, Hata Y, Takeuchi M et al. (1997) Binding of neuroligins to PSD-95. Science 277:1511–1515

Jamain S, Quach H, Betancur C et al. (2003) Mutations of the X-linked genes encoding neuroligins NLGN3 and NLGN4 are associated with autism, Nat Genet 34:27–29

Jamain S, Radyushkin K, Hammerschmidt K et al. (2008) Reduced social interaction and ultrasonic communication in a mouse model of monogenic heritable autism, Proc Natl Acad Sci USA 105:1710–1715

Kim HG, Kishikawa S, Higgins AW et al. (2008a) Disruption of neurexin 1 associated with autism spectrum disorder, Am J Hum Genet 82:199–207

Kim J, Jung SY, Lee YK et al. (2008b) Neuroligin-1 is required for normal expression of LTP and associative fear memory in the amygdala of adult animals, Proc Natl Acad Sci USA

Koehnke J, Jin X, Budreck EC et al. (2008) Crystal structure of the extracellular cholinesterase-like domain from neuroligin-2. Proc Natl Acad Sci USA 105:1873–1878

Kootz JP, Marinelli B and Cohen DJ (1981) Sensory receptor sensitivity in autistic children: response times to proximal and distal stimulation, Arch Gen Psychiatry 38:271–273

Laumonnier F, Bonnet-Brilhault F, Gomot M et al. (2004) X-linked mental retardation and autism are associated with a mutation in the NLGN4 gene, a member of the neuroligin family, Am J Hum Genet 74:552–557

Leonoudakis D, Conti LR, Radeke CM et al. (2004) A multiprotein trafficking complex composed of SAP97, CASK, Veli, and Mint1 is associated with inward rectifier Kir2 potassium channels, J Biol Chem 279:19051–19063

Marble DD, Hegle AP, Snyder ED, 2nd et al. (2005) Camguk/CASK enhances Ether-a-go-go potassium current by a phosphorylation-dependent mechanism, J Neurosci 25:4898–4907

Maximov A, Südhof TC and Bezprozvanny I (1999) Association of neuronal calcium channels with modular adaptor proteins, J Biol Chem 274:24453–24456

Meyer G, Varoqueaux F, Neeb A et al. (2004) The complexity of PDZ domain-mediated interactions at glutamatergic synapses: a case study on neuroligin, Neuropharmacology 47:724–733

Missler M, Zhang W, Rohlmann A et al. (2003) Alpha-neurexins couple Ca2+ channels to synaptic vesicle exocytosis, Nature 423:939–948

Moretti P, Levenson JM, Battaglia F et al. (2006) Learning and memory and synaptic plasticity are impaired in a mouse model of Rett syndrome, J Neurosci 26:319–327

Mukherjee K, Sharma M, Urlaub H et al. (2008) CASK Functions as a Mg2+-independent neurexin kinase, Cell 133:328–339

Nam CI and Chen L (2005) Postsynaptic assembly induced by neurexin-neuroligin interaction and neurotransmitter, Proc Natl Acad Sci USA 102:6137–6142

Oertel WH and Mugnaini E (1984) Immunocytochemical studies of GABAergic neurons in rat basal ganglia and their relations to other neuronal systems, Neurosci Lett 47:233–238
Petrenko AG, Ullrich B, Missler M et al. (1996) Structure and evolution of neurexophilin, J Neurosci 16:4360–4369
Piluso G, Carella M, D'Avanzo M et al. (2003) Genetic heterogeneity of FG syndrome: a fourth locus (FGS4) maps to Xp11.4-p11.3 in an Italian family, Hum Genet 112:124–130
Prange O, Wong TP, Gerrow K et al. (2004) A balance between excitatory and inhibitory synapses is controlled by PSD-95 and neuroligin, Proc Natl Acad Sci USA 101:13915–13920
Rowen L, Young J, Birditt B et al. (2002) Analysis of the human neurexin genes: alternative splicing and the generation of protein diversity, Genomics 79:587–597
Rudenko G, Hohenester E and Muller YA (2001) LG/LNS domains: multiple functions – one business end? Trends Biochem Sci 26:363–368
Rudenko G, Nguyen T, Chelliah Y et al. (1999) The structure of the ligand-binding domain of neurexin Ibeta: regulation of LNS domain function by alternative splicing, Cell 99:93–101
Sara Y, Biederer T, Atasoy D et al. (2005) Selective capability of SynCAM and neuroligin for functional synapse assembly, J Neurosci 25:260–270
Scheiffele P, Fan J, Choih J et al. (2000) Neuroligin expressed in nonneuronal cells triggers presynaptic development in contacting axons, Cell 101:657–669
Sebat J, Lakshmi B, Malhotra D et al. (2007) Strong association of de novo copy number mutations with autism, Science 316:445–449
Serajee FJ, Zhong H, Nabi R et al. (2003) The metabotropic glutamate receptor 8 gene at 7q31: partial duplication and possible association with autism, J Med Genet 40:e42
Shen KC, Kuczynska DA, Wu IJ et al. (2008) Regulation of neurexin 1beta tertiary structure and ligand binding through alternative splicing, Structure 16:422–431
Sheng M and Kim E (2000) The Shank family of scaffold proteins, J Cell Sci 113 (Pt 11):1851–1856
Song JY, Ichtchenko K, Südhof TC et al. (1999) Neuroligin 1 is a postsynaptic cell-adhesion molecule of excitatory synapses, Proc Natl Acad Sci USA 96:1100–1105
Squire LR and Zola SM (1996) Structure and function of declarative and nondeclarative memory systems, Proc Natl Acad Sci USA 93:13515–13522
Steffenburg S, Gillberg C, Hellgren L et al. (1989) A twin study of autism in Denmark, Finland, Iceland, Norway and Sweden, J Child Psychol Psychiatry 30:405–416
Sugita S, Saito F, Tang J et al. (2001) A stoichiometric complex of neurexins and dystroglycan in brain, J Cell Biol 154:435–445
Szatmari P, Paterson AD, Zwaigenbaum L et al. (2007) Mapping autism risk loci using genetic linkage and chromosomal rearrangements, Nat Genet 39:319–328
Tabuchi K, Blundell J, Etherton MR et al. (2007) A neuroligin-3 mutation implicated in autism increases inhibitory synaptic transmission in mice, Science 318:71–76
Tabuchi K and Südhof TC (2002) Structure and evolution of neurexin genes: insight into the mechanism of alternative splicing, Genomics 79:849–859
Ullrich B, Ushkaryov YA and Südhof TC (1995) Cartography of neurexins: more than 1000 isoforms generated by alternative splicing and expressed in distinct subsets of neurons, Neuron 14:497–507
Ushkaryov YA, Petrenko AG, Geppert M et al. (1992) Neurexins: synaptic cell surface proteins related to the alpha-latrotoxin receptor and laminin, Science 257:50–56
Varoqueaux F, Aramuni G, Rawson RL et al. (2006) Neuroligins determine synapse maturation and function, Neuron 51:741–754
Varoqueaux F, Jamain S and Brose N (2004) Neuroligin 2 is exclusively localized to inhibitory synapses, Eur J Cell Biol 83:449–456
Zhao X, Leotta A, Kustanovich V et al. (2007) A unified genetic theory for sporadic and inherited autism, Proc Natl Acad Sci USA 104:12831–12836
Zoghbi HY (2003) Postnatal neurodevelopmental disorders: meeting at the synapse? Science 302:826–830

Chapter 18
Synaptic Adhesion-Like Molecules (SALMs)

Philip Y. Wang and Robert J. Wenthold

Abstract The synaptic adhesion-like molecules (SALMs) are a newly discovered family of cell adhesion molecules that have a variety of functions in neuronal development, including aspects of neurite outgrowth and synapse formation (Ko et al. Neuron 50:233–245, 2006, Morimura et al. Gene 380:72–83, 2006, Wang et al. J Neurosci 26:2174–2183, 2006, Seabold et al. J Biol Chem 283:8395–8405, 2008, Wang et al. Mol Cell Neurosci, 39:83–94, 2008). Also known as Lrfn (leucine-rich and fibronectin III domain-containing), five family members have been identified thus far: SALM1/Lrfn2, SALM2/Lrfn1, SALM3/Lrfn4, SALM4/Lrfn3, and SALM5/Lrfn5. The SALMs have been shown to interact with NMDA receptors and the PSD-95 family of MAGUK proteins. Recent studies also indicate that the individual SALMs, while similar in structure, play distinct roles in heteromeric and homomeric protein interactions and neurite outgrowth (Seabold et al. J Biol Chem 283:8395–8405, 2008, Wang et al. Mol Cell Neurosci, 39:83–94, 2008). Neurite outgrowth and synapse formation are fundamental mechanisms in the development of the nervous system. While a considerable amount of information is known about both phenomena, the mechanism connecting the two is still enigmatic. SALMs join a growing mosaic of synaptic proteins that contribute to both neurite outgrowth and synapse formation during the course of development. Investigating SALMs and related proteins is essential for addressing fundamental questions of neuronal development.

Keywords Synapse · Dendrite · Axon · PDZ proteins · LRR · Growth cone · Heteromeric · Homomeric · Neurite outgrowth

---

R.J. Wenthold (✉)
Laboratory of Neurochemistry, National Institute on Deafness and Other Communication Disorders, National Institutes of Health, Bethesda, MD 20892, USA
e-mail: wenthold@nidcd.nih.gov

M. Hortsch, H. Umemori (eds.), *The Sticky Synapse*,
DOI 10.1007/978-0-387-92708-4_18, © Springer Science+Business Media, LLC 2009

18.1 SALM Family Structure and Expression

The synaptic adhesion-like molecules (SALMs) form a family of five adhesion molecules. In the mouse, they range from 626 to 788 amino acid residues in length. All SALMs contain a characteristic domain structure including six leucine-rich repeat (LRR) regions, an immunoglobulin C2-like domain (IgC2), a fibronectin type 3 domain (FN3), and a transmembrane region (TM) (Fig. 18.1). Additionally, SALMs 1–3 contain a PDZ-binding domain (PDZ-BD) at their distal C-termini, while SALMs 4–5 do not. The SALMs share a considerable amount of sequence similarity, though there are regions of high variability in both the N- and C-termini that could lend distinct characteristics to their individual functions (Fig. 18.2). Studies on SALMs and their functions thus far have focused on the rat and mouse representatives of this gene family. However, SALMs are present in a variety of mammalian species, as well as in fish and amphibians (Morimura et al. 2006). Interestingly, additional sequence analysis using the Ensembl database (www.ensembl.org) indicates that the protein structure of the SALMs is quite conserved across these species. As depicted in Fig. 18.3, the LRRs, IgC2, FN3, and PDZ-BD regions are all conserved across various SALM1 sequences, including those of human, mouse, platypus, chicken, and medaka (Japanese killifish). Analysis of the *Drosophila* genome indicates that SALMs share the closest sequence identity with a family of LRR and Ig-like domain containing transmembrane proteins called Kekkon (Kekkon 1–5), which function in the developmental regulation of EGF receptor activity during oogenesis (Ghiglione et al. 1999, 2003). Phylogenic analysis reveals that SALMs are closely related to a variety of leucine-rich molecules including the AMIGO, LINGO, NGL, PAL, and FLRT family of proteins (for review of these proteins, see Chen et al. 2006). As we describe later in this chapter, SALMs and these highly related proteins have one major characteristic in common: their role in regulating neurite outgrowth.

Northern blot analysis indicates that all SALM transcripts are found in the mouse and the rat brain, and transcripts for SALMs 2, 3, and 4 are seen to some extent in testis (Ko et al. 2006, Morimura et al. 2006). Additionally, SALM4

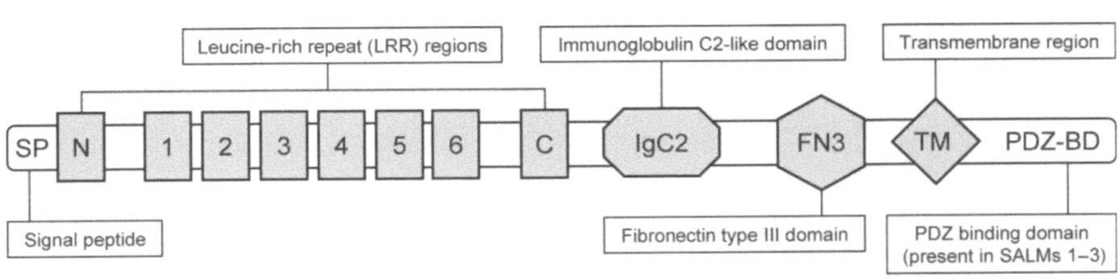

Fig. 18.1 SALM protein domain structure. Schematic diagram illustrating the domain structure of the SALM family of proteins. The SALMs contain an N-terminal signal peptide, six extracellular LRR regions flanked by N- and C-terminal LRR regions, an IgC2 domain, an FN3 domain, a single TM region, and a PDZ-BD (present in SALMs 1–3)

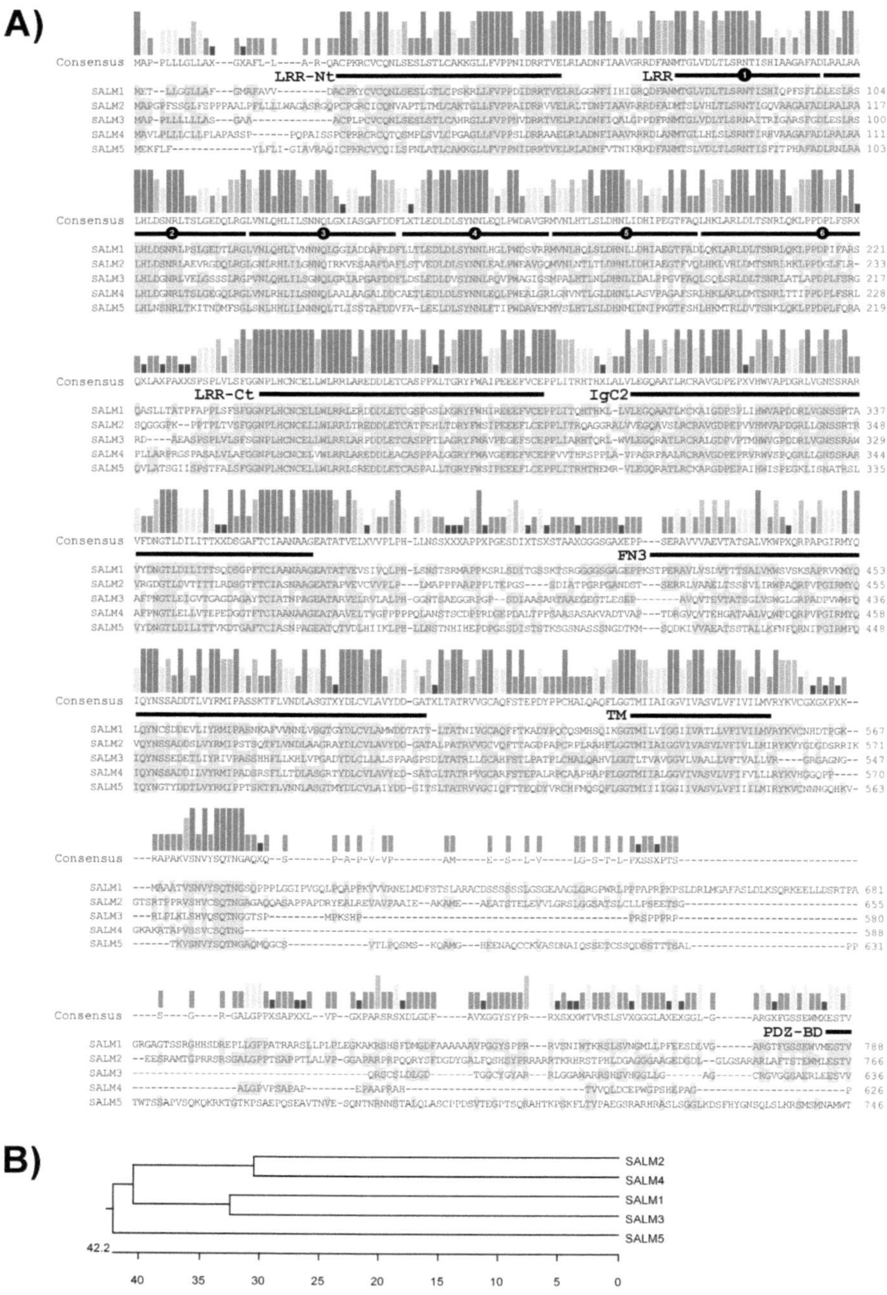

Fig. 18.2 SALM family sequence comparison. (**A**) Sequence analysis reveals that mouse SALMs contain six LRR regions (flanked by N- and C-terminal LRR sequences), an IgC2

transcripts are found in the gastrointenstinal tract and kidney (Morimura et al. 2006). Temporal expression profile blots indicate that transcript levels for SALMs 2–4 show an incremental increase starting from E10.5, while SALM1 and SALM5 increase from around E11.5–E12.5 (Morimura et al. 2006). In situ hybridization reveals that SALM transcripts are distinctly expressed in a variety of brain regions, including the cerebral cortex, hippocampus, dentate gyrus, and olfactory bulb (Ko et al. 2006, Morimura et al. 2006).

Western blot and subcellular fractionation experiments showed that SALM proteins are highly expressed in the rat brain, and enriched in synaptosomal and postsynaptic density fractions (Ko et al. 2006, Wang et al. 2006). SALM protein expression levels exhibit some differentiation among family members. Protein levels for SALM1 are detectable from E18, display high expression from P1 to P21, and then decrease at P28 (Wang et al. 2006). Protein levels for SALM2 increase from P21 and remain high at 6 weeks (Ko et al. 2006). SALM2 protein is widely distributed in brain, and for example, has been detected in cortical pyramidal neurons, hippocampal CA3 and CA1 neurons, and cerebellar Purkinje cells (Ko et al. 2006). At the subcellular level, SALM2 proteins localize to cell bodies, neurites, and punctate structures that co-localize with the presynaptic protein, synapsin I (Ko et al. 2006). Ultrastructural analysis using immunogold electron microscopy shows that native SALM4 is present at a variety of presynaptic, postsynaptic, and extrasynaptic sites in hippocampus, olfactory bulb, and cerebellar cortex (Seabold et al. 2008). Overexpressed SALMs are localized throughout the cell in the soma, axons, dendrites, and growth cones in young (<DIV7, days in vitro 7) neuronal cultures, both on the cell surface and intracellularly (Wang et al. 2006, 2008). In older (>DIV14) neuronal cultures, overexpressed SALMs are localized throughout the cell, on the cell surface, and at synapses (Ko et al. 2006, Wang et al. 2006, and unpublished observations). Additionally, Ko et al. (2006) demonstrated that SALM2 is localized to excitatory, but not inhibitory synapses, and that perturbation of SALM2 expression leads to aberrations in excitatory synaptic formation.

Fig 18.2 (continued) domain, an FN3 domain, a TM region, and a PDZ-BD at the distal C-termini (present in SALMs 1–3). Illustrated domain locations based on the SALM1 sequence. Similar amino acid residues are highlighted by the gray background, and the consensus sequence/strength is shown above the alignment. Consensus strength correlates with the length of the individual bars. Protein sequences for SALMs 1–5 (accession numbers NM_027452, NM_030562, NM_153388, NM_175478, and NM_178714, respectively) were aligned by the Clustal V method using MegAlign computer software. (**B**) Phylogenic tree comparing the mouse SALMs was constructed with MegAlign based on the alignment produced in (**A**)

Fig. 18.3 SALM1 species comparison. SALM1 protein sequences from a variety of species were acquired using the Ensembl database (www.ensembl.org). (**A**) The SALM1 sequence structure is highly conserved among a variety of mammalian, avian, and fish species including

18.2 SALM-Associated Proteins and Functional Significance

The known binding partners for the SALMs include the PSD-95 family of membrane-associated guanylate kinase (MAGUK) proteins, the NR1 subunit of the *N*-methyl-D-aspartate receptor (NMDAR), and the SALMs themselves – through homomeric and heteromeric interactions (Ko et al. 2006, Morimura et al. 2006, Wang et al. 2006, Seabold et al. 2008). The SALMs were independently identified in two different laboratories through yeast two-hybrid screens using the PDZ domains of MAGUKs as bait (SAP97 in Wang et al. 2006, and PSD-95 in Ko et al. 2006). Classically described as scaffolding proteins that assist in the tethering of receptors and associated proteins at the postsynaptic density (PSD), MAGUKs have been linked to a variety of functions in the CNS, including the trafficking of NMDARs to the synapse and neurite outgrowth (Kim and Sheng 2004, Charych et al. 2006). Overexpression of PSD-95 decreases dendritic branching in immature neurons, while knocking down PSD-95 increases it (Charych et al. 2006). Overexpression of SALMs in young neurons promotes neurite outgrowth, while overexpression of SALM 1–3 constructs lacking the PDZ-BD do not (Wang et al. 2008). This suggests that there may be a direct link in the process of neurite outgrowth and SALM–MAGUK associations, at least for SALMs 1–3. Interestingly, SALMs 4, 5 lack a PDZ-BD, but still promote neurite outgrowth. This may indicate that they act via a different mechanism or through heteromeric associations with other SALMs to induce MAGUK-associated neurite outgrowth. In mature neurons, deletion of the PDZ-BD has effects on synapse formation and morphology (Wang et al. 2006), further emphasizing the importance of associated PDZ proteins in SALM function.

Additional evidence suggests that SALMs may interact, indirectly or directly, with various other postsynaptic proteins. For example, bead-induced aggregation experiments revealed co-clustering of AMPARs and GKAP with SALM2 (Ko et al. 2006). SALM1 has also been shown to interact directly with NR1 in heterologous cells, and co-immunoprecipitates with NR1 and NR2 subunits of NMDARs from brain (Wang et al. 2006). The various protein domains that SALMs possess offer potential binding sites for other proteins.

Fig 18.3 (continued) human (*Homo sapiens*), mouse (*Mus musculus*), platypus (*Ornithorhynchus anatinus*), chicken (*Gallus gallus*), and Japanese killifish/medaka (*Oryzias latipes*). (**B**) Phylogenic tree comparing the SALM1 sequences of the various species. SALM1 proteins sequences were aligned by the Clustal V method using MegAlign. Ensembl gene IDs used for SALM1 human, mouse, platypus, chicken, and killifish were ENSG00000156564, ENSOANG00000014625, ENSGALG00000010050, and ENSORLG00000018222, respectively. Accession number NM_027452 was used for the SALM1 mouse sequence

18.3 Homomeric and Heteromeric SALM Interactions

Cell adhesion molecules (CAMs) are usually transmembrane proteins, and participate in the formation of cellular junctions through interactions between their extracellular protein domains. These CAM interactions can be *cis*, within the same membrane, or *trans*, across the cell–cell junctions of adjacent cells. These *cis* or *trans* interactions can be heteromeric (forming interactions with other distinct proteins), or homomeric (forming dimers or higher order multimers with a single protein type). For example, nectins form homomeric *cis* interactions (Takai and Nakanishi 2003), while L1-type CAMs and axonin-1 form heteromeric *cis* interactions (Buchstaller et al. 1996, Stoeckli et al. 1996). Cadherins form homomeric *trans* interactions (Tepass et al. 2000), while neuroligin and neurexin form heteromeric *trans* interactions (Ichtchenko et al. 1995, Scheiffele et al. 2000).

Seabold et al. (2008) demonstrated that SALMs are able to form heteromeric and homomeric interactions with each other, though with some distinctions among the individual SALM family members. From brain extracts, SALMs 1–3 co-immunoprecipitate with each other, indicating the formation of heteromeric interactions, while SALMs 4 and 5 do not. When individually expressed and co-plated in heterologous cells, only SALMs 4 and 5 are able to form homomeric *trans* interactions. These interactions are due to the extracellular N-terminus, as demonstrated through the use of chimera constructs made of the N- and C-termini of SALMs 2 and 4. When transfected into heterologous cells, a construct containing the N-terminus of SALM4 and the C-terminus of SALM2 (SALM4/2) forms homomeric *trans* interactions, while the reverse chimera construct (SALM2/4) does not. Application of antibodies directed to the extracellular LRR of SALMs blocks this interaction, indicating a function of the LRR region in these *trans* associations. Furthermore, when HeLa cells and primary hippocampal neurons overexpressing SALM4 are co-cultured, SALM4 is recruited to points of contact between the two different cell types. These SALM4 accumulations are seen at both axonal and dendritic adhesions between neurons and HeLa cells.

Interestingly, when co-transfected into heterologous cells in pairs, all five SALMs appear to form both heteromeric and homomeric complexes, indicating that the potential for such complexes may exist in vivo under certain conditions. The mechanism of SALM trafficking and the dynamics of their early interactions in the secretory pathway are not yet known. A hypothetical model of SALM interactions is depicted in Fig. 18.4. Early in the ER, SALMs may form combinations of dimers or higher order multimers with each other or other proteins. The constitution of such SALM multimers could contribute to the distinctiveness of SALM function, including their *cis*/*trans* interactions. The potential interplay between *cis* and *trans* interactions is highly complex, and may contribute to the variability among the SALM function. Formation of *cis* complexes may regulate the formation of *trans* complexes. For example,

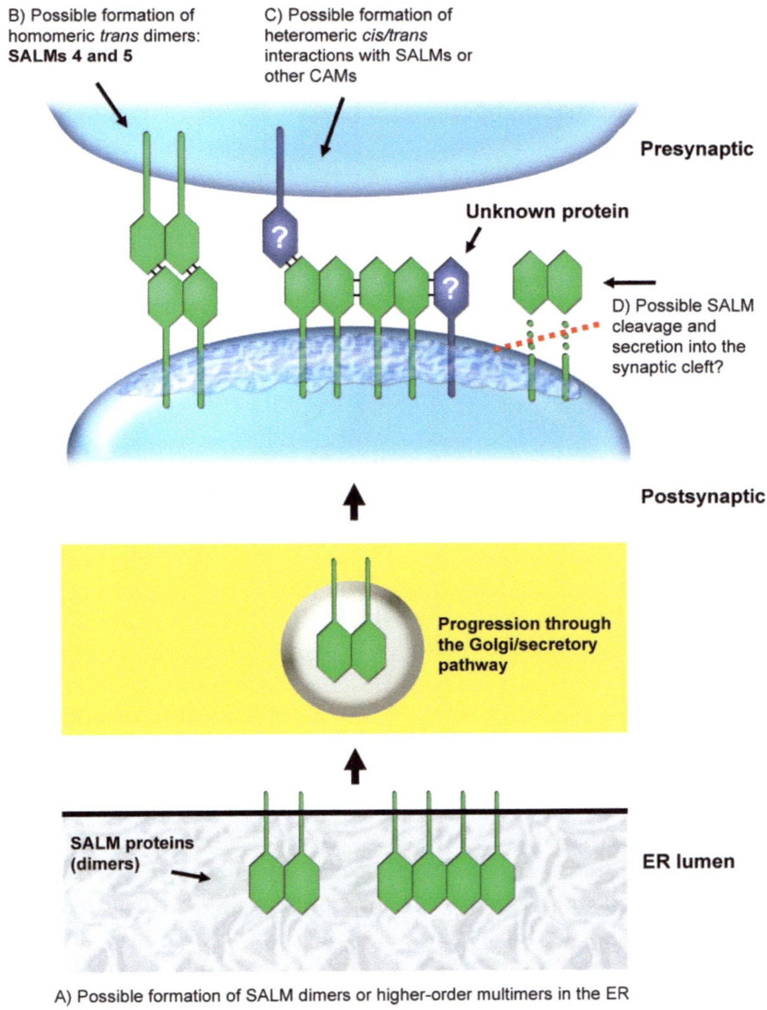

Fig. 18.4 Hypothetical model of SALM multimerization and *cis/trans*-synaptic interactions. (**A**) SALMs may assemble as dimers or higher order multimers within the ER, and traffic through the secretory pathway to the synapse. At the synapse, SALMs (specifically SALMs 4 and 5) may form *trans* complexes (**B**), or heteromeric *cis* and/or *trans* interactions with SALMs or other CAMs/unknown proteins (**C**). Another possible mechanism of regulating SALM function may involve proteolytic cleavage and release of the extracellular N-terminus into the synaptic cleft (**D**)

heteromeric *cis* interactions between axonin-1 and NgCAM (chicken L1-CAM) prevent the ability of axonin-1 to form homomeric *trans* interactions (Kunz et al. 1998, Sonderegger et al. 1998). Additionally, the formation of postsynaptic neuroligin-1/neurexin *cis* complexes inhibits the ability of postsynaptic neuroligin-1 to form *trans* interactions with presynaptic neurexin (Taniguchi et al. 2007). The *cis*/*trans* interactions of the various SALMs may work in concert to regulate various functions including intracellular signaling cascades and neurite outgrowth phenotypes of the SALMs. Further exploration of the trafficking, multimerization, and establishment of SALM interactions on the cell surface could be essential to understanding SALM function.

18.4 SALMs Promote Neurite Outgrowth

Neurite outgrowth is a fundamental event in the development of the nervous system. Through highly regulated mechanisms, neurons undergo axonal/dendritic polarization, and the subsequent branching and elongation of these neurites leads to the establishment of synaptic connections (da Silva and Dotti 2002). A vast collection of CAMs and other synaptic proteins is required for proper neurite outgrowth. For example, NCAM, N-cadherin, and L1-CAM regulate neurite outgrowth via MAPK signaling cascades, FGFR (fibroblast growth factor receptor) interactions, and changes in intracellular calcium levels (for review, see Doherty et al. 2000). CAMs regulate neurite outgrowth in various ways, as positive and negative regulators of outgrowth. As mentioned earlier, a set of closely related adhesion molecules with similar domain structures to SALMs all mediate neurite outgrowth, but with differential effects (Chen et al. 2006). Like SALMs, these proteins (including the AMIGO, LINGO, NGL, PAL, and FLRT families) have extracellular LRRs, a TM region, and often contain Ig-like and/or FN3 domains. AMIGO proteins contain six LRRs and one IgC2 domain, and promote neurite outgrowth in hippocampal neurons through their N-terminus (Kuja-Panula et al. 2003). Proteins in the LINGO family contain twelve LRRs and one IgC2, and negatively regulate neurite outgrowth by participating in Nogo-66 receptor signaling and modulating myelin-mediated inhibition (Mi et al. 2004). The nature of the outgrowth that is mediated by these structurally related CAMs is variable. Specific combinations of the various extracellular protein-interaction domains (X number of LRR $\pm$ IgC2 $\pm$ FN3) likely contribute to the resulting neurite outgrowth activity.

When overexpressed in young (DIV4) primary hippocampal neurons, all SALMs promote neurite outgrowth and neurite branching. However, they also have distinct effects from each other. SALM1 promotes an increase in mean process length. SALM2 promotes an increase in axon number, and SALM4 dramatically increases the number of short primary dendritic neurites. The mechanism of SALM-mediated neurite outgrowth involves both the C-termini and the N-termini of the SALMs. Regarding the C-terminus, SALM function is partially dependent upon PDZ interactions, as deletion of the PDZ-BD from SALMs 1–3 prevents their ability to promote neurite outgrowth. It is likely that SALMs 1–3 relay signaling cascades through PDZ interactions with PSD-95 and/or other PDZ domain proteins (for review, see Kim and Sheng 2004).

SALMs 4 and 5 do not contain PDZ-BDs, yet promote increases in neurite outgrowth, indicating that these SALMs act by a different mechanism, possibly involving alternate protein interactions at their C-termini (a region of considerable sequence variability between all SALMs).

Regarding the N-terminus, transfection of the SALM4/2 chimera construct promotes increases in the number of primary neurites, much like that of the full-length SALM4 phenotype (see Table 18.1 for a summary of known SALM-mediated neurite outgrowth characteristics). Numerous factors could contribute to these differential effects of SALM function. For example, while the N-termini of the SALMs show a large degree of similarity, there are distinct regions of variability that could lead to differential interactions with N-terminal

Table 18. 1 Summary of neurite outgrowth characteristics

	Total outgrowth	Mean process length	Number of branches	Number of processes
(A) SALM1	↑	↑	↑	–
(B) SALM2	↑	–	↑	↑
(C) SALM3	↑	–	↑	–
(D) SALM4	↑	–	↑	↑
(E) SALM5	↑	–	↑	–
(F) SALM2 (axons only)	↑	–	↑	↑
(G) αSALM-LRR	↓	↓	↓	–
(H) SALM12345 RNAi	↓	↓	–	–
(I) SALM1ΔPDZ	–	–	–	–
(J) SALM2ΔPDZ	–	–	–	–
(K) SALM3ΔPDZ	–	–	–	–
(L) SALM2/4	↑	–	–	–
(M) SALM4/2	↑	–	↑	↑

(A–E) The individual SALMs promote neurite outgrowth and various differentiation when transfected into young hippocampal cultures, as described in Wang et al. (2008). (F) When axons of SALM2-transfected neurons were digitally isolated and analyzed for neurite outgrowth characteristics, the resulting "axon-only" images show increases in total axonal outgrowth, axonal branches, and number of axons. (G, H) Application of function-blocking antibodies generated to the extracellular LRR region of SALMs, or knocking down SALM expression by RNA interference (quintuple transfection of SALMs 1–5 RNAi constructs) inhibits neurite outgrowth. (I–K) Transfection of SALM 1–3 constructs lacking the PDZ-BD has no significant effect on neurite outgrowth, indicating a role of PDZ interactions in SALM-mediated outgrowth. (L, M) Transfection of chimera constructs in which the N- and C-termini of SALM2 and SALM4 were switched indicates roles for both the N- and C-termini in SALM-mediated outgrowth. (M) Transfection of SALM4/2 (containing the N-terminus of SALM4 and the C-terminus of SALM2) resembles the SALM4 results (D), indicating that the SALM4 N-terminus contributes to the outgrowth effects. (L) While the SALM2/4 chimera promotes a statistically significant increase in total outgrowth, the magnitude of the increase is less than that of either full-length SALM2 (B) or SALM4 (D). In addition, there is no increase in neurite branching. This construct (containing the N-terminus of SALM2 and C-terminus of SALM4, which lacks a PDZ-BD) resembles SALM2ΔPDZ in both structure and outgrowth effects (J), further indicating the role of the C-terminus in SALM-mediated neurite outgrowth

binding partners. Additionally, sequence analysis reveals that there are distinct sites of potential N-linked glycosylation among the individual SALMs (unpublished observations). Glycosylation is a co-translational/post-translational modification that has implications for numerous mechanisms including cell adhesion, signal transduction, and protein transport (for review, see Scheiffele and Fullekrug 2000). The function of adhesion molecules may also be regulated by cleavage products and secreted molecules. For example, the cleaved N-terminal region of Slit can induce axonal branching of dorsal root ganglion neurons; however, this effect is antagonized by the presence of uncleaved Slit protein (Brose et al. 1999, Wang et al. 1999, Nguyen Ba-Charvet et al. 2001, and for review, see Brose and Tessier-Lavigne 2000). It is possible that the SALMs are themselves cleaved and act as soluble signaling molecules, or that individual SALMs differentially interact with soluble proteins.

18.5 SALMs at the Synapse

Synapse formation is a highly organized process that includes numerous variables of temporal and mechanical complexity. Establishment of a stable synapse includes the guidance and targeting of outgrowing neurites, trans-synaptic adhesion of pre- and postsynaptic sites, the recruitment of synaptic scaffolding proteins, and the insertion of neurotransmitter receptors into the synaptic membrane. Adhesive interactions that are mediated by CAMs facilitate several steps in this process. For example, cadherins are critical to proper synapse formation. Loss of cadherin junctions leads to delays in synapse formation, and those that are formed are reduced in size (Bozdagi et al. 2004). Neuroligins bind presynaptic neurexins to form functional synaptic terminals, and induce synaptic differentiation at both excitatory and inhibitory contacts (for review, see Dean and Dresbach 2006, Lise and El-Husseini 2006, Craig and Kang 2007). Presynaptic α–neurexins also participate in the clustering of Ca^{2+} channels (Missler et al. 2003, Zhang et al. 2005). SynCAM/nectins forms homomeric and heteromeric *trans* interactions to form functional synaptic terminals (Biederer et al. 2002, Sara et al. 2005, Fogel et al. 2007). NCAM promotes synapse stability through the binding of heparan sulfate proteoglycans (Dityatev et al. 2004).

While there is still much to be investigated, the SALMs have already been demonstrated to participate in many of these critical stages of synapse formation. SALMs are enriched in synaptic membranes and PSD subcellular fractions (Ko et al. 2006, Wang et al. 2006). As mentioned earlier, immunogold electron microscopy reveals that native SALM4 is present at both pre- and postsynaptic membranes of processes in the brain (Seabold et al. 2008). Additionally, the data indicate that SALM4 may be present at both excitatory and inhibitory synapses (Seabold et al. 2008). In contrast, SALM2 is reported to distribute exclusively to excitatory synapses (Ko et al. 2006). Overexpression of SALM2 increases the number of excitatory synapses and dendritic spines, while mislocalization of SALM2 decreases them. Knockdown of SALM2 reduces the frequency but not amplitude of miniature EPSCs, indicating a reduction of synapse number. This variability in localization among the various SALM family members could have important implications on synaptic development, and brings forth potentially interesting parallels between SALM and neuroligin function. Neuroligin-1 is present at excitatory synapses (Song et al. 1999), while neuroligin-2 is largely localized at inhibitory synapses (Varoqueaux et al. 2004). Neuroligins directly interact with PSD-95 via PDZ interactions (Irie et al. 1997), and overexpression of PSD-95 results in a change in the distribution of neuroligin-2 from inhibitory to excitatory synapses (Levinson et al. 2005). As mentioned earlier, neuroligins promote excitatory and inhibitory synapse formation, thereby mediating the delicate balance of inputs in the nervous system. The balance of neuronal excitation to inhibition is critical to normal brain function, and disruption of this balance has been implicated in underlying a variety of neurological disorders including autism (Rubenstein and Merzenich 2003). Additionally, several studies showed that various abnormalities in neuroligin genes are found in autistic individuals (for review, see Dean and Dresbach 2006, Lise and El-Husseini 2006, Craig and Kang 2007, and see Chapters 6 and 17). Localization of SALM proteins at inhibitory synapses has not yet been thoroughly investigated. The possibility that SALMs may function to regulate the balance of excitation to inhibition is highly intriguing and deserves further study.

SALMs also function to recruit neurotransmitter receptors and scaffolding proteins to the synapse. Wang et al. (2006) show that overexpression of SALM1 in DIV14 neurons recruits NMDARs and PSD-95 to dendritic puncta, and that this recruitment is dependent upon the PDZ-BD of SALM1. SALM1 also enhances surface expression of transfected NR2A, and co-immunoprecipitates with NR1 and NR2 subunits in brain. Additionally, Ko et al., (2006) demonstrate that bead aggregation of SALM2 induces the co-clustering of PSD-95 and other postsynaptic proteins, including GKAP and AMPARs. Finally, SALM1 modulates synapse morphology, as transfection of a SALM1 construct lacking the C-terminal region (post TM, SALM1ΔCT) into DIV14 primary hippocampal cultures induces a dramatic increase in the appearance of thin filopodia-like structures (Wang et al. 2006).

18.6 Dual Functions for SALMs

Neurite outgrowth and synapse formation are two distinct, but highly interconnected mechanisms that provide a framework for brain function in the CNS. While the intricacies of the molecular cues guiding these processes are continually being deciphered, the transition from neurite outgrowth to synapse formation is still a topic of great speculation. One strategy to elucidate this connection is to investigate the factors that are involved in both processes. SALMs join a growing list of synaptic proteins that have distinct roles in both processes (Fig. 18.5

illustrates the known roles of SALMs throughout development). For example, in young neurons, N-cadherin mediates growth cone migration (Letourneau et al. 1990) and neurite outgrowth (Bixby et al. 1987, Matsunaga et al. 1988, Bixby and Zhang 1990). In mature neurons, N-cadherin interacts with NMDAR complexes (Husi et al. 2000), and functions in long-term potentiation (LTP) (Tang et al. 1998), and synapse formation/maintenance (Fannon and Colman 1996). Early in development, L1-type proteins function in various processes, including axonal fasciculation, neurite outgrowth, and growth cone motility (Kamiguchi and Lemmon 1998, and for review, Kamiguchi and Lemmon 1997, Hortsch 2000). L1-type proteins are involved in numerous synaptic phenomena later in development, including LTP (Luthl et al. 1994, Matsumoto-Miyai et al. 2003) and synapse formation/organization (Saghatelyan et al. 2004, Godenschwege et al. 2006, Triana-Baltzer et al. 2006). DASM1 (dendrite arborization and synapse maturation 1) is an Ig, FN3, and PDZ-BD containing transmembrane protein and is involved in promoting both dendritic arborization early in development, and synapse maturation later in development (Shi et al. 2004a, b). DASM1 functions in silent synapse formation by regulating AMPARs-mediated synaptic transmission through PDZ interactions with Shank and S-CAM, two PDZ domain proteins that are involved in spine maturation (Shi et al. 2004a).

PDZ domain proteins are known to be involved in both neurite outgrowth and synapse formation (Charych et al. 2006, Hoogenraad et al. 2005). GRIP1 (glutamate receptor interacting protein 1) is an AMPA receptor-associated multi-PDZ domain protein that mediates the formation and outgrowth of dendrites in young neurons by regulating EphB receptor trafficking

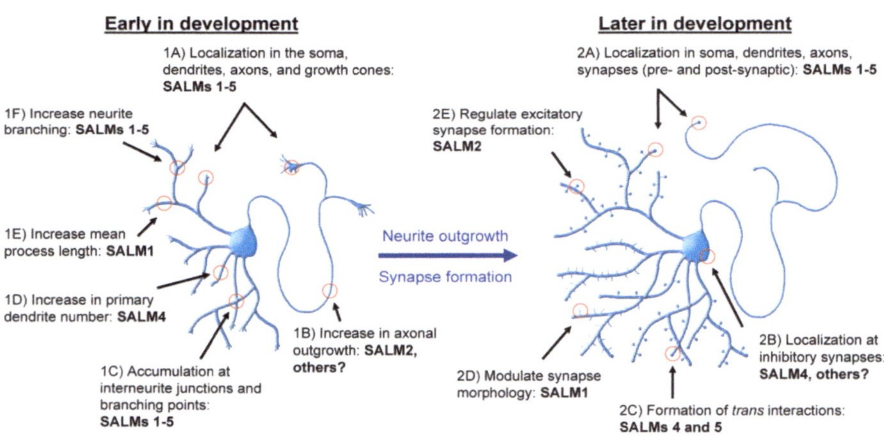

Fig. 18.5 Functional characteristics of SALMs. A schematic diagram illustrating the known functions of SALMs. The various SALMs have distinct roles throughout development. Early in development, SALMs promote neurite outgrowth and various kinds of neurite differentiation (1**A**–**F**). Later in development, SALMs are involved in synapse formation/stabilization (2**A**–**E**)

(Hoogenraad et al. 2005). In mature neurons, GRIP1 is involved in synaptic trafficking and stabilization of AMPARs (for review, see Song and Huganir 2002, Bredt and Nicoll 2003). In young neurons, PSD-95 regulates dendritic branching (Charych et al. 2006). Overexpression of PSD-95 decreases dendritic branching in primary hippocampal cultures, while knockdown of PSD-95 increases it (Charych et al. 2006). Among its many diverse synaptic roles, PSD-95 regulates synaptic localization of membrane proteins (Cline 2005, Han and Kim 2008), contributes to the formation and remodeling of the PSD (Marrs et al. 2001), and PSD-95 overexpression increases excitatory synapse formation (Ehrlich and Malinow 2004). In young neurons, SALMs may interact with PSD-95 and other MAGUKs to regulate neurite outgrowth.

Taken together, this suggests that some of the key players involved in synapse formation/stabilization and neurite outgrowth are shared, but raises questions about how these two processes are differentially regulated. For example, what factors determine the proportions of SALMs dedicated to synapse formation compared to neurite outgrowth? What developmental change signals their switch in functions? Do different protein interactions regulate outgrowth earlier in development, while others regulate synapse formation? Specifically, do SALMs have different PDZ interactions for neurite outgrowth and synapse formation? SALMs are enriched on the surface of axonal and dendritic growth cones (Wang et al. 2008). Does the contact of a growth cone to an impending postsynaptic surface signal a change in SALM function, thereby switching modalities to initiate synapse formation? Alternatively, perhaps the mechanistic switch from neurite outgrowth to synapse formation is not so distinct, but rather a fluid transition in the process of development that utilizes the same underlying machinery. Thus, there is an abundance of questions regarding both SALM function and the connection between neurite outgrowth and synapse formation, and striving to find the answers will lead to a greater understanding of adhesion molecules at synapses, and the development of the nervous system.

Acknowledgments We would like to thank Drs. Ron Petralia, Gail Seabold, and Stephan Brenowitz for critical reading of this manuscript. Philip Wang was supported by the Department of Biology, College of Chemical and Life Sciences, and the Neuroscience and Cognitive Science (NACS) program at the University of Maryland, College Park. This work was supported by the National Institute on Deafness and Other Communication Disorders (NIDCD) Intramural Research Program.

References

Biederer T, Sara Y, Mozhayeva M et al. (2002) SynCAM, a synaptic adhesion molecule that drives synapse assembly. Science 297:1525–1531

Bixby JL, Pratt RS, Lilien J et al. (1987) Neurite outgrowth on muscle cell surfaces involves extracellular matrix receptors as well as Ca2+-dependent and -independent cell adhesion molecules. Proc Natl Acad Sci USA 84:2555–2559

Bixby JL and Zhang R (1990) Purified N-cadherin is a potent substrate for the rapid induction of neurite outgrowth. J Cell Biol 110:1253–1260
Bozdagi O, Valcin M, Poskanzer K et al. (2004) Temporally distinct demands for classic cadherins in synapse formation and maturation. Mol Cell Neurosci 27:509–521
Bredt DS and Nicoll RA (2003) AMPA receptor trafficking at excitatory synapses. Neuron 40:361–379
Brose K, Bland KS, Wang KH et al. (1999) Slit proteins bind Robo receptors and have an evolutionarily conserved role in repulsive axon guidance. Cell 96:795–806
Brose K and Tessier-Lavigne M (2000) Slit proteins: key regulators of axon guidance, axonal branching, and cell migration. Curr Opin Neurobiol 10:95–102
Buchstaller A, Kunz S, Berger P et al. (1996) Cell adhesion molecules NgCAM and axonin-1 form heterodimers in the neuronal membrane and cooperate in neurite outgrowth promotion. J Cell Biol 135:1593–1607
Charych EI, Akum BF, Goldberg JS et al. (2006) Activity-independent regulation of dendrite patterning by postsynaptic density protein PSD-95. J Neurosci 26:10164–10176
Chen Y, Aulia S, Li L et al. (2006) AMIGO and friends: an emerging family of brain-enriched, neuronal growth modulating, type I transmembrane proteins with leucine-rich repeats (LRR) and cell adhesion molecule motifs. Brain Res Rev 51:265–274
Cline H (2005) Synaptogenesis: a balancing act between excitation and inhibition. Curr Biol 15:R203–205
Craig AM and Kang Y (2007) Neurexin-neuroligin signaling in synapse development. Curr Opin Neurobiol 17:43–52
da Silva JS and Dotti CG (2002) Breaking the neuronal sphere: regulation of the actin cytoskeleton in neuritogenesis. Nat Rev Neurosci 3:694–704
Dean C and Dresbach T (2006) Neuroligins and neurexins: linking cell adhesion, synapse formation and cognitive function. Trends Neurosci 29:21–29
Dityatev A, Dityateva G, Sytnyk V et al. (2004) Polysialylated neural cell adhesion molecule promotes remodeling and formation of hippocampal synapses. J Neurosci 24:9372–9382
Doherty P, Williams G and Williams EJ (2000) CAMs and axonal growth: a critical evaluation of the role of calcium and the MAPK cascade. Mol Cell Neurosci 16:283–295
Ehrlich I and Malinow R (2004) Postsynaptic density 95 controls AMPA receptor incorporation during long-term potentiation and experience-driven synaptic plasticity. J Neurosci 24:916–927
Fannon AM and Colman DR (1996) A model for central synaptic junctional complex formation based on the differential adhesive specificities of the cadherins. Neuron 17:423–434
Fogel AI, Akins MR, Krupp AJ et al. (2007) SynCAMs organize synapses through heterophilic adhesion. J Neurosci 27:12516–12530
Ghiglione C, Amundadottir L, Andresdottir M et al. (2003) Mechanism of inhibition of the Drosophila and mammalian EGF receptors by the transmembrane protein Kekkon 1. Development 130:4483–4493
Ghiglione C, Carraway KL 3rd, Amundadottir LT et al. (1999) The transmembrane molecule kekkon 1 acts in a feedback loop to negatively regulate the activity of the Drosophila EGF receptor during oogenesis. Cell 96:847–856
Godenschwege TA, Kristiansen LV, Uthaman SB et al. (2006) A conserved role for Drosophila Neuroglian and human L1-CAM in central-synapse formation. Curr Biol 16:12–23
Han K and Kim E (2008) Synaptic adhesion molecules and PSD-95. Prog Neurobiol 84:263–283
Hoogenraad CC, Milstein AD, Ethell IM et al. (2005) GRIP1 controls dendrite morphogenesis by regulating EphB receptor trafficking. Nat Neurosci 8:906–915
Hortsch M (2000) Structural and functional evolution of the L1 family: are four adhesion molecules better than one? Mol Cell Neurosci 15:1–10

Husi H, Ward MA, Choudhary JS et al. (2000) Proteomic analysis of NMDA receptor-adhesion protein signaling complexes. Nat Neurosci 3:661–669

Ichtchenko K, Hata Y, Nguyen T et al. (1995) Neuroligin 1: a splice site-specific ligand for beta-neurexins. Cell 81:435–443

Irie M, Hata Y, Takeuchi M et al. (1997) Binding of neuroligins to PSD-95. Science 277:1511–1515

Kamiguchi H and Lemmon V (1997) Neural cell adhesion molecule L1: signaling pathways and growth cone motility. J Neurosci Res 49:1–8

Kamiguchi H and Lemmon V (1998) A neuronal form of the cell adhesion molecule L1 contains a tyrosine-based signal required for sorting to the axonal growth cone. J Neurosci 18:3749–3756

Kim E and Sheng M (2004) PDZ domain proteins of synapses. Nat Rev Neurosci 5:771–781

Ko J, Kim S, Chung HS et al. (2006) SALM synaptic cell adhesion-like molecules regulate the differentiation of excitatory synapses. Neuron 50:233–245

Kuja-Panula J, Kiiltomaki M, Yamashiro T et al. (2003) AMIGO, a transmembrane protein implicated in axon tract development, defines a novel protein family with leucine-rich repeats. J Cell Biol 160:963–973

Kunz S, Spirig M, Ginsburg C et al. (1998) Neurite fasciculation mediated by complexes of axonin-1 and Ng cell adhesion molecule. J Cell Biol 143:1673–1690

Letourneau PC, Shattuck TA, Roche FK et al. (1990) Nerve growth cone migration onto Schwann cells involves the calcium-dependent adhesion molecule, N-cadherin. Dev Biol 138:430–442

Levinson JN, Chery N, Huang K et al. (2005) Neuroligins mediate excitatory and inhibitory synapse formation: involvement of PSD-95 and neurexin-1beta in neuroligin-induced synaptic specificity. J Biol Chem 280:17312–17319

Lise MF and El-Husseini A (2006) The neuroligin and neurexin families: from structure to function at the synapse. Cell Mol Life Sci 63:1833–1849

Luthl A, Laurent JP, Figurov A et al. (1994) Hippocampal long-term potentiation and neural cell adhesion molecules L1 and NCAM. Nature 372:777–779

Marrs GS, Green SH and Dailey ME (2001) Rapid formation and remodeling of postsynaptic densities in developing dendrites. Nat Neurosci 4:1006–1013

Matsumoto-Miyai K, Ninomiya A, Yamasaki H et al. (2003) NMDA-dependent proteolysis of presynaptic adhesion molecule L1 in the hippocampus by neuropsin. J Neurosci 23:7727–7736

Matsunaga M, Hatta K, Nagafuchi A et al. (1988) Guidance of optic nerve fibres by N-cadherin adhesion molecules. Nature 334:62–64

Mi S, Lee X, Shao Z et al. (2004) LINGO-1 is a component of the Nogo-66 receptor/p75 signaling complex. Nat Neurosci 7:221–228

Missler M, Zhang W, Rohlmann A et al. (2003) Alpha-neurexins couple Ca2+ channels to synaptic vesicle exocytosis. Nature 423:939–948

Morimura N, Inoue T, Katayama K et al. (2006) Comparative analysis of structure, expression and PSD95-binding capacity of Lrfn, a novel family of neuronal transmembrane proteins. Gene 380:72–83

Nguyen Ba-Charvet KT, Brose K, Ma L et al. (2001) Diversity and specificity of actions of Slit2 proteolytic fragments in axon guidance. J Neurosci 21:4281–4289

Rubenstein JL and Merzenich MM (2003) Model of autism: increased ratio of excitation/inhibition in key neural systems. Genes Brain Behav 2:255–267

Saghatelyan AK, Nikonenko AG, Sun M et al. (2004) Reduced GABAergic transmission and number of hippocampal perisomatic inhibitory synapses in juvenile mice deficient in the neural cell adhesion molecule L1. Mol Cell Neurosci 26:191–203

Sara Y, Biederer T, Atasoy D et al. (2005) Selective capability of SynCAM and neuroligin for functional synapse assembly. J Neurosci 25:260–270

Scheiffele P, Fan J, Choih J et al. (2000) Neuroligin expressed in nonneuronal cells triggers presynaptic development in contacting axons. Cell 101:657–669

Scheiffele P and Fullekrug J (2000) Glycosylation and protein transport. Essays Biochem 36:27–35

Seabold GK, Wang PY, Chang K et al. (2008) The SALM family of adhesion-like molecules forms heteromeric and homomeric complexes. J Biol Chem 283:8395–8405

Shi SH, Cheng T, Jan LY et al. (2004a) The immunoglobulin family member dendrite arborization and synapse maturation 1 (Dasm1) controls excitatory synapse maturation. Proc Natl Acad Sci USA 101:13346–13351

Shi SH, Cox DN, Wang D et al. (2004b) Control of dendrite arborization by an Ig family member, dendrite arborization and synapse maturation 1 (Dasm1). Proc Natl Acad Sci USA 101:13341–13345

Sonderegger P, Kunz S, Rader C et al. (1998) Discrete clusters of axonin-1 and NgCAM at neuronal contact sites: facts and speculations on the regulation of axonal fasciculation. Prog Brain Res 117:93–104

Song I and Huganir RL (2002) Regulation of AMPA receptors during synaptic plasticity. Trends Neurosci 25:578–588

Song JY, Ichtchenko K, Südhof TC et al. (1999) Neuroligin 1 is a postsynaptic cell-adhesion molecule of excitatory synapses. Proc Natl Acad Sci USA 96:1100–1105

Stoeckli ET, Ziegler U, Bleiker AJ et al. (1996) Clustering and functional cooperation of Ng-CAM and axonin-1 in the substratum-contact area of growth cones. Dev Biol 177:15–29

Takai Y and Nakanishi H (2003) Nectin and afadin: novel organizers of intercellular junctions. J Cell Sci 116:17–27

Tang L, Hung CP and Schuman EM (1998) A role for the cadherin family of cell adhesion molecules in hippocampal long-term potentiation. Neuron 20:1165–1175

Taniguchi H, Gollan L, Scholl FG et al. (2007) Silencing of neuroligin function by postsynaptic neurexins. J Neurosci 27:2815–2824

Tepass U, Truong K, Godt D et al. (2000) Cadherins in embryonic and neural morphogenesis. Nat Rev Mol Cell Biol 1:91–100

Triana-Baltzer GB, Liu Z and Berg DK (2006) Pre- and postsynaptic actions of L1-CAM in nicotinic pathways. Mol Cell Neurosci 33:214–226

Varoqueaux F, Jamain S and Brose N (2004) Neuroligin 2 is exclusively localized to inhibitory synapses. Eur J Cell Biol 83:449–456

Wang CY, Chang K, Petralia RS et al. (2006) A novel family of adhesion-like molecules that interacts with the NMDA receptor. J Neurosci 26:2174–2183

Wang KH, Brose K, Arnott D et al. (1999) Biochemical purification of a mammalian slit protein as a positive regulator of sensory axon elongation and branching. Cell 96:771–784

Wang PY, Seabold GK and Wenthold RJ (2008) Synaptic adhesion-like molecules (SALMs) promote neurite outgrowth. Mol Cell Neurosci 39:83–94

Zhang W, Rohlmann A, Sargsyan V et al. (2005) Extracellular domains of alpha-neurexins participate in regulating synaptic transmission by selectively affecting N- and P/Q-type $Ca2+$ channels. J Neurosci 25:4330–4342

Chapter 19
The Role of Integrins at Synapses

Devi Majumdar and Donna J. Webb

Abstract Integrins are transmembrane, adhesive receptors that serve as functional links between the extracellular matrix and the intracellular environment. Integrin engagement of a ligand can activate intracellular signaling pathways that regulate many cellular processes. The adhesive and signaling properties of integrins make them ideal for modulating connections at specialized adhesive junctions, such as synapses. Indeed, integrins are emerging as critical regulators of synaptic connections in the brain and the peripheral nervous system.

Keywords Integrins · Central nervous system synapses · Dendritic spines · Synaptic plasticity · Neuromuscular junction

19.1 Introduction

Integrins are a large family of heterodimeric (αβ) transmembrane receptors that mediate cell–cell and cell–matrix interactions. To date, 18 α and 8 β subunits have been identified in mammals that form 24 different integrins (Luo et al. 2007) (Fig. 19.1). Each integrin subunit contains a single transmembrane domain that connects a large, extracellular, ligand binding domain to a short cytoplasmic tail, which interacts with components of the actin cytoskeleton. The bidirectional linkage of integrins provides a physical connection between the extracellular environment and the interior of cells, making these receptors ideal for transmitting signals across the cell membrane. Indeed, integrin binding to a ligand can activate intracellular signaling cascades (outside-in signaling) that control many aspects of cellular behavior, including migration, adhesion, differentiation, proliferation, survival, and polarity (Hynes 2002). In turn,

D.J. Webb (✉)
Department of Biological Sciences and Vanderbilt Kennedy Center for Research on Human Development, Vanderbilt University, VU station B, Box 35-1634, Nashville, TN 37235, USA
e-mail: donna.webb@vanderbilt.edu

M. Hortsch, H. Umemori (eds.), *The Sticky Synapse*,
DOI 10.1007/978-0-387-92708-4_19,

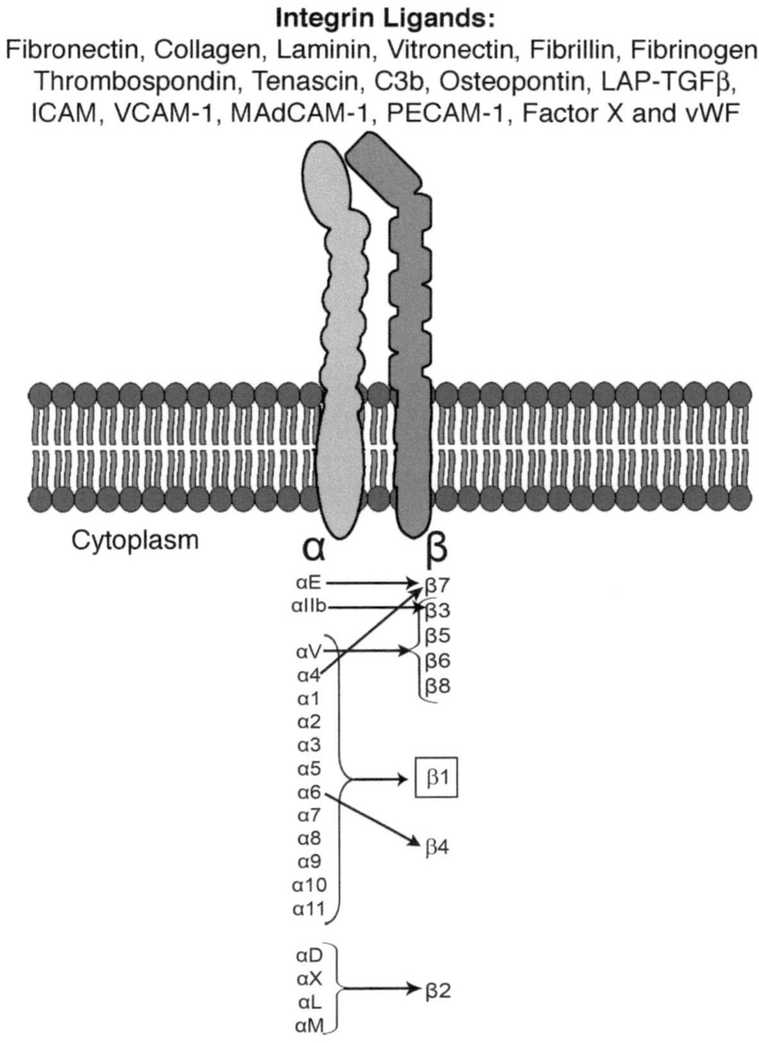

Fig. 19.1 Schematic representation of an integrin heterodimer. Integrins are heterodimeric, transmembrane receptors comprised of an α and β subunit. The subunits contain a large extracellular ligand binding domain, a single transmembrane domain, and a short cytoplasmic tail. To date, 18 α and 8 β subunits that form 24 different integrin heterodimers have been identified in mammals. Some of the extracellular integrin ligands are listed above the diagram

signaling events inside the cell can regulate the affinity of integrins for their ligands (inside-out signaling). Integrins can also associate with other receptors, such as growth factor receptors, and can function coordinately with these receptors to control cellular behavior (Schwartz and Shattil 2000, Hynes 2002, Miranti and Brugge 2002). In addition, integrins play an important role in cell–cell adhesion through their interaction with members of the Immunoglobulin superfamily, including cell adhesion molecules (CAMs).

The unique features of integrins position them well to contribute to the structure and function of specialized adhesive sites, such as synapses, in the nervous system. This chapter will focus on the role of integrins at central nervous system (CNS) synapses and at the neuromuscular junction.

19.2 Integrins at CNS Synapses

Neurons in the CNS are functionally connected by specialized cell–cell junctions, synapses, which allow for rapid transmission of signals between these cells. The known adhesive and signaling transduction properties of integrins make them attractive candidates to localize to these sites and contribute to synaptic function. Indeed, early studies showed that αvβ3 and α5β1 integrin epitopes were concentrated in brain synaptosomal membranes (Bahr et al. 1991, Bahr and Lynch 1992). Another integrin subunit, α8, was subsequently found at high concentrations in dendritic spines and at the postsynaptic density of neurons in the rat hippocampus (Einheber et al. 1996). Since that time, additional integrin subunits, including α3, β3, β5, and β8, have been found in CNS synapses (Nishimura et al. 1998, Chavis and Westbrook 2001, Kawaguchi and Hirano 2006, Shi and Ethell 2006). The presence of integrins at synapses raised the exciting possibility that they may be involved in the initiation and maintenance of synaptic connections.

19.2.1 Integrins and Synaptic Plasticity

A hint as to a role for integrins at CNS synapses came from integrin blocking experiments in rat hippocampal slices. Application of Arg-Gly-Asp (RGD)-containing peptides, which bind to integrins and inhibit their interaction with some ligands, affected long-term potentiation (LTP) (Staubli et al. 1990, Xiao et al. 1991). LTP, which is an increase in synaptic strength, is a widely studied form of synaptic plasticity in CNS synapses (Bliss and Lomo 1973). It includes an early phase, which begins immediately after LTP induction and lasts for several hours, and a late phase that starts a few hours after induction (Huang 1998). Disruption of integrin-mediated adhesion by application of RGD-containing peptides within 10 min after induction of LTP effectively blocked LTP, while application of the peptides 25 min after induction did not significantly affect LTP (Staubli et al. 1998). This implies that integrins are required for the stabilization of a very early stage of LTP. As RGD-containing peptides bind to multiple integrins and inhibit their binding to RGD-containing ligands, the specific integrin subunits, which contributed to this process, could not be identified in these studies. Additionally, the function of some integrins is not affected by RGD-containing peptides and their contribution to synaptic plasticity would not be detected by this approach, pointing to a need for alternative approaches to tackle this issue.

In this regard, pharmacological and genetic manipulations of integrin function and expression were used to specifically associate several α- and β-integrin subunits with the stabilization and modulation of LTP (Kramar et al. 2002, Chan et al. 2003, 2007). In rat hippocampal slices, an infusion of a function-blocking antibody to α3 integrin causes a slow and gradual decay of potentiation (Anderson and Ferreira 2004). Likewise, genetically altered mice with reduced expression of α3 integrin failed to maintain hippocampal LTP (Chan et al. 2003). α3 integrin has an additional role in the induction of LTP since hippocampal LTP was severely impaired in mice with a forebrain-specific loss of α3 integrin (Chan et al. 2007). In a similar way, the function of two other α-integrin subunits in synaptic plasticity was elucidated. The function of α5 integrin subunit in modulating synaptic plasticity is less clear although it does appear to be involved in this process. Neutralization of α5 integrin results in a gradual decline of LTP, suggesting that α5 integrin is involved in LTP stabilization (Chun et al. 2001). However, a reduced expression of α5 integrin in mice did not significantly affect hippocampal LTP (Chan et al. 2003). It should be pointed out that the level of α5 integrin was reduced by only 50% in these genetically altered mice, which may partially account for the conflicting results between these two studies. α8 integrin also plays a role in the induction and/or stability of hippocampal LTP (Chan et al. 2003). Although α5 and α8 integrin subunits are involved in regulating synaptic plasticity, the genetic studies suggest that α3 plays a more prominent role in this process.

β1 integrin is another attractive candidate to regulate synaptic plasticity because it is the only known subunit partner for α3, α5, and α8. Indeed, mice with a forebrain-specific knockout of β1 integrin exhibited a deficit in hippocampal LTP (Chan et al. 2006, Huang et al. 2006). As indicated by these studies, several different integrins appear to be involved in modulating synaptic plasticity. However, it is presently not clear whether there is a redundancy in their synaptic function or whether they contribute to different aspects of plasticity.

19.2.2 Integrins and Memory

Since changes in synaptic strength (synaptic plasticity) are thought to underlie learning and memory, integrins are prime candidates to contribute to memory formation. Work in *Drosophila* has pointed to the importance of integrins in short-term memory formation (Grotewiel et al. 1998). Disruption of the *Volado* (*Vol*) gene in *Drosophila*, which codes for two isoforms of an α-integrin, dominantly impairs olfactory memories within a few minutes of training. Such short-term memory impairment could be rescued by conditional expression of the *Vol* transgene in adult flies. The *Vol* mutant is also defective in forms of calcium and activity-dependent synaptic plasticity at NMJs (Rohrbough et al. 2000). The defects can be rescued by the conditional expression of a *Vol* transgene, implicating integrins in synaptic plasticity and memory formation. Recent studies indicate that integrins also contribute to memory processes in mammals. Mice deficient in α3 or β1 integrin subunits display a significant impairment in performing hippocampal-dependent working memory tasks (Chan et al. 2006, 2007). In addition, several integrins are implicated in spatial memory as mice with a reduced expression of three α-integrin subunits, α3, α5, and α8, are impaired in their ability to learn spatial information (Chan et al. 2003). Therefore, accumulating evidence clearly points to a critical role for integrins in synaptic plasticity and memory formation.

19.2.3 Integrins Modulate Neurotransmitter Receptors

Carefully coordinated signaling between pre- and postsynaptic terminals is essential for establishing and maintaining synaptic function. For the most part, rapid excitatory synaptic transmission takes place through two types of glutamate receptors: α-amino-3-hydroxy-5-methyl-4-isoxazole propionic acid (AMPA)-type receptor and *N*-methyl-D-asparate (NMDA)-type receptor. Evidence indicating that integrins are important modulators of synaptic plasticity suggests that these adhesion receptors can regulate the activities of the glutamate receptors. In hippocampal slices, integrin activation influences NMDA receptor-mediated synaptic currents (Lin et al. 2003). Src kinases, which phosphorylate the NR2 subunit of NMDA receptors (Hisatsune et al. 1999, Grosshans and Browning 2001), are thought to be mediators through which integrins modulate NMDA receptor activity. A study by the same group provided evidence that ligand-induced activation of β1 integrin subunits promotes phosphorylation of NMDA receptors through kinases of the focal adhesion kinase (FAK) and Src family members and thereby increases NMDA receptor responses (Bernard-Trifilo et al. 2005). There is also a functional link between integrin signaling and NMDA receptor responses in cortical neurons (Watson et al. 2007). NMDA receptor antagonists block integrin-induced increases in intracellular calcium levels and the activation of extracellular signal-regulated kinase (ERK) 1 and 2. Interestingly, activation of a specific integrin, α5β1, caused a similar increase in intracellular calcium (Lin et al. 2008). Integrin modulation of NMDA receptor activity implies a close relationship between these adhesion receptors and synaptic transmission.

Work in hippocampal neurons found that peptides containing the RGD motif block an activity-dependent reduction in glutamate release and a switch in the subunit composition of NMDA receptors (NR2B to NR2A subunits) (Chavis and Westbrook 2001). Both of these events are dependent on the β3 integrin subunit since they were able to be inhibited by a function-blocking antibody to β3 integrin. These events are associated with the maturation of

CNS excitatory synapses and provide evidence that integrin-mediated signaling is important for the orchestrated maturation of synapses. Interestingly, application of a broad-spectrum tyrosine kinase inhibitor mimics the effects of the blockade of β3 integrin function, suggesting that tyrosine kinase activity is coupled to β3 integrin signaling in this process.

The regulated addition and removal of AMPA receptors is the basis for long-lasting forms of synaptic plasticity, which allow for changes in synaptic strength (Bredt and Nicoll 2003, Shepherd and Huganir 2007). Emerging data indicates that AMPA receptors are stabilized by β3 integrin (Cingolani et al. 2008). Perturbation of β3 integrin promotes the internalization of GluR2-containing AMPA receptors, leading to a decrease in AMPA receptor synaptic currents. These findings show a critical role for integrins in modulating synaptic strength. Activation of AMPA receptors may in turn regulate expression of cell-surface integrins (Lin et al. 2005). Stimulation of AMPA receptors increases cell-surface levels of α5β1 integrin via a protein kinase C (PKC)-dependent mechanism (Lin et al. 2005). Consistent with this, glutamate stimulation increases the insertion of α5β1 into the synaptic plasma membrane (Webb et al. 2007). It is tempting to speculate that induction of LTP activates signaling pathways that drive the insertion of integrins into synaptic membranes, which subsequently contribute to the stabilization of LTP.

19.2.4 Dendritic Spines and Integrins

Postsynaptic terminals of excitatory CNS synapses are composed of dendritic spines (Matus et al. 1982), which are small, actin-rich extensions of the dendrite that form connections with axonal terminals (Matus et al. 1982). Dendritic spines receive the majority of the excitatory synaptic inputs in the CNS and are thought to mediate synaptic plasticity. Not surprisingly, several integrins have been found to localize to dendritic spines and regulate their development. In hippocampal neurons, α5β1 integrin localizes to dendritic spines and regulates spine and synapse formation via a signaling mechanism, which includes Src kinase, Rac, and the signaling adaptor protein GIT1 (Webb et al. 2007). The localization of α5β1 integrin to spines is activity-dependent since glutamate stimulation increases the concentration of this integrin in dendritic spines. β3 integrin subunits also localize to dendritic spines and regulate spine dynamics (Shi and Ethell 2006). Ligand binding to integrins induces spine elongation and promotes the formation of new dendrite protrusions, which are thought to be precursors for spines. β1 and β3 integrin subunits are critical to these processes, which are mediated through NMDA receptor and CaMKII (Huang et al. 2006). Other molecules also function through integrins to mediate their effects on dendritic spines. Stimulation of acute hippocampal slices with ephrin-A3 ligand, which binds EphA4, reduces the length and density of dendritic spines by inactivating β1 integrin signaling (Bourgin et al. 2007), indicating a potential mechanism for regulation for integrin signaling pathways.

19.3 Integrins in the Neuromuscular Junction Synapses

The NMJ is a specialized synaptic site by which motor neurons communicate with muscle fibers. As with CNS synapses, communication requires precise release of a neurotransmitter (acetylcholine) from the motor axon and binding to neurotransmitter receptors (acetylcholine receptors) in the postsynaptic cell membrane. A specialized extracellular matrix, the basal lamina, occupies the synaptic cleft between the muscle fiber and the nerve terminal and is thought to be an important organizer of postsynaptic differentiation (see Chapter 3). Since integrins are receptors for components of the basal lamina, they are attractive candidates to regulate synaptic function at the NMJ. In *Xenopus*, β1 integrin subunits are concentrated at NMJs. α3β1 integrin is enriched at active zones of motor nerve terminals (Anderson et al. 1996, Cohen et al. 2000). In human muscle, α1 and α7 integrin subunits are found in synaptic membranes at the NMJ (Martin et al. 1996). α7β1 integrin associates with acetylcholine receptors (AChRs) early during the formation of the postsynaptic membrane and together with laminin and agrin regulates clustering of AChRs, which is critical to the development of the NMJ (Burkin et al. 1998, 2000). αv and β1 integrin subunits appear to be necessary for agrin signaling in primary cultures of chicken muscle (Martin and Sanes 1997). Inactivation of the β1 integrin gene in mice results in defects in presynaptic nerve terminals and in postsynaptic differentiation (Schwander et al. 2004). Similar defects were observed in agrin null mice (Gautam et al. 1996, Lin et al. 2001), suggesting that agrin signaling is altered in the absence of β1 integrin.

As in CNS synapses, integrins regulate synaptic transmission at the NMJ. At the frog NMJ, muscle stretch increases the spontaneous and evoked neurotransmitter release from motor nerve terminals by a β1 integrin-dependent mechanism (Chen and Grinnell 1995). Similarly, integrins contribute to the enhanced neurotransmitter release induced by hypertonic solutions at frog nerve terminals and at *Drosophila* NMJs (Kashani et al. 2001, Suzuki et al. 2002). In *Drosophila*, three α-integrins, αPS1, αPS2, and αPS3 (Volado), and their β-binding partner, βPS, are found at NMJs (Beumer et al. 1999). The synaptic integrins act upstream of CaMKII in the regulation of NMJ architecture (Beumer et al. 2002). αPS3 integrin, which is encoded by the *Volado* gene, has additional roles in the regulation of synaptic transmission, activity-dependent synaptic plasticity, and short-term memory as described above (Grotewiel et al. 1998, Rohrbough et al. 2000). Collectively, these studies show that integrins have both structural and functional roles at the NMJ.

19.4 Role of Integrins in Synaptic Neuropathology

Various developmental and neurological disorders, including Alzheimer's disease, mental retardation, schizophrenia, and epilepsy, are associated with alterations in the density, morphology, and size of dendritic spines and synapses (Fiala et al. 2002). This raises the intriguing question of whether integrins contribute to these pathological changes through their function in regulating synaptic connections. Several lines of evidence suggest a possible link between integrins and these disorders. $\alpha5\beta1$ integrin signaling through the small GTPase Rac regulates the development of dendritic spines and synapses (Webb et al. 2007). Mutations in Rac activators and effectors, including αPIX and p21-activated kinase (PAK)3, which alter this signaling pathway has been found in patients with nonsyndromic mental retardation (Allen et al. 1998, Kutsche et al. 2000, Ramakers 2002). Integrins along with NMDA receptors may cooperate to regulate the uptake of amyloid-β (Aβ) peptides, which are the major neurotoxic component of senile plaques in Alzheimer's disease (AD) (Bi et al. 2002). The formation of Aβ deposits impairs synaptic transmission and synaptic plasticity in vivo (Stephan et al. 2001), pointing to a possible connection between integrins and the procession of AD. It is interesting to note that recent studies indicate loss of synapses in the hippocampus is an early event in AD caused by oligomerization of Aβ (Coleman et al. 2004).

19.5 Concluding Remarks

With their adhesive and signaling properties, integrins are poised to modulate structural and functional aspects of synaptic connections. Indeed, over the last 15 years, integrins have been shown to be involved in multifaceted functions at CNS synapses and NMJs (Fig. 19.2). Elegant genetic and pharmacological

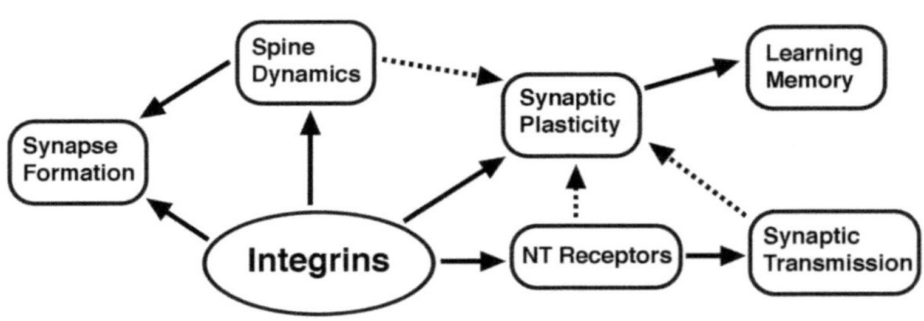

Fig. 19.2 Proposed functions for integrins at synapses. Integrins are critical regulators of synaptic connections in the central and peripheral nervous system. Integrins are involved in the modulation of synaptic plasticity, learning and memory processes, synaptic transmission, and neurotransmitter receptor responses. Integrins also play important roles in the development of dendritic spine and synapses

studies have clearly shown the importance of integrins in modulating synaptic plasticity, synaptic transmission, and memory processes. Their ability to regulate neurotransmitter receptor activities is perhaps one way in which integrins contribute to these processes. Integrins are also critical for the formation of new synapses and dendritic spines. In all, integrins are emerging as an important class of molecules that are essential for communication between neurons. However, many questions still remain regarding the function of integrins in regulating synaptic connections and the next 15 years should prove to be an exciting time as we more clearly elucidate the function of integrins in modulating the complex circuitry in the brain and peripheral nervous system.

Acknowledgments This work was supported by grant MH071674 from NIH to D.J.W.

References

Allen KM, Gleeson JG, Bagrodia S et al. (1998) PAK3 mutation in nonsyndromic X-linked mental retardation. Nat Genet 20:25–30

Anderson KL and Ferreira A (2004) alpha1 Integrin activation: a link between beta-amyloid deposition and neuronal death in aging hippocampal neurons. J Neurosci Res 75:688–697

Anderson MJ, Shi ZQ and Zackson SL (1996) Proteolytic disruption of laminin-integrin complexes on muscle cells during synapse formation. Mol Cell Biol 16:4972–4984

Bahr BA and Lynch G (1992) Purification of an Arg-Gly-Asp selective matrix receptor from brain synaptic plasma membranes. Biochem J 281 (Pt 1):137–142

Bahr BA, Sheppard A and Lynch G (1991) Fibronectin binding by brain synaptosomal membranes may not involve conventional integrins. Neuroreport 2:13–16

Bernard-Trifilo JA, Kramar EA, Torp R et al. (2005) Integrin signaling cascades are operational in adult hippocampal synapses and modulate NMDA receptor physiology. J Neurochem 93:834–849

Beumer K, Matthies HJ, Bradshaw A et al. (2002) Integrins regulate DLG/FAS2 via a CaM kinase II-dependent pathway to mediate synapse elaboration and stabilization during postembryonic development. Development 129:3381–3391

Beumer KJ, Rohrbough J, Prokop A et al. (1999) A role for PS integrins in morphological growth and synaptic function at the postembryonic neuromuscular junction of Drosophila. Development 126:5833–5846

Bi X, Gall CM, Zhou J et al. (2002) Uptake and pathogenic effects of amyloid beta peptide 1–42 are enhanced by integrin antagonists and blocked by NMDA receptor antagonists. Neuroscience 112:827–840

Bliss TV and Lomo T (1973) Long-lasting potentiation of synaptic transmission in the dentate area of the anaesthetized rabbit following stimulation of the perforant path. J Physiol 232:331–356

Bourgin C, Murai KK, Richter M et al. (2007) The EphA4 receptor regulates dendritic spine remodeling by affecting beta1-integrin signaling pathways. J Cell Biol 178:1295–1307

Bredt DS and Nicoll RA (2003) AMPA receptor trafficking at excitatory synapses. Neuron 40:361–379

Burkin DJ, Gu M, Hodges BL et al. (1998) A functional role for specific spliced variants of the alpha7beta1 integrin in acetylcholine receptor clustering. J Cell Biol 143:1067–1075

Burkin DJ, Kim JE, Gu M et al. (2000) Laminin and alpha7beta1 integrin regulate agrin-induced clustering of acetylcholine receptors. J Cell Sci 113 (Pt 16):2877–2886

Chan CS, Levenson JM, Mukhopadhyay PS et al. (2007) Alpha3-integrins are required for hippocampal long-term potentiation and working memory. Learn Mem 14:606–615
Chan CS, Weeber EJ, Kurup S et al. (2003) Integrin requirement for hippocampal synaptic plasticity and spatial memory. J Neurosci 23:7107–7116
Chan CS, Weeber EJ, Zong L et al. (2006) Beta 1-integrins are required for hippocampal AMPA receptor-dependent synaptic transmission, synaptic plasticity, and working memory. J Neurosci 26:223–232
Chavis P and Westbrook G (2001) Integrins mediate functional pre- and postsynaptic maturation at a hippocampal synapse. Nature 411:317–321
Chen BM and Grinnell AD (1995) Integrins and modulation of transmitter release from motor nerve terminals by stretch. Science 269:1578–1580
Chun D, Gall CM, Bi X et al. (2001) Evidence that integrins contribute to multiple stages in the consolidation of long term potentiation in rat hippocampus. Neuroscience 105:815–829
Cingolani LA, Thalhammer A, Yu LM et al. (2008) Activity-dependent regulation of synaptic AMPA receptor composition and abundance by beta3 integrins. Neuron 58:749–762
Cohen MW, Hoffstrom BG and DeSimone DW (2000) Active zones on motor nerve terminals contain alpha 3beta 1 integrin. J Neurosci 20:4912–4921
Coleman P, Federoff H and Kurlan R (2004) A focus on the synapse for neuroprotection in Alzheimer disease and other dementias. Neurology 63:1155–1162
Einheber S, Schnapp LM, Salzer JL et al. (1996) Regional and ultrastructural distribution of the alpha 8 integrin subunit in developing and adult rat brain suggests a role in synaptic function. J Comp Neurol 370:105–134
Fiala JC, Spacek J and Harris KM (2002) Dendritic spine pathology: cause or consequence of neurological disorders? Brain Res. Brain Res. Rev. 39:29–54
Fox MA (this volume) Development of the vertebrate neuromuscular junction. In Hortsch M and Umemori H (eds) The sticky synapse, Springer, New York
Gautam M, Noakes PG, Moscoso L et al. (1996) Defective neuromuscular synaptogenesis in agrin-deficient mutant mice. Cell 85:525–535
Grosshans DR and Browning MD (2001) Protein kinase C activation induces tyrosine phosphorylation of the NR2A and NR2B subunits of the NMDA receptor. J Neurochem 76:737–744
Grotewiel MS, Beck CD, Wu KH et al. (1998) Integrin-mediated short-term memory in Drosophila. Nature 391:455–460
Hisatsune C, Umemori H, Mishina M et al. (1999) Phosphorylation-dependent interaction of the N-methyl-D-aspartate receptor epsilon 2 subunit with phosphatidylinositol 3-kinase. Genes Cells 4:657–666
Huang EP (1998) Synaptic plasticity: going through phases with LTP. Curr Biol 8:R350–352
Huang Z, Shimazu K, Woo NH et al. (2006) Distinct roles of the beta 1-class integrins at the developing and the mature hippocampal excitatory synapse. J Neurosci 26:11208–11219
Hynes RO (2002) Integrins: bidirectional, allosteric signaling machines. Cell 110:673–687
Kashani AH, Chen BM and Grinnell AD (2001) Hypertonic enhancement of transmitter release from frog motor nerve terminals: Ca2+ independence and role of integrins. J Physiol 530:243–252
Kawaguchi SY and Hirano T (2006) Integrin alpha3beta1 suppresses long-term potentiation at inhibitory synapses on the cerebellar Purkinje neuron. Mol Cell Neurosci 31:416–426
Kramar EA, Bernard JA, Gall CM et al. (2002) Alpha3 integrin receptors contribute to the consolidation of long-term potentiation. Neuroscience 110:29–39
Kutsche K, Yntema H, Brandt A et al. (2000) Mutations in ARHGEF6, encoding a guanine nucleotide exchange factor for Rho GTPases, in patients with X-linked mental retardation. Nat Genet 26:247–250
Lin B, Arai AC, Lynch G et al. (2003) Integrins regulate NMDA receptor-mediated synaptic currents. J Neurophysiol 89:2874–2878

Lin CY, Hilgenberg LG, Smith MA et al. (2008) Integrin regulation of cytoplasmic calcium in excitatory neurons depends upon glutamate receptors and release from intracellular stores. Mol Cell Neurosci 37:770–780

Lin CY, Lynch G and Gall CM (2005) AMPA receptor stimulation increases alpha5beta1 integrin surface expression, adhesive function and signaling. J Neurochem 94:531–546

Lin W, Burgess RW, Dominguez B et al. (2001) Distinct roles of nerve and muscle in postsynaptic differentiation of the neuromuscular synapse. Nature 410:1057–1064

Luo BH, Carman CV and Springer TA (2007) Structural basis of integrin regulation and signaling. Annu Rev Immunol 25:619–647

Martin PT, Kaufman SJ, Kramer RH et al. (1996) Synaptic integrins in developing, adult, and mutant muscle: selective association of alpha1, alpha7A, and alpha7B integrins with the neuromuscular junction. Dev Biol 174:125–139

Martin PT and Sanes JR (1997) Integrins mediate adhesion to agrin and modulate agrin signaling. Development 124:3909–3917

Matus A, Ackermann M, Pehling G et al. (1982) High actin concentrations in brain dendritic spines and postsynaptic densities. Proc Natl Acad Sci USA 79:7590–7594

Miranti CK and Brugge JS (2002) Sensing the environment: a historical perspective on integrin signal transduction. Nat Cell Biol 4:E83–E90

Nishimura SL, Boylen KP, Einheber S et al. (1998) Synaptic and glial localization of the integrin alphavbeta8 in mouse and rat brain. Brain Res 791:271–282

Ramakers GJ (2002) Rho proteins, mental retardation and the cellular basis of cognition. Trends Neurosci 25:191–199

Rohrbough J, Grotewiel MS, Davis RL et al. (2000) Integrin-mediated regulation of synaptic morphology, transmission, and plasticity. J Neurosci 20:6868–6878

Schwander M, Shirasaki R, Pfaff SL et al. (2004) Beta1 integrins in muscle, but not in motor neurons, are required for skeletal muscle innervation. J Neurosci 24:8181–8191

Schwartz MA and Shattil SJ (2000) Signaling networks linking integrins and rho family GTPases. Trends Biochem Sci 25:388–391

Shepherd JD and Huganir RL (2007) The cell biology of synaptic plasticity: AMPA receptor trafficking. Annu Rev Cell Dev Biol 23:613–643

Shi Y and Ethell IM (2006) Integrins control dendritic spine plasticity in hippocampal neurons through NMDA receptor and Ca2+/calmodulin-dependent protein kinase II-mediated actin reorganization. J Neurosci 26:1813–1822

Staubli U, Chun D and Lynch G (1998) Time-dependent reversal of long-term potentiation by an integrin antagonist. J Neurosci 18:3460–3469

Staubli U, Vanderklish P and Lynch G (1990) An inhibitor of integrin receptors blocks long-term potentiation. Behav Neural Biol 53:1–5

Stephan A, Laroche S and Davis S (2001) Generation of aggregated beta-amyloid in the rat hippocampus impairs synaptic transmission and plasticity and causes memory deficits. J Neurosci 21:5703–5714

Suzuki K, Grinnell AD and Kidokoro Y (2002) Hypertonicity-induced transmitter release at Drosophila neuromuscular junctions is partly mediated by integrins and cAMP/protein kinase A. J Physiol 538:103–119

Watson PM, Humphries MJ, Relton J et al. (2007) Integrin-binding RGD peptides induce rapid intracellular calcium increases and MAPK signaling in cortical neurons. Mol Cell Neurosci 34:147–154

Webb DJ, Zhang H, Majumdar D et al. (2007) alpha5 integrin signaling regulates the formation of spines and synapses in hippocampal neurons. J Biol Chem 282:6929–6935

Xiao P, Bahr BA, Staubli U et al. (1991) Evidence that matrix recognition contributes to stabilization but not induction of LTP. Neuroreport 2:461–464

Chapter 20
Extracellular Matrix Molecules in Neuromuscular Junctions and Central Nervous System Synapses

Laurent Bogdanik and Robert W. Burgess

Abstract The formation and function of chemical synapses require the precise apposition of pre- and postsynaptic specializations. This alignment process requires trans-synaptic signaling, often mediated by transmembrane or matrix-bound molecules. In addition to the pre- and postsynaptic cells, the surrounding glial cells also contribute factors that are important in synapse formation and maintenance. This chapter will focus on the extracellular matrix molecules of the synapse, including those derived from the presynaptic terminal, the postsynaptic cell, and glial cells. We will also compare the composition and function of the matrix present in the cleft of the neuromuscular junction to that of central nervous system synapses and discuss the structural and signaling properties of these components.

Keywords Agrin · Laminin · Collagen · Proteoglycans · Thrombospondins

20.1 Introduction

This chapter begins with a discussion of the extracellular matrix (ECM) in neuromuscular junction (NMJ) formation. The advantages of the NMJ as an experimental model system for synapse formation have been introduced in a previous chapter of this book (see Chapter 3), and the depth of our knowledge about the molecular events of NMJ formation is testimony to its utility. The importance of the ECM in NMJ formation has been evident for 30 years, since experiments lesioning the nerve, the muscle, or both demonstrated that the regeneration of synaptic sites was directed by signals

L. Bogdanik (✉)
The Jackson Laboratory, 600 Main St, Bar Harbor, ME 04609, USA
e-mail: laurent.bogdanik@jax.org

R.W. Burgess (✉)
The Jackson Laboratory, 600 Main St, Bar Harbor, ME 04609, USA
e-mail: Robert.burgess@jax.org

M. Hortsch, H. Umemori (eds.), *The Sticky Synapse*,
DOI 10.1007/978-0-387-92708-4_20,

embedded in the basal lamina of the synapse (Letinsky et al. 1976, Marshall et al. 1977, Sanes et al. 1978, Burden et al. 1979). Subsequently, we will discuss the ECM in central nervous system synapse formation, where many of the same principles apply, but additional or different molecular components are involved.

The ECM is composed of secreted proteins that are often glycosylated, and in the case of the central nervous system, even free carbohydrates that are not associated with a core protein. These assemble into a cross-linked network that is a fibrous lamina in the periphery or a looser gel-like matrix in the central nervous system. In the periphery, the major components of the ECM are laminins and collagens that form a polymerized network of fibrils (Yurchenco and Schittny 1990). In addition, proteoglycans and other components are also present and serve both signaling and structural roles. In the central nervous system, the ECM is much different, with very little laminin or collagen, but a major contribution of proteoglycans and associated proteins that cross-link them into a mesh (Ruoslahti 1996).

Cells "sense" the ECM through a variety of cell-surface-associated receptors, and together the matrix and its receptors serve many diverse roles including signaling, adhesion, structure/architecture of the synapse, and even functional roles in synaptic transmission. Examples of proteins mediating each of these functions, and sometimes multiple functions, will be discussed. The examples provided are meant to be illustrative and not a comprehensive list of active components of the synaptic matrix. We have chosen both well-studied and emerging areas in the biology of the synaptic ECM, although many more topics will go unmentioned.

20.2 The Extracellular Matrix of the NMJ

20.2.1 Agrin

One of the best studied proteins involved in synaptogenesis is agrin. Agrin was first identified as a factor able to induce sites of acetylcholine receptor (AChR) clusters that resemble nascent sites of postsynaptic differentiation in cultured muscle fibers, and in 1990 U. J. McMahan proposed that agrin was the nerve-derived organizer of postsynaptic differentiation at the NMJ, an idea termed "The Agrin Hypothesis"(McMahan 1990). Agrin's in vitro AChR clustering activity allowed the purification of the agrin protein, the generation of antibodies, and eventually the cloning of the gene (Godfrey et al. 1984, Nitkin et al. 1987, Smith et al. 1987, Rupp et al. 1991, Tsim et al. 1992b). These analyses revealed that agrin is associated with the ECM of the NMJ and many other basement membranes throughout the body. At the NMJ, agrin is initially found throughout the basal lamina of the muscle and Schwann cells, as well as the synapse, but with age it becomes increasingly localized to the synaptic cleft. The protein itself is approximately 2000 amino acids in length, and the domain structure is indicated in Fig. 20.1A.

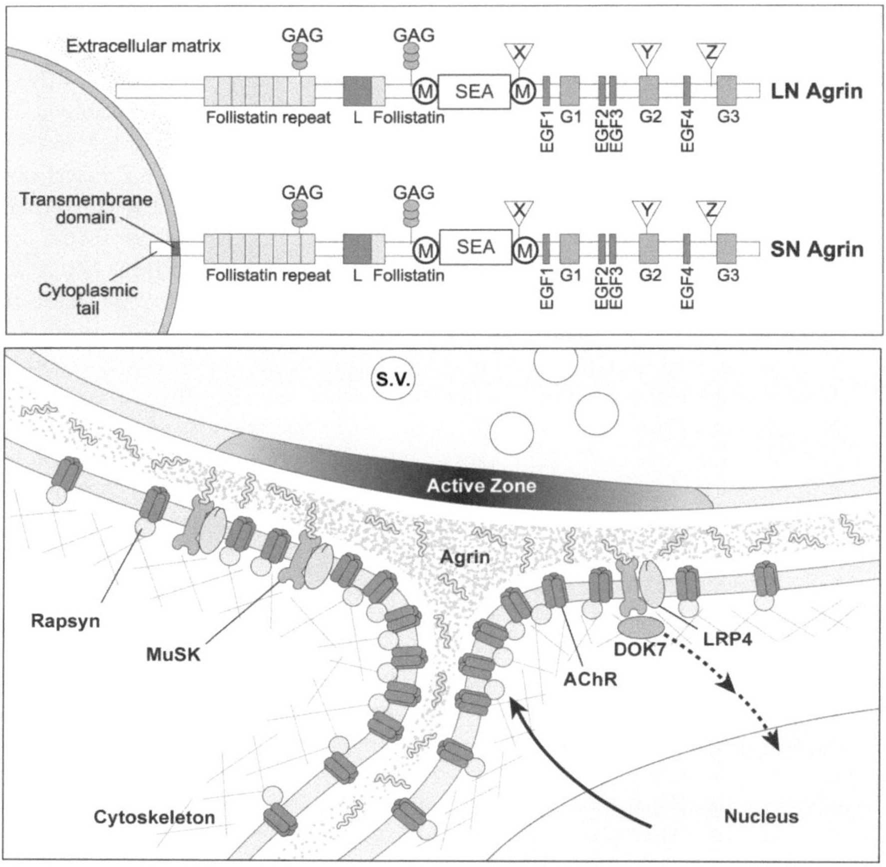

Fig. 20.1 Agrin protein structure and synaptic association. (**A**) Agrin is a heparan sulfate proteoglycan of approximately 2000 amino acids, depending on the alternative splicing. There are two amino-termini that arise from distinct transcriptional and translational start sites. The longer form (LN) has a signal peptide for secretion and binds laminin in the ECM. The shorter form (SN) converts agrin to a type II transmembrane protein and is the predominant isoform in neurons of the brain. There are nine follistatin-like repeats (F, also resembling Kazal protease inhibitor domains), a laminin-like domain (L), and multiple sites of glycosaminoglycan addition (GAG) in the amino-terminal half of the protein. The central domain of agrin is an SEA module (*s*ea urchin sperm protein-*e*nterokinase-*a*grin) that is flanked by mucin-like serine/threonine-rich repeats (M). The carboxy-terminal half of the protein contains four EGF-like repeats and three laminin-type globular domains (G). The carboxy-terminus also contains the Z-alternative splice that is crucial for agrin's activity in AChR clustering, as well as two other alternative splice sites of unknown functional

20.2.1.1 Alternative Splicing Controls Agrin Activity

The analysis of agrin expression also raised a dilemma: If agrin is to be the nerve-derived factor directing postsynaptic differentiation, how can it also be produced by Schwann cells and even by muscles? The answer was discovered in an examination of the alternative splicing of agrin (Ferns et al. 1992, Tsim et al. 1992a, Ferns et al. 1993, Hoch et al. 1993). In the nervous system, two exons (32 and 33 in the numbering of Rupp et al. (1992)) are alternatively spliced, and only transcripts found in the nervous system include either one or both of these exons, resulting in agrin proteins with insertions of 8, 11, or the combined 19 amino acids at a site immediately before the C-terminal G-domain termed the "Z" splice site (see Fig. 20.1A). Isoforms of agrin that include the Z insertions (Z8, Z11, or Z19, or generically Z+) are active in inducing AChR clustering, with the Z8 isoform conferring the greatest potency, and these isoforms are made by motor neurons. Isoforms that do not include these insertions (Z0 or Z- isoforms) are inactive in AChR clustering, and these are the forms made in other tissues, including muscle. These conclusions were made based on expression analysis and in vitro studies and were confirmed in vivo with the targeted deletion of just these alternative exons from the mouse genome (Burgess et al. 1999). Deletion of the Z exons results in a nearly complete loss of postsynaptic differentiation by late embryonic ages in mice, a phenotype that is identical to the complete loss of agrin from the NMJ. Interestingly, however, the earliest stages of postsynaptic differentiation are normal in the absence of agrin, as discussed below.

Agrin's residence in the ECM was also examined. Studies using immunocytochemistry are consistent with an extracellular matrix localization in the synaptic cleft of the NMJ. However, agrin was determined to have two alternative transcriptional and translational start sites, one conferring a signal peptide for secretion and a laminin-binding domain, the other generating a type 2 transmembrane protein that is the predominant isoform in the central nervous system (Neumann et al. 2001, Burgess et al. 2002) (Fig. 20.1A). A gene trap insertion allele in mice that eliminates the secreted isoform without disrupting the transmembrane isoform results in a complete loss of agrin from the NMJ and the same phenotype described above with an almost complete loss in postsynaptic differentiation and neonatal lethality (Burgess et al. 2000). In contrast, mice lacking the transmembrane isoform of agrin have normal

Fig. 20.1 (continued) consequence (X, Y). (**B**) At the NMJ, Z+ agrin from motor neurons is secreted into the synaptic cleft, where it binds other matrix components including laminins and dystroglycan. Agrin activates MuSK through the co-receptor LRP4. This signaling cascade results in AChR clustering mediated by the intracellular scaffolding protein rapsyn. Other molecules associate with the intracellular domain of MuSK, including DOK7, and ultimately influence not only protein localization to the NMJ, but also transcription from the myo-nuclei that lie directly beneath the synapse

NMJs (R.W.B., unpublished). Thus, the Z+ secreted form of the suppress "the" agrin is the form that is active and necessary for NMJ formation.

20.2.1.2 Agrin Signal Transduction

Agrin signals by activating a tyrosine kinase receptor, MuSK (muscle-specific kinase), on the muscle cell surface (Fig. 20.1B). All data are completely consistent with MuSK being critical for agrin signal transduction. Mice lacking MuSK have a phenotype similar to, or slightly more severe than, mice lacking agrin (DeChiara et al. 1996). Similarly, myotubes cultured from MuSK knockout mice are insensitive to agrin treatment, and agrin rapidly induces MuSK phosphorylation in wild-type myotubes (Glass et al. 1996). However, a direct interaction of MuSK and agrin has not been shown, leading to the hypothesized existence of a third accessory factor termed MASC (myotube-specific accessory component). This hypothetical factor also accounts for another oddity of agrin/MuSK signaling, which is that agrin is only able to activate MuSK in muscle cells, and not in heterologous cell types transfected with MuSK. This suggests that this postulated third factor must also be muscle specific.

The identity of MASC appears to have been recently solved. The LDL receptor-related protein4 (LRP4) is expressed in the endplate band of muscle and has a phenotype very similar to MuSK when mutated in mice (Weatherbee et al. 2006). Additional studies indicate that LRP4 is indeed a direct receptor for agrin that forms a complex with MuSK and mediates MuSK activation by agrin, even in transfected non-muscle cells (Kim et al. 2008, Zhang et al. 2008). Therefore, LRP4 appears to be the long-sought link between the agrin signal and the MuSK activation, which promotes postsynaptic differentiation at the NMJ.

Additional components of the MuSK downstream signaling cascade in muscle have been identified (Fig. 20.1B). First, members of the Src kinase family are critical (Smith et al. 2001). In addition, an adaptor protein, DOK7, binds directly to MuSK, and mutations in DOK7 in humans cause severe congenital myasthenias. Complete knockouts in mice resemble the MuSK or agrin phenotypes (Beeson et al. 2006, Okada et al. 2006). Ultimately, these signaling events result in the aggregation of acetylcholine receptors by the scaffolding protein rapsyn, the accumulation of other synaptically specialized proteins including components of the dystroglycan–glycoprotein complex (DGC), and a reprogramming of the myo-nuclei immediately below the nerve terminal to express a different set of synapse-specific genes such as AChR subunits.

20.2.1.3 Inducing Versus Stabilizing Postsynaptic Sites

The fact that postsynaptic differentiation is initially normal in the absence of agrin raised the question of whether agrin actually induces postsynaptic differentiation or instead stabilizes sites established by an intrinsic program of muscle differentiation. Indeed, prior to, or in the earliest stages at which motor axons reach the developing muscle (embryonic day 12–13 in the mouse), there is a region of postsynaptic specializations (AChR clusters) in the presumptive endplate band of the muscle. The formation of these earliest AChR clusters requires MuSK, but is independent of agrin and even independent of nerve contact (Lin et al. 2001, Yang et al. 2001). Therefore, this zone of postsynaptic differentiation is intrinsic to the muscle and is referred to as "Prepatterning." The prepatterning of muscle does help define the site of motor neuron innervation, and delaying prepatterning by deleting the gene encoding the embryonic γ subunit of the AChR results in a broader than normal endplate band (Liu et al. 2008). Agrin from the ingrowing nerve may therefore be stabilizing the sites of postsynaptic differentiation established by the muscle rather than inducing new sites de novo at the point of motor axon contact. However, the ability of agrin to induce AChR clustering in vivo and in vitro cannot be overlooked (Cohen et al. 1997, Meier et al. 1997). In addition, studies in both zebrafish, which followed the time course of motor innervation, and mouse, which compared medial versus lateral synaptic sites within an endplate band, indicated that motor neurons were initially stabilizing pre-existing sites, whereas later arriving terminals were inducing new sites of postsynaptic differentiation (Flanagan-Steet et al. 2005, Lin et al. 2008a). Therefore, agrin is likely to both stabilize prepatterned sites and to induce new synaptic sites, depending on the timing of synaptogenesis relative to initial nerve contact.

20.2.1.4 The Antagonistic Role of Cholinergic Transmission

The prepatterning of muscle is normal in the absence of agrin, but most postsynaptic sites are lost within a few days of nerve contact in agrin mutant mice (very sparse by E15–16, and virtually gone by E18). Interestingly, the prepattern persists in muscles that are never innervated. This indicates that the nerve is actively dispersing sites that are not stabilized by agrin, and from the perspective of postsynaptic differentiation, muscles are better off with no nerve than with a nerve that lacks agrin. So what may these dispersal signals be? One such signal appears to be cholinergic activity itself. In mice that lack both agrin and choline acetyltransferase (ChAT), the enzyme that synthesizes the neurotransmitter acetylcholine, there are more persistent sites of postsynaptic differentiation and innervation than in muscles lacking just agrin (Lin et al. 2005, Misgeld et al. 2005). Such a mechanism is logical during normal development, as it would allow the stabilization of innervated sites by agrin from the motor neuron, but would also promote the dispersal of non-innervated sites left over from the prepatterning of the muscle. Therefore, agrin is necessary to maintain postsynaptic differentiation against other disruptive neuronal activities such as cholinergic transmission.

As with most signaling pathways, agrin can be bypassed to some extent. The transgenic expression of Z+ agrin exclusively in muscle is able to restore not only postsynaptic differentiation but also motor neuron innervation in agrin mutant mice, rescuing the lethal phenotype (Lin et al. 2008b). The overexpression of MuSK in mice is sufficient both to induce relatively stable sites of postsynaptic differentiation and to confer stable innervation, even rescuing the lethality of agrin mutations for a few weeks (Kim and Burden 2008). The fact that this rescue is ultimately incomplete and the agrin mutant phenotype in a background of normal MuSK expression and cholinergic activity indicate that physiologically, Z+, matrix-bound agrin is essential for the formation of stable NMJs. These studies also indicate that innervation by motor neurons is maintained in the presence of normal postsynaptic differentiation, regardless of whether that differentiation is stabilized by Z+ agrin from the motor neuron, the muscle, or the artificial activation of MuSK by overexpression.

20.2.2 Laminins at the NMJ

In addition to agrin, the synaptic basal lamina of the neuromuscular junction contains other components that are common to many matrices, notably laminins and collagen IVs. However, the synaptic forms of these proteins are specialized isoforms that are often distinct from those found more widely in other tissues, and even from those found in the extrasynaptic regions of muscle. These proteins are generally produced in the muscle, but have effects on nerve terminal differentiation and synaptic structure, in addition to postsynaptic organization, making them trans-synaptic organizers that combine signaling and structural functions.

The laminins are trimeric molecules that are comprised of an α, β, and γ subunit, or "chain" (Fig. 20.2A). Each of these chains is encoded by its own gene, and there are five α chains, four β chains, and three γ chains known in mammalian genomes (Colognato and Yurchenco 2000). The structure of the trimer resembles a cross, with the N-terminal domains of each chain remaining separate, while the C-terminal halves associate in a three-part helix. The C-terminus of the α chain typically extends beyond this coiled structure and contains globular domains that interact with both integrins and dystroglycan. The N-terminal domains of the β and γ chains interact with other matrix components such as nidogen, collagen, and heparan sulfate proteoglycans. The N-terminal domains of many α chains mediate polymerization of laminin trimers into a net-like meshwork in basement membranes. This polymerization is likely to be essential for the spacing and structural organization of the basement membrane, and therefore its signaling properties to cell surface receptors. Laminin α4 is a notable exception, as it lacks an N-terminal domain and α4 containing trimers do not polymerize.

The laminin composition of the muscle extracellular matrix is different at the neuromuscular junction than in extrasynaptic regions of the muscle (Fig. 20.2). Extrasynaptic regions contain primarily laminin α2, β1, and γ1 (laminin 211 in the newest nomenclature or laminin 2 historically). The importance of laminins

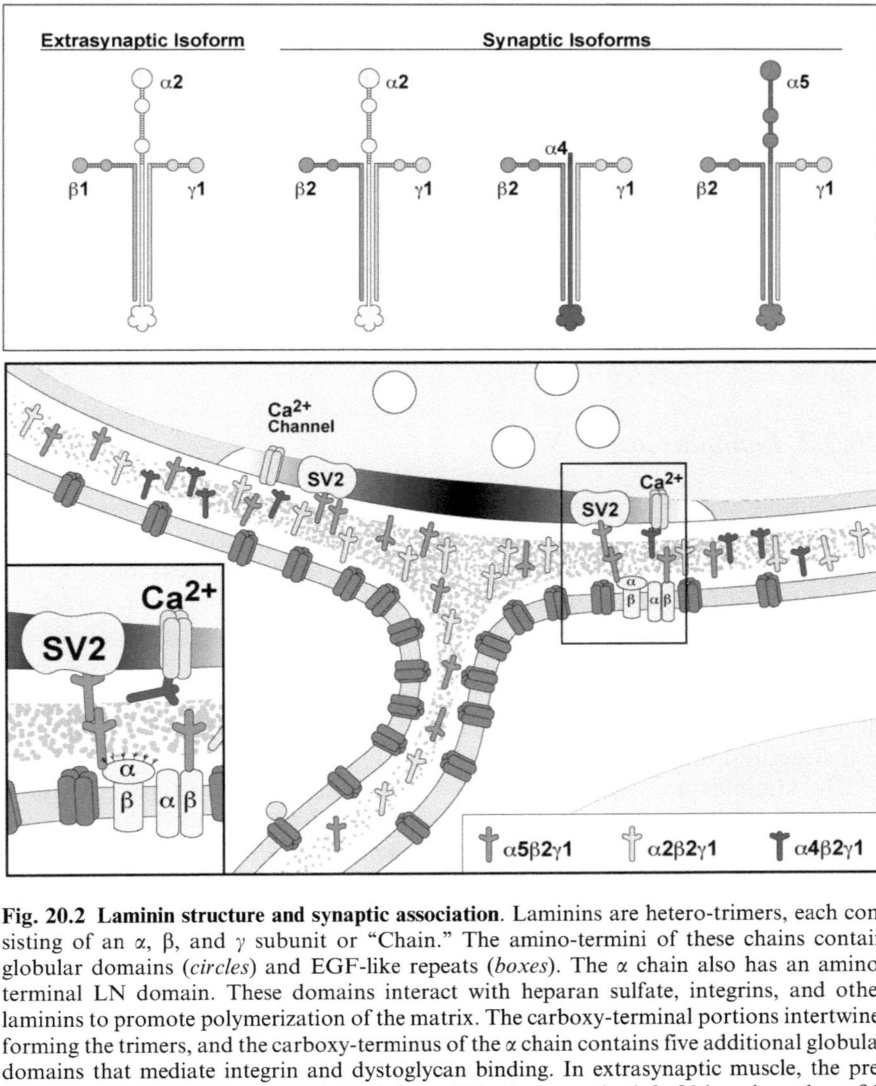

Fig. 20.2 Laminin structure and synaptic association. Laminins are hetero-trimers, each consisting of an α, β, and γ subunit or "Chain." The amino-termini of these chains contain globular domains (*circles*) and EGF-like repeats (*boxes*). The α chain also has an amino-terminal LN domain. These domains interact with heparan sulfate, integrins, and other laminins to promote polymerization of the matrix. The carboxy-terminal portions intertwine, forming the trimers, and the carboxy-terminus of the α chain contains five additional globular domains that mediate integrin and dystoglycan binding. In extrasynaptic muscle, the predominant trimer is α2β1γ1 (laminin 211), but within the synaptic cleft, β2 largely replaces β1, α5 is present throughout the cleft, and the non-polymerizing α4 surrounds the active zones. Laminin α5 interacts with SV2 presynaptically and integrins and dystroglycan postsynaptically. Laminin α4 interacts with presynaptic N-type calcium channels. Thus laminin localization and binding are central to establishing trans-synaptic specificity in aligning components of the pre- and postsynaptic transmission machinery

for maintaining muscle fiber integrity is exemplified by mutations in the laminin α2 gene (*LAMA2*), which cause congenital muscular dystrophy type 1A in humans and mice (Xu et al. 1994, Helbling-Leclerc et al. 1995, Sunada et al. 1995). Within the synapse, the α2 chain persists, but the α5 chain is also present

(Miner et al. 1997, Patton et al. 1997). The β2 chain also largely replaces the β1 chain (Hunter et al. 1989). Therefore, the predominant synaptic laminin is laminin 521. Laminin α4 is also found at the neuromuscular junction, though with a more restricted localization, as described below. The α4 and 5 chains are initially more broadly localized in the muscle and become restricted to the NMJ as it matures, whereas β2 is selectively synaptic throughout development.

The functions of these NMJ-specific laminins have been examined in vitro and in vivo using knockout mice, and the results indicate both pre- and postsynaptic effects. Mice lacking laminin β2 (*Lamb2*) fail to thrive and die at about 1 month of age. The NMJs arrest at an immature state and do not form elaborate postsynaptic specializations such as junctional folds in the muscle membrane, nor are presynaptic terminals fully polarized, with vesicles spread throughout the terminal and ill-defined presynaptic active zones in the mutant mice (Noakes et al. 1995). The terminal Schwann cell also invades the synaptic cleft of β2 knockout mice, impairing pre- and postsynaptic apposition and presumably synaptic transmission (Patton et al. 1998). This phenotype raised the possibility that β2-containing laminins are repellent signals for Schwann cells or an adhesive signal for motor nerve terminals. Indeed, results in culture suggest that both effects may be true, Schwann cells do not adhere as well to laminin 521 as to other β1-containing laminins, and motor neurons have shorter processes with more signs of presynaptic differentiation on laminin 521 than other laminins with the β1 chain (Cho et al. 1998). The β2 chain of laminin interacts with N-type voltage-gated calcium channels in the presynaptic membrane, and this may contribute to the molecular complexes required to form fully functional presynaptic active zones for cholinergic vesicle release, although this interaction may involve laminin 421 as described below (Sunderland et al. 2000, Nishimune et al. 2004). Laminin α5-containing trimers also bind to presynaptic components such as SV2, further supporting the hypothesis that laminins are at least in part responsible for organizing presynaptic structures (Son et al. 2000) (Fig. 20.2B).

Mice lacking laminin α4 also have striking, though subtle, synaptic defects. These mice are overtly normal, but have a misalignment of the pre- and postsynaptic active zones at the NMJ (Patton et al. 2001). Normally, there is a thickening of the presynaptic membrane and an accumulation of synaptic vesicles in the nerve terminal, visible by transmission electron microscopy, that are the sites of synaptic vesicle fusion and neurotransmitter release (active zones). These active zones always align directly across the synaptic cleft from junctional folds, and the highest densities of acetylcholine receptors are present at the mouths of these fold. In the laminin α4 knockout mice, these pre- and postsynaptic specializations are present, but out of register, no longer aligning directly across the cleft from one another. Interestingly, laminin α4 has a very discrete localization within the synaptic cleft, being found immediately flanking the active zones. This localization and the association with N-type calcium channels mentioned above again suggest that the organization of laminins within the basal lamina of the synapse is directing the organization of pre- and postsynaptic specializations needed for efficient synaptic transmission. Though the defects associated with the loss of laminin $\alpha4$ are not overly deleterious to mice housed under laboratory conditions, it will be interesting to see if more severe phenotypes are observed if stresses such as aging, increased activity, or denervation/reinnervation are introduced.

20.2.3 Collagens

The collagens are another family of matrix molecules that often colocalize with laminins, but are biochemically and structurally distinct. Collagens are also trimeric protein complexes (trimers), with long tightly twisted C-terminal domains that form "collagenous" fibrils (see Fig. 3.13). At the NMJ, the primary collagens are of the collagen 4 family, which consists of six genes in mammals (Col4α1–6) (Khoshnoodi et al. 2008). The collagen 4 trimers do not assemble at random, but instead are always found in one of the following combinations: $(\alpha1)_2(\alpha2)$, $(\alpha3)(\alpha4)(\alpha5)$, and $(\alpha5)_2(\alpha6)$. The elimination or perturbation of a single chain of the trimer results in a complete loss of assembly and absence of the protein complex from the basement membrane. In vitro collagen 4 peptides have an activity as presynaptic differentiation factors, and in vivo the collagen 4s act sequentially in NMJ formation (Fox et al. 2007). In early development, collagen 4 $(\alpha1)_2(\alpha2)$ is found throughout the muscle ECM. A mutation that deletes exon 40 of Col4A1 in the mouse genome causes a partial loss-of-function allele (Gould et al. 2005), and these mice have defects in early postnatal motor nerve terminal morphology, which disappear by 3 weeks of age. At this age, the other collagen 4 trimers [$(\alpha3)(\alpha4)(\alpha5)$ and $(\alpha5)_2(\alpha6)$] become specifically localized to the NMJ. Mouse mutations eliminating Col4A3 or Col4A6, which affect only one of the two trimers, do not have defects at the NMJ. However, mice lacking Col4A5, which affects both of these trimers, have NMJ defects that are apparent by 1 month of age. These defects include partial occupancy of the postsynaptic site by the nerve terminal, dystrophic axons and terminals, abnormal neurofilament-positive loops in the nerves, and fragmentation of the postsynaptic receptor field in the muscle cell membrane. Therefore, collagen 4 $(\alpha1)_2(\alpha2)$ appears to function early in the formation of the NMJ, and the collagen 4 $(\alpha3)(\alpha4)(\alpha5)$ and $(\alpha5)_2(\alpha6)$ function in the maintenance of the NMJ after it is reaches its mature morphology.

20.2.4 Matrix Components Involved in Synaptic Function

Thus far we have described agrin as a nerve-derived signaling molecule that effects postsynaptic differentiation, whereas laminins and collagens are muscle-derived structural molecules that are important for organizing the architecture of the NMJ. All of these molecules reside in the ECM and interact with cell surface receptors to mediate their effects. However, other proteins in the extracellular matrix are also important for synaptic function. An obvious example of this is acetylcholine esterase (AChE), the enzyme that degrades acetylcholine and thereby terminates cholinergic synaptic transmission. At the NMJ, AChE is found in asymmetric forms that are anchored to the matrix by a second protein, ColQ, which shares structural and sequence homology with collagen tails (Krejci et al. 1997). ColQ mediates the interaction of AChE enzymatic subunits with heparan sulfate proteoglycans in the matrix and MuSK in the muscle membrane, thus localizing the enzyme to the synapse (Deprez et al. 2003, Cartaud et al. 2004). AChE itself has many alternative splice forms that have different properties, including GPI-linked forms to monomers, but the ColQ linked forms are the predominant species at the NMJ. Most cases of congenital myasthenia resulting from endplate AChE deficiency are caused by mutations in the ColQ gene, which result in a failure to properly anchor the enzyme in the synaptic cleft. The pathophysiology of this disease indicates the essential functions of both proteins (AChE and ColQ) in synaptic transmission at the NMJ, but also the ability of the NMJ synapse to compensate for their loss by downregulating neurotransmitter release and/or AChR density (Donger et al. 1998, Ohno et al. 1998, Feng et al. 1999, Adler et al. 2004).

There are a number of other matrix specializations that are specific to the NMJ. These include synapse-specific proteases, carbohydrate moieties and possibly related glycosyl transferases, and growth factors that are localized to or specifically presented at the NMJ. The functions of these NMJ components are still very much an area of active research, but many studies clearly indicate their significance.

20.2.5 Proteases

The remodeling of the extracellular matrix by proteases is an important area of cell biology. These enzymes can activate proteins by releasing active fragments that are otherwise tethered to the ECM or plasma membrane. Conversely, they may also inactivate proteins by removing them from the cell membrane or ECM, and thus eliminate their signaling ability. Ectodomain shedding is an example of the second possibility. There is some evidence that agrin may be subject to such regulation. Following denervation, postsynaptic sites on the muscle are relatively stable over a time period of weeks awaiting reinnervation by the regenerating axon. If reinnervation is directed ectopically, the previously stable denervated site disintegrates very quickly. This process requires matrix metalloprotease 3 (MMP3) and suggests that MMPs are involved in the destabilization of NMJs, probably through the degradation of agrin (Stanco and Werle 1997, Van Saun and Werle 2000, Van Saun et al. 2003).

Other proteases also target agrin. Neurotrypsin is a recently identified protease in the brain, whose only known substrate is agrin (Reif et al. 2007).

Mutations in humans in neurotrypsin cause profound mental retardation, suggesting a critical role for this enzyme in CNS function (Molinari et al. 2002). Neurotrypsin is localized to CNS synapses and this is regulated by neuronal activity (Stephan et al. 2008). Whether the same will be true at the NMJ and how many of neurotrypsin's biological effects may be mediated through the cleavage of agrin remain to be determined.

20.2.6 Synapse-Specific Carbohydrates

In addition to synaptically specialized protein isoforms, the matrix of the NMJ also contains specialized carbohydrate moieties. A survey of carbohydrates present on muscle revealed some to be present only at the NMJ. For instance, N-acetyl galactosamine (GalNAc) at the terminus of a polysaccharide chain was found to be concentrated in the NMJ basal lamina, but absent from the other areas of the muscle (Sanes and Cheney 1982). In contrast, non-reducing terminal α-D-glucose, α-D-mannose, or sialic acid are not specifically localized (Scott et al. 1988). The CT antigens, two GalNAc-containing carbohydrates recognized by two distinct antibodies directed against a cytotoxic lymphocyte-specific phosphatase (CD45), are similarly enriched at the NMJ and are produced by both the motor neuron and the muscle (Martin et al. 1999). Of more than 15 proteins carrying these carbohydrate moieties, some have been identified as the synapse-specific isoform of AChE and agrin, as well as dystroglycan (Martin and Sanes 1995, Xia et al. 2002). A possible function of these carbohydrates is to modulate the binding of their core proteins. For instance, GalNAc-terminal carbohydrates are required for the anchoring of AChE in the synaptic cleft. When O-linked to the mucin-like domains of agrin, this carbohydrate modification increases agrin's AChR clustering activity (Parkhomovskiy et al. 2000).

Further evidence for a signaling role of carbohydrates comes from in vitro studies. Treatment of cultured myotubes with either neuraminidase, which removes sialic acid modifications, or the lectin VVA (*vicia villosa* agglutinin) causes AChR clustering that is distinct from agrin-induced clustering (Grow et al. 1999b, a).

In addition to the carbohydrates themselves, several glycosyltransferases have been identified as important enzymes in muscle. Mutations in these enzymes often cause a complex phenotype of muscular dystrophy, accompanied by defects in the eye and central nervous system (Table 20.1). A common substrate of these enzymes is dystroglycan, and defects in dystroglycan function can account for many of the pathological changes observed. One enzyme identified in a spontaneous mutation in *myodystrophy* mice (*myd*) with muscular dystrophy and various neurological phenotypes is LARGE (Grewal et al. 2001). LARGE is required for the glycosylation of dystroglycan's mucin domain and thereby regulates its affinity for laminins and agrin (Michele

Table 20.1 Glycosyltransferase genes and human diseases associated with their dysfunction

Gene	Human disease
FKTN, fukutin	Fukuyama congenital muscular dystrophy, limb-girdle MD type 2 M, Walker–Warburg syndrome, dilated cardiomyopathy 1X
FKRP, fukutin-related protein	Congenital muscular dystrophy 1C, limb-girdle muscular dystrophy 2I, muscle–eye–brain disease, Walker–Warburg syndrome
POMT1, protein *O*-mannosyltransferase 1	Limb-girdle muscular dystrophy type 2 K, congenital muscular dystrophy plus mental retardation, Walker–Warburg syndrome
POMT2, protein *O*-mannosyltransferase 2	Walker–Warburg syndrome
LARGE, acetylglucosaminyltransferase-like protein	Congenital muscular dystrophy type 1D, Walker–Warburg syndrome
POMGNT1, protein *O*-mannose beta-1,2-N-acetylglucosaminyltransferase	Muscle–eye–brain disease

et al. 2002, Kanagawa et al. 2004). LARGE mutant mice also exhibit a reduced aggregation of AChRs and have reduced levels of agrin, perlecan, and MuSK at the NMJ, but not in the extrasynaptic muscle basal lamina (Kanagawa et al. 2005). Interestingly, there is not a dramatic disruption of the basal lamina integrity (Levedakou et al. 2005). Another enzyme, fukutin, also regulates the glycosylation of the dystroglycan mucin domain, and consequently its binding to laminins. Mice lacking fukutin recapitulate the LARGE phenotype at the NMJ, with additional reductions in the presence of laminin and AChE (Saito et al. 2007). Several human diseases are associated with glycosyltransferases. These are generally categorized as Walker–Warburg syndromes, but also include congenital and limb-girdle muscular dystrophies, muscle–eye–brain disease, and Fukuyama's muscular dystrophy [POMGNT1 (Yoshida et al. 2001); fukutin (Kobayashi et al. 1998, Hayashi et al. 2001, Takeda et al. 2003); fukutin-related protein (Brockington et al. 2001a, b); LARGE (Longman et al. 2003, van Reeuwijk et al. 2007); POMT1 (Beltran-Valero de Bernabe et al. 2002); POMT2 (van Reeuwijk et al. 2005)] (Table 20.1). While these diseases are not synapse-specific, synaptic defects are likely to contribute to their pathophysiology, and mouse models do show neuromuscular junction defects (Levedakou et al. 2005).

20.2.7 Dystroglycan

The extent to which a failure to glycosylate dystroglycan can account for the disease phenotypes of glycosyltransferases is an interesting question. Indeed, deletion of dystroglycan from developing mouse epiblasts recapitulates many of the CNS and eye phenotypes seen in Walker–Warburg-like conditions (Satz et al. 2008). Dystroglycan is an important cell surface receptor at the NMJ and throughout muscle. It is able to bind laminins and agrin and is therefore central to cell surface/matrix interactions. The glycosylation of dystroglycan accounts for half of its apparent molecular weight, most of which is composed of O-linked additions on its mucin domain. These post-translational modifications are important for the binding properties of dystroglycan, but may have regional specificity on muscle cell surface, differing between synaptic and extrasynaptic regions or costameres for example (Ibraghimov-Beskrovnaya et al. 1992). Chimeric mice with mixed dystroglycan mutant and wild-type muscle fibers have fragmented NMJs (Cote et al. 1999). Cultured myotubes from dystroglycan mutant mice do form AChRs aggregates in response to agrin, but these clusters of receptors are fragmented and lack laminin $\alpha 5$ and $\gamma 1$ found on normal myotubes (Grady et al. 2000). This suggests that part of the postsynaptic defects observed in absence of dystroglycan could relate to specific defects in ECM composition. Dystroglycan also regulates NMJ structure and functions at the somewhat distant model of the *Drosophila* NMJ. Like at the mammalian NMJ, dystroglycan controls the localization of laminin, and both dystroglycan and laminin surround active zones, but are excluded from the actual sites of neurotransmitter release. This perisynaptic localization of sarcolemma-tethered dystroglycan controls presynaptic neurotransmitter release and defects in the glycosylation of dystroglycan have the same effects as its muscle-specific suppression (Bogdanik et al. 2008, Wairkar et al. 2008). Noticeably, these defects in synaptic function were found in the absence of any major muscle damage and illustrate the trans-synaptic communication mediated by dystroglycan and its dependence on glycosylation.

20.2.8 Growth Factors

Another important role for the ECM, and particularly the glycosaminoglycans embedded in it, is the presentation of growth factors. The importance of growth factors at the NMJ was first suggested by in vitro experiments, which focused on presynaptic differentiation in cultured neurons, and is supported by more recent genetic manipulations in mice. The requirement of the interaction with heparan sulfate proteoglycan (HSPGs) for the function of fibroblast growth factors (FGFs) is well established (review in Ornitz 2000). HSPGs protect FGFs against proteolysis and limit their diffusion. HSPG, FGF, and FGF receptor form a ternary complex that is required for the FGF Receptor dimerization and activation. Evidence of the presynaptic role for FGFs came from experiments on *Xenopus* motor neurons in culture, where localized exposure to bFGF (or FGF2)-coated latex beads induced the accumulation of synaptic vesicles and synaptic proteins, as well as the development of excitation–secretion properties of the nearby plasma membrane (Dai and Peng 1995). These hallmarks of a presynaptic differentiation were dependent on a receptor-tyrosine kinase, consistent with the FGF receptor. Recent work has extended the understanding of FGFs at the NMJ. Mouse myotubes in culture express FGF7, -10, and -22, which can activate FGFR2b expressed by motor neurons in culture. When they contact culture myotubes, motoneurons accumulate synaptic vesicles in their terminal, a process that is impaired when binding of FGFs to their neuronal receptor is blocked. In vivo, the same system is present, with muscles expressing FGFs and nerve terminal containing FGFR2b. In FGFR2b-mutant mice, which die at birth, accumulation of synaptic vesicles in the nerve terminals is impaired. When FGFR2 is genetically suppressed from the motor neurons only, mice are viable at birth, and the defect in synaptic vesicle accumulation is only transiently observed after birth, before resolving by 3 weeks of age. Double mutants for FGFR2 and laminin β2 displayed more severe neonatal phenotypes, suggesting that FGF and laminin exert partially redundant presynaptic differentiation functions, with laminin β2 maintaining differentiation of the NMJ after FGF secretion by the muscle is gradually downregulated after birth (Fox et al. 2007).

In addition to FGFs, other neurotrophic factors also influence the NMJ. For example, glial cell-derived neurotrophic factor (GDNF), which is expressed by muscles, causes a hyperinnervation of muscles, with more motor axons innervating endplates in GDNF overexpressing transgenic mice. This effect correlated with increasing expression of the GDNF, but also showed specificity, as GDNF expressed by glial cells is ineffective, as are neurotrophin-3 and -4 expressed by muscle (Nguyen et al. 1998). Like FGFs, GDNF is regulated by its interaction with heparan sulfate proteoglycans (Ai et al. 2007). Thus, growth factors are key regulators of NMJ innervation and synaptic differentiation, and the extracellular matrix is a key regulator of growth factor activity.

20.3 The Extracellular Matrix and CNS Synapses

Central nervous system synapse formation shares many conceptual features with NMJ formation, but only a few of the proteins involved are shared. In CNS synapses, there is also a precise alignment of pre- and postsynaptic components, and establishing this structure clearly requires the same sorts of trans-synaptic signaling events that are required for pre- and postsynaptic differentiation at the NMJ. However, CNS synaptic sites are much smaller than NMJs and may therefore require less of an intracellular signaling cascade to promote synaptic differentiation. Perhaps it is therefore more accurate to think of these synapses as highly specialized sites of cell adhesion, and consistent with this, one of the most striking differences in the CNS synapses versus the NMJ is the tight contact of pre- and postsynaptic membranes and the almost complete absence of a basal lamina in the synaptic cleft in the CNS. This close proximity allows many of the proposed CNS synaptogenic factors to be transmembrane molecules that span the cleft from the pre- to the postsynaptic cell, as described in other chapters of this book.

20.3.1 Agrin in the CNS

Agrin, the well-characterized signal from motor neurons to muscle fibers that resides in the basal lamina of the NMJ, also exists as an alternative isoform that is a type 2 transmembrane protein in the CNS (Fig. 20.1) (Neumann et al. 2001, Burgess et al. 2002). Indeed, this transmembrane form accounts for approximately 70–80% of the agrin in the brain. However, despite its abundant expression in the brain, agrin's role in inter-neuronal synapse formation is more subtle than its role in NMJ formation. Neurons in culture respond to the application of soluble agrin, which induces C-fos expression and CREB phosphorylation, but these neurons do not form sites of postsynaptic (or presynaptic) differentiation in response to agrin as muscle fibers do (Ji et al. 1998, Hilgenberg et al. 2002). The identity of the receptor for agrin on CNS neurons is also controversial, but it is probably not MuSK, which is found only at low levels in the brain. Dystroglycan, a prominent binding partner of agrin at the NMJ, is found widely throughout the CNS, and agrin was recently shown to bind a subunit of the Na-K-ATPase; however, the physiological consequences of these interactions remain to be determined (Hilgenberg et al. 2006). Mice lacking agrin in the brain have a smaller number of synapses in a given volume of cortical neuropil than control mice. Other defects such as a reduced number and size of synapses between the pre- and postganglionic neurons in the sympathetic nervous system have also been reported (Gingras et al. 2002, Ksiazek et al. 2007). However, the catastrophic failure of synaptogenesis that is observed at the NMJ in the absence of agrin is not evident. This may reflect a less crucial role for agrin in the CNS, or it may be the result of redundancy with other synaptogenic mechanisms. Nonetheless, it is indicative that other synaptogenic mechanisms are at play in the brain.

Other NMJ proteins are clearly absent from the CNS but have clear functional counterparts. For example, the postsynaptic scaffolding protein rapsyn is necessary for aggregating AChRs and anchoring them to other components of the postsynaptic apparatus and the actin cytoskeleton at the NMJ. In the CNS, this role is filled by gephryn for inhibitory neurotransmitter receptors such as GABAa or glycine receptors and multi-PDZ-domain-containing proteins at many excitatory synapses (Prior et al. 1992, Feng et al. 1998).

20.3.2 Laminins in the CNS

Interestingly, laminins are largely absent from the CNS. Again this is consistent with the idea that most trans-synaptic organization in the CNS is mediated by cell–cell contacts through transmembrane proteins and not through the organization of the basal lamina and extracellular matrix. Laminin expression in the brain is more discrete than in other tissues, with the highest expression in basal laminae of the meninges and blood vessels. It is largely of either glial or endothelial origin, and reports on laminin expression in the neurons have been conflicting (Yin et al. 2003, Lein et al. 2007). Laminins are clearly present in basement membrane structures in the CNS, such as the inner limiting membrane of the retina, where laminin α4 is expressed with laminin β2 and γ3. Synaptic sites in the outer plexiform layer of the retina, where photoreceptors form ribbon synapses onto bipolar and horizontal cells, also contain laminins. Laminin β2-mutant mice have "floating ribbons," fully differentiated, vesicle-rich presynaptic photoreceptors terminals without any postsynaptic element facing them (Libby et al. 1999).

20.3.3 Proteoglycans in the CNS Extracellular Matrix

Differences in the matrix of the NMJ and CNS have implications even for those proteins that have shared functions in both locations. For example, AChE is anchored in the brain by a transmembrane protein PRiMA, which is itself glycosylated (Perrier et al. 2002). The extent to which the brain is suited to PRiMA anchoring, and the NMJ to ColQ anchoring, is likely dependent in part on differences in the ECM and the greater prevalence of fibrilar laminin/ collagen-containing matrices at the NMJ.

The matrix of the central nervous system is largely a meshwork of proteoglycans and non-protein carbohydrates (Fig. 20.3). A variety of glycosaminoglycans are present, most prominently chondroitin sulfate, but also heparan sulfate, keratin sulfate, and dermatin sulfate. There are many chondroitin

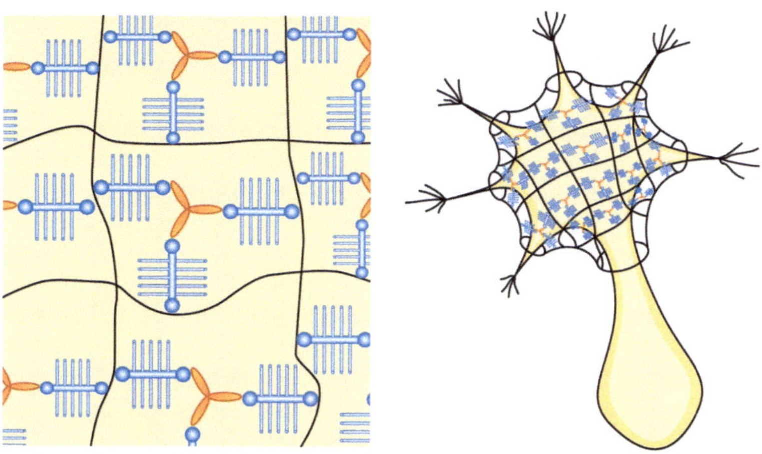

Fig. 20.3 The extracellular matrix of the brain. The CNS matrix is molecularly very distinctive and rich in carbohydrates. Chondroitin sulfate proteoglycans such as aggrecan (*blue*) bind both tenascin (*brown*) and hyaluronin (*black lines*), a carbohydrate with no associated core protein. This network can ensheath neuronal cell bodies and processes, forming perineuronal nets that may serve to cement synaptic contacts into place in the mature nervous system

sulfate proteoglycans, but members of the aggrecan family are particularly key in the formation of CNS proteoglycan meshwork (Schwartz et al. 1999). In addition, hyaluronan, which does not have an associated core protein, is also abundant. Tenascins interact with proteoglycans, which in turn interact with hyaluronin, creating a lattice that is analogous to the laminin lattice of other basement membranes.

Perineuronal nets (PNNs) are one interesting manifestation of this lattice (Galtrey and Fawcett 2007) (Fig. 20.3). PNNs were first described by Golgi over 100 years ago, but dismissed as a fixation artifact by Ramon y Cajal. Persistent studies have revealed that these structures do exist around neurons in the brain, and they are most commonly detected using the lectin *Wisteria floribunda* agglutinin. Chemically, they are particularly rich in chondroitin sulfate, and they are found surrounding the soma and proximal dendrites of many cell types, especially parvalbumin-positive cells, although estimates of types and the percentage of cells with PNNs vary widely. PNNs form as the nervous system matures, and their appearance coincides with the end of critical periods of plasticity. Enzymatic treatment with chondroitinase ABC or similar enzymes can temporarily restore plasticity and may influence axonal regeneration in the brain (Pizzorusso et al. 2002). Thus it appears that PNNs effectively glue a neuron's synaptic connections in place once synaptogenesis and the subsequent activity-dependent pruning of synaptic connections is complete. The ability to regulate the integrity of PNNs, either through the normally occurring upregulation of enzymes that allow remodeling or through exogenous application of such enzymes following injury or disease is an interesting avenue of research.

20.3.4 Thrombospondins

Of the many synaptogenic factors mentioned in previous chapters, most are cell-associated proteins that often span the plasma membrane and mediate cell adhesion. One matrix-associated family of adhesive proteins that has recently been implicated in CNS synapse formation and maintenance is the thrombospondins.

The thrombospondins are matrix-associated, multi-domain proteins that assemble into trimers or pentamers in vertebrates. Like many matrix molecules, they are glycosylated and interact with chondroitin- and heparan sulfate proteoglycans and integrins through a C-terminal RGD sequence. They have been extensively studied for their role in cell adhesion, especially in platelets and during angiogenesis. However, they have also been characterized in the nervous system. Thrombospondin-4 (TSP4) was identified at the NMJ as a factor that is upregulated following denervation and was a preferred substrate for neurite outgrowth in cultured neurons. In addition to enrichment at the NMJ, the TSP4 protein was also found in synapse-rich regions of the retina and cerebellum (Arber and Caroni 1995). Studies on TSP1 and 2 also suggest a role for thrombospondins in synapse formation. TSP1 and 2 are astrocyte-derived signals that promote synaptogenesis in cultured retinal ganglion cells (Christopherson et al. 2005). These synapses were structurally normal and were active presynaptically, but silent postsynaptically. This suggests that there are additional signals required for the formation of completely functional synaptic connections, but that TSP1 and 2 are capable of promoting the alignment and adhesion of pre- and postsynaptic sites in neurons. Furthermore, TSP1 and 2 double-knockout mice have a reduced number of synapses, again consistent with these proteins supporting the coordination of synaptogenesis.

Thrombospondin type 1 repeats are a site of glycosylation that may be of particular functional importance. These repeats are found in many proteins besides the thrombospondins, and a failure to glycosylate these domains may contribute to human diseases such as Peter's plus syndrome (Hess et al. 2008). The extent to which such disease phenotypes can be directly attributed to effects mediated by thrombospondins, as opposed to other proteins containing this domain, remains to be determined.

The thrombospondins therefore exemplify many of the principles discussed in this chapter. They are matrix molecules that influence both pre- and postsynaptic differentiation, though not to equal degrees. They may play a role both at the NMJ and in the brain, although different family members appear to function at different sites, and these functional forms are most likely derived from glial cells.

20.4 Conclusions

The ECM plays a critical role in synapse formation. This is particularly evident at the peripheral NMJ, where a clear basal lamina runs through the synaptic cleft, but is true at CNS synapses as well. The importance of these proteins becomes obvious based on human diseases caused by their dysfunction. Myasthenias are caused by mutations in MuSK, rapsyn, and DOK7, all proteins in the agrin signaling pathway, and MuSK is the primary autoimmune target in serum-negative myasthenia gravis (those cases that do not have autoimmunity against the AChR themselves). Mutations in matrix-associated proteins that are essential for proper NMJ function also result in congenital myasthenic syndromes, including ColQ, the matrix anchor for AChE. Diseases associated with agrin have not been identified, and diseases, which are associated with collagens and laminins, have defects primarily associated with the kidney. However, this may stem in part from a lack of identifying patients or properly phenotyping neuromuscular deficits in a background of other pathologies.

The role of these proteins in the brain is undoubtedly more subtle than their role at the NMJ. This may reflect the greater diversity of synaptic types in the CNS, and therefore imply that only a subset of synapses is affected.

Alternatively, it may be the result of more redundancy in CNS synaptogenesis. However, some CNS functions are postulated for matrix-associated proteins as well. An interesting case is agrin; CNS defects have been reported in agrin-deficient mice, but the consequences of these changes for cognition have not yet been tested. In humans, mutations in a trypsin-like protease, neurotrypsin, cause profound mental retardation and the only known substrate for this protease is agrin, suggesting a role for agrin in the human CNS. In addition, mutations in glycosyltransferases cause severe developmental defects in the CNS, exemplified by muscle–eye–brain disease. These enzymes are known to glycosylate dystroglycan and the mis-glycosylated protein no longer properly interacts with its matrix-associated ligands such as agrin and laminins. Whether this is the sole function of these enzymes, and whether the failure to glycosylate dystroglycan also underlies the CNS defects observed in these patients, remains to be determined. However, it is highly suggestive that the carbohydrate-mediated interaction of cell surface proteins and the matrix is essential for normal brain development.

Another principle that is clearly established is that synapse formation and maintenance require trans-synaptic signaling between the pre- and the post-synaptic cell, even at synapses like the NMJ where these cells are not in direct contact. In addition, the surrounding glial cells strongly influence the synapse. These effects include the promotion of synaptogenesis by glial factors such as thrombospondin, and terminal Schwann cells at the NMJ, which guide rein-nervation, but can also impede synaptic transmission if they invade the synaptic cleft.

Thus, the extracellular matrix mediates trans-synaptic signaling, affects both pre- and postsynaptic cells, and influences the formation, plasticity, and maintenance of synaptic connections in both the peripheral and the central nervous system.

References

Adler M, Manley HA, Purcell AL et al. (2004) Reduced acetylcholine receptor density, morphological remodeling, and butyrylcholinesterase activity can sustain muscle function in acetylcholinesterase knockout mice. Muscle Nerve 30:317–327

Ai X, Kitazawa T, Do AT et al. (2007) SULF1 and SULF2 regulate heparan sulfate-mediated GDNF signaling for esophageal innervation. Development (Cambridge, England) 134:3327–3338

Arber S and Caroni P (1995) Thrombospondin-4, an extracellular matrix protein expressed in the developing and adult nervous system promotes neurite outgrowth. J Cell Biol 131:1083–1094

Beeson D, Higuchi O, Palace J et al. (2006) Dok-7 mutations underlie a neuromuscular junction synaptopathy. Science (New York, NY) 313:1975–1978

Beltran-Valero de Bernabe D, Currier S, Steinbrecher A et al. (2002) Mutations in the O-mannosyltransferase gene POMT1 give rise to the severe neuronal migration disorder Walker-Warburg syndrome. Am J Human Genet 71:1033–1043

Bogdanik L, Framery B, Frolich A et al. (2008) Muscle dystroglycan organizes the postsynapse and regulates presynaptic neurotransmitter release at the Drosophila neuromuscular junction. PLoS ONE 3:e2084

Brockington M, Blake DJ, Prandini P et al. (2001a) Mutations in the fukutin-related protein gene (FKRP) cause a form of congenital muscular dystrophy with secondary laminin alpha2 deficiency and abnormal glycosylation of alpha-dystroglycan. Am J Human Genet 69:1198–1209

Brockington M, Yuva Y, Prandini P et al. (2001b) Mutations in the fukutin-related protein gene (FKRP) identify limb girdle muscular dystrophy 2I as a milder allelic variant of congenital muscular dystrophy MDC1C. Human Mol Genet 10:2851–2859

Burden SJ, Sargent PB and McMahan UJ (1979) Acetylcholine receptors in regenerating muscle accumulate at original synaptic sites in the absence of the nerve. J Cell Biol 82:412–425

Burgess RW, Dickman DK, Nunez L et al. (2002) Mapping sites responsible for interactions of agrin with neurons. J Neurochem 83:271–284

Burgess RW, Nguyen QT, Son YJ et al. (1999) Alternatively spliced isoforms of nerve- and muscle-derived agrin: their roles at the neuromuscular junction. Neuron 23:33–44

Burgess RW, Skarnes WC and Sanes JR (2000) Agrin isoforms with distinct amino termini: differential expression, localization, and function. J Cell Biol 151:41–52

Cartaud A, Strochlic L, Guerra M et al. (2004) MuSK is required for anchoring acetylcholinesterase at the neuromuscular junction. J Cell Biol 165:505–515

Cho SI, Ko J, Patton BL et al. (1998) Motor neurons and Schwann cells distinguish between synaptic and extrasynaptic isoforms of laminin. J Neurobiol 37:339–358

Christopherson KS, Ullian EM, Stokes CC et al. (2005) Thrombospondins are astrocyte-secreted proteins that promote CNS synaptogenesis. Cell 120:421–433

Cohen I, Rimer M, Lomo T et al. (1997) Agrin-induced postsynaptic-like apparatus in skeletal muscle fibers in vivo. Mol Cell Neurosci 9:237–253

Colognato H and Yurchenco PD (2000) Form and function: the laminin family of heterotrimers. Dev Dyn 218:213–234

Cote PD, Moukhles H, Lindenbaum M et al. (1999) Chimaeric mice deficient in dystroglycans develop muscular dystrophy and have disrupted myoneural synapses. Nature Genet 23:338–342

Dai Z and Peng HB (1995) Presynaptic differentiation induced in cultured neurons by local application of basic fibroblast growth factor. J Neurosci 15:5466–5475

DeChiara TM, Bowen DC, Valenzuela DM et al. (1996) The receptor tyrosine kinase MuSK is required for neuromuscular junction formation in vivo. Cell 85:501–512

Deprez P, Inestrosa NC and Krejci E (2003) Two different heparin-binding domains in the triple-helical domain of ColQ, the collagen tail subunit of synaptic acetylcholinesterase. J Biol Chem 278:23233–23242

Donger C, Krejci E, Serradell AP et al. (1998) Mutation in the human acetylcholinesterase-associated collagen gene, COLQ, is responsible for congenital myasthenic syndrome with end-plate acetylcholinesterase deficiency (type Ic). Am J Hum Genet 63:967–75

Feng G, Krejci E, Molgo J et al. (1999) Genetic analysis of collagen Q: roles in acetylcholinesterase and butyrylcholinesterase assembly and in synaptic structure and function. J Cell Biol 144:1349–1360

Feng G, Tintrup H, Kirsch J et al. (1998) Dual requirement for gephyrin in glycine receptor clustering and molybdoenzyme activity. Science (New York, NY) 282:1321–1324

Ferns M, Hoch W, Campanelli JT et al. (1992) RNA splicing regulates agrin-mediated acetylcholine receptor clustering activity on cultured myotubes. Neuron 8:1079–1086

Ferns MJ, Campanelli JT, Hoch W et al. (1993) The ability of agrin to cluster AChRs depends on alternative splicing and on cell surface proteoglycans. Neuron 11:491–502

Flanagan-Steet H, Fox MA, Meyer D et al. (2005) Neuromuscular synapses can form in vivo by incorporation of initially aneural postsynaptic specializations. Development (Cambridge, England) 132:4471–4481

Fox MA, Sanes JR, Borza DB et al. (2007) Distinct target-derived signals organize formation, maturation, and maintenance of motor nerve terminals. Cell 129:179–193
Galtrey CM and Fawcett JW (2007) The role of chondroitin sulfate proteoglycans in regeneration and plasticity in the central nervous system. Brain Res Rev 54:1–18
Gingras J, Rassadi S, Cooper E et al. (2002) Agrin plays an organizing role in the formation of sympathetic synapses. J Cell Biol 158:1109–1118
Glass DJ, Bowen DC, Stitt TN et al. (1996) Agrin acts via a MuSK receptor complex. Cell 85:513–523
Godfrey EW, Nitkin RM, Wallace BG et al. (1984) Components of Torpedo electric organ and muscle that cause aggregation of acetylcholine receptors on cultured muscle cells. J Cell Biol 99:615–627
Gould DB, Phalan FC, Breedveld GJ et al. (2005) Mutations in Col4a1 cause perinatal cerebral hemorrhage and porencephaly. Science (New York, NY) 308:1167–1171
Grady RM, Zhou H, Cunningham JM et al. (2000) Maturation and maintenance of the neuromuscular synapse: genetic evidence for roles of the dystrophin–glycoprotein complex. Neuron 25:279–293
Grewal PK, Holzfeind PJ, Bittner RE et al. (2001) Mutant glycosyltransferase and altered glycosylation of alpha-dystroglycan in the myodystrophy mouse. Nature Genet 28:151–154
Grow WA, Ferns M and Gordon H (1999a) A mechanism for acetylcholine receptor clustering distinct from agrin signaling. Dev Neurosci 21:436–443
Grow WA, Ferns M and Gordon H (1999b) Agrin-independent activation of the agrin signal transduction pathway. J Neurobiol 40:356–365
Hayashi YK, Ogawa M, Tagawa K et al. (2001) Selective deficiency of alpha-dystroglycan in Fukuyama-type congenital muscular dystrophy. Neurology 57:115–121
Helbling-Leclerc A, Zhang X, Topaloglu H et al. (1995) Mutations in the laminin α-chain gene (LAMA2) cause merosin-deficient congenital muscular dystrophy. Nature Genet 11:216–218
Hess D, Keusch JJ, Oberstein SA et al. (2008) Peters Plus syndrome is a new congenital disorder of glycosylation and involves defective Omicron-glycosylation of thrombospondin type 1 repeats. J Biol Chem 283:7354–7360
Hilgenberg LG, Ho KD, Lee D et al. (2002) Agrin regulates neuronal responses to excitatory neurotransmitters in vitro and in vivo. Mol Cell Neurosci 19:97–110
Hilgenberg LG, Su H, Gu H et al. (2006) Alpha3Na + /K + -ATPase is a neuronal receptor for agrin. Cell 125:359–369
Hoch W, Ferns M, Campanelli JT et al. (1993) Developmental regulation of highly active alternatively spliced forms of agrin. Neuron 11:479–490
Hunter DD, Shah V, Merlie JP et al. (1989) Laminin-like adhesive protein concentrated in the synaptic cleft of the neuromuscular junction. Nature 338:229–234
Ibraghimov-Beskrovnaya O, Ervasti JM, Leveille CJ et al. (1992) Primary structure of dystrophin-associated glycoproteins linking dystrophin to the extracellular matrix. Nature 355:696–702
Ji RR, Bose CM, Lesuisse C et al. (1998) Specific agrin isoforms induce cAMP response element binding protein phosphorylation in hippocampal neurons. J Neurosci 18:9695–9702
Kanagawa M, Michele DE, Satz JS et al. (2005) Disruption of perlecan binding and matrix assembly by post-translational or genetic disruption of dystroglycan function. FEBS Lett 579:4792–4796
Kanagawa M, Saito F, Kunz S et al. (2004) Molecular recognition by LARGE is essential for expression of functional dystroglycan. Cell 117:953–964
Khoshnoodi J, Pedchenko V and Hudson BG (2008) Mammalian collagen IV. Microscopy research and technique 71:357–370
Kim N and Burden SJ (2008) MuSK controls where motor axons grow and form synapses. Nature Neurosci 11:19–27

Kim N, Stiegler AL, Cameron TO et al. (2008) Lrp4 Is a Receptor for Agrin and Forms a Complex with MuSK. Cell 135:334–42

Kobayashi K, Nakahori Y, Miyake M et al. (1998) An ancient retrotransposal insertion causes Fukuyama-type congenital muscular dystrophy. Nature 394:388–392

Krejci E, Thomine S, Boschetti N et al. (1997) The mammalian gene of acetylcholinesterase-associated collagen. J Biol Chem 272:22840–22847

Ksiazek I, Burkhardt C, Lin S et al. (2007) Synapse loss in cortex of agrin-deficient mice after genetic rescue of perinatal death. J Neurosci 27:7183–7195

Lein ES, Hawrylycz MJ, Ao N et al. (2007) Genome-wide atlas of gene expression in the adult mouse brain. Nature 445:168–176

Letinsky MS, Fischbeck KH and McMahan UJ (1976) Precision of reinnervation of original postsynaptic sites in frog muscle after a nerve crush. J Neurocytol 5:691–718

Levedakou EN, Chen XJ, Soliven B et al. (2005) Disruption of the mouse Large gene in the enr and myd mutants results in nerve, muscle, and neuromuscular junction defects. Mol Cell Neurosci 28:757–769

Libby RT, Lavallee CR, Balkema GW et al. (1999) Disruption of laminin beta2 chain production causes alterations in morphology and function in the CNS. J Neurosci 19:9399–9411

Lin S, Landmann L, Ruegg MA et al. (2008a) The role of nerve- versus muscle-derived factors in mammalian neuromuscular junction formation. J Neurosci 28:3333–3340

Lin S, Maj M, Bezakova G et al. (2008b) Muscle-wide secretion of a miniaturized form of neural agrin rescues focal neuromuscular innervation in agrin mutant mice. Proc Natl Acad Sci USA 105:11406–11411

Lin W, Burgess RW, Dominguez B et al. (2001) Distinct roles of nerve and muscle in postsynaptic differentiation of the neuromuscular synapse. Nature 410:1057–1064

Lin W, Dominguez B, Yang J et al. (2005) Neurotransmitter acetylcholine negatively regulates neuromuscular synapse formation by a Cdk5-dependent mechanism. Neuron 46:569–579

Liu Y, Padgett D, Takahashi M et al. (2008) Essential roles of the acetylcholine receptor gamma-subunit in neuromuscular synaptic patterning. Development (Cambridge, England) 135:1957–1967

Longman C, Brockington M, Torelli S et al. (2003) Mutations in the human LARGE gene cause MDC1D, a novel form of congenital muscular dystrophy with severe mental retardation and abnormal glycosylation of alpha-dystroglycan. Human Mol Genet 12:2853–2861

Marshall LM, Sanes JR and McMahan UJ (1977) Reinnervation of original synaptic sites on muscle fiber basement membrane after disruption of the muscle cells. Proc Natl Acad Sci USA 74:3073–3077

Martin PT and Sanes JR (1995) Role for a synapse-specific carbohydrate in agrin-induced clustering of acetylcholine receptors. Neuron 14:743–754

Martin PT, Scott LJ, Porter BE et al. (1999) Distinct structures and functions of related pre- and postsynaptic carbohydrates at the mammalian neuromuscular junction. Mol Cell Neurosci 13:105–118

McMahan UJ (1990) The agrin hypothesis. Cold Spring Harb Symp Quant Biol 55:407–418

Meier T, Hauser DM, Chiquet M et al. (1997) Neural agrin induces ectopic postsynaptic specializations in innervated muscle fibers. J Neurosci 17:6534–6544

Michele DE, Barresi R, Kanagawa M et al. (2002) Post-translational disruption of dystroglycan-ligand interactions in congenital muscular dystrophies. Nature 418:417–422

Miner JH, Patton BL, Lentz SI et al. (1997) The laminin α chains: expression, developmental transitions, and chromosomal locations of α1–5, identification of heterotrimeric laminins 8–11, and cloning of a novel α3 isoform. J Cell Biol 137:685–701

Misgeld T, Kummer TT, Lichtman JW et al. (2005) Agrin promotes synaptic differentiation by counteracting an inhibitory effect of neurotransmitter. Proc Natl Acad Sci USA 102:11088–11093

Molinari F, Rio M, Meskenaite V et al. (2002) Truncating neurotrypsin mutation in autosomal recessive nonsyndromic mental retardation. Science (New York, NY) 298:1779–1781
Neumann FR, Bittcher G, Annies M et al. (2001) An alternative amino-terminus expressed in the central nervous system converts agrin to a type II transmembrane protein. Mol Cell Neurosci 17:208–225
Nguyen QT, Parsadanian AS, Snider WD et al. (1998) Hyperinnervation of neuromuscular junctions caused by GDNF overexpression in muscle. Science (New York, NY) 279:1725–1729
Nishimune H, Sanes JR and Carlson SS (2004) A synaptic laminin-calcium channel interaction organizes active zones in motor nerve terminals. Nature 432:580–587
Nitkin RM, Smith MA, Magill C et al. (1987) Identification of agrin, a synaptic organizing protein from Torpedo electric organ. J Cell Biol 105:2471–2478
Noakes PG, Gautam M, Mudd J et al. (1995) Aberrant differentiation of neuromuscular junctions in mice lacking s-laminin/laminin β 2. Nature 374:258–262
Ohno K, Brengman J, Tsujino A et al. (1998) Human endplate acetylcholinesterase deficiency caused by mutations in the collagen-like tail subunit (COLQ) of the asymmetric enzyme. Proc Natl Acad USA 95:9654–9659
Okada K, Inoue A, Okada M et al. (2006) The muscle protein Dok-7 is essential for neuromuscular synaptogenesis. Science (New York, NY) 312:1802–1805
Ornitz DM (2000) FGFs, heparan sulfate and FGFRs: complex interactions essential for development. Bioessays 22:108–112
Parkhomovskiy N, Kammesheidt A and Martin PT (2000) N-acetyllactosamine and the CT carbohydrate antigen mediate agrin-dependent activation of MuSK and acetylcholine receptor clustering in skeletal muscle. Mol Cell Neurosci 15:380–397
Patton BL, Chiu AY and Sanes JR (1998) Synaptic laminin prevents glial entry into the synaptic cleft. Nature 393:698–701
Patton BL, Cunningham JM, Thyboll J et al. (2001) Properly formed but improperly localized synaptic specializations in the absence of laminin alpha4. Nature Neurosci 4:597–604
Patton BL, Miner JH, Chiu AY et al. (1997) Distribution and function of laminins in the neuromuscular system of developing, adult, and mutant mice. J Cell Biol 139:1507–1521
Perrier AL, Massoulie J and Krejci E (2002) PRiMA: the membrane anchor of acetylcholinesterase in the brain. Neuron 33:275–285
Pizzorusso T, Medini P, Berardi N et al. (2002) Reactivation of ocular dominance plasticity in the adult visual cortex. Science (New York, NY) 298:1248–1251
Prior P, Schmitt B, Grenningloh G et al. (1992) Primary structure and alternative splice variants of gephyrin, a putative glycine receptor-tubulin linker protein. Neuron 8:1161–1170
Reif R, Sales S, Hettwer S et al. (2007) Specific cleavage of agrin by neurotrypsin, a synaptic protease linked to mental retardation. FASEB J 21:3468–3478
Ruoslahti E (1996) Brain extracellular matrix. Glycobiology 6:489–492
Rupp F, Ozcelik T, Linial M et al. (1992) Structure and chromosomal localization of the mammalian agrin gene. J Neurosci 12:3535–3544
Rupp F, Payan DG, Magill-Solc C et al. (1991) Structure and expression of a rat agrin. Neuron 6:811–823
Saito F, Masaki T, Saito Y et al. (2007) Defective peripheral nerve myelination and neuromuscular junction formation in fukutin-deficient chimeric mice. J Neurochem 101:1712–1722
Sanes JR and Cheney JM (1982) Lectin binding reveals a synapse-specific carbohydrate in skeletal muscle. Nature 300:646–647
Sanes JR, Marshall LM and McMahan UJ (1978) Reinnervation of muscle fiber basal lamina after removal of myofibers. Differentiation of regenerating axons at original synaptic sites. J Cell Biol 78:176–198
Satz JS, Barresi R, Durbeej M et al. (2008) Brain and eye malformations resembling Walker-Warburg syndrome are recapitulated in mice by dystroglycan deletion in the epiblast. J Neurosci 28:10567–10575

Schwartz NB, Pirok EW, 3rd, Mensch JR, Jr. et al. (1999) Domain organization, genomic structure, evolution, and regulation of expression of the aggrecan gene family. Prog Nucleic Acid Res Mol Biol 62:177–225

Scott LJ, Bacou F and Sanes JR (1988) A synapse-specific carbohydrate at the neuromuscular junction: association with both acetylcholinesterase and a glycolipid. J Neurosci 8:932–944

Smith CL, Mittaud P, Prescott ED et al. (2001) Src, Fyn, and Yes are not required for neuromuscular synapse formation but are necessary for stabilization of agrin-induced clusters of acetylcholine receptors. J Neurosci 21:3151–3160

Smith MA, Yao YM, Reist NE et al. (1987) Identification of agrin in electric organ extracts and localization of agrin-like molecules in muscle and central nervous system. J Exp Biol 132:223–230

Son YJ, Scranton TW, Sunderland WJ et al. (2000) The synaptic vesicle protein SV2 is complexed with an alpha5-containing laminin on the nerve terminal surface. J Biol Chem 275:451–460

Stanco AM and Werle MJ (1997) Agrin and acetylcholine receptors are removed from abandoned synaptic sites at reinnervated frog neuromuscular junctions. J Neurobiol 999–1018

Stephan A, Mateos JM, Kozlov SV et al. (2008) Neurotrypsin cleaves agrin locally at the synapse. FASEB J 22:1861–1873

Sunada Y, Bernier SM, Utani A et al. (1995) Identification of a novel mutant transcript of laminin $\alpha 2$ chain gene responsible for muscular dystrophy and dysmyelination in dy2J mice. Human Mol Genet 4:1055–1061

Sunderland WJ, Son YJ, Miner JH et al. (2000) The presynaptic calcium channel is part of a transmembrane complex linking a synaptic laminin (alpha4beta2gamma1) with non-erythroid spectrin. J Neurosci 20:1009–1019

Takeda S, Kondo M, Sasaki J et al. (2003) Fukutin is required for maintenance of muscle integrity, cortical histiogenesis and normal eye development. Human Mol Genet 12:1449–1459

Tsim KW, Ruegg MA, Escher G et al. (1992a) cDNA that encodes active agrin. Neuron 8:677–689

Tsim KW, Ruegg MA, Escher G et al. (1992b) cDNA that encodes active agrin (published erratum appears in Neuron 1992 Aug. 9(2): following 381). Neuron 8:677–689

van Reeuwijk J, Grewal PK, Salih MA et al. (2007) Intragenic deletion in the LARGE gene causes Walker-Warburg syndrome. Human Genet 121:685–690

van Reeuwijk J, Janssen M, van den Elzen C et al. (2005) POMT2 mutations cause alpha-dystroglycan hypoglycosylation and Walker-Warburg syndrome. J Med genetics 42:907–912

Van Saun M, Herrera AA and Werle MJ (2003) Structural alterations at the neuromuscular junctions of matrix metalloproteinase 3 null mutant mice. J Neurocytol 32:1129–1142

Van Saun M and Werle MJ (2000) Matrix metalloproteinase-3 removes agrin from synaptic basal lamina. J Neurobiol 44:369

Wairkar YP, Fradkin LG, Noordermeer JN et al. (2008) Synaptic defects in a Drosophila model of congenital muscular dystrophy. J Neurosci 28:3781–3789

Weatherbee SD, Anderson KV and Niswander LA (2006) LDL-receptor-related protein 4 is crucial for formation of the neuromuscular junction. Development (Cambridge, England) 133:4993–5000

Xia B, Hoyte K, Kammesheidt A et al. (2002) Overexpression of the CT GalNAc transferase in skeletal muscle alters myofiber growth, neuromuscular structure, and laminin expression. Dev Biol 242:58–73

Xu H, Wu XR, Wewer UM et al. (1994) Murine muscular dystrophy caused by a mutation in the laminin alpha 2 (Lama2) gene. Nature Genet 8:297–302

Yang X, Arber S, William C et al. (2001) Patterning of muscle acetylcholine receptor gene expression in the absence of motor innervation. Neuron 30:399–410

Yin Y, Kikkawa Y, Mudd JL et al. (2003) Expression of laminin chains by central neurons: analysis with gene and protein trapping techniques. Genesis 36:114–127
Yoshida A, Kobayashi K, Manya H et al. (2001) Muscular dystrophy and neuronal migration disorder caused by mutations in a glycosyltransferase, POMGnT1. Dev Cell 1:717–724
Yurchenco PD and Schittny JC (1990) Molecular architecture of basement membranes. FASEB J 4:1577–1590
Zhang B, Luo S, Wang Q et al. (2008) LRP4 serves as a coreceptor of agrin. Neuron 60:285–297

Chapter 21
Gap Junctions as Electrical Synapses

Juan Mauricio Garré and Michael V.L. Bennett

Abstract Gap junctions form between many cell types. Between neurons, they constitute one class of electrical synapses. Gap junctions are aggregates of membrane channels between the conjoined cells and in mammals they are comprised of connexins, which are encoded by a gene family that has 21 members in humans. Each of the coupled cells contributes a hemichannel to each cell–cell channel. Channel turnover can occur within hours, or channels may last a lifetime. Not all connexins will form channels with every other connexin, and connexin compatibility is one limit on junction formation. Other mechanisms including cell attachment and recognition molecules contribute to specificity of gap junction formation. Electrical synapses are characterized by specificity, but mistakes, i.e., inappropriate connections, are sometimes made. Pannexins/innexins form gap junctions in invertebrates, but apparently only hemichannels in mammals.

Keywords Gap junction · Connexin · Pannexin · Innexin · Electrical coupling · Dye coupling · Synchronization · Synaptic delay

21.1 Introduction

In this chapter we consider formation of gap junctions, which are electrical synapses when they are between neurons and are also common between other cell types, both electrically excitable and inexcitable. Gap junctions are aggregates of cell–cell channels that connect the cytoplasm of the coupled cells. They mediate electrical transmission, which is the flow of current, largely carried by K^+ ions and driven by the potential difference between cells. They also allow movement of small molecules, charged and uncharged, by simple diffusion.

M.V.L. Bennett (✉)
Dominick P Purpura Department of Neuroscience, Albert Einstein College of Medicine, Bronx, NY 10804, USA
e-mail: mbennett@aecom.yu.edu

M. Hortsch, H. Umemori (eds.), *The Sticky Synapse*,
DOI 10.1007/978-0-387-92708-4_21,

Gap junctions in vertebrates are formed by connexins, a protein family encoded by 21 genes in humans (Söhl and Willecke 2004). A brief discussion of invertebrate gap junction forming proteins, innexins, and their vertebrate homologs, pannexins, appears at the end of this chapter; for now gap junctions refers only to those comprised of connexins. A single gap junction channel is composed of two connexin hexamers, one in each of the apposed membranes that are docked together across the eponymic gap between the cells; there is essentially no leak from the channel to the extracellular space of the gap. The hexamers are called hemichannels or connexons.

Not uncommonly, cells express multiple connexins, and a connexin hexamer can be homomeric, i.e., made of a single connexin, or heteromeric, i.e., made of more than one connexin. A junction can be homotypic, as between hemichannels of the same kind, or heterotypic, when formed by different connexins in the two hemichannels. A further nomenclatural nicety is whether junctions are between cells of the same or different kind, homocellular or heterocellular. Another pair of terms, homologous and heterologous, is sometimes used for these latter junctions. This nomenclature is suboptimal, because "homologous" is generally used to refer to molecules or structures that have a common evolutionary origin.

In this chapter we will not consider other forms of electrical interaction, which include the electrical inhibitory synapse on the Mauthner cell of teleosts and mutual excitation by electrical coupling across extracellular space. The former has a complex anatomy involving several cell types (Nakajima 1974), and the latter is generally an unspecialized area of apposition with little morphological specialization (see Bennett 1972, 1977).

21.2 Life History of a Gap Junction

To form a gap junction, each cell must express a connexin capable of making a channel with a connexin in the other cell. Pairings that have been shown to allow or not to allow junction formation are shown in Table 21.1. This table has many blank entries, and for connexins that are expressed in few cell types, most possible heterotypic combinations are not physiologically relevant.

Connexins are cotranslationally inserted into membrane of the endoplasmic reticulum (ER) and can be assembled into hexamers in the ER (Koval 2006) or in a post-Golgi compartment (Musil and Goodenough 1993). [Cx26 can be synthesized in the absence of membrane and inserted into membranes post-translationally (Zhang et al. 1996, Ahmad and Evans 2002).] No connexin is known to be glycosylated, so this function of the Golgi complex is not required. Newly synthesized hexamers are transported to the plasma membrane in vesicles that are ultimately inserted into the membrane, the luminal aspect of the hemichannel in the vesicle now facing the external milieu (Fig. 21.1). Trafficking may be along microtubules and vectorial for delivery close to tight junctions (Shaw et al. 2007). [Cx26 may again be an exception as it is transported along actin

Table 21.1 Connexin compatibility, or which (homomeric) connexin hemichannels will form gap junctions with which?

Cx	26	30	30.2	32	36	37	40	43	45	46	47	50
26	+	+		+	–	–	–	–	+	+		+
30	+	+		+	–	+	+	+	+			
30.2			+	+			+	+	+			
32	+	+	+	+	–	–	–	–	–	+	+	+
36	–	–		–	+	+	–	+	+	–	+	–
37	–	+		–	+	+	+	+	+	+		–
40	–	+	+	–	–	+	+	–	+	–		–
43	–	+	+	–	+	+	–	+	+	+	+	–
45	+	+	+	–	+	+	+	+	+	–	+	
46	+			+	–	+	–	+	–	+		+
47				+	+			+	+		+	
50	+			+	–	–	–	–		+		+

In most cases, recombinant connexins were expressed in cell lines. Although there are a number of studies of structural requirements for compatibility, there is no clear relation between connexin group and compatibility. +: junctions do form. –: junctions do not form. blank: not examined. (from Bukauskas 2001 and his unpublished work).

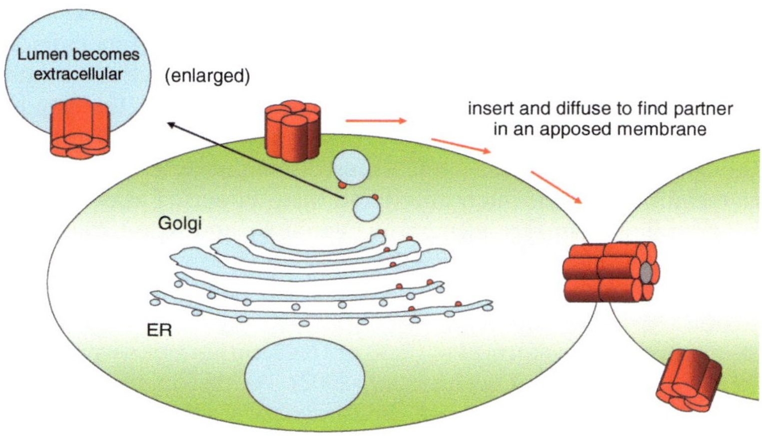

Fig. 21.1 Connexin hemichannels are cotranslationally inserted into membrane of the endoplasmic reticulum and assembled into hemichannels, either in the endoplasmic reticulum or in a post-Golgi compartment. Vesicles with hemichannels are transported to the plasma membrane and inserted. Hemichannels can then diffuse laterally to find and dock with hemichannels in an apposed membrane. A common site of docking is along the edge of a preexisting junction where the membranes closely approach one another (diagram by Jorge E Contreras modified from Bennett 2008)

filaments (Thomas et al. 2001).] Many epithelia are characterized by apical and basolateral faces with different molecular compositions, and hemichannel insertion can be polarized. Alternatively, at least in cell lines and possibly unpolarized cells such as fibroblasts and many types of astrocytes, insertion may not be particularly localized. Once inserted hemichannels appear free to diffuse in the surface membrane. However, some connexins have PDZ and other binding motifs, which might restrict them to specific cell regions (Giepmans 2004). Furthermore, participation in a gap junction leads to reduced mobility because of increased mass and attachment to the apposed cell. If there is a preexisting gap junction, the hemichannels in non-junctional membrane can diffuse to the edge of the junction where the membranes are closely apposed and they are able to dock with a hemichannel in the apposed membrane. The intercellular gap of nominally 20 nm contains cell adhesion molecules, glycoproteins, and other molecules that are anchored in the membranes lining the gap. This space is much too large for connexin hemichannels on either side to interact with each other.

Timed labeling of new channels shows that they are added at the edge of existing gap junctions (Gaietta et al. 2002, Segretain and Falk 2004). Moreover, in a gap junction that remains of a constant size, channels near the center of the gap junction are internalized at a rate equal to that at which they are added at the edge. The internalization process is remarkable. The entire junction thickness, including both membranes and channels, where each cell contributes a hemichannel, are taken up into one or the other cell, along with a little cytoplasm from the donor (or loser) cell (Fig. 21.2). In this process the surface membrane of each cell is broken and resealed, and the corresponding membranes in the internalized junction are also resealed, and the cytoplasm in the vesicle also retains its contents (see Gumpert et al. 2008). Subsequently, the internalized vesicles are transported to lysosomes, where they are degraded. There is no indication that gap junction components are recycled (other than by reentry into cell metabolism).

Although cell–cell channels are not reused, unapposed hemichannels in the surface membrane can be internalized and then reinserted as part of the constitutive recycling of surface membrane (VanSlyke and Musil 2005). Reduced degradation of the internalized hemichannels by a variety of metabolic stressors increases surface expression and formation of gap junctions, possibly by simple mass action.

In some gap junctions formed of Cx32 or Cx43, turnover is surprisingly rapid, a matter of a few hours (Gaietta et al. 2002). At the opposite extreme, gap junctions in the lens nucleus, which are formed of Cx46 and Cx50 (Zampighi et al. 2000), cannot turn over, as there is no capacity for protein synthesis in the mature lens fibers, and the junctions last a lifetime. The significance of the rapid turnover is obscure, but does allow changes in cell–cell coupling by alterations in the rate of internalization over reasonable time frames. For most connexins, turnover times have not been determined, and we do not know the extent to which they vary among different cells and tissues expressing the same connexins.

Until relatively recently, it was thought that hemichannels in the surface membrane would not open until docking with another hemichannel in an apposed membrane; the high permeability suggested by the permeability of cell–cell channels would lead to excess Ca^{2+} and Na^{+} influx and loss of metabolites. Now it is clear that under certain conditions hemichannels can open, although with low probability, and that open hemichannels may mediate

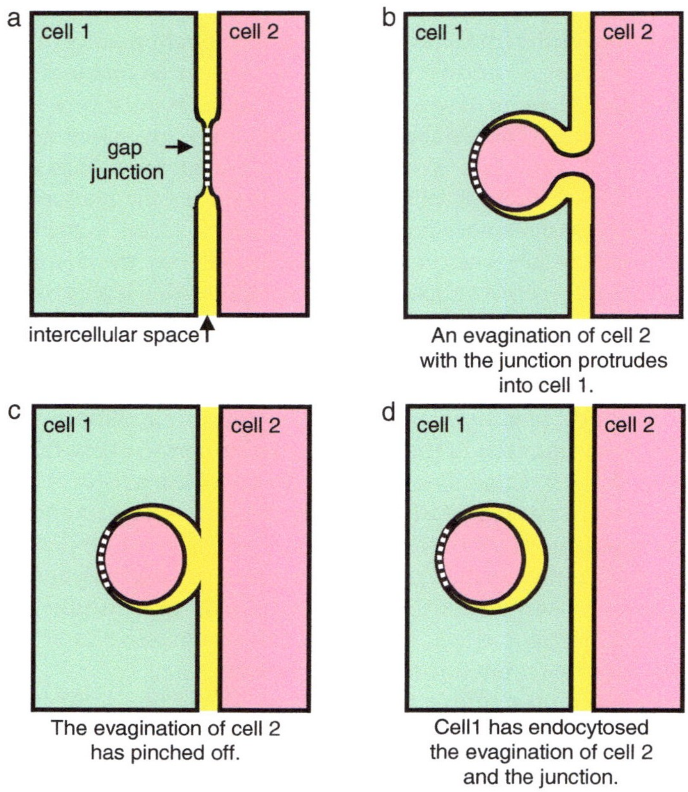

Fig. 21.2 Steps during the internalization of a gap junction. Internalization of small central areas of a gap junction is the more common mode, but this diagram clarifies the breaking and resealing of the two plasma membranes. The internalized junction is destined for lysosomal degradation (modified from Bennett 2009)

autocrine and paracrine cell signaling, as well as an increase in cell stress (Bennett et al. 2003)

21.3 The First Opening

Using dual whole cell patch clamp recording it is possible to detect the first gap junction channel opening when two cells are brought together. If cells that do not express connexins are transfected with Cx43-EGFP (Cx43 with enhanced green fluorescent protein attached to its C-terminal), one can visualize the formation of gap junctions (Bukauskas et al. 2000). Not surprisingly, if there is a large fluorescent aggregate of gap junction channels (or a junctional plaque) between the cells, there is a large junctional conductance between them, and if there is no apparent junction, there is no coupling. Now one asks how big a junction must be for the first channel to open. One might predict that no visible aggregate of channels would be necessary, if an isolated channel could open. Surprisingly, an aggregate of several hundred channels must be present for the first channel to open. [The number of channels can be estimated from the total fluorescence of the aggregate. Whether these aggregates are docked hemichannels is uncertain (see Preus et al. 1981).] Moreover, as the plaque increases in size, junctional conductance increases, but only a small percentage of the channels are open at any one time. It is unclear, whether the channels that are opening (and closing) are a subpopulation opening with a high probability or a larger fraction opening with a low probability. [Estimates from gap junctions at the club endings on the Mauthner cell also suggest that only a small fraction of the channels are open at any one time (Tuttle et al. 1986, Pereda et al. 2004).]

We do not know the mechanism(s) underlying the requirement for many channels (or hemichannels) in the apposed membranes to allow the opening of even a single channel. Candidates are tension along the axis of the channel or lateral compression associated with forces in the surface membrane causing the channels to aggregate. The glycocalyx molecules that must be pushed aside to allow the hemichannels in the apposed membranes to make contact provide a force analogous to surface tension that would tend to cause the channels to aggregate (the Cheerio® effect, Fig. 21.3). The compressive force aggregating the channels may be required for channel opening.

The dependence on multiple channels for a single one to open would not appear to apply to very small junctions such as those formed of strings of particles in the retina (Raviola and Gilula 1973) or to isolated, unapposed hemichannels in the surface but not at junctions (Bennett et al. 2003). Small junctions containing only tens of particles have also been described (e.g., Rash et al. 2007b). Moreover, the question arises of how the first junction is formed. Cells may be able to put out filopodia that can overcome the structural barrier provided by adhesion molecules (Yamane et al. 1999).

Tight junctions are another site where membranes approach closely enough for gap junction hemichannels to touch each other. At these structures the extracellular space is completely occluded, and aggregates of gap junction channels are often found alongside tight junction strands. Facilitated cell–cell channel formation at tight junctions may account for targeted transport of hemichannel containing vesicles to regions where there are tight junctions (Shaw et al. 2007).

21.4 Gap Junctions as Sites of Attachment

Using appropriate detergents gap junctions can be isolated more or less intact. Thus it appears that the two cells are mechanically held together by gap junctions. Low Ca^{2+} solutions that allow cell dissociation by disrupting Ca^{2+}-dependent attachments at desmosomal and adherens junctions do not

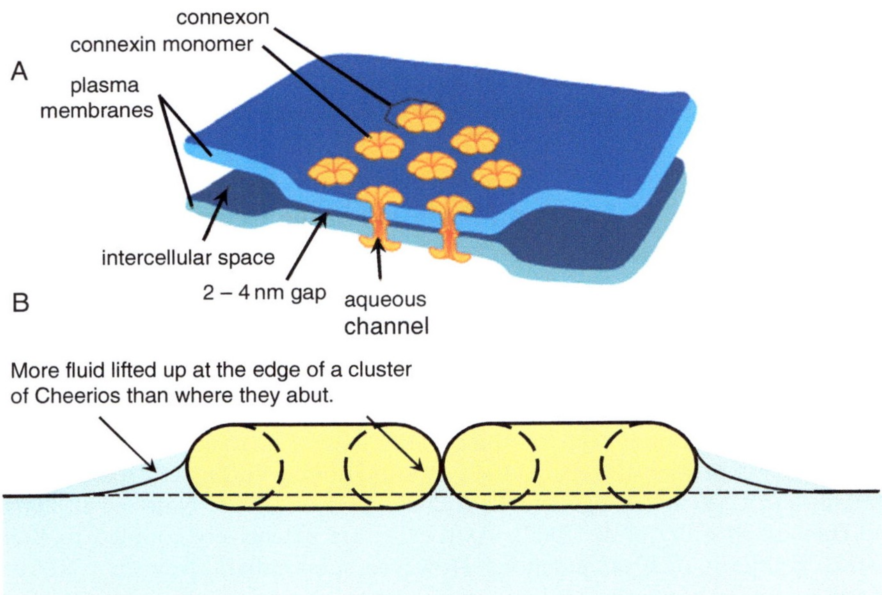

Fig. 21.3 A mechanism for clustering of gap junction channels, the Cheerio® effect. (**A**) The membranes of the coupled cells are much closer at a gap junction than the normal separation between them, a space into which cell adhesion and glycocalyx molecules extend. Energy is required to push these molecules aside to narrow the extracellular space to a separation across which gap junction channels can form. The energy requirement is less if channels are clustered. (**B**) Similarly, Cheerios® cluster on the surface of milk or water because less fluid is raised where the Cheerios® are in contact than at the outer edge of a cluster (gap junction diagram from Wikipedia)

separate gap junctions. When tissues are mechanically dissociated in low Ca^{2+} solutions, intact gap junctions go with one or the other cell, often with a vesicle of membrane from the previously coupled cell (Mazet et al. 1985). Isolated junctions can have one membrane stripped away with the probe of an atomic force microscope to expose the extracellular aspect of the hemichannels in the other membrane (Müller et al. 2002), so a potential plane of cleavage does appear to be present at the center of the channels. Conventional methods of cell dissociation do not separate the channels at this site. Although gap junctions do not have the cytoskeletal connections of desmosomes, interactions with cytoskeletal elements such as ZO-1 do occur (Thomas et al. 2002, Giepmans 2004). In at least some cases gap junction internalization is mediated by clathrin (Gumpert et al. 2008), which implies a mechanical connection, although possibly a weak one compared to that of the classical adhering junctions.

Cx43 is necessary for neuroblast migration along radial glia. However, this cell migration appears not to require the formation of patent intercellular channels, and a mechanical function for Cx43 was inferred (Elias et al. 2007).

Gap junctions also have a role in invasion of glioblastoma cells. This function may be adhesive or may be mediated by some other aspect of gap junctional communication (Oliveira et al. 2005, Cotrina et al. 2008). The adhesive roles of gap junctions are reviewed in more detail by Prochnow and Dermietzel (2008).

21.5 Specificity of Junction Formation in the Central Nervous System

For cells to form heterocellular gap junctions they must express connexins that will form gap junctions with each other (Table 21.1). In the central nervous system, neurons, astrocytes, and oligodendrocytes all live together and frequently contact one another. Each cell type expresses different connexins, which to some extent will limit the formation of heterocellular junctions. Astrocytes express Cx26, Cx30, and Cx43 and oligodendrocytes express Cx29, Cx32, and Cx47 (Altevogt and Paul 2004, Li et al. 2004, Nagy et al. 2004, Orthmann-Murphy et al. 2007). Astrocytes are extensively coupled to each other, primarily by Cx43 junctions. However, subpopulations in the CNS can form restricted networks implying specificity mechanisms in gap junction formation between astrocytes (Houades et al. 2008). Oligodendrocytes only rarely couple to each other, but do couple to astrocytes through (Cx26 and/or Cx30)/Cx32 and Cx43/Cx47 junctions on their cell bodies (Nagy et al. 2003). Cx29 is apposed to the axonal membrane in the paranodal region and probably does not form gap junctions (Kleopa et al. 2004).

Neurons express primarily Cx36 and variably Cx45 (and Cx57 in the horizontal cells of the retina) (Nagy et al. 2004). A new neuronal connexin is Cx30.2 that is found in inhibitory interneurons, often coexpressed with Cx36 (Kreuzberg et al. 2008), and also in the atrioventricular conduction system of the heart (Bukauskas et al. 2006). Although Cx36 will form gap junctions with Cx43, Cx45, and Cx47 in expression systems, it does not appear to do so in the CNS, and there is no evidence of glial/neuronal junctions in the adult nervous system (Rash et al. 2000, Nagy et al. 2004, Rash et al. 2005). Cx30.2 is compatible with Cx40, Cx43, and Cx45 (Table 21.1), but as yet there are no data on its colocalization with other connexins at gap junctions in the nervous system.

21.6 Specificity of Gap Junction Formation Between Neurons

Connexin incompatibility clearly does not account for the absence of glial/neuronal coupling. One must appeal to other cell adhesion and recognition molecules, such as those that have demonstrated roles in the specificity of formation of chemical synapses. For a description of these other adhesive molecules see Chapters 7–20. An early example of the importance of adhesion molecules for junction formation was the greatly increased incidence of gap junctions between mouse sarcoma cells after transfection with a cell adhesion molecule (Mege et al. 1988). Other adhesion molecules are implicated in gap junction formation in studies of other non-neuronal tissues (Meyer et al. 1992, Prowse et al. 1997, Yamane et al. 1999, Wei et al. 2005). Here we cite a few examples of gap junction synapses, where specific contacts are made and gap junction formation is not a simple outcome of cell proximity. As for coupling between glial cells and the absence of coupling between glia and neurons, specificity is likely to be determined by the same or similar molecules as those that determine the formation of chemical synapses. (There remains the conundrum that these mechanisms of adhesion usually do not get the membranes in close enough proximity to form gap junctions.)

There are two main classes of electrical synapses between neurons, gap junctions at axodendritic or axosomatic contacts, and gap junctions between dendrites or somata (Fig. 21.4A, B). Coupling of dendrites and cell bodies tends synchronize impulse activity of the coupled cells. Axosomatic and axodendritic gap junctions can shorten propagation time between cells, particularly in cold blooded animals where synaptic delays tend to be longer. These junctions can also couple and synchronize the postsynaptic cells, if the axons have short branches forming gap junctions on multiple cells (Fig. 21.4C).

Inhibitory interneurons are commonly coupled by dendrodendritic junctions (Gibson et al. 1999, Fukuda and Kosaka 2000, Mann-Metzer and Yarom 2000,

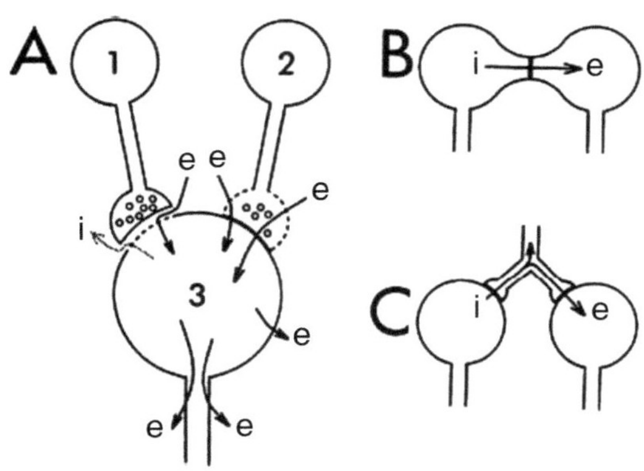

Fig. 21.4 Types of synapse. (**A**) Axosomatic synapses. Cell 1 forms a chemical synapse on cell 3. Arrows indicate current flow if the synapse is excitatory (**e**) or inhibitory (**i**). Cell 2 forms an excitatory electrical synapse on cell 3. During the presynaptic impulse Na^+ flows into the terminal and K^+, the main charge carrier in the cytoplasm, flows through gap junction(s) into the postsynaptic cell. The initial segment, where the impulse arises, has difficulty distinguishing the mode of excitatory transmission. (**B**) Dendrodendritic synapse mediating electrotonic coupling and synchronous activity of neuronal somata. If the arrow indicates the direction of current flow, the cell at the arrow head is excited (**e**) and the cell at the tail is inhibited (**i**); thus, the synapses are both excitatory and inhibitory. (**C**) Coupling of cell bodies by way of presynaptic fibers can serve the same synchronizing function as dendrodendritic junctions (modified from Bennett 1972)

Bennett and Zukin 2004, Connors and Long 2004, Cruikshank et al. 2005). Coupling is not very close, but is sufficient to loosely synchronize cell firing. This activity is implicated in generation of gamma rhythms (30–60 Hz oscillations in the electroencephalogram that are implicated in perceptual processes, see Bartos et al. 2007). In cortical areas, there are multiple types of interneurons, e.g., fast spiking (FS) and low-threshold spiking (LTS). Each class of neurons tends to be coupled to like neurons, with a small fraction being coupled across classes. These cases may arise from the failure of the specificity mechanisms. Possible developmental errors have been described in the retina, where horizontal cells are coupled to occasional bipolar cells, which are not their functional target (Trexler et al. 2001). In other sites, such as cerebellum and thalamic reticular nuclei, coupling is between the inhibitory neurons, but the principle cells do not appear to express a connexin and thus are not "coupling competent". As well as inhibiting principle cells, these inhibitory interneurons may form chemical inhibitory synapses with each other, and dendrodendritic inhibitory synapses can be found alongside dendrodendritic gap junctions (Fukuda et al. 2006).

Coupling of projection neurons is also well established at several sites. Examples are the inferior olive, olfactory glomeruli (Rash et al. 2005), locus coeruleus (Rash et al. 2007a), suprachiasmatic nuclei (Rash et al. 2007b), and various hypothalamic nuclei (see also reviews cited in preceding paragraph). Electrical coupling of hippocampal pyramidal cells has been reported (MacVicar and Ducek 1981, Mercer et al. 2006), although there remains some controversy (Bennett and Pereda 2006). There is no obvious incompatibility between hippocampal pyramidal cells that do and do not form junctions with each other or between the inhibitory neurons that make chemical synapses on them and also form gap junctions with each other. Thus, neurons that express Cx36 do not necessarily form gap junctions with each other, although they come in close enough contact for chemical transmission.

In mammals cortical pyramidal cells that project to the spinal cord can form dual chemical and electrical synapses on spinal neurons (Rash et al. 2000), but do not couple in the cortex (Mercer et al. 2006). Conversely, in the electromotor system of mormyrid fishes, medullary relay cells are closely coupled by dendrodendritic gap junctions, but their axons transmit chemically to the electromotor neurons in the spinal cord, which are closely coupled to each other by dendrodendritic junctions (Bennett et al. 1967). In the electric eel, the medullary neurons are coupled to each other and transmit electrically to the spinal electromotor neurons, which are coupled by way of presynaptic fibers (as in Fig. 21.4C) and apparently not by way of dendrodendritic connections (Meszler et al. 1974). Coupling of postsynaptic neurons by way of presynaptic fibers in the absence of direct dendrodendritic junctions may also occur in the vestibular nucleus of the rat (Korn et al. 1973). In all of these examples regulatory mechanisms in addition to connexin compatibility must be operative.

At most axodendritic and axosomatic synapses with gap junctions, there are also structures suggestive of chemical transmission, and these appositions are often called "mixed synapses." Where the actual mode of transmission has not been determined, a more precise characterization would be "*morphologically* mixed." Generally, mixed synapses in electromotor systems of fishes transmit only electrically (Bennett 1972, 1977), and with weak stimulation of the VIIIth nerve forming the club endings on the Mauthner cell, there may be electrical transmission with little or no chemical component (Pereda et al. 2004). Stimulating a larger fraction of the fibers afferent to the Mauthner cell can increase the fraction of synapses that transmit chemically (apparently depolarization in the dendrite spreads antidromically into the terminals and increases action potential amplitude), and tetanic stimulation can induce long-term potentiation of both components, even when initially there is no chemical component in response to weak stimulation. Given the low open probability inferred for the gap junction channels, potentiation could result from change in channel open probability, rather than the insertion of new channels as suggested for glutamatergic synapses (Isaac 2003, Kerchner and Nicoll 2008). Since the fraction of open channels at a gap junction can be quite small, the possibility should be considered that a morphologically described gap junction, whether or not at a mixed synapse, may not have any open channels. However, no examples have been reported to date.

21.7 Why Electrical Coupling?

Most of the classical properties of synapses can be observed at both chemical and electrical synapses (Bennett 1972, 1977). An electrical connection is thought of as faster, and transmission starts without delay. However, it takes time to charge the postsynaptic membrane capacity, and gap junctions act as low-pass filters. The range of observed delays overlaps for the two classes of synapses and in mammals delay may be of little use in determining the mode of transmission. With the possible exception of the vestibular nuclei of rat, a greater speed of transmission does not appear to be important for the function of gap junctions in the mammalian CNS (Korn et al. 1973). There may be an advantage to electrical synapses, because their reciprocity and linearity permit more effective synchronization of impulse activity. Subthreshold as well as impulse signals can be transmitted, but the degree of precision does not require the rapidity of electrical transmission. The reciprocal glutamatergic excitation between olfactory mitral cells does demonstrate the possibility of synchronization by chemical synapses (Christie and Westbrook 2006, Pimentel and Margrie 2008).

21.8 Gap Junctions in Development

For some days after birth, most cortical neurons are dye coupled (Peinado et al. 1993), probably via Cx45 junctions (Maxeiner et al. 2003). However, over time coupling declines, and few if any pyramidal cells are coupled by day 14 (Yuste et al. 1995). Synchronous firing appears to operate in the specification of neuronal circuits (Yuste et al. 1995, Wong 1999, Butts et al. 2007). Metabolic communication through gap junctions may also be important in establishing developmental gradients of signaling molecues and allowing cells to choose different developmental paths.

21.9 A Fixation Artifact?

Although measurements of dye and electrical coupling are relatively straightforward, artifacts are possible. For example, reports of dye coupling between fixed cells (Coleman and Sengelaub 2002) are suspect, because fixation is known to block gap junctions in several tissues (Bennett 1973, Bushong et al. 2002, Ogata and Kosaka 2002). However, fixation can promote cell fusion on a microscopic scale that leads to dye and electrical coupling; cell fusion can be excluded by application of a gap junction blocking agent or the injection of large molecular weight tracers, such as fluoresceinated dextrans or proteins (Bennett 1973). If cell fusion is produced by fixation, it implies that the cells are closely apposed in vivo, probably without separation by astrocyte (glial) processes.

21.10 Pannexins/Innexins

Gap junctions in Cnidaria and protostomes are formed by a protein family initially termed innexins. When homologous genes were identified in the human genome, the term pannexin was proposed for both vertebrate and invertebrate homologs (Shestopalov and Panchin 2008). This gene family is unrelated to connexins (Yen and Saier 2007, Shestopalov and Panchin 2008). It can be stated with confidence that connexins are absent in *Drosophila*, *Caenorhabditis elegans*, *Nematostella* (an anthozoan), and *Hydra*, since the genomes of these species have been completely sequenced. Where in the deuterostome lineage connexin genes first appeared has not been determined (see Putnam et al. 2007). Connexins are found in ascidians (Sasakura et al. 2003, White et al. 2004), and the genome of *Amphioxus* has been sequenced, but the types of gap junction genes have not been reported (Putnam et al. 2008, Toyoda et al. 2008). Pannexins/innexins and connexins are stated to be absent in the genome of an echinoderm, the sea urchin *Strongylocentrus purpuratus* (Sea Urchin Genome Sequencing Consortium 2006). There are reports of electrical and dye coupling in echinoderms, so it is likely that there are either analogs or homologs of the gap junction forming proteins.

Although not homologous with connexins, pannexins/innexins share many structural and functional properties in common with them (Bruzzone et al. 2005, Scemes et al. 2007). Both classes are tetraspan proteins with cytoplasmic N- and C-terminals. Both form hexameric hemichannels. Both have cysteine groups in the two extracellular loops where formation of intramolecular disulphide bonds is likely required for structural stability. Both are sensitive to similar blocking agents, including long chain alcohols (heptanol, octanol), high intracellular Ca^{2+} and low intracellular pH. In both protein families, some junctions are quite voltage insensitive to either or both transjunctional voltage (V_j) and potential between the inside and outside of the cells, whereas others show significant voltage dependence. Voltage dependence can be very fast, probably due to rectification at the single channel level. Other junctions show slower changes associated with changes in channel open probability, and there can be distinct channel subconductance states. Both can open as hemichannels, and in mammals, pannexins appear not to form cell–cell channels and to function only as hemichannels in the surface membrane. This failure to form gap junctions may be true in other vertebrates as well, although mammalian pannexins will mediate electrical coupling when expressed in *Xenopus* oocytes.

There are a few functional differences between pannexins/innexins and connexins. Pannexin/innexin junctions are permeable to somewhat larger molecules than connexin junctions (Simpson et al. 1977), and the intramembrane particles seen in freeze fracture replicas are a bit larger. There is some indication that, unlike connexin channels, pannexin/innexin channels can split in the plane of the gap leaving hemichannels in each membrane that can then be reused (Pappas et al. 1971, Lane and Swales 1978). Modern labeling techniques should be applied to verify this inference. Innexins are in general encoded by multiple exons. Most vertebrate connexins are encoded by a single exon; Cx36 is encoded by two. Multiple exons encode connexins in ascidians. Identification of earlier arising connexins may clarify whether introns have been lost or gained in evolution.

References

Ahmad S and Evans WH (2002) Post-translational integration and oligomerization of connexin 26 in plasma membranes and evidence of formation of membrane pores: implications for the assembly of gap junctions. Biochem J 365:693–699

Altevogt BM and Paul DL (2004) Four classes of intercellular channels between glial cells in the CNS. J Neurosci 24:4313–4323

Bartos M, Vida I and Jonas P (2007) Synaptic mechanisms of synchronized gamma oscillations in inhibitory interneuron networks. Nat Rev Neurosci 8:45–56

Bennett MVL (1972) Electrical vs. chemical neurotransmission. In: Neurotransmitters. A.R. N.M.D 50:58–89

Bennett MVL (1973) Permeability and structure of electrotonic junctions and intercellular movement of tracers. In: Kater SB and Nicholson C (eds) Intracellular staining techniques in neurobiology, Springer, New York, pp. 115–133

Bennett MVL (1977) Electrical transmission: a functional analysis and comparison with chemical transmission. In: Kandel ER (ed) Cellular biology of neurons Vol. I, Sec. I, Handbook of Physiology. The nervous system, Williams and Wilkins, Baltimore, pp. 357–416

Bennett MVL (2009) Gap junctions and electrical synapses. In: Squire LR (ed.) Encyclopedia of Neuroscience, vol. 4, pp. 529–548, Oxford, Academic Press

Bennett MVL, Contreras JE, Bukauskas FF et al. (2003) New roles for astrocytes: gap junction hemichannels have something to communicate. Trends Neurosci 26:610–617

Bennett MV, Pappas GD, Aljure E et al. (1967) Physiology and ultrastructure of electrotonic junctions. II. Spinal and medullary electromotor nuclei in mormyrid fish. J Neurophysiol 30:180–208

Bennett MVL and Pereda A (2006) Pyramid power: principal cells of the hippocampus unite! Brain Cell Biol 35:5–11

Bennett MVL and Zukin RS (2004) Electrical coupling and neuronal synchronization in the mammalian brain. Neuron 41:495–511

Bruzzone R, Barbe MT, Jakob NJ et al. (2005) Pharmacological properties of homomeric and heteromeric pannexin hemichannels expressed in Xenopus oocytes. J Neurochem 92:1033–1043

Bukauskas FF (2001) Inducing de novo formation of gap junction channels. Methods Mol Biol 154:379–393

Bukauskas FF, Jordan K, Bukauskiene A et al. (2000) Clustering of connexin 43-enhanced green fluorescent protein gap junction channels and functional coupling in living cells. Proc Natl Acad Sci USA 97:2556–2561

Bukauskas FF, Kreuzberg MM, Rackauskas M et al. (2006) Properties of mouse connexin 30. 2 and human connexin 31.9 hemichannels: implications for atrioventricular conduction in the heart. Proc Natl Acad Sci USA 103:9726–9731

Bushong EA, Martone ME, Jones YZ et al. (2002) Protoplasmic astrocytes in CA1 stratum radiatum occupy separate anatomical domains. J Neurosci 22:183–192

Butts DA, Kanold PO and Shatz CJ (2007) A burst-based "Hebbian" learning rule at retinogeniculate synapses links retinal waves to activity-dependent refinement. PLoS Biol 5:e61

Christie JM and Westbrook GL (2006) Lateral excitation within the olfactory bulb. J Neurosci 26:2269–2277

Coleman AM and Sengelaub DR (2002) Patterns of dye coupling in lumbar motor nuclei of the rat. J Comp Neurol 454:34–41

Connors BW and Long MA (2004) Electrical synapses in the mammalian brain. Annu Rev Neurosci 27:393–418

Cotrina ML, Lin JH and Nedergaard M (2008) Adhesive properties of connexin hemichannels. Glia 56:1791–1798

Cruikshank SJ, Landisman CE, Mancilla JG et al. (2005) Connexon connexions in the thalamocortical system. Prog Brain Res 149:41–57

Elias LA, Wang DD and Kriegstein AR (2007) Gap junction adhesion is necessary for radial migration in the neocortex. Nature 448:901–907

Fukuda T, Kosaka T, Singer W et al. (2006) Gap junctions among dendrites of cortical GABAergic neurons establish a dense and widespread intercolumnar network. J Neurosci 26:3434–3443

Fukuda T and Kosaka T (2000) The dual network of GABAergic interneurons linked by both chemical and electrical synapses: a possible infrastructure of the cerebral cortex. Neurosci Res 38:123–130

Gaietta G, Deerinck TJ, Adams SR et al. (2002) Multicolor and electron microscopic imaging of connexin trafficking. Science 296:503–507

Gibson JR, Beierlein M and Connors BW (1999) Two networks of electrically coupled inhibitory neurons in neocortex. Nature 402:75–79

Giepmans BN (2004) Gap junctions and connexin-interacting proteins. Cardiovasc Res 62:233–245

Gumpert AM, Varco JS, Baker SM et al. (2008) Double-membrane gap junction internalization requires the clathrin-mediated endocytic machinery. FEBS Lett 582:2887–2892

Houades V, Koulakoff A, Ezan P et al. (2008) Gap junction-mediated astrocytic networks in the mouse barrel cortex. J Neurosci 28:5207–5217

Isaac JT (2003) Postsynaptic silent synapses: evidence and mechanisms. Neuropharmacology 45:450–460

Kerchner GA and Nicoll RA (2008) Silent synapses and the emergence of a postsynaptic mechanism for LTP. Nat Rev Neurosci 9:813–825

Kleopa KA, Orthmann JL, Enriquez A et al. (2004) Unique distributions of the gap junction proteins connexin29, connexin32, and connexin47 in oligodendrocytes. Glia 47:346–357

Korn H, Sotelo C and Crepel F (1973) Electronic coupling between neurons in the rat lateral vestibular nucleus. Exp Brain Res 16:255–275

Koval M (2006) Pathways and control of connexin oligomerization. Trends Cell Biol 16:159–166

Kreuzberg MM, Deuchars J, Weiss E et al. (2008) Expression of connexin30.2 in interneurons of the central nervous system in the mouse. Mol Cell Neurosci 37:119–134

Lane NJ and Swales LS (1978) Changes in the blood-brain barrier of the central nervous system in the blowfly during development, with special reference to the formation and disaggregation of gap and tight junctions. Dev Biol 62:415–431

Li X, Ionescu AV, Lynn BD et al. (2004) Connexin47, connexin29 and connexin32 co-expression in oligodendrocytes and Cx47 association with zonula occludens-1 (ZO-1) in mouse brain. Neuroscience 126:611–630

MacVicar BA and Dudek FE (1981) Electrotonic coupling between pyramidal cells: a direct demonstration in rat hippocampal slices. Science 213:782–785

Mann-Metzer P and Yarom Y. (2000) Electrotonic coupling synchronizes interneuron activity in the cerebellar cortex. Prog Brain Res 124:115–122

Maxeiner S, Krüger O, Schilling K et al. (2003) Spatiotemporal transcription of connexin45 during brain development results in neuronal expression in adult mice. Neuroscience 119:689–700

Mazet F, Wittenberg BA and Spray DC (1985) Fate of intercellular junctions in isolated adult rat cardiac cells. Circ Res 56:195–204

Maeda S, Nakagawa S, Suga M et al. (2009) Structure of the connexin 26 gap junction channel at 3.5 A resolution. Nature 458:597–602

Mege RM, Matsuzaki F, Gallin WJ et al. (1988) Construction of epithelioid sheets by transfection of mouse sarcoma cells with cDNAs for chicken cell adhesion molecules. Proc Natl Acad Sci USA 85:7274–7278

Mercer A, Bannister AP and Thomson AM (2006) Electrical coupling between pyramidal cells in adult cortical regions. Brain Cell Biol 35:13–27

Meszler RM, Pappas GD and Bennett VL (1974) Morphology of the electromotor system in the spinal cord of the electric eel, *Electrophorus electricus*. J Neurocytol 3:251–261

Meyer RA, Laird DW, Revel JP et al. (1992) Inhibition of gap junction and adherens junction assembly by connexin and A-CAM antibodies. J Cell Biol 119:179–189

Müller DJ, Hand GM, Engel A et al. (2002) Conformational changes in surface structures of isolated connexin 26 gap junctions. EMBO J 21:3598–3607

Musil LS and Goodenough DA (1993) Multisubunit assembly of an integral plasma membrane channel protein, gap junction connexin43, occurs after exit from the ER. Cell 74:1065–1077

Nagy JI, Dudek FE and Rash JE (2004) Update on connexins and gap junctions in neurons and glia in the mammalian nervous system. Brain Res Brain Res Rev 47(1–3):191–215
Nagy JI, Ionescu AV, Lynn BD et al. (2003) Coupling of astrocyte connexins Cx26, Cx30, Cx43 to oligodendrocyte Cx29, Cx32, Cx47: implications from normal and connexin32 knockout mice. Glia 44:205–218
Nakajima Y (1974) Fine structure of the synaptic endings on the Mauthner cell of the goldfish. J Comp Neurol 156:379–402
Ogata K and Kosaka T (2002) Structural and quantitative analysis of astrocytes in the mouse hippocampus. Neuroscience 113:221–233
Oliveira R, Christov C, Guillamo JS et al. (2005) Contribution of gap junctional communication between tumor cells and astroglia to the invasion of the brain parenchyma by human glioblastomas. BMC Cell Biol 6:7
Orthmann-Murphy JL, Freidin M, Fischer E et al. (2007) Two distinct heterotypic channels mediate gap junction coupling between astrocyte and oligodendrocyte connexins. J Neurosci 27:13949–13957
Pappas GD, Asada Y and Bennett MVL (1971) Morphological correlates of increased coupling resistance at an electrotonic synapse. J Cell Biol 49:173–188
Peinado A, Yuste R and Katz LC (1993) Gap junctional communication and the development of local circuits in neocortex. Cereb Cortex 3:488–498
Pereda AE, Rash JE, Nagy JI et al. (2004) Dynamics of electrical transmission at club endings on the Mauthner cells. Brain Res Brain Res Rev 47:227–244
Pimentel DO and Margrie TW (2008) Glutamatergic transmission and plasticity between olfactory bulb mitral cells. J Physiol 586:2107–2119
Preus D, Johnson R, Sheridan J et al. (1981) Analysis of gap junctions and formation plaques between reaggregating Novikoff hepatoma cells. J Ultrastruct Res 77:263–276
Prochnow N and Dermietzel R (2008) Connexons and cell adhesion: a romantic phase. Histochem Cell Biol 130:71–77
Prowse DM, Cadwallader GP and Pitts JD (1997) E-cadherin expression can alter the specificity of gap junction formation. Cell Biol Int 21:833–843
Putnam NH, Butts T, Ferrier DE et al. (2008) The amphioxus genome and the evolution of the chordate karyotype. Nature 453:1064–1071
Putnam NH, Srivastava M, Hellsten U et al. (2007) Sea anemone genome reveals ancestral eumetazoan gene repertoire and genomic organization. Science 317:86–94
Rash JE, Staines WA, Yasumura T et al. (2000) Immunogold evidence that neuronal gap junctions in adult rat brain and spinal cord contain connexin-36 but not connexin-32 or connexin-43. Proc Natl Acad Sci USA 97:7573–7578
Rash JE, Davidson KG, Kamasawa N et al. (2005) Ultrastructural localization of connexins (Cx36, Cx43, Cx45), glutamate receptors and aquaporin-4 in rodent olfactory mucosa, olfactory nerve and olfactory bulb. J Neurocytol 34:307–341
Rash JE, Olson CO, Davidson KG et al. (2007a) Identification of connexin36 in gap junctions between neurons in rodent locus coeruleus. Neuroscience 147:938–956
Rash JE, Olson CO, Pouliot WA et al. (2007b) Connexin36 vs. connexin32, "miniature" neuronal gap junctions, and limited electrotonic coupling in rodent suprachiasmatic nucleus. Neuroscience 149:350–371
Raviola E and Gilula NB (1973) Gap junctions between photoreceptor cells in the vertebrate retina. Proc Natl Acad Sci USA 70:1677–1681
Sasakura Y, Shoguchi E, Takatori N et al. (2003) A genomewide survey of developmentally relevant genes in Ciona intestinalis. X. Genes for cell junctions and extracellular matrix. Dev Genes Evol 213:303–313
Scemes E, Suadicani SO, Dahl G et al. (2007) Connexin and pannexin mediated cell-cell communication. Neuron Glia Biol 3:199–208

Sea Urchin Genome Sequencing Consortium (2006) The genome of the sea urchin *Strongylocentrotus purpuratus*. Science 314: 941–952

Segretain D and Falk MM (2004) Regulation of connexin biosynthesis, assembly, gap junction formation, and removal. Biochim Biophys Acta 1662:3–21

Shaw RM, Fay AJ, Puthenveedu MA et al. (2007) Microtubule plus-end-tracking proteins target gap junctions directly from the cell interior to adherens junctions. Cell 128:547–560

Shestopalov VI and Panchin Y (2008) Pannexins and gap junction protein diversity. Cell Mol Life Sci 65:376–394

Simpson I, Rose B and Loewenstein WR (1977) Size limit of molecules permeating the junctional membrane channels. Science 195:294–296

Söhl G and Willecke K (2004) Gap junctions and the connexin protein family. Cardiovasc Res 62:228–232

Thomas MA, Huang S, Cokoja A et al. (2002) Interaction of connexins with protein partners in the control of channel turnover and gating. Biol Cell 94:445–456

Thomas T, Jordan K and Laird DW (2001) Role of cytoskeletal elements in the recruitment of Cx43-GFP and Cx26-YFP into gap junctions. Cell Commun Adhes 8:231–236

Toyoda A, Bronner-Fraser M, Fujiyama A et al. (2008) The amphioxus genome and the evolution of the chordate karyotype Nature 453:1064–1071

Trexler EB, Li W, Mills SL et al. (2001) Coupling from AII amacrine cells to ON cone bipolar cells is bidirectional. J Comp Neurol 437:408–422

Tuttle R, Masuko S and Nakajima Y (1986) Freeze-fracture study of the large myelinated club ending synapse on the goldfish Mauthner cell: special reference to the quantitative analysis of gap junctions. J Comp Neurol 246:202–211

VanSlyke JK and Musil LS (2005) Cytosolic stress reduces degradation of connexin43 internalized from the cell surface and enhances gap junction formation and function. Mol Biol Cell 16:5247–5257

Wei CJ, Francis R, Xu X et al. (2005) Connexin43 associated with an N-cadherin-containing multiprotein complex is required for gap junction formation in NIH3T3 cells. J Biol Chem 280:19925–19936

White TW, Wang H, Mui R et al. (2004) Cloning and functional expression of invertebrate connexins from *Halocynthia pyriformis*. FEBS Lett 577:42–48

Wong RO (1999) Retinal waves and visual system development. Annu Rev Neurosci 22:29–47

Yamane Y, Shiga H, Asou H et al. (1999) Dynamics of astrocyte adhesion as analyzed by a combination of atomic force microscopy and immuno-cytochemistry: the involvement of actin filaments and connexin 43 in the early stage of adhesion. Arch Histol Cytol 6:355–361

Yen MR and Saier MH Jr (2007) Gap junctional proteins of animals: the innexin/pannexin superfamily. Prog Biophys Mol Biol 94:5–14

Yuste R, Nelson DA, Rubin WW et al. (1995) Neuronal domains in developing neocortex: mechanisms of coactivation. Neuron 14:7–17

Zhang JT, Chen M, Foote CI et al. (1996) Membrane integration of in vitro-translated gap junctional proteins: co- and post-translational mechanisms. Mol Biol Cell 7:471–482

Zampighi GA, Eskandari S and Kreman M (2000) Epithelial organization of the mammalian lens. Exp Eye Res 71:415–435

Zhang JT, Chen M, Foote CI et al. (1996) Membrane integration of in vitro-translated gap junctional proteins: co- and post-translational mechanisms. Mol Biol Cell 7:471–482

Index

D

H

I

K

L

N

P

Q

R

W

X

Z

MIX
Papier aus verantwortungsvollen Quellen
Paper from responsible sources
FSC® C105338

If you have any concerns about our products,
you can contact us on
ProductSafety@springernature.com

In case Publisher is established outside the EU,
the EU authorized representative is:
Springer Nature Customer Service Center GmbH
Europaplatz 3, 69115 Heidelberg, Germany

Printed by Libri Plureos GmbH
in Hamburg, Germany